Remote Sensing Based Building Extraction

Remote Sensing Based Building Extraction

Special Issue Editors

Mohammad Awrangjeb
Xiangyun Hu
Bisheng Yang
Jiaojiao Tian

MDPI • Basel • Beijing • Wuhan • Barcelona • Belgrade • Manchester • Tokyo • Cluj • Tianjin

Special Issue Editors

Mohammad Awrangjeb
Griffith University
Australia

Xiangyun Hu
Wuhan University
China

Bisheng Yang
Wuhan University
China

Jiaojiao Tian
German Aerospace Center (DLR)
Germany

Editorial Office
MDPI
St. Alban-Anlage 66
4052 Basel, Switzerland

This is a reprint of articles from the Special Issue published online in the open access journal *Remote Sensing* (ISSN 2072-4292) (available at: https://www.mdpi.com/journal/remotesensing/special_issues/Building_Detection).

For citation purposes, cite each article independently as indicated on the article page online and as indicated below:

LastName, A.A.; LastName, B.B.; LastName, C.C. Article Title. *Journal Name* **Year**, *Article Number*, Page Range.

ISBN 978-3-03928-382-8 (Pbk)
ISBN 978-3-03928-383-5 (PDF)

© 2020 by the authors. Articles in this book are Open Access and distributed under the Creative Commons Attribution (CC BY) license, which allows users to download, copy and build upon published articles, as long as the author and publisher are properly credited, which ensures maximum dissemination and a wider impact of our publications.

The book as a whole is distributed by MDPI under the terms and conditions of the Creative Commons license CC BY-NC-ND.

Contents

About the Special Issue Editors vii

Mohammad Awrangjeb, Xiangyun Hu, Bisheng Yang and Jiaojiao Tian
Editorial for Special Issue: "Remote Sensing based Building Extraction"
Reprinted from: *Remote Sensing* 2020, 12, 549, doi:10.3390/rs12030549 1

Wenchao Kang, Yuming Xiang, Feng Wang and Hongjian You
EU-Net: An Efficient Fully Convolutional Network for Building Extraction from Optical Remote Sensing Images
Reprinted from: *Remote Sensing* 2019, 11, 2813, doi:10.3390/rs11232813 5

Yang Cui, Qingquan Li and Zhen Dong
Structural 3D Reconstruction of Indoor Space for 5G Signal Simulation with Mobile Laser Scanning Point Clouds
Reprinted from: *Remote Sensing* 2019, 11, 2262, doi:10.3390/rs11192262 31

Yan Zhang, Weiguo Gong, Jingxi Sun and Weihong Li
Web-Net: A Novel Nest Networks with Ultra-Hierarchical Sampling for Building Extraction from Aerial Imageries
Reprinted from: *Remote Sensing* 2019, 11, 1897, doi:10.3390/rs11161897 61

Yaning Yi, Zhijie Zhang, Wanchang Zhang, Chuanrong Zhang, Weidong Li and Tian Zhao
Semantic Segmentation of Urban Buildings from VHR Remote Sensing Imagery Using a Deep Convolutional Neural Network
Reprinted from: *Remote Sensing* 2019, 11, 1774, doi:10.3390/rs11151774 85

Xudong Lai, Jingru Yang, Yongxu Li and Mingwei Wang
A Building Extraction Approach Based on the Fusion of LiDAR Point Cloud and Elevation Map Texture Features
Reprinted from: *Remote Sensing* 2019, 11, 1636, doi:10.3390/rs11141636 105

Shiran Song, Jianhua Liu, Heng Pu, Yuan Liu and Jingyan Luo
The Comparison of Fusion Methods for HSRRSI Considering the Effectiveness of Land Cover (Features) Object Recognition Based on Deep Learning
Reprinted from: *Remote Sensing* 2019, 11, 1435, doi:10.3390/rs11121435 123

Haiqing He, Junchao Zhou, Min Chen, Ting Chen, Dajun Li and Penggen Cheng
Building Extraction from UAV Images Jointly Using 6D-SLIC and Multiscale Siamese Convolutional Networks
Reprinted from: *Remote Sensing* 2019, 11, 1040, doi:10.3390/rs11091040 153

Xuran Pan, Fan Yang, Lianru Gao, Zhengchao Chen, Bing Zhang, Hairui Fan and Jinchang Ren
Building Extraction from High-Resolution Aerial Imagery Using a Generative Adversarial Network with Spatial and Channel Attention Mechanisms
Reprinted from: *Remote Sensing* 2019, 11, 917, doi:10.3390/rs11080917 187

Weijia Li, Conghui He, Jiarui Fang, Juepeng Zheng, Haohuan Fu and Le Yu
Semantic Segmentation-Based Building Footprint Extraction Using Very High-Resolution Satellite Images and Multi-Source GIS Data
Reprinted from: *Remote Sensing* 2019, 11, 403, doi:10.3390/rs11040403 205

Weixuan Ma, Youchuan Wan, Jiayi Li, Sa Zhu and Mingwei Wang
An Automatic Morphological Attribute Building Extraction Approach for Satellite High Spatial Resolution Imagery
Reprinted from: *Remote Sensing* **2019**, *11*, 337, doi:10.3390/rs11030337 225

Prakhar Misra, Ram Avtar and Wataru Takeuchi
Comparison of Digital Building Height Models Extracted from AW3D, TanDEM-X, ASTER, and SRTM Digital Surface Models over Yangon City
Reprinted from: *Remote Sensing* **2018**, *10*, 2008, doi:10.3390/rs10122008 249

Linfu Xie, Qing Zhu, Han Hu, Bo Wu, Yuan Li, Yeting Zhang and Ruofei Zhong
Hierarchical Regularization of Building Boundaries in Noisy Aerial Laser Scanning and Photogrammetric Point Clouds
Reprinted from: *Remote Sensing* **2018**, *10*, 1996, doi:10.3390/rs10121996 275

Youqiang Dong, Li Zhang, Ximin Cui, Haibin Ai and Biao Xu
Extraction of Buildings from Multiple-View Aerial Images Using a Feature-Level-Fusion Strategy
Reprinted from: *Remote Sensing* **2018**, *10*, 1947, doi:10.3390/rs10121947 297

Hui Yang, Penghai Wu, Xuedong Yao, Yanlan Wu, Biao Wang and Yongyang Xu
Building Extraction in Very High Resolution Imagery by Dense-Attention Networks
Reprinted from: *Remote Sensing* **2018**, *10*, 1768, doi:10.3390/rs10111768 327

Mohammad Awrangjeb, Syed Gilani and Fasahat Siddiqui
An Effective Data-Driven Method for 3-D Building Roof Reconstruction and Robust Change Detection
Reprinted from: *Remote Sensing* **2018**, *10*, 1512, doi:10.3390/rs10101512 343

Tingting Lu, Dongping Ming, Xiangguo Lin, Zhaoli Hong, Xueding Bai and Ju Fang
Detecting Building Edges from High Spatial Resolution Remote Sensing Imagery Using Richer Convolution Features Network
Reprinted from: *Remote Sensing* **2018**, *10*, 1496, doi:10.3390/rs10091496 375

Ying Sun, Xinchang Zhang, Xiaoyang Zhao and Qinchuan Xin
Extracting Building Boundaries from High Resolution Optical Images and LiDAR Data by Integrating the Convolutional Neural Network and the Active Contour Model
Reprinted from: *Remote Sensing* **2018**, *10*, 1459, doi:10.3390/rs10091459 395

Guangming Wu, Zhiling Guo and Xiaowei Shao
A Boundary Regulated Network for Accurate Roof Segmentation and Outline Extraction
Reprinted from: *Remote Sensing* **2018**, *10*, 1195, doi:10.3390/rs10081195 413

About the Special Issue Editors

Mohammad Awrangjeb (Dr) received his Ph.D. degree from Monash University. He is currently working as a Senior Lecturer at Griffith University, since Jan 2020. Before joining Griffith University as a Lecturer in 2016, he worked as a (Senior) Research Fellow at the Federation University Australia, Monash University, and the University of Melbourne. His research interests include automatic feature extraction and matching, multimedia security and image processing, automatic building extraction and 3D modeling from remote sensing data, fusion of hyperspectral imagery, and point cloud data for forest vegetation modeling and biomass estimation.

Xiangyun Hu received a Ph.D. degree in photogrammetry and remote sensing from Wuhan University, Wuhan, China, in 2001. From 2002 to 2005, he was a Postdoctoral Research Fellow with the Department of Earth and Space Science and Engineering, Lassonde School of Engineering, York University, Canada. He developed semiautomatic feature extraction technology SmartDigitizer acquired by PCI Geomatics, Leica Geosystems, and Microsoft. From 2005 to 2010, he was a Senior Engineer with ERDAS Inc., USA. Since 2010, he has been a professor at Wuhan University. From 2018 to 2019, he was a Visiting Scholar/Professor at Purdue University, USA. He is currently the Head of the Department of Photogrammetry, School of Remote Sensing and Information Engineering, Wuhan University, where he is also the Founder and the Director of the Earthvision Laboratory. He is the author or co-author of more than 40 articles published in journals and conferences in intelligent analysis and feature extraction of remotely sensed data.

Bisheng Yang received a B.S. degree in engineering survey, and M.S. and Ph.D. degrees in photogrammetry and remote sensing from Wuhan University, Wuhan, China, in 1996, 1999, and 2002, respectively. From 2002 to 2006, he was a Post-Doctoral Research Fellow with the University of Zurich, Zurich, Switzerland. Since 2007, he has been a Professor with the State Key Laboratory of Information Engineering in Surveying, Mapping, and Remote Sensing, Wuhan University, where he is currently the Director of the 3S and Network Communication Laboratory. He has hosted a project of the National High Technology Research and Development Program, a key project of the Ministry of Education, and four National Scientific Research Foundation Projects of China. His current research interests include 3-D geographic information systems, urban modeling, and digital city. He was a Guest Editor of the ISPRS Journal of Photogrammetry and Remote Sensing and Computers and Geosciences.

Jiaojiao Tian (Dr) is a scientist at the Photogrammetry and Image Analysis department, Remote Sensing Technology Institute, German Aerospace Center, Germany, where she is currently heading the 3D modeling team. She received her Ph.D. degree in Mathematics and Computer Sciences from Osnabrueck University in 2013. Her research interests include 3D Change Detection, building reconstruction, 3D point cloud segmentation, DSM generation, and DSM-assisted object extraction and classification.

Editorial

Editorial for Special Issue: "Remote Sensing based Building Extraction"

Mohammad Awrangjeb [1,*], Xiangyun Hu [2], Bisheng Yang [3] and Jiaojiao Tian [4]

1. Institute for Integrated and Intelligent Systems, Griffith University, Nathan QLD 4111, Australia
2. School of Remote Sensing and Information Engineering, Wuhan University, 129 Luoyu Road, Wuhan, Hubei 430079, China; huxy@whu.edu.cn
3. State Key Laboratory of Information Engineering in Surveying, Mapping and Remote Sensing (LIESMARS), Wuhan University, Wuhan, Hubei 430072, China; bshyang@whu.edu.cn
4. Remote Sensing Technology Institute, German Aerospace Center (DLR), Muenchener Strasse 20, 82234 Wessling, Germany; jiaojiao.tian@dlr.de
* Correspondence: m.awrangjeb@griffith.edu.au

Received: 21 January 2020; Accepted: 4 February 2020; Published: 7 February 2020

Building extraction from remote sensing data plays an important role in urban planning, disaster management, navigation, updating geographic databases, and several other geospatial applications [1]. Even though significant research has been carried out for more than two decades, the success of automatic building extraction and modelling is still largely impeded by scene complexity, incomplete cue extraction and sensor dependency of data. Most recently, deep neural networks (DNN) have been widely applied for high classification accuracy in various areas including land-cover and land-use classification [2]. Therefore, intelligent and innovative algorithms are in dire need for high success of automatic building extraction and modelling. This Special Issue focuses on the newly-developed methods for classification and feature extraction from remote sensing data for automatic building extraction and 3D roof modelling.

In the Special Issue, the published papers cover a wide range of related topics including building detection [3], boundary extraction [4] and regularization [5], 3D indoor space (room) modelling [6], land cover classification [7], building height model extraction [8], 3D roof modelling [6,9] and change detection [9].

In terms of datasets, some of the published works use publicly available benchmark datasets, e.g., ISPRS (International Society for Photogrammetry and Remote Sensing) urban object extraction and modelling datasets [4,5,10]; ISPRS 2D semantic labelling datasets [1]; Inria aerial image labelling benchmark datasets [11–13]; and IEEE (Institute of Electrical and Electronics Engineers) DeepGlobe Satellite Challenge datasets [14].

The proposed methods fall into two main categories depending the use of the input data sources: Methods based on single source data, and methods that use multi-source data. Methods based on single source data can use point cloud data [9], aerial imagery [4] and digital surface models (DSM) [8]. The multi-source data-based methods can use the same types of data, e.g., panchromatic band and multispectral imagery [7], optical imagery and light detection and ranging (LiDAR) data [4].

Recently, the rapid development of DNNs has been focused in remote sensing, and the networks have achieved remarkable progress in image classification and segmentation tasks [11]. The majority of the articles published in the Special Issue propose classification based on the DNN [1–6,8,11–13]. There are also a small number of methods based on segmentation [6] and morphological filtering [15].

Using aerial LiDAR data, Awrangjeb et al. [16] introduce a new 3D roof reconstruction technique that constructs an adjacency matrix to define the topological relationships among the roof planes. This method then uses the generated building models to detect 3D changes in buildings.

Among the methods that integrate data from multiple sources, Lai et al. [16] apply a particle swarm optimization algorithm for building extraction based on the fusion of LiDAR point cloud and texture features from the elevation map which is generated from the LiDAR point cloud. Ying et al. [1] combine the optical imagery and LiDAR data in a robust classification framework using the convolutional neural networks (CNN) and active contour model (ACM) to overcome the current limitations (e.g., salt and pepper artefacts) in algorithms for building boundary extraction. The influence of vegetation and salt and pepper artefacts in the extracted buildings is reduced. Li et al. [14] propose a DNN to fuse high-resolution satellite images and multi-source GIS data for building footprint extraction. This method offers better results than the top three solutions in the SpaceNet building detection competition. Dong et al. [10] present a framework for detecting and regularizing the boundary of individual buildings using a feature-level-fusion strategy based on features from dense image matching point clouds, orthophoto and original aerial images. Song et al. [7] present a comparative study on image fusion methods, that achieves the complementarity information of the panchromatic band and multispectral bands in high spatial resolution remote sensing images.

By using optical imagery only, Lu et al. [3] propose a building edge detection model using a richer convolutional features (RCF) network. The RCF-building model can detect building edges accurately and completely, with at least 5% better performance than the baseline methods. Wu et al. [17] present a boundary regulated network called BR-Net for accurate aerial image segmentation and building outline extraction. The BR-Net achieves significantly higher performance than the state-of-the-art U-Net model. Yang et al. [1] propose a novel deep network based on DenseNets and the attention mechanism, called the dense-attention network (DAN), to overcome the difficulty with using both high-level and low-level feature maps in the same network. The results show that DAN offers better performance than other deep networks. Yi et al. [14] effectively perform urban building segmentation from high resolution imagery using a DNN and generate accurate segmentation results. This method outperforms the six existing methods and particularly shows better results for irregular-shaped and small-sized buildings. Zhang et al. [18] use a nested network architecture for building extraction from aerial imageries. It can even extract the building areas covered by shadows. Kang et al. [13] design a dense spatial pyramid pooling to extract dense and multi-scale features simultaneously, to facilitate the extraction of buildings at all scales. He et al. [18] present an effective approach to extracting buildings from Unmanned Aerial Vehicle (UAV) images through the incorporation of superpixel segmentation and semantic recognition. Pan et al. [13] propose a generative adversarial network with spatial and channel attention mechanisms (GAN-SCA) for the robust segmentation of buildings in remote sensing images. Experimental results show that the proposed GAN-SCA achieves a higher accuracy than several state-of-the-art approaches.

Among the other published papers, Cui et al. [6] present a novel method coupling linear structures with three-dimensional geometric surfaces to automatically reconstruct 3D models using point cloud data from mobile laser scanning [6]. A new morphological attribute building index (MABI) and shadow index (MASI) are proposed in Ma et al. [15] for automatically extracting building features from high-resolution remote sensing satellite images. In experiments, this method shows better performance than the two widely used supervised classifiers, namely the support vector machine (SVM) and random forest (RF). Misra et al. [8] compare the digital building height models extracted from four freely available but coarse-resolution global DSMs. Thus, these DSMs can help to cost effectively analyse the vertical urban growth of rapidly growing cities. Xie et al. [5] propose a hierarchical regularization method for noisy building boundary points, through fusion of aerial laser scanning or photogrammetric point clouds. This is formulated as a Markov random field and solved efficiently via graph cut.

Acknowledgments: We want to thank the authors who contributed towards this Special Issue on "Remote Sensing based Building Extraction", as well as the reviewers who provided the authors with comments and very constructive feedback.

Conflicts of Interest: The authors declare no conflict of interest.

References

1. Yang, H.; Wu, P.; Yao, X.; Wu, Y.; Wang, B.; Xu, Y. Building Extraction in Very High Resolution Imagery by Dense-Attention Networks. *Remote Sens.* **2018**, *10*, 1768. [CrossRef]
2. Jahan, F.; Zhou, J.; Awrangjeb, M.; Gao, Y. Fusion of Hyperspectral and LiDAR Data Using Discriminant Correlation Analysis for Land Cover Classification. *IEEE J. Select. Topics Appl. Earth Obs. Remote Sens.* **2018**, *11*, 3905–3917. [CrossRef]
3. Lu, T.; Ming, D.; Lin, X.; Hong, Z.; Bai, X.; Fang, J. Detecting Building Edges from High Spatial Resolution Remote Sensing Imagery Using Richer Convolution Features Network. *Remote Sens.* **2018**, *10*, 1496. [CrossRef]
4. Sun, Y.; Zhang, X.; Zhao, X.; Xin, Q. Extracting Building Boundaries from High Resolution Optical Images and LiDAR Data by Integrating the Convolutional Neural Network and the Active Contour Model. *Remote Sens.* **2018**, *10*, 1459. [CrossRef]
5. Xie, L.; Zhu, Q.; Hu, H.; Wu, B.; Li, Y.; Zhang, Y.; Zhong, R. Hierarchical Regularization of Building Boundaries in Noisy Aerial Laser Scanning and Photogrammetric Point Clouds. *Remote Sens.* **2018**, *10*, 1996. [CrossRef]
6. Cui, Y.; Li, Q.; Dong, Z. Structural 3D Reconstruction of Indoor Space for 5G Signal Simulation with Mobile Laser Scanning Point Clouds. *Remote Sens.* **2019**, *11*, 2262. [CrossRef]
7. Song, S.; Liu, J.; Pu, H.; Liu, Y.; Luo, J. The Comparison of Fusion Methods for HSRRSI Considering the Effectiveness of Land Cover (Features) Object Recognition Based on Deep Learning. *Remote Sens.* **2019**, *11*, 1435. [CrossRef]
8. Misra, P.; Avtar, R.; Takeuchi, W. Comparison of Digital Building Height Models Extracted from AW3D, TanDEM-X, ASTER, and SRTM Digital Surface Models over Yangon City. *Remote Sens.* **2018**, *10*, 2008. [CrossRef]
9. Dong, Y.; Zhang, L.; Cui, X.; Ai, H.; Xu, B. Extraction of Buildings from Multiple-View Aerial Images Using a Feature-Level-Fusion Strategy. *Remote Sens.* **2018**, *10*, 1947. [CrossRef]
10. Awrangjeb, M.; Gilani, S.A.N.; Siddiqui, F.U. An Effective Data-Driven Method for 3-D Building Roof Reconstruction and Robust Change Detection. *Remote Sens.* **2018**, *10*, 1512. [CrossRef]
11. Zhang, Y.; Gong, W.; Sun, J.; Li, W. Web-Net: A Novel Nest Networks with Ultra-Hierarchical Sampling for Building Extraction from Aerial Imageries. *Remote Sens.* **2019**, *11*, 1897. [CrossRef]
12. Kang, W.; Xiang, Y.; Wang, F.; You, H. EU-Net: An Efficient Fully Convolutional Network for Building Extraction from Optical Remote Sensing Images. *Remote Sens.* **2019**, *11*, 2813. [CrossRef]
13. Pan, X.; Yang, F.; Gao, L.; Chen, Z.; Zhang, B.; Fan, H.; Ren, J. Building Extraction from High-Resolution Aerial Imagery Using a Generative Adversarial Network with Spatial and Channel Attention Mechanisms. *Remote Sens.* **2019**, *11*, 917. [CrossRef]
14. Yi, Y.; Zhang, Z.; Zhang, W.; Zhang, C.; Li, W.; Zhao, T. Semantic Segmentation of Urban Buildings from VHR Remote Sensing Imagery Using a Deep Convolutional Neural Network. *Remote Sens.* **2019**, *11*, 1774. [CrossRef]
15. Ma, W.; Wan, Y.; Li, J.; Zhu, S.; Wang, M. An Automatic Morphological Attribute Building Extraction Approach for Satellite High Spatial Resolution Imagery. *Remote Sens.* **2019**, *11*, 337. [CrossRef]
16. Lai, X.; Yang, J.; Li, Y.; Wang, M. A Building Extraction Approach Based on the Fusion of LiDAR Point Cloud and Elevation Map Texture Features. *Remote Sens.* **2019**, *11*, 1636. [CrossRef]
17. Wu, G.; Guo, Z.; Shi, X.; Chen, Q.; Xu, Y.; Shibasaki, R.; Shao, X. A Boundary Regulated Network for Accurate Roof Segmentation and Outline Extraction. *Remote Sens.* **2018**, *10*, 1195. [CrossRef]
18. He, H.; Zhou, J.; Chen, M.; Chen, T.; Li, D.; Cheng, P. Building Extraction from UAV Images Jointly Using 6D-SLIC and Multiscale Siamese Convolutional Networks. *Remote Sens.* **2019**, *11*, 1040. [CrossRef]

© 2020 by the authors. Licensee MDPI, Basel, Switzerland. This article is an open access article distributed under the terms and conditions of the Creative Commons Attribution (CC BY) license (http://creativecommons.org/licenses/by/4.0/).

Article

EU-Net: An Efficient Fully Convolutional Network for Building Extraction from Optical Remote Sensing Images

Wenchao Kang [1,2,3,*], Yuming Xiang [1,2,3], Feng Wang [1,3] and Hongjian You [1,2,3]

1. Aerospace Information Research Institute, Chinese Academy of Sciences, Beijing 100190, China; ymxiang@mail.ie.ac.cn (Y.X.); wangfeng003020@aircas.ac.cn (F.W.); hjyou@mail.ie.ac.cn (H.Y.)
2. School of Electronic, Electrical and Communication Engineering, University of Chinese Academy of Sciences, Huairou District, Beijing 101408, China
3. Key Laboratory of Technology in Geo-spatial Information Processing and Application System, Chinese Academy of Sciences, Beijing 100190, China
* Correspondence: kangwenchao16@mails.ucas.ac.cn; Tel.: +86-1371-810-5044

Received: 12 October 2019; Accepted: 25 November 2019; Published: 27 November 2019

Abstract: Automatic building extraction from high-resolution remote sensing images has many practical applications, such as urban planning and supervision. However, fine details and various scales of building structures in high-resolution images bring new challenges to building extraction. An increasing number of neural network-based models have been proposed to handle these issues, while they are not efficient enough, and still suffer from the error ground truth labels. To this end, we propose an efficient end-to-end model, EU-Net, in this paper. We first design the dense spatial pyramid pooling (DSPP) to extract dense and multi-scale features simultaneously, which facilitate the extraction of buildings at all scales. Then, the focal loss is used in reverse to suppress the impact of the error labels in ground truth, making the training stage more stable. To assess the universality of the proposed model, we tested it on three public aerial remote sensing datasets: WHU aerial imagery dataset, Massachusetts buildings dataset, and Inria aerial image labeling dataset. Experimental results show that the proposed EU-Net is superior to the state-of-the-art models of all three datasets and increases the prediction efficiency by two to four times.

Keywords: building extraction; high-resolution aerial imagery; fully convolutional network; semantic segmentation

1. Introduction

Land cover and land use (LCLU) classification is the fundamental task in remote sensing image interpretation, with the goal of assigning a category label to each pixel of an image [1]. It provides the opportunity to monitor and analyze the evolution of global earth and key regions, and has spawned many new applications, e.g., precision agriculture [1], population density estimation [2], location information service [3]. Among these applications, automatic building extraction with optical remote sensing (ORS) images is one of the most popular research directions [4–7], owing to its convenience and feasibility.

As the resolution of ORS images has reached the decimeter level, more and more elaborate structure, texture and spectral information of buildings has become available. Meanwhile, the increasing intra-class variance and decreasing inter-class variance in VHR images make it more difficult to manually design classification features [8,9]. Therefore, the traditional methods based on hand-crafted features are no longer suitable for building extraction in VHR images [10–13]. Fortunately, the rise of deep learning, especially the convolutional neural network (CNN), has brought us new solutions, as it can automatically learn effective classification features. In recent years, with the development of semantic segmentation technology, building extraction from ORS images has been continuously improved.

In earlier studies, semantic labels have been independently determined pixel by pixel using patch-based CNN models, which predict the label relying on only a small patch around the target pixel and ignores the inherent relationship between patches. Patch-based CNN models have achieved remarkable performance in building extraction, while they cannot guarantee the spatial continuity and integrity of the building structures [14,15]. Moreover, patch-based CNN methods are time consuming.

To overcome the problems of patch-based CNNs, Long et al. [16] proposed the fully convolutional networks (FCNs), which have become a new paradigm for semantic segmentation. FCNs replace the fully connected layers in traditional CNNs with convolutional layers and upsampling layers. Based on the basic FCN8 model [16], several modifications of FCNs have been proposed. For example, DeconvNet [17], SegNet [18] and U-net [19] used the encoder-decoder structure to improve the segmentation accuracy, FastFCN [20] proposed Joint Pyramid Upsampling (JPU) to extract high-resolution feature maps and DeepLab [21] used the dilated convolution to enlarge model receptive field.

To train supervised neural network models, datasets with large number of tagged samples are necessary. In recent years, as more and more remote sensing datasets have become available, FCNs have drawn increasing attention in building extraction research and demonstrated remarkable classification ability on different datasets such as the WHU dataset [22,23], the Massachusetts dataset [6,24], and the Inria Aerial Image Labeling dataset [7,25,26].

Compared with the natural image semantic segmentation tasks, there are two challenges for building extraction from high-resolution ORS images. One is how to accurately extract the regularized contours of buildings. The other one is that buildings in different areas show complex shapes and diverse scales. The scales of different buildings may vary by dozens of times.

Regardless of the diversity of building shapes, they have clear contours. The most commonly used loss function in building extraction is cross-entropy loss function, but it only focuses on the accuracy of single pixel classification. Therefore, the spatial continuity of building shapes is entirely dependent on the features extracted by the models. In order to get accurate contours, some researchers choose to use post-processing methods. A common post-processing method to capture fine edge details is conditional random fields (CRFs) [27]. Shrestha et al. [28] proposed a ELU-FCN model, which replaced the rectified linear unit (ReLU) in FCN8 with the exponential linear unit (ELU) [29], to extract preliminary building maps and then used CRFs to recover accurate building boundaries. Alshehhi et al. [30] extracted features with a single patch-based CNN architecture and integrated them with low-level features of adjacent regions during the post-processing stage to improve the performance. Another branch to solve the contour problem is to use the generative adversarial networks (GANs)[31]. GANs have achieved great success in image conversion [32,33] and super-resolution reconstruction [34,35]. A GAN model consists of two parts: a generator network and a discriminator network. These two networks are trained with adversarial strategy alternatively until the discriminator cannot distinguish the generated image from the real one. Many researchers believe that the GANs can enforce spatial label continuity to refine the building boundaries [8,25]. Li et al. [25] adopted a stable learning strategy to train a GAN model and tested it on the Inria dataset and the Massachusetts dataset. Although this model gave the state-of-the-art results on the Inria dataset, it needed 21 days to train even on a NVIDIA K80 GPU. In addition, the GAN model is prone to collapse, leading to an extremely unstable training procedure. In contrast to the above methods only using RGB images, some studies improved the extraction accuracy of building boundaries by introducing additional geographic information (digital elevation model (DEM), digital surface model (DSM), etc.) [36,37].

Besides the contour problem in building extraction, the buiding sizes can vary greatly, even in one remote sensing image. To deal with the multi-scale problem, one way is to keep the network model unchanged and train the model with input images of different scales. Ji et al. [5] used the original images and double down-sampled images to train one Siamese U-Net (SiU-Net) model simultaneously and share weights between two networks. Although the model can simultaneously learn multi-scale features of the buildings, the training resources are double, which greatly reduces the training efficiency. In order to improve efficiency, researchers reuse single-scale inputs and hope that deep networks can simultaneously exploit multi-scale features extracted by different layers. There are two branches here. The first is to study multi-scale feature extraction block and output one building extraction map [22,24,38]. The JointNet [24] gave a new dense atrous convolution block (DACB), which used dense connectivity block and atrous convolution to acquire multi-scale features. Through extensive use of the DACB modules, the JointNet has achieved the best results on the Massachusetts dataset, while consuming a large amount of GPU memory. Another branch uses the multiple outputs of the middle layer to constrain the model [39–42]. Ji et al. [40] proposed a scale-robust FCN (SR-FCN) and trained it with five outputs of two atrous spatial pyramid pooling (ASPP) structures.

On the one hand, high-resolution focal detail features are indispensable to improve the accuracy of building contour extraction. On the other hand, in order to extract buildings with varying morphologies and scales, global semantic features are the key. Furthermore, to use more context information, models tend to enlarge receptive fields [24,40], which will further increase the difficulty of extracting accurate contours. To solve this conflict, Liu et al. [22] proposed an SRI-Net model to handle the balance between discrimination and detail-preservation abilities. To this end, the SRI-Net used large kernel convolution and the spatial residual inception (SRI) module to preserve detail information while obtaining a large receptive field. These strategies made SRI-Net achieve the best results on the WHU dataset, but also made it computationally expensive.

Although FCNs-based models have achieved great success in remote sensing building extraction task, the accuracies of existing results are still not satisfactory due to the poor prediction of boundaries. Moreover, the state-of-the-art models are complex and inefficient, i.e., they are difficult to train and time-consuming to forecast. Furthermore, they are not versatile and can only achieve good results on a single dataset. To solve the foregoing problems, we proposed a simple but efficient U-Net for building extraction, named EU-Net. The main contributions of this paper can be summarized as follows.

1. A simple but efficient model EU-Net is proposed for optical remote sensing image building extraction. It can be trained efficiently with large learning rate and large batch size.
2. By applying the dense spatial pyramid pooling (DSPP) structure, multi-scale dense features can be extracted simultaneously from more compact receptive field and then buildings of different scales can be better detected. By using the focal loss in reverse, we reduced the impact of error labels in the datasets on model training, leading to a significant improvement of the accuracy.
3. Exhaustive experiments were performed for evaluation and comparison using three public remote sensing building datasets. Compared with the state-of-the-art models on each dataset, the results have demonstrated the universality of the proposed model for building extraction task.

This paper is organized as follows. Some preliminary concepts of neural network are introduced in Section 2. Section 3 details the proposed EU-Net and the loss function used in this paper. Then, the datasets, implementation settings and experiment results are illustrated in Section 4. A series of comparative experiments are discussed in Section 5. Finally, a conclusion is made in Section 6.

2. Preliminaries

A standard CNN model consists of convolutional layers, nonlinear layers, pooling layers and fully connected layers. To build the encoder-decoder structure, FCNs-based models replace the fully connected layers with different kinds of upsampling layers, such as the transposed convolutional layers and the uppooling layers. In addition, many other kinds of layers such as dropout layers [43] and SoftMax layers [44], etc. are also often used in these models. In the following, we only introduce each of the basic layers used in the proposed model.

- **Standard convolutional layer:** The standard convolutional layers are usually used for different purposes with different convolution kernel sizes. For example, convolution with 3*3 kernel is the most common choice for feature extraction and convolution with 1*1 kernel is always used to reintegrate features from different sources in concatenate layer or reduce feature channels. In order to use the spatial context information around the pixel, the convolution kernel is at least a 3*3 convolution kernel. Compared with larger kernel convolution, cascade 3*3 convolutions can get the same receptive field with fewer parameters and introduce more nonlinear functions at the same time. As for 1*1 kernel, the fewest parameters can be used to reduce feature channels.
- **ReLU layer:** The rectified linear unit (ReLU) [45] is the preferred nonlinear activation function for most neural network models. The function of ReLU is very simple, keeping the positive values while setting negative values to zero, i.e., max(0,x).
- **Pooling layer:** Pooling is a general option for downsampling feature maps along the spatial dimension. The max-pooling is adopted by most models and we also use it in our model.
- **Dilated convolution layer:** By adjusting the dilated rate, the dilated convolution can change the receptive field without changing the number of parameters. Therefore, the dilated convolution is used to expand receptive field and simultaneously acquire features of different scales.
- **Transposed convolution layer:** Transposed convolution is used to recover the resolution of feature maps and implement pixel-to-pixel prediction. Different with the uppooling used in U-Net or SegNet, the transposed convolution is trainable and more flexible.
- **Batch normalization layer:** Batch normalization (BN) is used to accelerate model training by normalizing layer inputs [46]. By doing this, the internal covariate shift can be suppressed, and much higher learning rate can be used.
- **Concatenate layer:** Concatenate layer is used to connect feature maps from different sources.

3. Methodology

In this paper, we designed a simple but effective FCNs-based model for remote sensing images building extraction. The complete network of our proposed model is illustrated in Figure 1. As shown in the figure, we divide the model into three parts, namely encoder, DSPP and decoder.

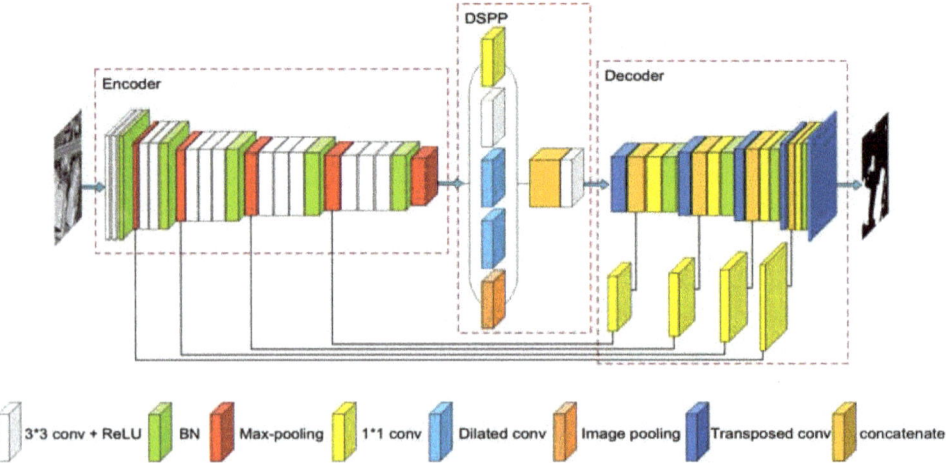

Figure 1. Architecture of the proposed EU-Net, which consists of three parts: encoder, DSPP and decoder.

3.1. Encoder

Currently, model training generally uses small sample slices. As the resolution of ORS images has reached the centimeter level, a large building can cover a large part of a slice. Moreover, background information plays an important role in building extraction. In order to detect large buildings, we use input images with larger size. Though large images will increase the training time, an efficient encoder is chosen to improve the efficiency of our model. We compared the floating-point operations per second (FLOPs) and training parameters of several commonly used encoders in Table 1. VGG16 has the fewest layers but the least efficiency. Xception41 seems to best meet our needs. However, due to the use of depthwise separable convolution, Xception41 needs to consume more GPU memory. Limited by the hardware resources, if we use the Xception41, we can only use a smaller batch size, which shows an adverse effect on BN layers [47]. We note that the parameters and computational cost of VGG16 are mainly concentrated in the last three convolutional layers. If we only use the first 13 layers of VGG16, we can get an encoder that meets both less parameters and low GPU memory cost requirements.

Table 1. Comparison of FLOPs and training parameters for different encoder structures.

Encoders	FLOPs (M)	Parameters (M)
VGG16 [16]	268.52	134.27
VGG16 (first 13 layers)	29.42	14.71
ResNet50 [48]	76.06	38.07
ResNet101 [48]	149.66	74.91
Xception41 [49]	62.54	31.33
Xception65 [49]	96.94	48.57

It should be noted that many researchers only use three [23,50] or four [49,51,52] downsampling layers to retain more detail features, while we keep all five pooling layers to reduce the need for hardware, mainly referring the consumption of GPU memory. As for the detail information, it is handled by the decoder. All convolutions in the encoder use 3*3 kernels and the ReLU activation function is used after each convolution. We only add a BN layer before each pooling layer rather than after each convolution layer. With this encoder, we can use larger input images and larger batch size at the same time.

3.2. DSPP

In our model, we use a dense spatial pyramid pooling (DSPP) structure after the encoder to increase the receptive field and acquire multi-scale features. The receptive field of our encoder is 212*212 which is smaller than the input image size we used. Considering that the effective receptive field was significantly smaller than the theoretical receptive field [53], we use two dilated convolution with rate = 3 (404*404) and rate = 6 (596*596) to make sure that the receptive field can cover the entire input image. However, we notice that the receptive field of DSPP is much bigger than the receptive field of encoder and the moderate features between them are ignored. Thus, we add a standard 3*3 convolution whose receptive field is 276*276. In addition, we add a 1*1 convolution to reintegrate the pooled features and an image pooling to get the importance of different feature channels. Consequently, dense and multi-scale features can be extracted by the proposed structure, and this is the first reason we call it DSPP.

Thereafter, the concatenation of the five outputs passes a 3*3 standard convolution. Compared with the 1*1 convolution in ASPP, a 3*3 convolution can better integrate more information while reducing the feature dimension. As shown in Figure 2, the numbers represent the rates of different dilated convolutions, where 0 represents the 1*1 convolution. The position of the number indicates which feature on the input image can be used by a single pixel on the output image. Through the comparison we can see, a 1*1 convolution can only integrate very sparse features in a 15*15 pixels area, but a 3*3 convolution can nearly cover all pixels. This is another reason we call it DSPP. Finally, we add a BN layer after the 3*3 convolutional layer.

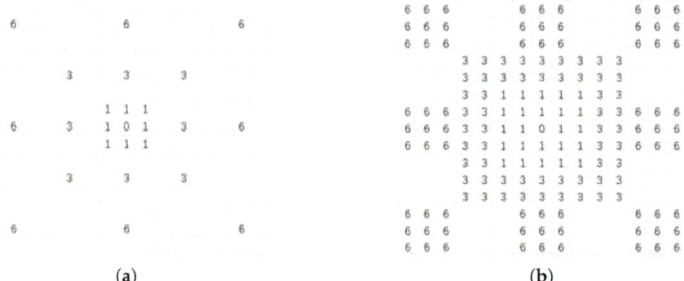

Figure 2. A comparison of which features on the input image can be used by a single pixel on the output image when a different size convolution kernel is used at the end of the DSPP. (**a**) Use a 1*1 convolution. (**b**) Use a 3*3 convolution.

3.3. Decoder

The decoder is used to restore output resolution. We use five transposed convolutions with an upsampling factor of 2 in our decoder. In contrast to the U-Net, we do not use extra 3*3 convolutions to refine the features as shown in Figure 1, since the decoder of U-Net cannot predict accurate building boundaries. To solve this problem, we use a short connection to concatenate the outputs of the first four transposed convolution layers with the corresponding low-level layers in the encoder. However, as mentioned in paper [49], it is not the best choice to directly use the low-level features with many channels, which may outweigh the high-level semantic features and affect the final classification accuracy. Therefore, we first apply a 1*1 convolution on the low-level features before the connection to reduce the number of channels. For the best channel ratio of low-level and high-level features, we will further discuss it in Section 5. After the short connection, another 1*1 convolution is used to reintegrate features and then a BN layer is applied to accelerate model training.

3.4. Loss Function

Cross-entropy loss (CE) is the most commonly used loss function in semantic segmentation task. For binary classification, CE loss function can be described as:

$$CE(p,y) = \begin{cases} -\log(p) & if\, y = 1 \\ -\log(1-p) & otherwise. \end{cases} \quad (1)$$

where $y \in \{0,1\}$ is the ground truth label and $p \in [0,1]$ is the prediction result. For notational convenience, here we define the probability p_t as:

$$p_t = \begin{cases} p & if\, y = 1 \\ 1-p & otherwise. \end{cases} \quad (2)$$

Then, we can rewrite the loss function in (3) as:

$$CE(p_t) = -\log(p_t) \quad (3)$$

Focal loss (FL) [54] is proposed to address extreme imbalances between foreground and background classes in dense object detection. Now, it is also often used to deal with category imbalances in semantic segmentation. Focal loss is improved from CE loss. To address class imbalance, an intuitive idea is to use weighting coefficients. The modified CE loss function can be expressed as:

$$CE(p_t) = -\alpha_t log(p_t) \qquad (4)$$

$\alpha_t \in [0,1]$ is the weighting factor for the two classes, which is defined analogously to p_t.

Based on this, FL adds a modulating factor $(1-p_t)^\gamma$ to further reduce the loss of the easy classification category. $\gamma \geq 0$ is a tunable focusing parameter. FL function can be described as:

$$FL(p_t) = -\alpha_t(1-p_t)^\gamma log(p_t). \qquad (5)$$

As $p_t \to 1$, the factor $(1-p_t)^\gamma \to 0$, i.e., samples that have been correctly classified tend to show a reduced impact on the model training. Conversely, samples that are difficult to classify will determine the subsequent model training. Through the focusing factor γ, the rate at which easy samples are downweighted can be smoothly adjusted. FL is equivalent to CE when $\gamma = 0$.

In our task, the error ground truth labels instead of the category imbalance are the most serious impact that constrains the effectiveness of the model training. We can regard the error labels as hard examples in FL function, and eliminate the influence of error labels by constraining its weight in the later stages of training. Therefore, our final loss function is:

$$Loss = CE(p_t) - \beta \cdot FL(p_t). \qquad (6)$$

β is the weighting factor to control the weight of FL in total loss. The larger the β, the smaller the impact of the difficult (error) sample on the model training. By using the FL in reverse, we can reduce the impact of error labels in CE. Lin et al. [54] pointed out the best choice was $\alpha = 0.25$ and $\gamma = 2$ and $\alpha = 0.5$ worked nearly as well. In our task, two typical error labels, the missing of buildings and the error classification of buildings, are considered. We think these two kinds of error labels should be treated equally, and thus we set $\alpha = 0.5$. As for the γ, we follow the setting in [54] as $\gamma = 2$ (Appendix A).

4. Experimental Results

In this section, we first present detailed dataset description, implementation settings, evaluation metrics and comparative methods. Then, extensive experiments were performed to evaluate the performance of the proposed EU-Net.

4.1. Dataset Description

In this paper, we evaluate the proposed EU-Net on three publicly available building datasets for semantic labeling. The metrics of subsequent experiments are evaluated on the test sets of three datasets.

WHU Building Dataset: This dataset is proposed in [5] and includes both aerial and satellite images. In this paper, we only use the aerial imagery dataset (0.3 m ground resolution) which has higher label accuracy. Therefore, we use this dataset to evaluate the accuracy of building extraction. The aerial dataset contains 8189 tiles with 512*512 pixels. Paper [5] divided the samples into three parts: a training set (4736 tiles with 130,500 buildings), a validation set (1036 tiles with 14,500 buildings) and a test set (2416 tiles with 42,000 buildings). An example is shown in Figure 3.

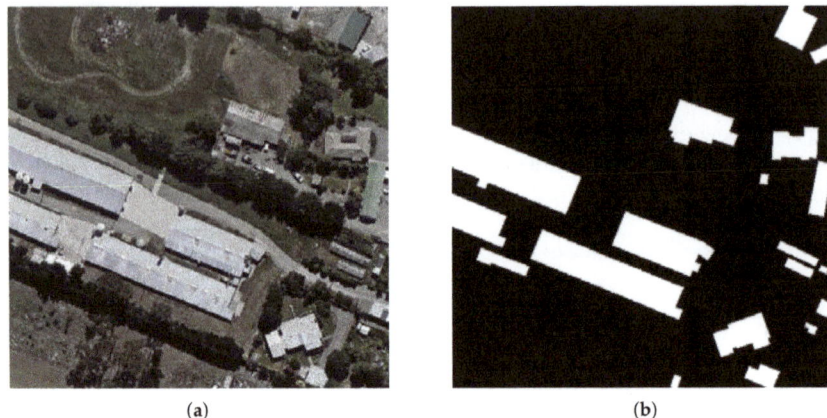

Figure 3. An example of the WHU dataset. (**a**) Original image. (**b**) Ground truth label.

Massachusetts Building Dataset: This dataset is proposed in [6]. Unlike the WHU dataset, the ground resolution of this dataset is 1m, which is relatively low. The label accuracy of this dataset is also lower than the WHU dataset. Thus, we use this dataset to evaluate the ability to handle fuzzy images. There are 151 images with 1500*1500 pixels and paper [6] divided them into three parts: a training set of 137 images, a validation set of 4 images and a test set of 10 images. An example is shown in Figure 4.

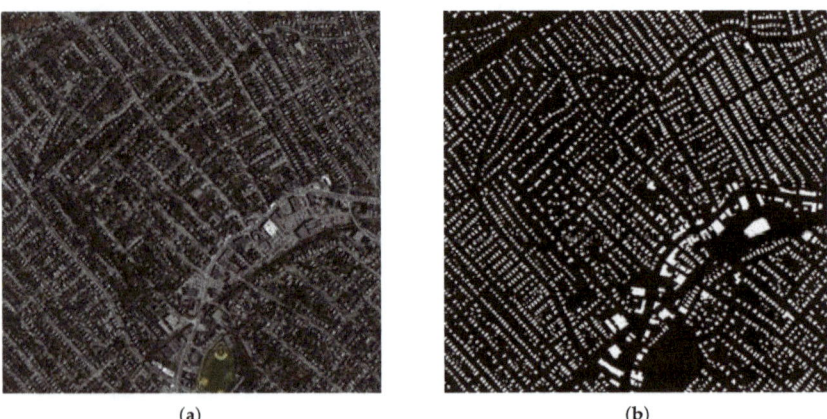

Figure 4. An example of the Massachusetts dataset. (**a**) Original image; (**b**) Ground truth label.

Inria Aerial Image Labeling Dataset: This dataset is proposed in [7] and includes 180 images with public labels and 180 images without public labels. For quantitative analysis, we only use the former in this paper. There are five dissimilar urban settlements (Austin, Chicago, Kitsap County, Western Tyrol and Vienna) with 36 images respectively, ranging from densely populated areas to alpine towns. Since the label accuracy of this dataset is lower than the first dataset, we use this dataset to evaluate the generalization ability of the model. The ground resolution of this dataset is also 0.3 m and image size is 5000*5000 pixels. The first five images of each city are set as test images. An example is shown in Figure 5.

Figure 5. An example of the Inria dataset. (a) Original image; (b) Ground truth label.

4.2. Implementation Settings

Due to hardware limitations, raw images are too large to be directly used for training. In this paper, the raw images are cropped into 512*512 patches in preprocessing with no overlap. The WHU dataset has 4736 512*512 patches, the Massachusetts dataset has 1065 512*512 patches, and the Inria dataset has 15,500 512*512 patches. Then, in each iteration, a batch is clipped to 256*256 pixels using the same random cropping to further increase the diversity of the training samples. Except for random cropping, we do not use other data augmentation tricks such as rotation and flip.

We implemented our EU-Net model based on the Keras API in TensorFlow framework. In the experiments, we did not use any pre-training parameters. The convolution kernels were initialized with Glorot uniform initializer [55] and the biases were initialized to 0. Our proposed network was trained from scratch using SGD optimizer with batch size 64, momentum 0.9. Unlike most literature, we did not use any learning rate adjustment strategies. For the WHU dataset and the Inria dataset, the learning rate was set to 0.2. And for the Massachusetts dataset, the learning rate was set to 0.5. As for the β in (6), we set 0 for the WHU dataset, and 2 for the Massachusetts dataset and the Inira dataset. This was because the WHU dataset had a high-precision label and therefore did not require the FL. The model was trained using two NVIDIA GeForce GTX 1080Ti and tested with one. We trained EU-Net 300 epochs with WHU dataset, 2000 epochs with Massachusetts dataset, and 400 epochs with Inria dataset.

4.3. Evaluation Metrics

To assess the quantitative performance, four benchmark metrics are used, i.e., recall (Rec), precision (Pre), F1 score (F1) and intersection over union (IoU). These four metrics are defined as:

$$Recall = \frac{TP}{TP+FN}. \tag{7}$$

$$Precision = \frac{TP}{TP+FP}. \tag{8}$$

$$F1 = \frac{2*Rec*Pre}{Rec+Pre}. \tag{9}$$

$$IoU = \frac{TP}{TP+FP+FN}. \tag{10}$$

where TP, FP and FN are the number of true positives, false positives and false negatives, respectively. In addition, we will give the normalized confusion matrix, following [56,57]. The form of normalized confusion matrix is shown in Figure 6. The indexes in the i_{th} row denote the rates of the pixels that are classified as each class from the i_{th} class.

		Predicted label	
		Building	Other
True label	Building	$\dfrac{TP}{TP+FN}$	$\dfrac{FN}{TP+FN}$
	Other	$\dfrac{FP}{TN+FP}$	$\dfrac{TN}{TN+FP}$

Figure 6. The form of normalized confusion matrix.

As we all know, every building has a clear boundary, no matter how the shape of the building changes. Therefore, in addition to using the original mask labels, we also create contour labels to evaluate the model. The criterion for judging whether a building pixel belongs to the contour is based on whether there are background pixels among its four adjacent pixels. If the judgment is true, then the pixel is a contour pixel, and vice versa. An example of the contour label in WHU dataset is shown in Figure 7.

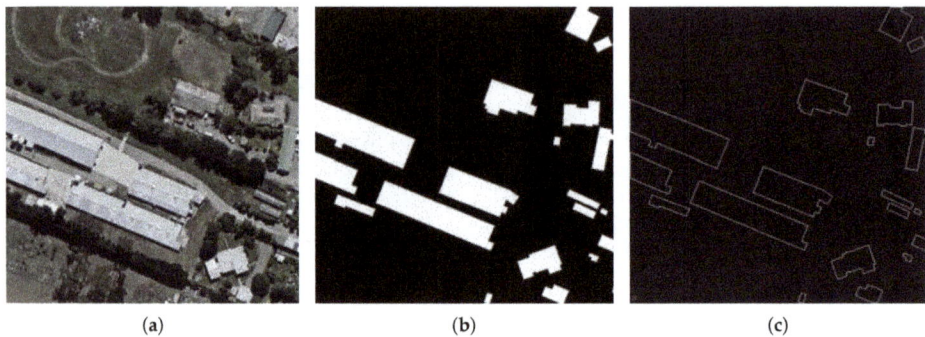

Figure 7. An example of the contour label extracted from the mask label. (**a**) Original image. (**b**) Mask label. (**c**) Contour label.

The four metrics based on mask labels and contour labels are both presented in the subsequent experiments.

In addition to extraction accuracy, the efficiency of the model is also our focus. Considering that the size of the original image in the WHU dataset is only 512*512 pixels, we use the number of images processed per second by the model as a metric. As for the other two datasets, we use the time processing an image as the metric.

4.4. Comparing Methods

To demonstrate its superior performance, the proposed EU-Net is compared with the state-of-the-art methods on each dataset. In this subsection, we will give a brief introduction of the best performing model on each dataset. In addition, we use the results of DeepLabv3+ and FastFCN as the benchmarks for all three datasets.

SRI-Net: Liu et al. [22] proposed SRI-Net for building detection, which was tested on the WHU dataset and the Inria dataset. According to our research, it achieved the best performance on the WHU dataset. We reproduced the SRI-Net with the Keras API and retrained it on the WHU dataset. We followed the training settings in [22]: an Adam optimizer was initialized with a learning rate of 1×10^{-4}, the learning rate was decayed at a rate of 0.9 per epoch, L2 regularization was introduced with a weight decay of 0.0001. Cross-entropy was used as loss function. We trained SRI-Net 300 epochs on WHU dataset.

JointNet: Zhang et al. [24] proposed JointNet for building detection, which was tested on the Massachusetts dataset. According to our research, it achieved the best performance on this dataset. We reproduced the JointNet with the Keras API and retrained it on the Massachusetts dataset. We followed the training settings in [24]: an Adam optimizer was initialized with a learning rate of 1×10^{-4}, and focal loss was used as loss function. We trained JointNet 400 epochs on Massachusetts dataset.

Web-Net: Zhang et al. [38] proposed a nested encoder-decoder deep network for building extraction, named Web-Net. To balance the local cues and the structural consistency, the Web-Net used the Ultra-Hierarchical Sampling (UHS) blocks to extract and fuse the inter-level features. According to our research, it achieved the best performance on the Inria dataset. In order to achieve the best result, Web-Net had to use the pretrained parameters from ImageNet.

FastFCN: The FastFCN model was proposed by Wu et al. [20] and achieved the state-of-the-art results on the ADE20K dataset and the PASCAL Context dataset. We reproduced the FastFCN with the Keras API. We trained FastFCN 300 epochs on WHU dataset, 1600 epochs on Massachusetts dataset, and 350 epochs on Inria dataset. FastFCN was trained with SGD, of which the momentum was set to 0.9 and the weight decay was set to 1×10^{-4}. We set the learning rate to 0.1 and reduced it following the 'poly' strategy. Loss function was kept same with EU-Net.

DeepLabv3+: The DeepLab networks were proposed by Chen et al. and have been improved several times, including v1 [21], v2 [58], v3 [59], and v3+ [49]. The DeepLabv3+ achieved the state-of-the-art results on the Cityscapes dataset and the PASCAL VOC 2012 dataset. We reproduced the DeepLabv3+ with the Keras API. We trained DeepLabv3+ 300 epochs on WHU dataset, 1600 epochs on Massachusetts dataset, and 400 epochs on Inria dataset. DeepLabv3+ was trained with SGD, of which the momentum was set to 0.9 and the weight decay was set to 1×10^{-4}. We set the learning rate to 0.01 for WHU dataset, 0.5 for Massachusetts dataset, and 0.01 for Inria dataset. We also reduced the learning rate following the 'poly' strategy. Loss function was kept same with EU-Net.

4.5. Comparison with Deep Models

4.5.1. WHU Dataset

Table 2 gives the quantitative evaluation indexes of different models on the test images of the WHU dataset. Our proposed EU-Net outperforms SRI-Net, FastFCN, and DeepLabv3 on all the four metrics, except for the precision of SRI-Net. However, the precision of our EU-Net is only 0.23% lower than that of SRI-Net, and our EU-Net makes a better balance between recall and precision, as can be intuitively seen from the F1. The mask IoU metrics are all higher than 85% and our EU-Net even exceeds 90%. This is because the WHU dataset is of higher quality and is easier to distinguish than the other two datasets [22]. Compared with the result of SRI-Net in [22], our EU-Net improved the mask IoU by 1.47%. Moreover, our model can process 16.78 images per second, while SRI-Net can only process 8.51 images per second. In other words, our model efficiency is approximately twice that of the latter. The FastFCN and the DeepLabv3+ are also much slower than our model. As a reference, we also presented the normalized confusion matrix for EU-Net in Figure 8.

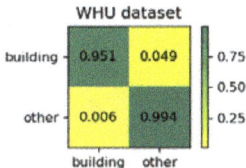

Figure 8. The normalized confusion matrix of EU-Net on WHU dataset.

Table 2. Evaluation results on the test set of WHU dataset. The best values are masked as bold.

		Recall (%)	Precision (%)	F1 (%)	IoU (%)	Images/s
Web-Net [38] (report)	mask	-	-	-	88.76	-
SRI-Net [22] (report)	mask	93.28	**95.21**	94.23	89.09	-
SRI-Net [22]	mask	91.92	92.75	92.33	85.75	8.51
	contour	36.65	37.36	37.00	22.70	
FastFCN [20]	mask	81.37	87.98	84.55	73.23	9.78
	contour	20.27	16.61	18.26	10.05	
DeepLabv3+ [49]	mask	92.99	93.11	93.05	87.00	9.44
	contour	40.49	40.67	40.58	25.45	
EU-Net	mask	**95.10**	94.98	**95.04**	**90.56**	**16.78**
	contour	**48.73**	**49.10**	**48.91**	**32.38**	

Figure 9 shows the visual performance of four different scenarios. The first scenario is an example of an oversized building. The result of FastFCN was very terrible where the oversized building was almost undetected. SRI-Net and DeepLabv3+ misclassified the cement floor in the lower middle area into buildings, while our EU-Net successfully avoided this error. The second scenario is sparsely distributed medium-sized buildings. Only our EU-Net successfully detected the building in the upper left corner and all four models misclassified the containers in the lower right corner into buildings. The third scenario is densely distributed small-sized buildings. All four models missed several buildings of very small size. The fourth scenario is a negative sample. Only our EU-Net gave the right prediction and the other three models had more or less misclassifications. In summary, our EU-Net gives the best results both on the integrity of building shapes and the accuracy of building contours.

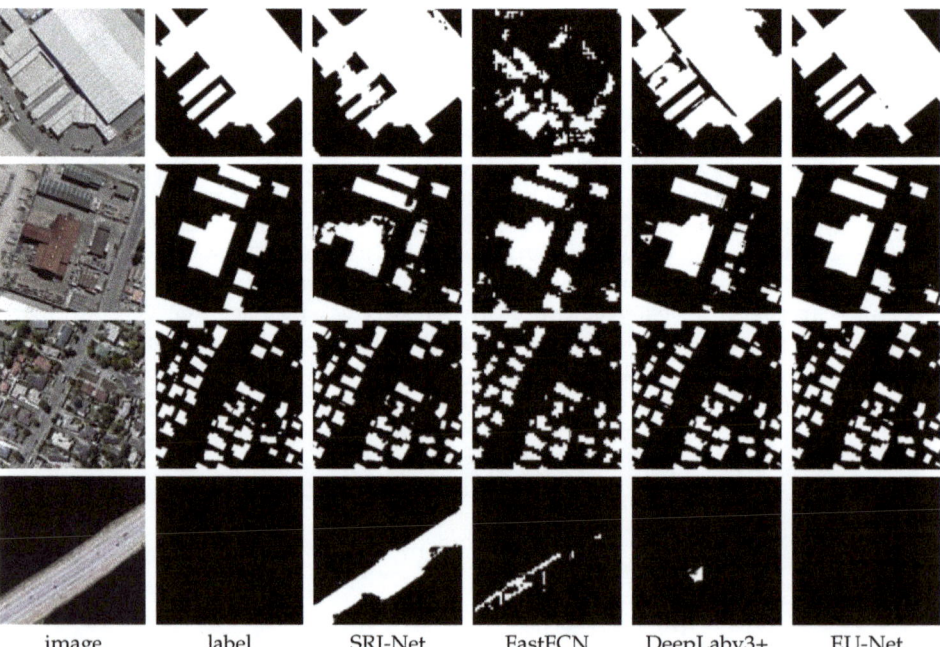

Figure 9. Examples of building extraction results produced by four models on the WHU dataset. The first three rows are examples of oversized building, medium-sized buildings and small-sized buildings, respectively. The last row is an example which has no buildings. Columns 2-6 are the ground truth labels and prediction maps from SRI-Net, FastFCN, DeepLabv3+, and EU-Net, respectively.

4.5.2. Massachusetts Dataset

In this dataset, we predicted the test images in blocks of 512*512 pixels, with a sliding stride of 256 pixels. Table 3 gives the quantitative evaluation indexes and time costs of processing one image for different models. It is clear that our proposed EU-Net outperforms JointNet, FastFCN, and DeepLabv3+. Compared with the results of the WHU dataset, all metrics are much lower for the four models. There are two reasons for the results. First, the scenes of the Massachusetts dataset are more complicated. Especially the shadows of high buildings bring great difficulties to the classification. Second, the Massachusetts dataset has a lower resolution and label quality. As shown in Table 3, recall metrics for all models are much lower than the precision metrics and our model has the smallest gap between the two metrics. Compared with the report result of JointNet in [24], our EU-Net improved the IoU by 1.94%. In terms of processing efficiency, the FastFCN is almost the same as the DeepLabv3+, and the time to process one image is nearly twice that of our model. The most time-consuming JointNet takes more than three times as much as our model. As a reference, we also presented the normalized confusion matrix for EU-Net in Figure 10.

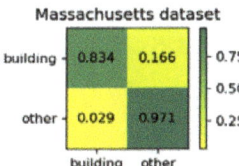

Figure 10. The normalized confusion matrix of EU-Net on Massachusetts dataset.

Table 3. Evaluation results on the test set of Massachusetts dataset. The best values are masked as bold.

		Recall (%)	Precision (%)	F1 (%)	IoU (%)	Time (s)
JointNet [24] (report)	mask	81.29	86.21	83.68	71.99	-
JointNet [24]	mask	79.85	85.21	82.44	70.13	4.16
	contour	27.30	27.32	27.31	15.82	
FastFCN [20]	mask	65.70	78.83	71.67	55.85	2.19
	contour	13.13	14.38	13.73	7.37	
DeepLabv3+ [49]	mask	69.90	83.21	75.98	61.26	2.20
	contour	21.38	22.50	21.92	12.31	
EU-Net	mask	**83.40**	**86.70**	**85.01**	**73.93**	1.13
	contour	**28.23**	**29.44**	**28.83**	**16.84**	

Figure 11 shows the visual performance of four different scenarios. The odd rows are original images and the even rows are the regions of interest selected by red boxes in the original images. The first scenario showed the impact of high-rise shadows on the accuracy of building extraction. It could be seen that the buildings predicted by four models were all incomplete and the integrity of DeepLabv3+ was the worst. Although the integrity of FastFCN was better than DeepLabv3+, there was a significant sawtooth effect on the building contours. Our EU-Net gave the most complete and accurate building extraction results. The main problem was that the shadows in the middle of the buildings cannot be accurately classified. As mentioned before, there exist obvious wrong labels in Massachusetts dataset. An example of error labels is shown in the 4th and 8th row of Figure 11, where the ground truth presents a wrong label on the grassland area. The last two scenarios were used to show the performance of four models for detecting buildings of different sizes. For small and medium-sized buildings, JointNet, DeepLabv3+ and EU-Net had similar performance. Meanwhile, for oversized buildings, only our EU-Net gave a relatively complete prediction.

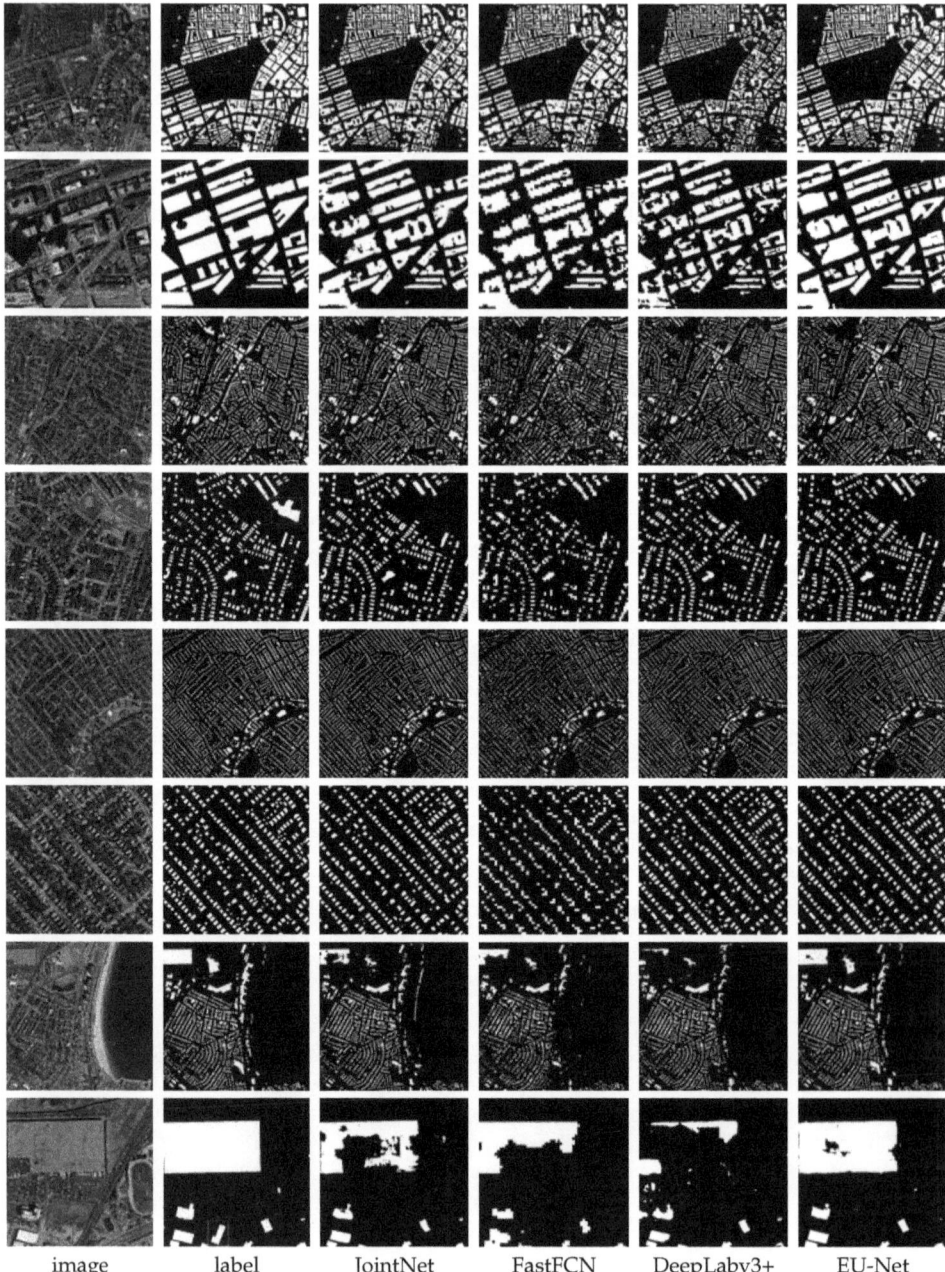

Figure 11. Examples of building extraction results produced by four models on the Massachusetts dataset. The even rows are the enlargements of the red box selected areas in the odd rows. The red box selected areas in the odd rows are error label examples. Columns 2–6 are the ground truth labels and prediction maps from JointNet, FastFCN, DeepLabv3+, and EU-Net, respectively.

4.5.3. Inria Dataset

The test image size is the same as the Massachusetts dataset. Table 4 gives the quantitative evaluation indexes and time costs of processing one image for different models. Similar conclusions can be drawn from Table 4. The results on the Inria dataset are also worse than the WHU dataset, but better than the Massachusetts dataset. The latter is because the resolution of the Inria dataset is higher than the resolution of the Massachusetts dataset. Compared with the WHU dataset, the Inria dataset has a lower quality where some obvious error labels can be found in the ground truth. Moreover, the scenes of the Inria dataset are more complicated, as there are five different cities in the dataset. These factors cause the results on the Inria dataset to be worse than the WHU dataset. Compared with the result of Web-Net reported in [38], our EU-Net improved the mask IoU by 0.4%. According to paper [38], the Web-Net takes 56.5 s to process one image, while the EU-Net only needs 14.79 s, i.e., our model has improved the efficiency four times compared with the Web-Net. As a reference, we also presented the normalized confusion matrix for EU-Net in Figure 12.

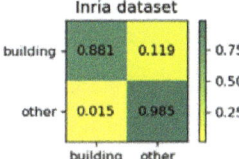

Figure 12. The normalized confusion matrix of EU-Net on Inria dataset.

Table 4. Evaluation results on the test set of Inria dataset. The best values are masked as bold.

		Recall (%)	Precision (%)	F1 (%)	IoU (%)	Time (s)
SRI-Net [22] (report)	mask	81.46	85.77	83.56	71.76	-
2-levels U-Nets [60] (report)	mask	-	-	-	74.55	208.8
Building-A-Net [25] (report)	mask	-	-	-	78.73	150.50
Web-Net [38] (report)	mask	-	-	-	80.10	56.50
FastFCN [20]	mask	83.55	87.51	85.48	74.64	29.12
	contour	11.31	11.07	11.18	5.92	
DeepLabv3+ [49]	mask	84.00	87.88	85.90	75.28	29.61
	contour	13.94	15.85	14.84	8.01	
EU-Net	mask	**88.14**	**90.28**	**89.20**	**80.50**	14.79
	contour	**19.81**	**21.18**	**20.47**	**11.40**	

We also test the performance of EU-Net on five cities respectively, showed in Table 5. Compared with Web-Net, the proposed EU-Net gained better performance on Austin, Chicago, and Vienna. But for Kitsap and Tyrol-w, Web-Net performed better. In general, the overall performance of the proposed EU-Net is slightly better than Web-Net.

Table 5. IoU metrics (%) for each city of test set in Inria dataset. The best values are masked as bold.

	Austin	Chicago	Kitsap	Tyrol-w	Vienna	Overall
Web-Net [38] (report)	82.49	73.90	**70.71**	**83.72**	83.49	80.10
FastFCN [20]	75.56	70.05	64.37	74.10	78.97	74.64
DeepLabv3+ [49]	78.89	69.93	66.11	73.09	79.24	75.28
EU-Net	**82.86**	**76.18**	70.68	80.83	**83.55**	**80.50**

We selected a representative image from each of the five cities, as shown in Figure 13. The buildings in Kitsap County (the third row in Figure 13) are randomly scattered throughout the image and the buildings in Western Tyrol (the fourth row in Figure 13) are concentrated in parts of the image. The buildings of the other three cities are densely distributed throughout the images. Among them, the buildings in Chicago (the second row in Figure 13) are neatly distributed, while the buildings in Austin (the first row in Figure 13) and Vienna (the last row in Figure 13) are irregular. Compared with Austin, the building size varies drastically in Vienna.

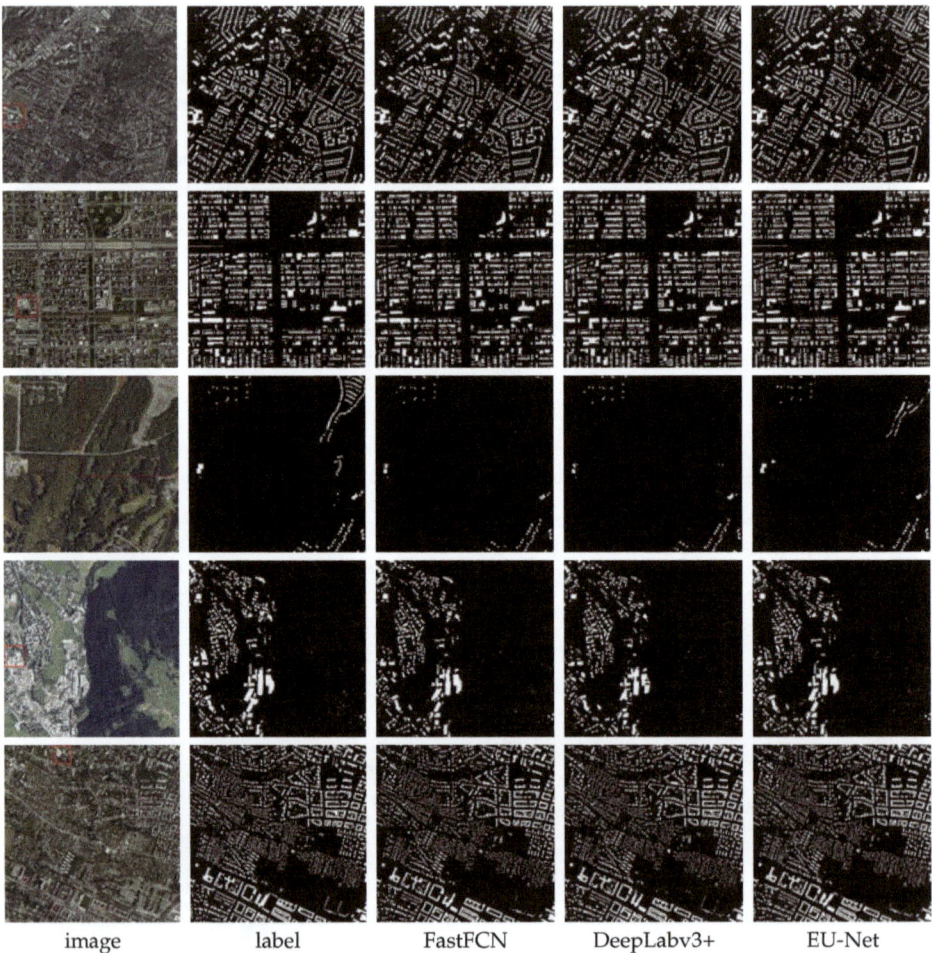

Figure 13. Examples of building extraction results produced by three models on the Inria dataset. Five scenarios from top to bottom are chosen from Austin, Chicago, Kitsap, Tyrol and Vienna. Columns 2–5 are the ground truth labels and prediction maps from FastFCN, DeepLabv3+, and EU-Net, respectively.

Roughly speaking, the three model predictions are visually similar. In order to compare the details, we enlarged the red boxes in Figure 13, shown in Figure 14. From top to bottom are Austin, Chicago, Kitsap, Tyrol and Vienna. For the large buildings in Austin, the predictions of FastFCN and our model were more complete than DeepLabv3+, and the contour accuracy of DeepLabv3+ and our model was better than that of FastFCN. Moreover, there were some misclassifications of small-size buildings in the predictions of FastFCN and DeepLabv3+. For the Chicago, there are many neatly arranged buildings. In the prediction of FastFCN, the adjacent buildings were treated as a whole. The prediction of DeepLabv3+ was slightly better, and only EU-Net successfully distinguished these buildings. In addition, the selected areas in Chicago showed some error labels. Although all three models gave correct predictions for these areas, they gave some false alarms as shown in the yellow rectangles. Moreover, the large building in the prediction of DeepLabv3+ obviously missed the right piece. As for the Kitsap, all buildings in the selected area were error labels. Our EU-Net showed a bad performance while the predictions of DeepLabv3+ and FastFCN only had a few noises in the selected area. We ascribed the bad performance to the simple encoder of EU-Net. Compared with the other four cities, there were much fewer buildings in Kitsap, and thus it was difficult for models to learn enough effective features to correctly extract the buildings in Kitsap, especially for our simple encoder. This conclusion can be also drawn from the quantitative indexes on Kitsap among the five cities. Because there were few buildings in the Kitsap, the error labels had a great influence on the evaluation metrics, leading to abnormally low metrics in this area. If we ignored these obviously error labels, the IoU metrics for EU-Net, DeepLabv3+ and FastFCN could be improved by 4.91%, 7.83% and 7.50%, which were obvious improvements. There are four large buildings in Tyrol. The prediction results of FastFCN and DeepLabv3+ missed two of them while the EU-Net extracted four buildings. As for the enlarged areas in Vienna, the buildings in yellow box were under construction. It is clear that the FastFCN and DeepLabv3+ both failed to extract the two buildings, and the proposed EU-Net successfully detected one of them.

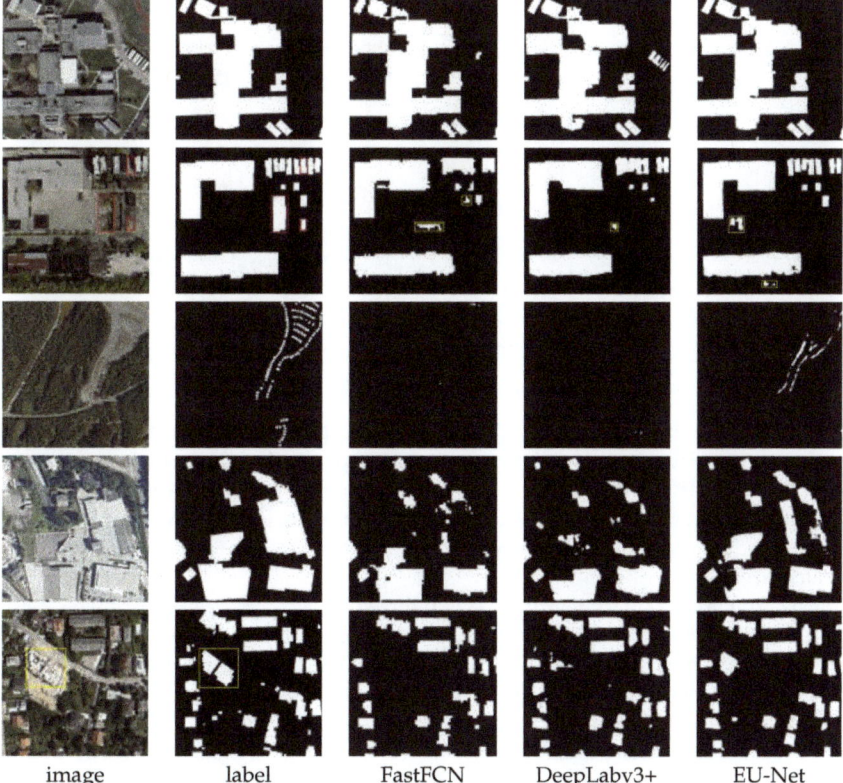

Figure 14. The enlargements of red box selected areas in Figure 13. Five scenarios from top to bottom are Austin, Chicago, Kitsap, Tyrol and Vienna. Columns 2–5 are the ground truth labels and prediction maps from FastFCN, DeepLabv3+, and EU-Net, respectively. The red box selected areas have error labels and correct predictions. The yellow box selected area has correct label and error prediction.

5. Discussion

5.1. Channel Ratio in Short Connection

Unlike natural objects, buildings are a type of artificial objects with clear boundaries. Therefore, we hope to obtain more accurate building contours when improving the overall classification accuracy. As we know, shallow focal features have more detailed information, while rich semantic information in deep global features is more conducive to overall classification. Consequently, we want to choose a best channel ratio between shallow features and deep features in the short connection. A comparative experiment of the channel ratios versus IoU is presented in Figure 15, where 0 means no short connection is used, and 1 means the shallow features have the same channels with deep features. Obviously, the result is much worse without using short connections. Basically, the contour IoU increases as the channel ratio increases and the mask IoU increases first and then decreases. When the ratio is less than 1/4, the addition of shallow features can provide more accurate position information for the semantic features, thereby improving the segmentation accuracy. When the ratio exceeds 1/4, the model will pay too much attention to detail accuracy and ignore the semantic integrity. When ratio is 1/4, the mask IoU achieves the maximum value and therefore the channel ratio is set as 1/4.

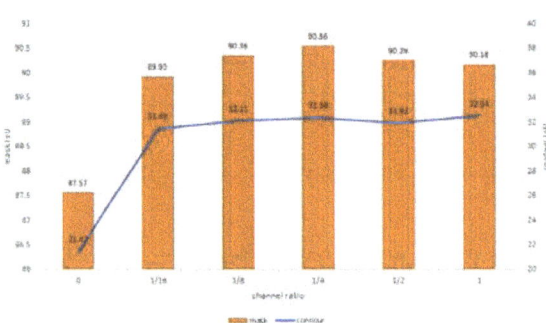

Figure 15. Mask IoU histogram and contour IoU line for different channel ratios of shallow features to deep features used in the short connections.

5.2. Larger Sample Size or Larger Batch Size

Training sample size has a strong influence on the training model since models can potentially capture more fine-grained patterns with higher resolution samples [61]. In the case of limited hardware resources, the batch size must be reduced to increase the sample size. However, as mentioned before, BN is sensitive to input batch size. Considering the performance of BN layers, the batch size should not be too small. So, which one is more important for building extraction, the larger sample size or the larger batch size?

To make full use of GPU memory, we set up three sets of comparison experiments for (sample size, batch size), which are (256*256, 64), (354*354, 32) and (512*512, 16). Table 6 lists the IoU metrics of each set on the three datasets. We can see that increasing the sample size and reducing the batch size make the mask IoU metrics worse on three datasets, i.e., for building extraction, a larger batch size is more important than a larger sample size. In addition, the metric gaps on the Inria dataset were the smallest. For the other two datasets, the results of (512*512, 16) had a significant drop. We think this is because there are no sufficient training samples for the first two datasets when sample size is 512*512. Specifically, the WHU dataset has 4737 samples and the Massachusetts dataset only has 1065 samples. Conversely, for the Inria dataset with sufficient samples, the contour IoU metric of (512*512, 16) is even better than that of (354*354, 32). Overall, the best parameter setting is (256*256, 64).

Table 6. IoU metrics (%) of EU-Net prediction results using different parameter pairs on the test set of three datasets.

		(256*256, 64)	(354*354, 32)	(512*512, 16)
WHU	mask	90.56	90.35	89.00
	contour	32.38	32.22	29.91
Massachusetts	mask	73.93	73.75	70.42
	contour	16.84	16.78	14.99
Inria	mask	80.50	80.24	80.20
	contour	11.40	11.27	11.30

To further verify the impact of batch size and sample size on model training, we used the Inria dataset to supplement three sets of contrast experiments (224*224, 64), (256*256, 32) and (256*256, 16). When training model with the latter two sets of parameters, only one NVIDIA GeForce GTX 1080Ti was used. The IoU metrics are list in Table 7. As expected, compared with the results in Table 6, the IoU continues to decrease as the batch size decrease when the sample size is fixed. And the IoU increases as the sample size increases when batch size keeps same. Consequently, we think that our EU-Net will work better when trained with large sample size and large batch size at the same time.

Table 7. IoU metrics (%) of EU-Net prediction results using different batch sizes on the test set of Inira dataset.

	(224*224, 64)	(256*256, 32)	(256*256, 16)
mask IoU	79.83	80.22	80.01
contour IoU	11.12	10.91	10.87

5.3. DSPP

In this subsection, we will verify the effect of DSPP block by ablation experiment. Training sample size and batch size are set to 256*256 and 64. The EU-Net model without DSPP block is denoted as EU-Net-simple. We tested the EU-Net-simple with three datasets and compared them with the previous experiment results. The IoU metrics are showed in Table 8, and we can see that the IoU metrics have improved on all three datasets by using the DSPP. The minimum increase is 0.35% on the WHU dataset and the maximum increase is 1.76% on the Inria dataset.

Table 8. Comparison of IoU metrics (%) with or without DSPP block on the test sets of three datasets.

		WHU	Massachusetts	Inria
EU-Net	mask	90.56	73.93	80.50
	contour	32.38	16.84	11.40
EU-Net-simple	mask	90.21	73.07	78.74
	contour	32.13	16.52	10.61

We propose to use DSPP to acquire multi-scale features, which can improve the extraction accuracy of buildings of different sizes, especially the medium-sized and oversized buildings. For a clearer visualization, we chose an obvious area from each dataset. The selected areas are shown in Figure 16. For the WHU block, part of the roof parking was misclassified in the result of EU-Net-simple. For the Massachusetts block with multi-scale buildings, the extraction accuracy of EU-Net was higher for all buildings of different sizes. As for the Inria block, the results of the small-sized buildings had no significant difference. EU-Net clearly performed better on the large building extractions. All the above results demonstrate that DSPP does play its intended role. The red box selected areas in the Massachusetts block and the Inria block are error ground truth labels, and both models gave the right predictions.

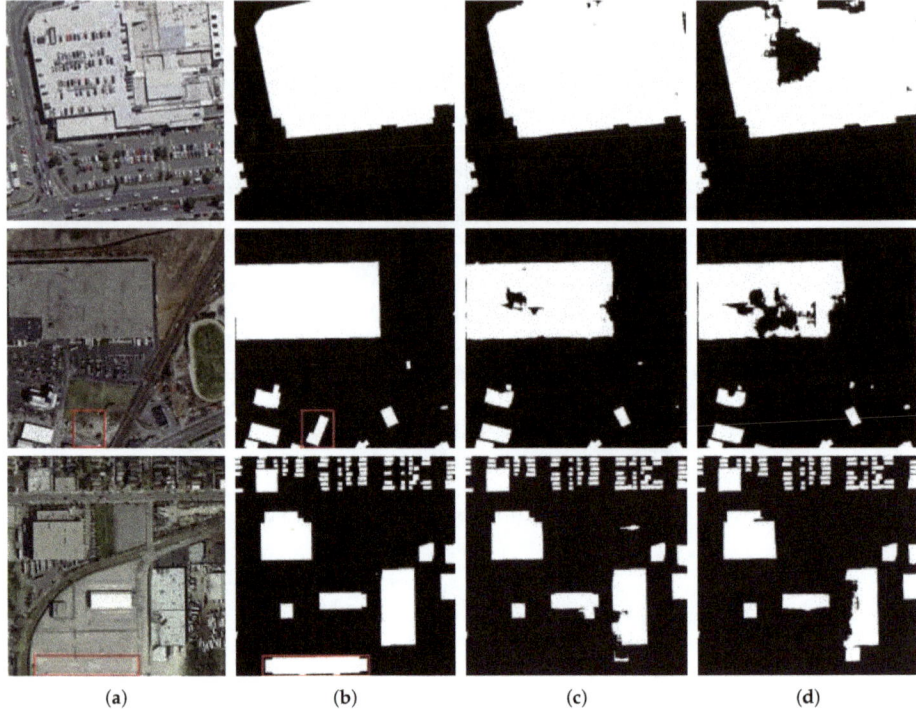

Figure 16. Sample comparison of prediction results with or without DSPP. Three scenarios from top to bottom are from WHU, Massachusetts and Inria. (**a**) Original image. (**b**) Ground truth label. (**c**) Prediction of EU-Net. (**d**) Prediction of EU-Net-simple. The red box selected areas have error labels.

Moreover, adding DSPP block does not significantly reduce the efficiency of the model. We compare the time of processing one image in the Inria dataset. The EU-Net-simple took 13.95 s, and the EU-Net took 14.79 s. After adding DSPP block, consuming time to process a 5000*5000-pixel image only increased 0.84 s. This time increase is worthwhile compared to the increase in metrics.

5.4. Loss Function

The error labels can hinder model training. To overcome this problem, we introduced FL in reverse to reduce the gradient generated by the error labels. We also verified the performance of FL through comparative experiments. We set $\beta = 0$ to remove the FL from loss function and performed the EU-Net on the Massachusetts dataset and the Inria dataset. Training sample size and batch size were still set to 256*256 and 64. The IoU metrics are listed in Table 9. The mask IoU has been improved on the Massachusetts dataset by 1.1% and improved on the Inria dataset by 0.05%. It turns out that negative FL does work. In addition, we can see that even without using FL, the results of our EU-Net are better than the state-of-the-art results on these two datasets.

Table 9 proved that negative FL does work. To analyze the influence of the FL weight, we changed β to 1 and 3, and retrained EU-Net on Massachusetts dataset. The IoU metrics are listed in Table 10. For both mask IoU and contour IoU, the maximums were obtained when β was 2. Therefore, we set β to 2 when training the model on Massachusetts dataset and Inria dataset.

Table 9. IoU metrics (%) of prediction results with or without FL on the test set of Massachusetts dataset and the test set of Inria dataset.

		Massachusetts	Inria
CE+FL	mask	73.93	80.50
	contour	16.84	11.40
CE	mask	72.83	80.45
	contour	16.53	11.32

Table 10. IoU metrics (%) of prediction results with different β on the test set of Massachusetts dataset.

β	1	2	3
mask	73.67	73.93	73.51
contour	16.62	16.84	16.65

5.5. Learning Rate and Epoch

The learning rate and number of epochs can greatly affect the model performance. In order to obtain the optimal hyperparameters, we conducted some comparative experiments. As we know, the learning rate affects the convergence speed of the model, and increasing the learning rate in a certain range can speed up the network convergence. Therefore, we used different learning rates and fixed the number of epochs to test the convergence speed of the network. Taking the Massachusetts dataset as an example, we started with the learning rate of 0.01 and then increased to 0.1, 0.2, 0.5 and 0.8. For different learning rates, we fixed the number of epochs to 500. As shown in Figure 17, the convergence speed continues to increase until the learning rate reaches 0.5. While continuing to increase to 0.8, the convergence speed has no obvious increasing. Consequently, we use learning rate of 0.5 to train EU-Net on Massachusetts dataset.

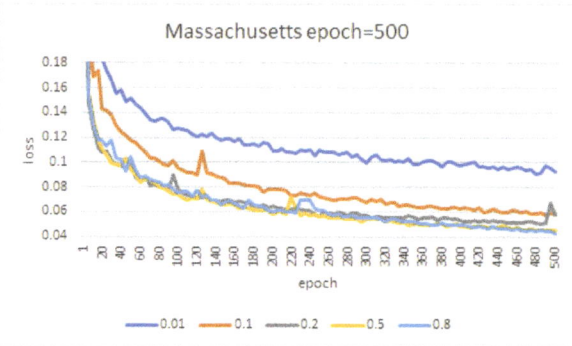

Figure 17. The EU-Net training loss curves for different learning rate on Massachusetts dataset.

After determining the appropriate learning rate, we trained our model with more epochs. As shown in Figure 18a, the valid_accuracy curve tends to be flat when the number of epochs increases to 300, and the valid_loss curve has no obvious upward trend, which means that model has no overfitting. Consequently, 300 epochs are sufficient to train EU-Net on WHU dataset. We can see from Figure 18b,c that although the valid_accuracy curves have not decreased, the valid_loss curves have begun to increase. In other words, continuing to train the model may lead to overfitting. Therefore, 2000 epochs and 400 epochs are sufficient to train EU-Net on Massachusetts dataset and Inria dataset.

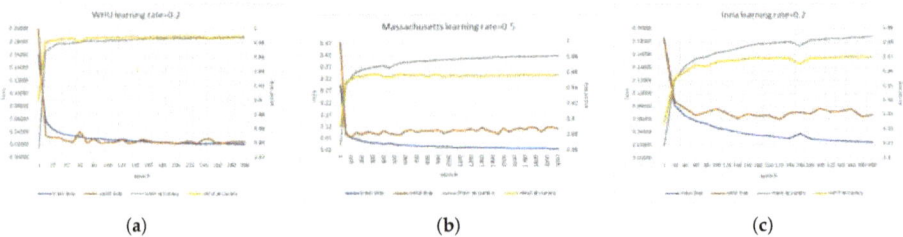

(a)   (b)   (c)

Figure 18. The loss curves and accuracy curves on training set and validation set for three datasets. (a) WHU dataset. (b) Massachusetts dataset. (c) Inria dataset.

6. Conclusions

In this paper, we proposed an effective FCN-based neural network model for building extraction from aerial remote sensing images. The proposed network consists of three parts: encoder, DSPP block and decoder. In order to save GPU memory cost, we chose the first 13 layers in VGG16 as our encoder. Then, we can train our model with large batch size and learning rate to improve training efficiency. The DSPP block was proposed to enlarge the receptive field and extract dense multi-scales features. In the decoder, by adjusting the channel ratio of the deep features and the shallow features, we can achieve higher building contour accuracy as much as possible while improving the semantic segmentation accuracy.

The experiments were conducted on three datasets. Experimental results show that the DSPP block can improve the extraction accuracy of multi-size buildings. The impact of error labels in training samples could be successfully suppressed by using the focal loss in reverse. Without any post-processing, our model refreshed the state-of-the-art performance on the three datasets simultaneously.

Although our model has achieved satisfactory results, the building boundary accuracy is still very low. In future studies, we will try to modify the loss function or adjust the network structure to improve the extraction accuracy of the building contours.

Author Contributions: Conceptualization, W.K.; Methodology, W.K.; Resources, F.W.; Supervision, F.W. and H.Y.; Validation, Y.X.; Writing—original draft, W.K.; Writing—review & editing, Y.X., F.W. and H.Y.

Funding: This research was funded by the From 0 to 1 Original Innovative Project of the Chinese Academy of Sciences Frontier Science Research Program under Grant ZDBS-LY-JSC036, and the National Natural Science Foundation of China under Grant 61901439.

Acknowledgments: The authors would like to thank the anonymous reviewers for their constructive comments and suggestions.

Conflicts of Interest: The authors declare no conflict of interest.

Appendix A. Derivatives

We use y = 1 to specify the building label and y = −1 to specify the others. A pixel in output map has two predictions, which we denote as x for label y = 1 and $\hat{x}$ for label y = −1. Before calculating the loss, we use SoftMax to calculate the probability p that a pixel belongs to building. p can define as:

$$p = \frac{e^x}{e^x + e^{\hat{x}}}. \tag{A1}$$

Derivative for p regarding x is:

$$\frac{\partial p}{\partial x} = p(1-p). \tag{A2}$$

Then, we use p_t to specify the pixel belongs to different classes. p_t can defined as:

$$p_t = \begin{cases} p & y = 1 \\ 1 - p & y = -1. \end{cases} \tag{A3}$$

This is compatible with Equation (2). According to the chain rule, derivative for p_t regarding x is:

$$\frac{\partial p_t}{\partial x} = \frac{\partial p_t}{\partial p}\frac{\partial p}{\partial x} = yp(1-p). \tag{A4}$$

According to Equations (3) and (5), we can get the derivatives for CE and FL regarding p_t are:

$$\frac{\partial CE}{\partial p_t} = -\frac{1}{p_t}. \tag{A5}$$

$$\frac{\partial FL}{\partial p_t} = \alpha_t(1-p_t)^\gamma (\frac{\gamma log(p_t)}{1-p_t} - \frac{1}{p_t}). \tag{A6}$$

According to the chain rule, derivatives for CE and FL regarding x are:

$$\frac{\partial CE}{\partial x} = y(p_t - 1), \tag{A7}$$

$$\frac{\partial FL}{\partial x} = y\alpha_t(1-p_t)^\gamma (\gamma p_t log(p_t) + p_t - 1). \tag{A8}$$

According the definition of loss function in Equation (6), derivative for our loss regarding x is:

$$\frac{\partial Loss}{\partial x} = y(p_t - 1) - \beta y\alpha_t(1-p_t)^\gamma (\gamma p_t log(p_t) + p_t - 1). \tag{A9}$$

It is obvious that the proposed loss function is globally continuous and differentiable.

References

1. Liu, Y.; Fan, B.; Wang, L.; Bai, J.; Xiang, S.; Pan, C. Semantic labeling in very high resolution images via a self-cascaded convolutional neural network. *ISPRS J. Photogramm. Remote Sens.* **2018**, *145*, 78–95. [CrossRef]
2. Mohammadimanesh, F.; Salehi, B.; Mahdianpari, M.; Gill, E.; Molinier, M. A new fully convolutional neural network for semantic segmentation of polarimetric SAR imagery in complex land cover ecosystem. *ISPRS J. Photogramm. Remote Sens.* **2019**, *151*, 223–236. [CrossRef]
3. Kemker, R.; Salvaggio, C.; Kanan, C. Algorithms for semantic segmentation of multispectral remote sensing imagery using deep learning. *ISPRS J. Photogramm. Remote Sens.* **2018**, *145*, 60–77. [CrossRef]
4. Huang, J.; Zhang, X.; Xin, Q.; Sun, Y.; Zhang, P. Automatic building extraction from high-resolution aerial images and LiDAR data using gated residual refinement network. *ISPRS J. Photogramm. Remote Sens.* **2019**, *151*, 91–105. [CrossRef]
5. Ji, S.; Wei, S.; Lu, M. Fully convolutional networks for multisource building extraction from an open aerial and satellite imagery data set. *IEEE Trans. Geosci. Remote Sens.* **2018**, *57*, 574–586. [CrossRef]
6. Mnih, V. *Machine Learning for Aerial Image Labeling*; University of Toronto: Toront, ON, Canada, 2013.
7. Maggiori, E.; Tarabalka, Y.; Charpiat, G.; Alliez, P. Can semantic labeling methods generalize to any city? the inria aerial image labeling benchmark. In Proceedings of the 2017 IEEE International Geoscience and Remote Sensing Symposium (IGARSS), Fort Worth, TX, USA, 23–28 July 2017; pp. 3226–3229.
8. Pan, X.; Yang, F.; Gao, L.; Chen, Z.; Zhang, B.; Fan, H.; Ren, J. Building Extraction from High-Resolution Aerial Imagery Using a Generative Adversarial Network with Spatial and Channel Attention Mechanisms. *Remote Sens.* **2019**, *11*, 917. [CrossRef]
9. Xu, Y.; Wu, L.; Xie, Z.; Chen, Z. Building extraction in very high resolution remote sensing imagery using deep learning and guided filters. *Remote Sens.* **2018**, *10*, 144. [CrossRef]

10. Ok, A.O.; Senaras, C.; Yuksel, B. Automated detection of arbitrarily shaped buildings in complex environments from monocular VHR optical satellite imagery. *IEEE Trans. Geosci. Remote Sens.* **2012**, *51*, 1701–1717. [CrossRef]
11. Huang, X.; Yuan, W.; Li, J.; Zhang, L. A new building extraction postprocessing framework for high-spatial-resolution remote-sensing imagery. *IEEE J. Sel. Top. Appl. Earth Obs. Remote Sens.* **2016**, *10*, 654–668. [CrossRef]
12. Chen, R.; Li, X.; Li, J. Object-based features for house detection from RGB high-resolution images. *Remote Sens.* **2018**, *10*, 451. [CrossRef]
13. Senaras, C.; Ozay, M.; Vural, F.T.Y. Building detection with decision fusion. *IEEE J. Sel. Top. Appl. Earth Obs. Remote Sens.* **2013**, *6*, 1295–1304. [CrossRef]
14. Saito, S.; Aoki, Y. Building and road detection from large aerial imagery. *Proc. SPIE* **2015**, *9405*, 94050K.
15. Vakalopoulou, M.; Karantzalos, K.; Komodakis, N.; Paragios, N. Building detection in very high resolution multispectral data with deep learning features. In Proceedings of the 2015 IEEE International Geoscience and Remote Sensing Symposium (IGARSS), Milan, Italy, 13–18 July 2015; pp. 1873–1876.
16. Long, J.; Shelhamer, E.; Darrell, T. Fully convolutional networks for semantic segmentation. In Proceedings of the IEEE Conference on Computer Vision and Pattern Recognition, Boston, MA, USA, 7–12 June 2015; pp. 3431–3440.
17. Noh, H.; Hong, S.; Han, B. Learning deconvolution network for semantic segmentation. In Proceedings of the IEEE International Conference on Computer Vision, Santiago, Chile, 7–13 December 2015; pp. 1520–1528.
18. Badrinarayanan, V.; Kendall, A.; Cipolla, R. Segnet: A deep convolutional encoder-decoder architecture for image segmentation. *IEEE Trans. Pattern Anal. Mach. Intell.* **2017**, *39*, 2481–2495. [CrossRef] [PubMed]
19. Ronneberger, O.; Fischer, P.; Brox, T. U-net: Convolutional networks for biomedical image segmentation. In Proceedings of the International Conference on Medical Image Computing and Computer-Assisted Intervention, Munich, Germany, 5–9 October 2015; pp. 234–241.
20. Wu, H.; Zhang, J.; Huang, K.; Liang, K.; Yu, Y. FastFCN: Rethinking Dilated Convolution in the Backbone for Semantic Segmentation. *arXiv* **2019**, arXiv:1903.11816.
21. Chen, L.C.; Papandreou, G.; Kokkinos, I.; Murphy, K.; Yuille, A.L. Semantic image segmentation with deep convolutional nets and fully connected crfs. *arXiv* **2014**, arXiv:1412.7062.
22. Liu, P.; Liu, X.; Liu, M.; Shi, Q.; Yang, J.; Xu, X.; Zhang, Y. Building Footprint Extraction from High-Resolution Images via Spatial Residual Inception Convolutional Neural Network. *Remote Sens.* **2019**, *11*, 830. [CrossRef]
23. Lin, J.; Jing, W.; Song, H.; Chen, G. ESFNet: Efficient Network for Building Extraction From High-Resolution Aerial Images. *IEEE Access* **2019**, *7*, 54285–54294. [CrossRef]
24. Zhang, Z.; Wang, Y. JointNet: A Common Neural Network for Road and Building Extraction. *Remote Sens.* **2019**, *11*, 696. [CrossRef]
25. Li, X.; Yao, X.; Fang, Y. Building-A-Nets: Robust Building Extraction From High-Resolution Remote Sensing Images With Adversarial Networks. *IEEE J. Sel. Top. Appl. Earth Obs. Remote Sens.* **2018**, *11*, 3680–3687. [CrossRef]
26. Mou, L.; Zhu, X.X. RiFCN: Recurrent network in fully convolutional network for semantic segmentation of high resolution remote sensing images. *arXiv* **2018**, arXiv:1805.02091.
27. Krähenbühl, P.; Koltun, V. Efficient inference in fully connected crfs with gaussian edge potentials. *Adv. Neural Inf. Process. Syst.* **2011**, *24*, 109–117.
28. Shrestha, S.; Vanneschi, L. Improved fully convolutional network with conditional random fields for building extraction. *Remote Sens.* **2018**, *10*, 1135. [CrossRef]
29. Clevert, D.A.; Unterthiner, T.; Hochreiter, S. Fast and accurate deep network learning by exponential linear units (elus). *arXiv* **2015**, arXiv:1511.07289.
30. Alshehhi, R.; Marpu, P.R.; Woon, W.L.; Dalla Mura, M. Simultaneous extraction of roads and buildings in remote sensing imagery with convolutional neural networks. *ISPRS J. Photogramm. Remote Sens.* **2017**, *130*, 139–149. [CrossRef]
31. Goodfellow, I. NIPS 2016 tutorial: Generative adversarial networks. *arXiv* **2016**, arXiv:1701.00160.
32. Zhu, J.Y.; Park, T.; Isola, P.; Efros, A.A. Unpaired image-to-image translation using cycle-consistent adversarial networks. In Proceedings of the IEEE International Conference on Computer Vision, Venice, Italy, 22–29 October 2017; pp. 2223–2232.

33. Liu, M.Y.; Breuel, T.; Kautz, J. Unsupervised image-to-image translation networks. *arXiv* **2017**, arXiv:1703.00848.
34. Ledig, C.; Theis, L.; Huszár, F.; Caballero, J.; Cunningham, A.; Acosta, A.; Aitken, A.; Tejani, A.; Totz, J.; Wang, Z.; et al. Photo-realistic single image super-resolution using a generative adversarial network. In Proceedings of the IEEE Conference on Computer Vision and Pattern Recognition, Honolulu, HI, USA, 21–26 July 2017; pp. 4681–4690.
35. Wang, X.; Yu, K.; Wu, S.; Gu, J.; Liu, Y.; Dong, C.; Qiao, Y.; Change Loy, C. Esrgan: Enhanced super-resolution generative adversarial networks. In Proceedings of the European Conference on Computer Vision (ECCV), Munich, Germany, 8–14 September 2018.
36. Pan, B.; Shi, Z.; Xu, X. MugNet: Deep learning for hyperspectral image classification using limited samples. *ISPRS J. Photogramm. Remote Sens.* **2018**, *145*, 108–119. [CrossRef]
37. Bittner, K.; Adam, F.; Cui, S.; Körner, M.; Reinartz, P. Building footprint extraction from VHR remote sensing images combined with normalized DSMs using fused fully convolutional networks. *IEEE J. Sel. Top. Appl. Earth Obs. Remote Sens.* **2018**, *11*, 2615–2629. [CrossRef]
38. Zhang, Y.; Gong, W.; Sun, J.; Li, W. Web-Net: A Novel Nest Networks with Ultra-Hierarchical Sampling for Building Extraction from Aerial Imageries. *Remote Sens.* **2019**, *11*, 1897. [CrossRef]
39. Wu, G.; Shao, X.; Guo, Z.; Chen, Q.; Yuan, W.; Shi, X.; Xu, Y.; Shibasaki, R. Automatic building segmentation of aerial imagery using multi-constraint fully convolutional networks. *Remote Sens.* **2018**, *10*, 407. [CrossRef]
40. Ji, S.; Wei, S.; Lu, M. A scale robust convolutional neural network for automatic building extraction from aerial and satellite imagery. *Int. J. Remote Sens.* **2019**, *40*, 3308–3322. [CrossRef]
41. Audebert, N.; Le Saux, B.; Lefèvre, S. Beyond RGB: Very high resolution urban remote sensing with multimodal deep networks. *ISPRS J. Photogramm. Remote Sens.* **2018**, *140*, 20–32. [CrossRef]
42. Bischke, B.; Helber, P.; Folz, J.; Borth, D.; Dengel, A. Multi-task learning for segmentation of building footprints with deep neural networks. In Proceedings of the 2019 IEEE International Conference on Image Processing (ICIP), Taipei, Taiwan, 22–29 September 2019; pp. 1480–1484.
43. Srivastava, N.; Hinton, G.; Krizhevsky, A.; Sutskever, I.; Salakhutdinov, R. Dropout: A simple way to prevent neural networks from overfitting. *J. Mach. Learn. Res.* **2014**, *15*, 1929–1958.
44. Bridle, J.S. Probabilistic interpretation of feedforward classification network outputs, with relationships to statistical pattern recognition. In *Neurocomputing*; Springer: Berlin/Heidelberg, Germany, 1990; pp. 227–236.
45. Nair, V.; Hinton, G.E. Rectified linear units improve restricted boltzmann machines. In Proceedings of the 27th International Conference on Machine Learning (ICML-10), Haifa, Israel, 21–25 June 2010; pp. 807–814.
46. Ioffe, S.; Szegedy, C. Batch normalization: Accelerating deep network training by reducing internal covariate shift. *arXiv* **2015**, arXiv:1502.03167.
47. Wu, Y.; He, K. Group normalization. In Proceedings of the European Conference on Computer Vision (ECCV), Munich, Germany, 8–14 September 2018; pp. 3–19.
48. He, K.; Zhang, X.; Ren, S.; Sun, J. Identity mappings in deep residual networks. In Proceedings of the European Conference on Computer Vision, Amsterdam, The Netherlands, 11–14 October 2016; pp. 630–645.
49. Chen, L.C.; Zhu, Y.; Papandreou, G.; Schroff, F.; Adam, H. Encoder-decoder with atrous separable convolution for semantic image segmentation. In Proceedings of the European Conference on Computer Vision (ECCV), Munich, Germany, 8–14 September 2018; pp. 801–818.
50. Marcu, A.; Costea, D.; Slusanschi, E.; Leordeanu, M. A multi-stage multi-task neural network for aerial scene interpretation and geolocalization. *arXiv* **2018**, arXiv:1804.01322.
51. Ruan, T.; Liu, T.; Huang, Z.; Wei, Y.; Wei, S.; Zhao, Y. Devil in the details: Towards accurate single and multiple human parsing. In Proceedings of the AAAI Conference on Artificial Intelligence, Honolulu, HI, USA, 27 January–1 February 2019; Volume 33, pp. 4814–4821.
52. Lu, T.; Ming, D.; Lin, X.; Hong, Z.; Bai, X.; Fang, J. Detecting building edges from high spatial resolution remote sensing imagery using richer convolution features network. *Remote Sens.* **2018**, *10*, 1496. [CrossRef]
53. Luo, W.; Li, Y.; Urtasun, R.; Zemel, R. Understanding the effective receptive field in deep convolutional neural networks. *arXiv* **2016**, arXiv:1701.04128.
54. Lin, T.Y.; Goyal, P.; Girshick, R.; He, K.; Dollár, P. Focal loss for dense object detection. In Proceedings of the IEEE International Conference on Computer Vision, Venice, Italy, 22–29 October 2017; pp. 2980–2988.

55. Glorot, X.; Bengio, Y. Understanding the difficulty of training deep feedforward neural networks. In Proceedings of the Thirteenth International Conference on Artificial Intelligence and Statistics, Sardinia, Italy, 13–15 May 2010; pp. 249–256.
56. Han, W.; Feng, R.; Wang, L.; Cheng, Y. A semi-supervised generative framework with deep learning features for high-resolution remote sensing image scene classification. *ISPRS J. Photogramm. Remote Sens.* **2018**, *145*, 23–43. [CrossRef]
57. Kang, J.; Körner, M.; Wang, Y.; Taubenböck, H.; Zhu, X.X. Building instance classification using street view images. *ISPRS J. Photogramm. Remote Sens.* **2018**, *145*, 44–59. [CrossRef]
58. Chen, L.C.; Papandreou, G.; Kokkinos, I.; Murphy, K.; Yuille, A.L. Deeplab: Semantic image segmentation with deep convolutional nets, atrous convolution, and fully connected crfs. *IEEE Trans. Pattern Anal. Mach. Intell.* **2017**, *40*, 834–848. [CrossRef]
59. Chen, L.C.; Papandreou, G.; Schroff, F.; Adam, H. Rethinking atrous convolution for semantic image segmentation. *arXiv* **2017**, arXiv:1706.05587.
60. Khalel, A.; El-Saban, M. Automatic pixelwise object labeling for aerial imagery using stacked u-nets. *arXiv* **2018**, arXiv:1803.04953.
61. Tan, M.; Le, Q.V. EfficientNet: Rethinking Model Scaling for Convolutional Neural Networks. *arXiv* **2019**, arXiv:1905.11946.

© 2019 by the authors. Licensee MDPI, Basel, Switzerland. This article is an open access article distributed under the terms and conditions of the Creative Commons Attribution (CC BY) license (http://creativecommons.org/licenses/by/4.0/).

Article

Structural 3D Reconstruction of Indoor Space for 5G Signal Simulation with Mobile Laser Scanning Point Clouds

Yang Cui [1], Qingquan Li [1,*] and Zhen Dong [2]

1. Shenzhen Key Laboratory of Spatial Smart Sensing and Services & The Key Laboratory for Geo-Environment Monitoring of Coastal Zone of the National Administration of Surveying, Mapping and GeoInformation & College of Information Engineering, Shenzhen University, Nanhai Road 3688, Shenzhen 518060, China; cuiyang@szu.edu.cn
2. State Key Laboratory of Information Engineering in Surveying, Mapping and Remote Sensing, Wuhan University, Wuhan 430079, China; dongzhenwhu@whu.edu.cn
* Correspondence: liqq@szu.edu.cn; Tel.: +86-755-2653-1934

Received: 2 September 2019; Accepted: 23 September 2019; Published: 27 September 2019

Abstract: 3D modelling of indoor environment is essential in smart city applications such as building information modelling (BIM), spatial location application, energy consumption estimation, and signal simulation, etc. Fast and stable reconstruction of 3D models from point clouds has already attracted considerable research interest. However, in the complex indoor environment, automated reconstruction of detailed 3D models still remains a serious challenge. To address these issues, this paper presents a novel method that couples linear structures with three-dimensional geometric surfaces to automatically reconstruct 3D models using point cloud data from mobile laser scanning. In our proposed approach, a fully automatic room segmentation is performed on the unstructured point clouds via multi-label graph cuts with semantic constraints, which can overcome the over-segmentation in the long corridor. Then, the horizontal slices of point clouds with individual room are projected onto the plane to form a binary image, which is followed by line extraction and regularization to generate floorplan lines. The 3D structured models are reconstructed by multi-label graph cuts, which is designed to combine segmented room, line and surface elements as semantic constraints. Finally, this paper proposed a novel application that 5G signal simulation based on the output structural model to aim at determining the optimal location of 5G small base station in a large-scale indoor scene for the future. Four datasets collected using handheld and backpack laser scanning systems in different locations were used to evaluate the proposed method. The results indicate our proposed methodology provides an accurate and efficient reconstruction of detailed structured models from complex indoor scenes.

Keywords: 3D reconstruction; indoor modelling; mobile laser scanning; point clouds; 5G signal simulation

1. Introduction

Three-dimensional (3D) reconstruction of the indoor environment has received significant attention due to the development of smart cities. However, the automation of generating high-quality models remains to be a challenging issue due to the complexities of the indoor environment. The industry foundation classes (IFC) defines building information modelling (BIM) as having rich semantic information, 3D structural information, spatial relationships, and interoperable geometry [1]. BIM has been used in a number of applications, such as indoor navigation [2,3], space management [4,5], energy simulation [6] and real-time emergency response [7,8]. Primary indoor locations utilizing BIM include indoor offices, parking lots, and commercial establishments, which are commonly comprised of basic elements, such as the ceilings, floors, walls, windows, doors, and pillars, and not objects, such as furniture. 3D Indoor models are often generated manually by creating geometric representations using point cloud data and commercial software. This often requires significant investment in time and in training personnel [9]. To accelerate the efficiency of data acquisition and improve the automation of reconstructed models, various laser scanning technologies and automated modelling methods have been developed. Over the past few years, terrestrial laser scanning (TLS) has been used to obtain high-quality data in the indoor scene, but often suffers from low mapping efficiency due to laborious scan station resetting, protracted registration procedures, and high costs. Thus, its application has excluded large-scale indoor data acquisition. RGBD panorama is acquired by a camera and a depth sensor [10], with added advantages of affordability and convenience of use. However, the image distorted and noisy that lead to difficultly build accurate models in large scale scenes. With the development of the simultaneous localization and mapping (SLAM), various types of mobile laser scanning (MLS) devices have been used for data acquisition, such as handheld, backpack, push-cart and robot mobile laser scanners. MLS systems can obtain point clouds by moving from different spaces and measure from different locations. The easy-to-use and affordable indoor MLS systems are mostly used for data acquisition of large indoor scenes [11]. While MLS ensures good coverage for indoor environment mapping, the data can be affected by a number of factors (e.g., moving objects, multiple reflections, and dynamic occlusions) resulting in quality losses, which present serious challenges in model reconstruction.

Recently, numerous studies have focused on modelling the indoor environment. For example, some research [10,12–22] segmented the unstructured points clouds into individual rooms to provide a prior knowledge in building the indoor model. The over-segmentation often occurs in the spatial partition of long corridors [13,14], creating a substantial challenge when segmenting long corridors. Neoteric methods [9,15,22–31] have focused on extracting piecewise planar surfaces to construct the model. Although the results of room segmentation have been satisfactory, indoor models still are difficult to accurately reconstruct because of high levels of occlusion and noisy point cloud data, while still requiring interaction [22]. Large scale scene modelling based on surface elements has significant computational challenges. In order to increase computational efficiency, many line-based reconstructed methods have been proposed [32–40]. Their results have shown that line-based methods are efficient and effective, which can produce accurate and complete line segments. However, the reconstructed models are only represented in vector line structure without room semantic information, even when these elements are unconnected. Furthermore, a satisfactory solution for indoor interior reconstruction has not been developed due to the complexity of the indoor environment and unfiltered data noise.

In this study, a novel method is developed that combines line structures and 3D surface geometry to automatically build a 3D indoor model with detailed structural and semantic information using MLS point clouds. A fully automatic room segmentation was performed on unstructured point clouds via multi-label graph cuts, which can solve the over-segmentation in long corridors. Point cloud slices of individual rooms are transformed as a binary image, from which extraction, and regularization of line elements, aimed at improving the computational efficiency and structural accuracy. 3D structured models were reconstructed using multi-label graph cuts, and room segmentation and lines were used as semantic and structural constraints of 2D floorplan, with surfaces providing 3D geometric information. Finally, an innovative approach was employed using 5G signal simulation based on the reconstructed model, where the basic structural elements (e.g., windows, doors, pillars, walls, ceilings, and floors) have a critical effect on the signal transmission.

This study offers two major contributions. First, room segmentation was employed to semantically label the unstructured point clouds using multi-label graph cuts with semantic constraints of openings that can overcome the over-segmentation in corridors. Point cloud slices of individual rooms are transformed into images, which are then extracted and regularized into the floorplan lines. This innovation contributes to improving the efficiency and precision of extracting structural elements. Second, 3D structured models were reconstructed using multi-label graph cuts with room segmentation and 2D floorplan lines, and with 3D surfaces used as constraints. The resulting structured models provided room adjacency relationship, geometric characteristics, and semantic information, which are then applied to the signal simulation to provide the optimal location of the 5G small base station in indoor environment for future.

2. Related Work

In the last decade, various approaches designed for indoor modelling using 3D laser scanning have been developed, which mostly consisted of three fundamental steps: (1) room segmentation, (2) reconstruction of indoor space, and (3) indoor model application.

2.1. Room Segmentation

Room segmentation is key semantic information used in model reconstruction. Ikehata et al. [10] proposed the segmentation method of rooms, which repeated k-medoids algorithm to cluster sub-sampled pixels, the distance metric for clustering is based on a binary visibility vector of scanning center. Mura et al. [12] presented an approach that established a global affinity measure between cells by diffusion maps, and partitioned rooms by clustering 2D cells iteratively. Ochmann et al. [13] proposed a method that segmented indoor point clouds into individual rooms by the visibility-based and class-conditional probabilities, which based on initial knowledge of scans and scan positions; however, the over-segmentation occurs in long corridors. Turner et al. [14] proposed an approach that triangulated a 2D sampling of wall positions and separated these triangles into interior and exterior domains. The room segmentation can be obtained by Graph-cut in the triangulated map. However, these methods of room partition are limited to depend on TLS the scanning position, but not for MLS point clouds. Wang et al. [17] employed the hierarchical clustering method for partitioning rooms, which established diffusion maps to merge the over-segmented spaces. The method uses scan trajectories instead of scanner positions. Díaz-Vilariño et al. [18] proposed a method that used the timestamp information to determine the visible point clouds of each trajectory point and constructed the energy minimization function for global spatial optimization to complete individual room segmentation. Their method relies heavily on data quality and integrity and has been shown effective in simple scenarios. Li et al. [19] proposed a comprehensive segmentation method that is created by a morphological erosion and connectivity analysis methods on the floor space, which overcomes over-segmentation in long corridors. Similarly, Ochmann et al. [22] proposed a fully automatic room segmentation that performed visibility tests by the ray casting between point patches on surfaces to build visibility graph, and then the nodes of this graph are clustered by the Markov Clustering method [21].

2.2. Reconstruction of Indoor Space

Current methods for the reconstruction of indoor spaces are mainly based on the extraction of surfaces [9,15,22–31] and lines [32–40].

(1) Surface-Based Reconstruction

The accuracy of reconstruction models mostly depends on the extraction of surfaces. Bassier and Vergauwen [9] proposed an innovative approach to segment walls using the Conditional Random Field and concluded that the generated wall clusters were better than traditional region growing. Other researchers have also extracted unconnected planes from 3D point clouds [23,24], but only enabled visualization and excluded the spatial topological relationship. Monszpart et al. [25] proposed an effective approach to extract Regular Arrangements of Planes (RAP) from unstructured point clouds in rebuilding man-made scenes. However, the method requires long computing time for the reconstruction of large scenes. Awrangjeb et al. [26] proposed a novel 3D roof reconstruction technique that constructs an adjacency matrix to define the topological relationships among the detected roof planes, in addition, used the generated building models to detect 3D change in buildings. Xiao and

Furukawa [27] employed constructive solid geometry (CSG) operations to generate volumetric wall model, which focused on the large-scale reconstruction without semantic information. To overcome this deficiency, Ochmann et al. [15] extracted the piece planar surfaces by the RANSAC approach [28], constructed partitions based on the wall surfaces, utilized the global optimization to reconstruct wall elements, and then finally built the volumetric model of single room by extruding the walls. However, the thickness of model walls was assigned a fixed threshold that led to significant errors. Mura et al. [29] extracted the permanent components used in constructing adjacent relations and partitions of 3D polyhedral cells. The final general three-dimensional interior model was reconstructed using multi-label to optimize cell selection; however, this method was only applied to small-scale scenes. Ochmann et al. [22] extracted wall candidates and formulated the optimization method to arrange volumetric wall entities to build the structural model. Reconstructing the model presents latent difficulties due to occlusion and clutter point clouds in the indoor environment. While this approach can reshape the model manually, its main limitation includes slanted walls, ceilings and floors, and detailed pillar reconstruction.

(2) Line-Based Reconstruction

Many researchers have also studied indoor reconstruction based on lines. Lin et al. [32] proposed a method where line segments can accurately be extracted from unorganized point clouds. However, the line elements remained completely isolated and devoid of information about topological relations. Similarly, Xia and Wang [33], Lu et al. [34] extracted unstructured line elements from point clouds. Liu et al. [35] proposed the FloorNet where a deep neural architecture can automatically reconstruct the floorplan from RGBD videos with camera poses. Extracting initial line structures from labeled points, Wang et al. [37] proposed a conditional Generative Adversarial Nets (cGAN) deep learning method to optimize the detected lines to rebuild line frameworks with structural representation in the cluttered indoor environment. Bauchet et al. [38] proposed an approach that extract flexibility on polygon shape, which better recover geometric patterns but still lacks topological information. Sui et al. [39] introduced an automatic method for extracting floorplans from slices that correct both normal vector and position to obtain accurate boundaries, which are then used in propagating to the other floors. However, the reconstructed models are only applicable for visualization and cannot be used for geometric manipulation. For the underground infrastructure, Novakovic et al. [40] extracted the 2D profile from the point cloud data of tunnel, built the spatial parameter model and simulated cargo tunnel pass.

2.3. Indoor Model Application

Previous studies have investigated the various applications of the BIM model. Díaz-Vilariño et al. [3] proposed an approach based on the BIM model that determined the optimal scan positions in planning the shortest route for an automatic robot visit. Boyes et al. [4] proposed the combined use of BIM and GIS for spatial data management (e.g., location queries). Tomasi et al. [5] introduced the use of the BIM model in computing for the optimal coverage of Wireless sensor networks (WSNs). Rafiee et al. [6] applied the methods transforming BIM model with geometric and semantic information into geo-referenced vector model for view and shadow analyses, which are useful in urban spatial planning. Tang and Kim [7] introduced a dynamic fire simulation based on the Fire Dynamics Simulator (FDS) and BIM model, which included simulation control, fire and smoke modelling, and occupant evacuation in the indoor environment. Boguslawski et al. [8] introduced that the route planning of indoor fire emergency based on BIM model. Thus, the indoor model reconstruction has become extremely valuable in urban development.

2.4. Summary

These surface-based reconstruction methods [9,15,22–31] mostly depend on the accuracy of surface extraction. In complex indoor scenes, the efficiency of surface extraction is low and contains excessive noise. Line-based reconstruction methods [32–40] can be used to completely represent the geometric information; however, these do not contain semantic information and adjacency relationship of rooms, leaving the reconstructed models to be useful only for visualization. In order to address the above shortcomings, we propose an innovative approach combining the rich structure of 2D lines with 3D geometry of surfaces to automatically build 3D structured models using MLS point cloud data. The output structural model presents novel applications in signal simulation, including the capability of providing optimal locations for 5G small base stations in the future.

3. Methodology

The complete flowchart is illustrated in Figure 1, showing the four key steps in the proposed methodology: room segmentation, floorplan extraction and regularization, structural model reconstruction, and 5G signal simulation. For the room segmentation, the door position and the simulated visible point clouds of sample trajectories are used to establish the initial space, while the global optimization of the indoor space is solved by the energy minimization function via multi-label graph cuts. For the floor extraction and regularization, the line elements are processed, which included the following steps: (1) the 3D point cloud slices are transformed into a binary image, and the line elements are extracted from the image; (2) the correction of line elements are based on global optimization; and, (3) similar lines are clustered to remove redundant line elements. The three-dimensional structural models are reconstructed via multi-label graph cuts, with room segmentation, 2D line elements, and 3D surfaces as semantic constraints. Finally, the signal intensity of 5G small base station is simulated using the structural models in the indoor environment.

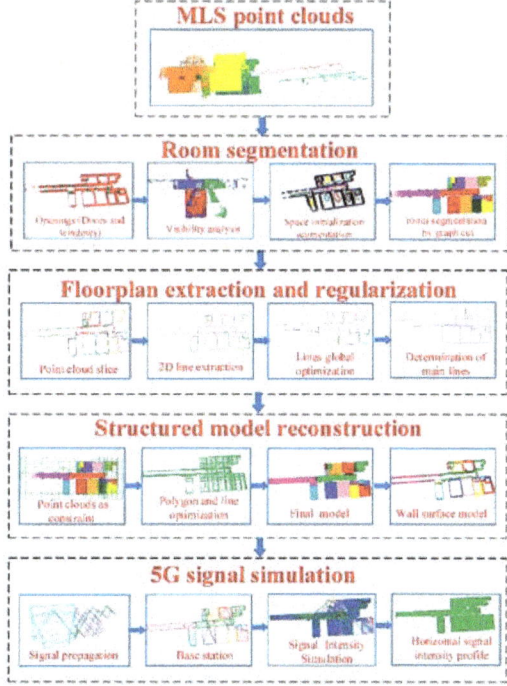

Figure 1. Flowchart of the proposed method.

3.1. Room Segmentation

The input of our approach, consists of unstructured point clouds and trajectories acquired from the mobile laser scanner system. In the indoor scene, every room has at least one door, representing a transition from one indoor space to another. For room segmentation, the position of detected doors and the simulated visible point clouds of sample trajectories are used to establish the initial space, while the global optimization of the indoor space is solved using the energy minimization function via multi-label graph cuts.

3.1.1. Detection of Openings

Since openings are generally attached to wall surfaces and as holes in the point clouds of wall surfaces, the extraction of doors and windows is based on the hierarchical relationship of plane-contour. The surfaces are first extracted based on the previous plane segmentation [31] (see wall surfaces illustrated in Figure 2a). The 3D point clouds of wall surfaces are projected into a 2D plane using the following conversion:

$$X_v = \frac{(0,0,1) \times (n_x, n_y, n_z)}{|(0,0,1) \times (n_x, n_y, n_z)|} \quad Y_v = \frac{X_d \times (n_x, n_y, n_z)}{|X_d \times (n_x, n_y, n_z)|} \quad Z_v = (n_x, n_y, n_z)^T$$
$$T = (X_v, Y_v, Z_v) \quad (x, y, z) \cdot T = (x_2, y_2, z_2) \tag{1}$$

where (n_x, n_y, n_z) are normal vectors of 3D plane; X_v, Y_v, Z_v are vectors of the 3D plane coordinate axis that constituted the transformation matrix T; (x, y, z) are coordinates of 3D plane; (x_2, y_2) are the 2D coordinates of the transformed plane; and, z_2 is depth.

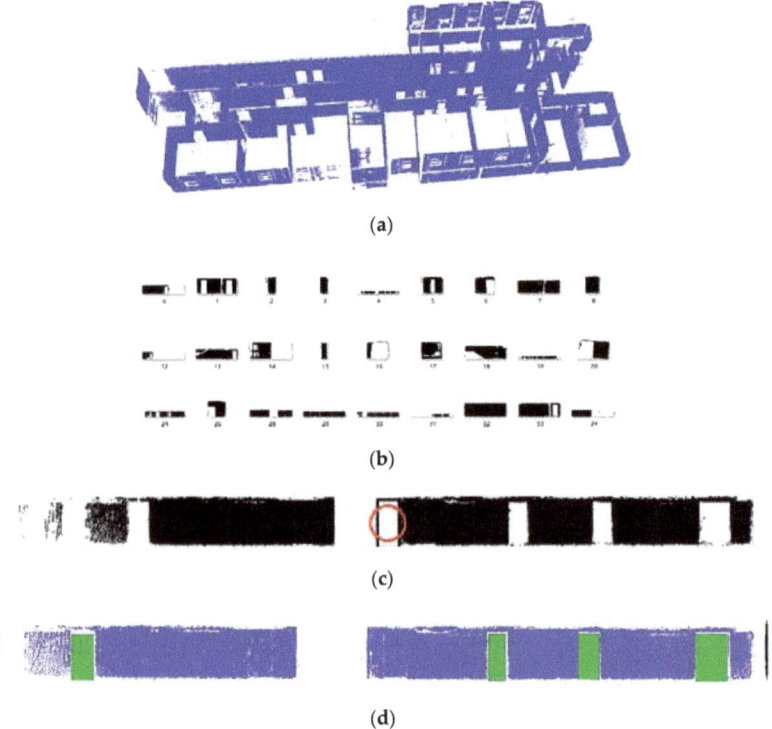

Figure 2. Detection of openings. (a) Extracted wall surfaces. (b) Wall surfaces converted into binary image. (c) Template match. (d) Detected doors.

The 2D projected points are converted into a binary image, as shown in Figure 2b. For every binary image, morphological corrosion transformation is used to remove noise. The find-contour method [41] is then applied during the extraction of plane outline to get sets of contour points, such that every contour is independent. Afterward, the bounding boxes of contours are calculated. Based on the size of the bounding box, the contours are categorized into doors, windows, invalid. The invalid consists of holes resulting from occlusion and undetected openings. In our study, the template match method, which is implemented in the OpenCV library [42], is applied to extract undetected openings. (In Figure 2c, an extracted opening as the template that encircled in red; in Figure 2d, the identified doors are shaded in green). Compared with the previous method [31], the extracted opening boundaries are visually more defined. As for the extraction of pillars, the technique is similar to extracting openings.

3.1.2. Room-Space Segmentation

The segmentation of rooms is accomplished by first simulating the visible point clouds from the MLS trajectory based on the line-of-sight. The position of doors is then used to limit the range of visible point clouds and to partition the trajectory segments in establishing the initial space. Finally, similar visible points between scanning trajectories are automatically clustering using global optimization based on multi-label graph cuts.

Inspired by the previous room segmentation [31], the visibility analysis is that simulating the visible point clouds along sample trajectory of the MLS and the grid cells' center point based on the line-of-sight. Instead of being dependent on segmented planes, the original point clouds are divided into uniform grids and the sampled point is one of every 200 trajectory points from the original trajectories. Figure 3 shows the flowchart, which illustrates the intersections between rays and all the other cells found along the line-of-sight. The points in the target cell are only visible if the point number of the cell where the ray passes through is within the threshold. Figure 4a shows the simulated visible point clouds of the three sample trajectories, some points are collected by different rooms due to the openings. Thus, the position of doors can be used to separate different spaces by limiting the range of visible points, as shown in Figure 4b.

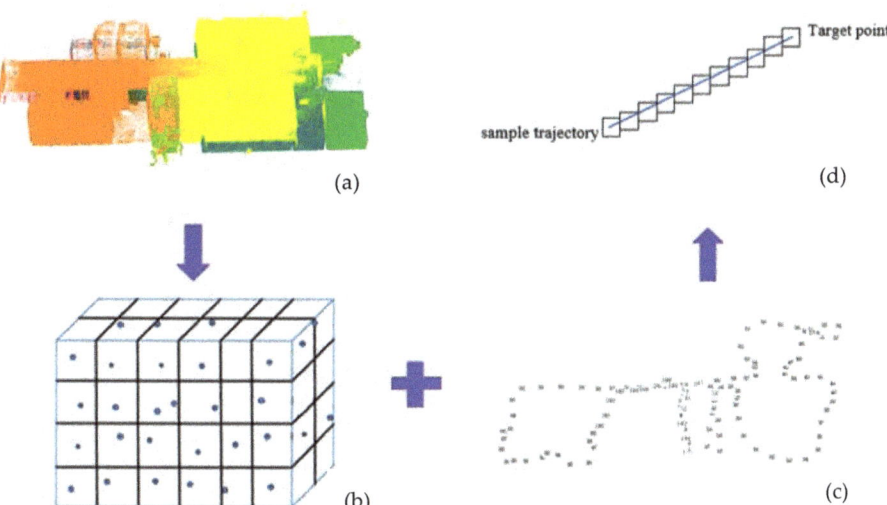

Figure 3. The diagram of visible point clouds of simulated trajectory points. (**a**) Original point clouds. (**b**) Original point clouds divided into uniform grids. (**c**) Sample trajectory points. (**d**) Visibility analysis based on line-of-sight.

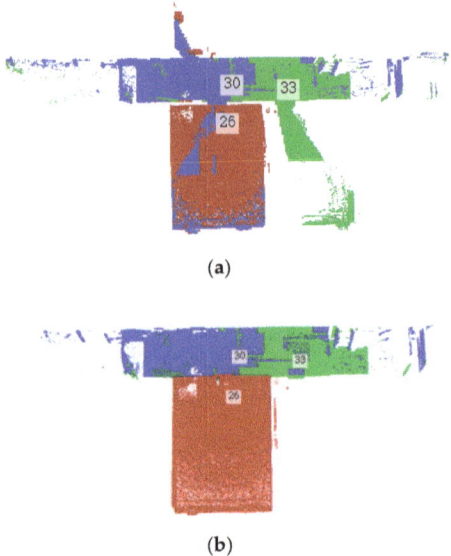

Figure 4. Simulating visible point clouds of sample trajectories. (**a**) The visible points of the three trajectory points 26, 30, and 33. (**b**) Visible points limited by door.

The location of doors plays an important role in room segmentation. In our proposed methodology, the location of doors is used to subdivide the trajectories into initialized subspaces. Figure 5b illustrates how the trajectories are segmented by each door. Each trajectory segment corresponds to only one room, but not all rooms are depicted by a single segment. Trajectories in the same space have similar visible point clouds, so the individual room can be segmented, which regard as automatic clustering of similar visible point clouds by the global optimized method.

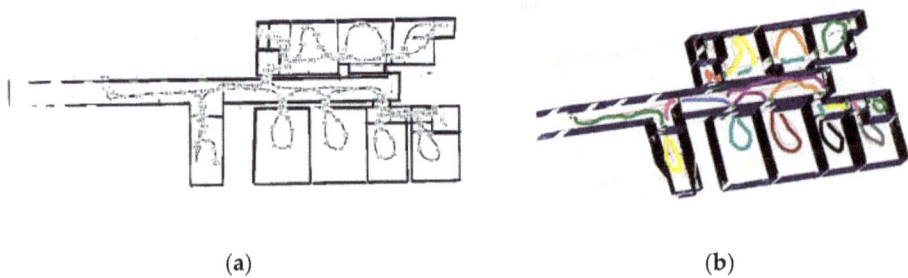

Figure 5. Position of doors subdivide trajectory segments. (**a**) Sample trajectory points. (**b**) Partition of trajectory segments.

The global optimization of indoor spaces is performed by solving an energy minimization function via multi-label graph cuts. The optimization function consists of a unary and smooth term, which is expressed as Equation (2), where weight parameters α, γ are used in balancing the data term and the smooth term in the energy function. The initial trajectory is first segmented using the doors as positional constraints, then, its corresponding clustering spaces are determined by minimizing a predefined energy function.

$$E = \min(E_D + E_S)$$
$$E(l) = \min(\sum_{v \in V} \alpha \cdot D_v(l_v) + \sum_{(v,w) \in E} \gamma \cdot B_{v,w}(l_v, l_w)) \quad (2)$$

Data term. E_D is the sum of unary functions; and, $D_v(l_v)$ is the difference in visual area between trajectory i and every trajectory segment ϕ_v, which is expressed in Equation (3):

$$r_o = \frac{o_i \cap G_{\phi_v}}{G_{\phi_v}}$$
$$D_v(l_v) = I_v - r_o \quad (3)$$

where l_v is the label for the trajectory i belonging to trajectory segment ϕ_v; I_v is the ideal value for the label l_v; r_o is the ratio of the overlapping area between trajectory i and each trajectory segment ϕ_v; o_i is the visual area of trajectory i, and G_{ϕ_v} is the visual area of trajectory segment ϕ_v; $\{\phi_1, \ldots \phi_v \ldots \phi_\theta\}$ are the set of initial trajectory segments, where $\phi_v, v \in \{1, \ldots, \theta\}$. Lower $D_v(l_v)$ values mean less penalty when assigning the sample trajectory point i to the trajectory segment ϕ_v. The overlap between the visible area of two trajectory points is calculated using the number of the same index for visible cells.

Smoothness term. E_S is the sum of binary functions $B_{v,w}(l_v, l_w)$ and is used to regularize label by penalizing the assignment of different labels to adjacent trajectories, as defined by Equation (4):

$$B_{v,w}(l_v, l_w) = \begin{cases} (\frac{1}{2}e^{(-\frac{dis(i,j)}{\Delta d})} + \frac{1}{2}e^{-(1-O_{(i,j)})}) & if \ (l_v \neq l_w) \\ 0 & otherwise \end{cases} \quad (4)$$
$$O_{(i,j)} = \frac{1}{2}(\frac{o_i \cap o_j}{o_i} + \frac{o_i \cap o_j}{o_j})$$

where i is the sampled trajectory point; and let i be as center point and get its k-nearest neighbor (KNN) trajectory points, $j \in K$; o_i, o_j are visible area of trajectory points i and j, respectively; $dis(i,j)$ is the distance between trajectory points i and j, $O_{(i,j)}$ is the overlapping ratio of visual area of trajectory points i and j; Δd is a distance threshold. The smoothness term indicates the penalty between adjacent trajectory points. If a pair of neighboring trajectory points belong to the same space, the smooth cost between them is 0; otherwise, this value is closer to 1 as the overlapping ratio is greater and the distance of the adjacent trajectory points is smaller. The smooth term can reduce the number of redundant spaces, thus solving the over-segmentation problem in long corridors for complex indoor environment. Again, the global optimization of the indoor space is solved by an energy minimization function via multi-label graph cuts [43–45] to automatically cluster similar visible point clouds. The room segmentation results are shown in Figure 6b.

Figure 6. Point clouds before and after space labeling. (a) Original point clouds. (b) Point clouds after space labeling.

3.2. Floorplan Extraction and Regularization

These methods [15,22,27,29] extracted mainly the piecewise planar surfaces to reconstruct indoor models. However, due to the complexity of the indoor environment, the quality of point clouds can suffer significantly from a number of factors such as moving objects, multiple reflections, and occlusions.

Building high-accuracy indoor model automatically becomes complex, which may require interaction. Since lines are commonly used in expressing key information in modelling, line-based reconstruction ensures the efficiency and precision of the model. In this study, our method combines lines and surfaces in building the 3D structured model of the indoor scene. The extraction and regularization of lines are conducted prior to the reconstruction of the vector model with more detailed features.

3.2.1. Floorplan Line Extraction

The horizontal slices of the point clouds with individual room are determined based on certain heights from the ceiling. Connectivity analysis is performed to filter outliers from the point cloud slicing (results are shown in Figure 7a). Line segments are then extracted based on the image gradient, which recovers detailed structural features and greatly improves computational efficiency. The point clouds of horizontal slices are converted into a binary image, as shown in Figure 7b, with a pixel size of 5 cm. We use the Line-Segment-Detector (LSD) [46] method to extract lines from the binary image, which region-growing method is applied to the image gradient clustering. The larger gradient is used as the seed point, while the given angle threshold is used as the growing condition. The line extraction results are shown in Figure 7c with the line elements of different rooms presented in varying colors.

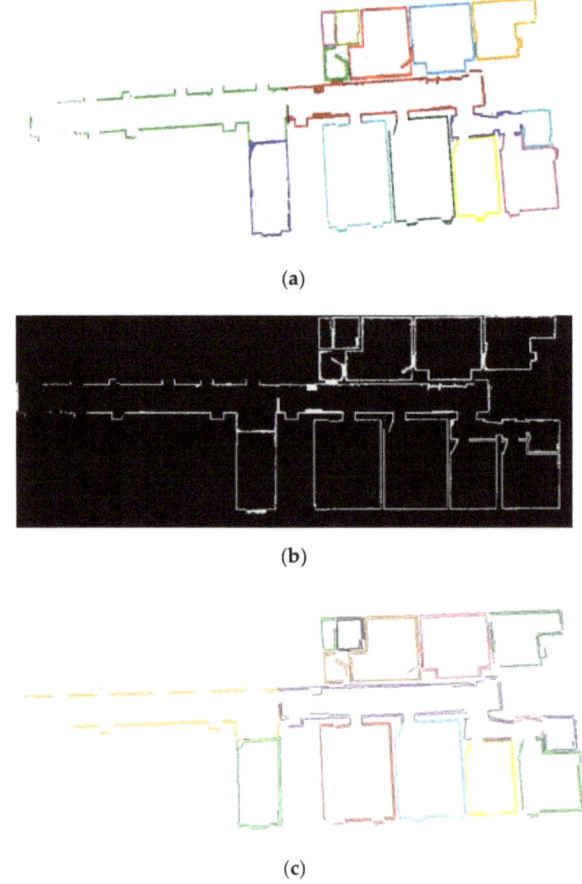

Figure 7. Floorplan Line extraction. (a) Split labeled point clouds from given height. (b) The conversion of projected points into a binary image. (c) Extraction of line elements with label information.

3.2.2. Line Global Optimization

Figure 7c shows that the extracted initial lines are able to preserve detailed features. However, due to laser point clouds with information loss, holes, noise, etc., the extracted results mainly have four kinds of errors: angle deviation, distance deviation, excessive redundancy, and incomplete boundaries. Inspired the method [38], the angle and distance deviations of the line-segments are corrected by global optimization, as shown in Figure 8a,b. The problem is expressed as an energy function, as shown in Equation (5), which is minimized by g2o that called a general framework for graph optimization [47]. The weight parameter λ is used for balancing the different terms.

$$E(x) = (1-\lambda)\cdot D(x) + \lambda \cdot B(x) \tag{5}$$

For angle correction, the data term $D(x)$ is used to correct the angle deviations corresponding to the initial orientation, as expressed by:

$$D(x) = \frac{1}{n}\sum_{i=1}^{n}\left(\frac{x_i}{\theta_{max}}\right)^2 \tag{6}$$

where the correction value $x_i \in [-\theta_{max}, \theta_{max}]$ can be added to the initial orientation of the line i with respect to its center; clockwise direction indicates positive value; θ_{max} is the angle threshold adjustable based on the quality of point clouds; and n is the number of extracted initial lines.

The smooth term $B(x)$ is used to correct the geometric relationship of the adjacent lines, as expressed by:

$$B(x) = \frac{1}{\sum_{i=1}^{n}\sum_{j=1}^{k} u_{ij}}\sum_{i=1}^{n}\sum_{j=1}^{k} u_{ij}\frac{\left|\theta_{ij} - (|x_j| + |x_i|)\right|}{4\theta_{max}} \tag{7}$$

$$\theta_{ij} = \begin{cases} \theta_{ij} \pm 2\pi & if\ (\frac{7}{4}\pi \leq |\theta_{ij}| \leq 2\pi) \\ \theta_{ij} \pm \frac{3}{2}\pi & if\ (\frac{5}{4}\pi \leq |\theta_{ij}| \leq \frac{7}{4}\pi) \\ \theta_{ij} \pm \pi & if\ (\frac{3}{4}\pi \leq |\theta_{ij}| \leq \frac{5}{4}\pi) \\ \theta_{ij} \pm \frac{1}{2}\pi & if\ (\frac{1}{4}\pi \leq |\theta_{ij}| \leq \frac{3}{4}\pi) \\ \theta_{ij} & otherwise \end{cases} \tag{8}$$

$$u_{ij} = \begin{cases} 1 & if\ (|\theta_{ij}| < 2\theta_{max}) \\ 0 & otherwise \end{cases}$$

where θ_{ij} is the angle between adjacent lines s_i, s_j (such that $\theta_{ij} \in [-2\pi, 2\pi]$). The adjacent lines are encouraged which are nearly-parallel or nearly-orthogonal or nearly-themselves. The angle θ_{ij} is adjusted to close to the coordinate axis, as expressed by Equation (8). In addition, the following conditions are satisfied: if $|\theta_{ij}| < 2*\theta_{max}$, $u_{ij} = 1$, otherwise $u_{ij} = 0$. k is the number of lines adjacent to line s_i; which can be obtained that s_i is as the center and search its KNN from all other lines. The distance correction is similar to the angle, as expressed by:

$$D(x) = \frac{1}{n}\sum_{i=1}^{n}\left(\frac{x_i}{d_{max}}\right)^2$$

$$B(x) = \frac{1}{\sum_{i=1}^{n}\sum_{j=1}^{k} u_{ij}}\sum_{i=1}^{n}\sum_{j=1}^{k} u_{ij}\frac{|d_{ij}-(|x_j|+|x_i|)|}{4d_{max}} \tag{9}$$

$$u_{ij} = \begin{cases} 1 & if\ (|d_{ij}| < 2d_{max}) \\ 0 & otherwise \end{cases}$$

where $x_i \in [-d_{max}, d_{max}]$ is added on the line s_i along its normal direction; and, d_{ij} is the distance between adjacent parallel lines s_i, s_j. If $|d_{ij}| < 2 * d_{max}$, then $u_{ij} = 1$, otherwise $u_{ij} = 0$. The global optimization results for the lines are shown in Figure 8c, which contains the correct geometric information.

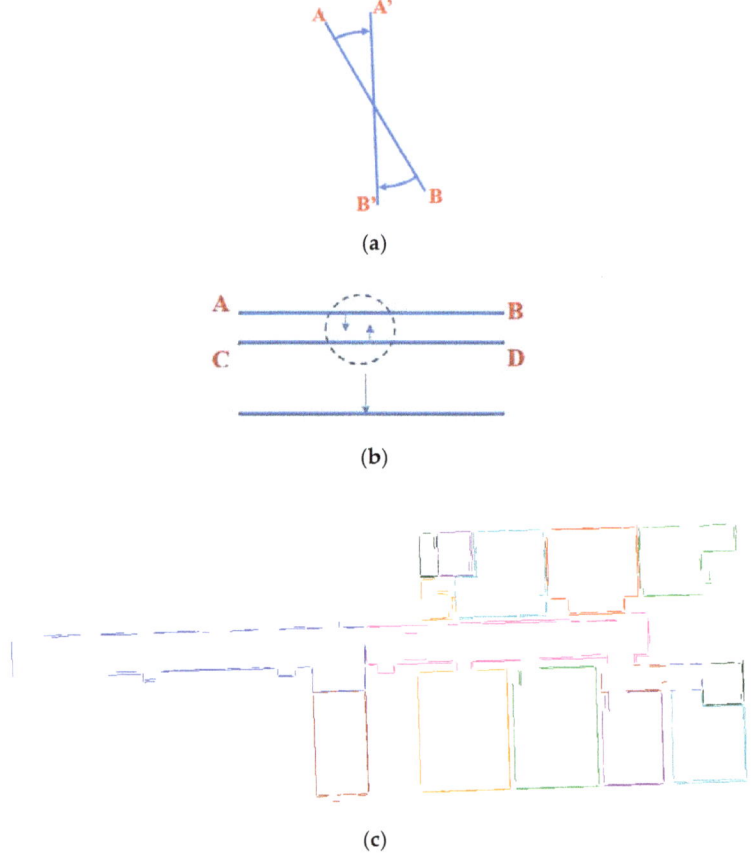

Figure 8. Global optimization of lines. (**a**) Correction of angle. (**b**) Correction of distance. (**c**) The global optimization results.

3.2.3. Clustering Similar Lines

After optimization, the whole scene consists of a set of small lines with different labels requiring further refinement. Inspired by [46], the following region growing algorithm incrementally merges adjacent basic units with similar features into a set of main lines. The number of line similar with each line is estimated, which is qualified to meet the user-defined angle and distance thresholds between lines. Seed lines with more similar line number are tested first as they are more likely to belong to the mainline. Each line region starts primarily with just a seed line. The orientation and distance of other lines from the seed line are estimated whether they meet the certain threshold, given by:

$$\begin{array}{ll} \Delta D = s_{nx} \cdot x_m + s_{ny} \cdot y_m + s_{offset} & \Delta D < d_{threshold} \\ \Delta A = a\cos\left(\frac{s_n \cdot s_{on}}{|s_n||s_{on}|}\right) & \Delta A < a_{threshold} \end{array} \quad (10)$$

where $(s_{nx}, s_{ny}, s_{offset})$ are the seed line parameters, s_n is its normal vector; (x_m, y_m) is the midpoint coordinate of other line, s_{on} is its normal vector. The lines that meet the threshold are then added to the seed line.

Lastly, the main lines consist of line groups. A mainline is determined at least by a starting point, an endpoint, an offset, and a normal vector. The normal vector of the seed line and the mean offset serve as the final parameters for the mainline. The line groups with different labels are then projected onto the mainline to find the endpoints and create the bounding boxes (see Figure 9a). The final main lines are presented in Figure 9b.

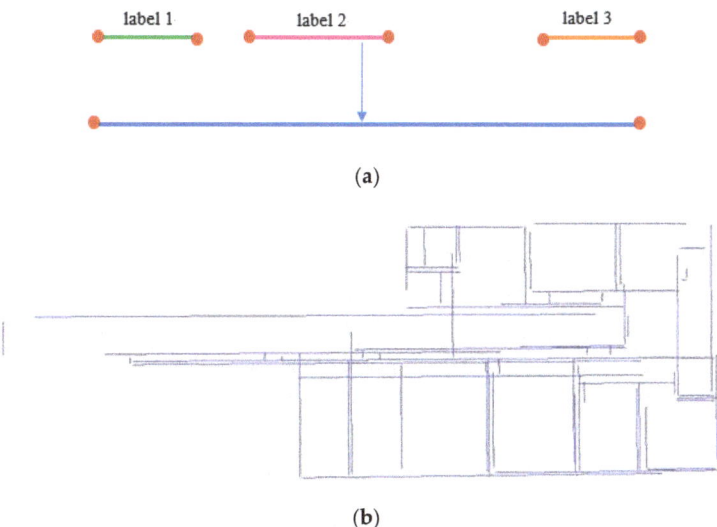

Figure 9. Clustering similar lines. (**a**) The line groups project onto the cluster line. (**b**) The final main lines.

3.3. Structured Model Reconstruction

3.3.1. Model Reconstruction

Figure 9b shows the line segments are incomplete due to missing point clouds. The lines are extended to form the enclosed floorplans (shown in green lines in Figure 10). Lines in the enclosed floorplans have topology and the point clouds in the segmented rooms have semantic information. Thus, the line segments and point clouds with individual rooms are can be used as constraint conditions in building the two-dimensional floorplan using multi-label graph cuts. The labelled Point clouds are projected on the 2D polygon floorplan is shown in Figure 10. Each cell is assigned a label from set $\{l_1, \ldots, l_{Nrooms}, l_{out}\}$, which includes one label for each room plus an additional label l_{out} for the outer space. Each line cell is assigned a label from the initial $\{l_1, \ldots, l_{Nrooms}, l_{out}\}$. The labelled line segments are used to segregate cells from adjacent region, of which cell labels should be different. Our approach differs from Ochmann's work [15,22] such that the line segments are projected directly on the floorplan and divided into 2D line cells to improve the accuracy and efficiency of the model. With the approach expressed as an energy minimization function [31], the 2D polygons and lines are globally optimized to build the floorplan model, which are then extended to the estimated the floor and ceiling heights from segmented surfaces to build 3D room models. Figure 11 shows that the reconstructed model can better retain the details of indoor scene.

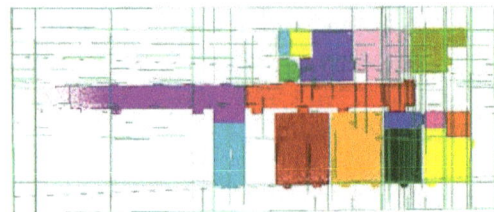

Figure 10. Line segments and point clouds as constraint conditions.

Figure 11. The reconstructed room models.

3.3.2. Room Structured Connection

The above reconstructed models are expressing room semantic and geometric information, but still lacks room topological connection types. In the last step, the room's structural connection is recreated. For indoor scenes, the space created by a door is a type of connection space. Thus, the door's position is used to analyze the room connection types. In this study, the doors were extracted based on the segmented surfaces, and the model reconstruction is based on the horizontal slice of the segmented rooms. To correct some distance errors introduced during reconstruction, the detected door are attached to the walls if the following conditions must be satisfied: (a) the door must be parallel to a wall; (b) the distance along the normal should be less than the threshold of 0.2 m; and, (c) the door completely overlaps with the walls. The door connecting adjacent rooms is a subspace with thickness. Figure 12 shows the structured model with the reconstituted doors.

Figure 12. The structured model with openings.

Every room is associated with geometry elements that ceiling, floor, walls, doors and windows. In our work, openings between neighboring rooms are detected to obtain a room connectivity graph, in addition, for each wall of the room, we search a matching, approximately parallel surface with opposing normal orientation within user-defined distance and angle thresholds. Each matching pair of wall surfaces forms adjacent rooms. According to these rules, a building's room topology graph is constructed. If a space that is connected to more than three rooms and has many doors that is labeled a corridor. The rooms with topological relationship are shown in Figure 13, which can be applied to a service application in an indoor environment.

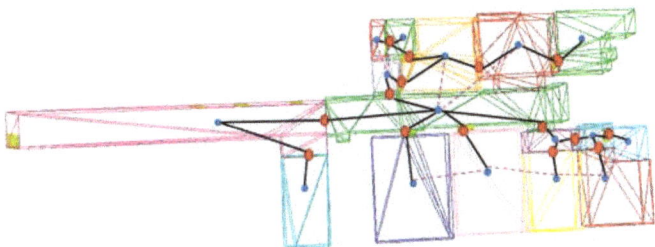

Figure 13. The rooms with topological relationship (the solid lines are connected by the doors, and the dotted lines are the connected by adjacent walls).

3.3.3. G Signal Intensity Simulation

In the 5G era, 5G devices have design features that support higher signal frequency and the shorter wavelength to generate faster transmission speed; however, this leads to diminished capability of penetrating through walls [48]. With 80% of today's businesses occurring indoors, setting up large-scale small base stations to increase signal intensity has become a common occurrence. However, due to the complexities of the indoor environment (e.g., occlusion problems), network construction has become a challenging undertaking.

In this study, the output structured models have three properties: semantic, geometric, and connection types. These models are made up of basic building elements, such as the ceilings, floors, walls, windows, doors and pillars, which have direct effect on signal propagation in the real world. Thus, the structured model can become an important tool in analyzing 5G signal simulation.

According to the standard of 3GPP [49], the non-line-of-sight signal propagation loss model of indoor space is expressed:

$$L_{fs,dB} = 32.4 + 31.9 \cdot \lg(d_p) + 20 \cdot \lg(f) \tag{11}$$

where $L_{fs,dB}$ is the propagation loss; d_p is the max propagation distance (100 m); and, f is frequency of the electromagnetic wave (0.5 GHz – 100 GHz). The formula suggests that greater propagation loss occurs with larger wave frequency or with longer propagation distance. In an ideal indoor environment (no attenuation losses), when the frequency remains constant, the propagation loss increases with increasing distance, which then decreases the power received by the receiver.

The indoor environment is comprised of open cubicles, walled offices, open areas, corridors, etc. In this study, the 5G base stations were assumed to be located at the height of 2 m, near the ceilings. The ray-tracing solution is adopted to provide a detailed multipath and accurately simulate the spatial variation. Figure 14 illustrates the principle of single signal propagation, where the intensity multipath results from the reflection of walls and transmission of openings.

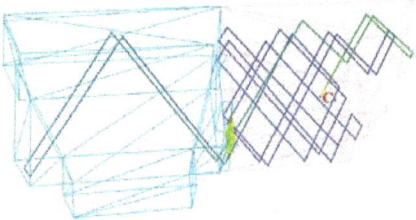

Figure 14. Principle of signal propagation (the signal intensity changes from strong to weak that corresponds to color from red to blue).

In our experiment, the signal propagation intensity was simulated in the structured model. Three base stations were mounted at the corridor and a room, as shown in Figure 15a. Every base station was assumed to be at the center of a sphere, randomly launching 150 rays. Using a frequency of 100 GHz, the intensity was calculated within a range of 100 m using the signal propagation loss model (see Figure 15b). The profile provided an effective means of measuring the changes of signal intensity and became a useful tool in visualization and inspection of 3D interpolation results. Therefore, the interpolation method of Inverse Distance Weight method (IDW) used in simulating the intensity profile can be calculated with:

$$P = \sum_{i=1}^{n} \varepsilon_i P_i \quad \varepsilon_i = \frac{\frac{1}{(D_i)^m}}{\sum_{i=1}^{n} \frac{1}{(D_i)^m}} \tag{12}$$

where the intensity value P of the interpolation point is defined as the weighted average value of known point intensity value P_i; D_i is the Euclidean distance from interpolation point to its nearest sampling point; and, m is the power exponent. Finally, the profile result is shown in Figure 15c.

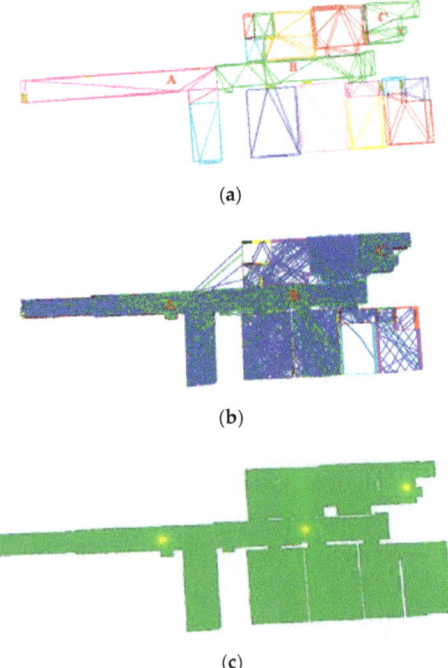

(a)

(b)

(c)

Figure 15. Signal intensity simulation. (**a**) Setting three base stations. (**b**) Multipath signal propagation. (**c**) Horizontal profile of signal intensity.

4. Experiment

4.1. Datasets

The proposed method was tested on four datasets acquired by MLS in different indoor scenes, as shown in Figure 16. Table 1 lists the technical specifications of system, and the Table 2 details the specifications of the point clouds. The first dataset, the ISPRS Benchmark Data [50], was captured using a handheld scanner, Zeb-Revo, in one of the buildings at the Technical University of Braunschweig, Germany. The data were acquired from across two floors connected via a staircase. The point clouds and trajectories are shown in Figure 16a. The walls had different thicknesses, the ceilings were of different heights, and the level of point cloud quality was high. The second dataset was captured in the 14th floor of the Technology Building of Shenzhen University, using our own developed backpack laser scanning (BLS), which contains a 16-beam 3D laser scanner. The location was in a corridor with glass walls and contains a number of moving objects; the collected point clouds had a high level of noise, as shown in Figure 16b. The third and fourth datasets were acquired on a corridor and a parking lot by the backpack mapping system of Xiamen University (shown in Figure 16c,d). This laser scanning system [37] contains two 16-beam laser scanners and can obtain higher quality 3D point cloud data.

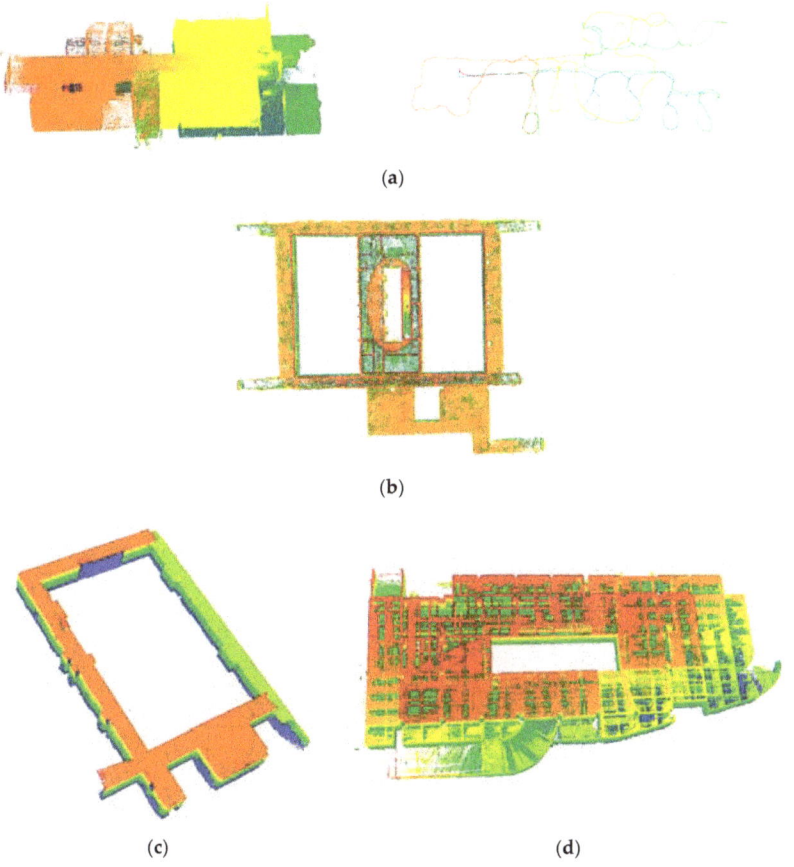

Figure 16. The experiment data. (**a**) Benchmark point clouds and trajectories acquired by handheld laser scanning (HLS), ZEB-REVO. (**b**) Point clouds acquired by Shenzhen University (BLS) system of Shenzhen University. (**c**) A closed-loop corridor by BLS system of Xiamen University. (**d**) Parking lot by BLS system of Xiamen University.

Table 1. Technical specifications of the laser scanning system.

Sensor	ZEB REVO	BLS (Shenzhen University)	BLS (Xiamen University)
Max range	30 m	100 m	100 m
Speed (points/sec)	43×10^3	300×10^3	300×10^3
Horizontal Angular Resolution	0.625°	0.1–0.4°	0.1–0.4°
Vertical Angular Resolution	1.8°	2.0°	2.0°
Angular FOV	270 × 360°	30 × 360°	2 × 30 × 360°

Table 2. Specifications of point clouds.

Dataset	Benchmark Data	Corridor (Shenzhen University)	Corridor (Xiamen University)	Parking lot (Xiamen University)
Number of points	21.560.263	1.980.911	2.098.634	7.683.766
Clutter	Low	High	Low	High

4.2. Parameters

The parameters of the proposed indoor structural model method for four datasets are listed in Table 3. Based on preliminary findings from the experiments, proposed methods showed robustness, and most of the parameters were insensitive to point cloud data in various indoor scenes and did not require manual modification. For opening extraction, the point clouds were transform into 2D image, where C_{2D} is the pixel size, w_d and h_d are the width and height of the regularized door; and, w_w and h_w are the width and height of the regularized window. For room segmentation, the point clouds were transformed into 3D grids, where C_{3D} is the size of 3D grid; and, α, γ are weight parameters used for balancing importance between data term and smooth term in the energy function. For the line global optimization, the θ_{max} and d_{max} were used to correct the angle and distance value of line; K is the nearest neighbor number of every line; and, λ is weight parameter used for balancing different terms in the energy function. For clustering similar lines, a_{thread} and d_{thread} were the angle and distance threshold. In 5G signal intensity simulation, d_p is the signal propagation distance, f is the frequency of the electromagnetic wave, and m is the power exponent by IDW interpolation method. These parameters can be depended on the point cloud data from different indoor scenes with similar characteristics.

Table 3. Parameters of the proposed indoor structural model method.

Parameters	Values	Descriptions
Extracting Openings		
C_{2D}	0.05 m	The size of the pixel (point clouds transform into image)
w_d/h_d	$0.7\,m \leq w_d \leq 1.5\,m$ $1.8\,m \leq h_d \leq 2.2\,m$	The width and height of regularized door
w_w/h_w	$0.5\,m \leq w_w \leq 1.5\,m$ $0.5\,m \leq h_w \leq 1.5\,m$	The width and height of the regularized window
Segmentation of Rooms		
C_{3D}	0.1 m	The size of the 3D grid (point clouds transform into 3D grid)
α, γ	1.0/0.5	Parameters of data term and smooth term of the energy function
Line Global Optimization		
θ_{max}	$0° \leq \theta_{max} \leq 45°$	Angle correction of lines
d_{max}	$0 \leq d_{max} \leq 0.1\,m$	Distance correction of lines
K	50	k-nearest of lines
λ	0.9	The weight parameter of line global optimization
Cluster Similar Lines		
a_{thread}	5°	Angle threshold of merging similar lines
d_{thread}	0.1 m	Distance threshold of merging similar lines
5G Signal Intensity Simulation		
d_p	100 m	The signal propagation distance
f	100 GHz	The frequency of the electromagnetic wave
m	1	The power exponent by IDW interpolation

4.3. Results

The algorithm was implemented in C++, edited by Microsoft Visual Studio 2017. All experiments were performed with a Window 10, 64-bit operating system with an Alienware Intel (R) Core (TM) i7-7700HQ CPU @ 2.80 GHz and a 16GB RAM.

Preliminary visual results of the structured model showed correctness and completeness of the model. For the benchmark data (shown in Figure 17a), the width and length of the extracted doors (green) and windows (yellow) were close to the true value. The room segmentation results of the first and second floors (as shown in Figure 17b) showed that the unstructured point clouds were correctly partitioned based on the multi-label graph cuts. In order to ensure the model accuracy, the structural model was reconstructed using visible point clouds, which eliminated the error from wall thickness estimation. Figure 17c,d show the reconstructed models to have detailed wall information. The doors (yellow) and windows (red) were correctly positioned and completely embedded within the walls, and the adjacent rooms had different heights. The structured model and original point clouds were well-matched, as shown in Figure 17e.

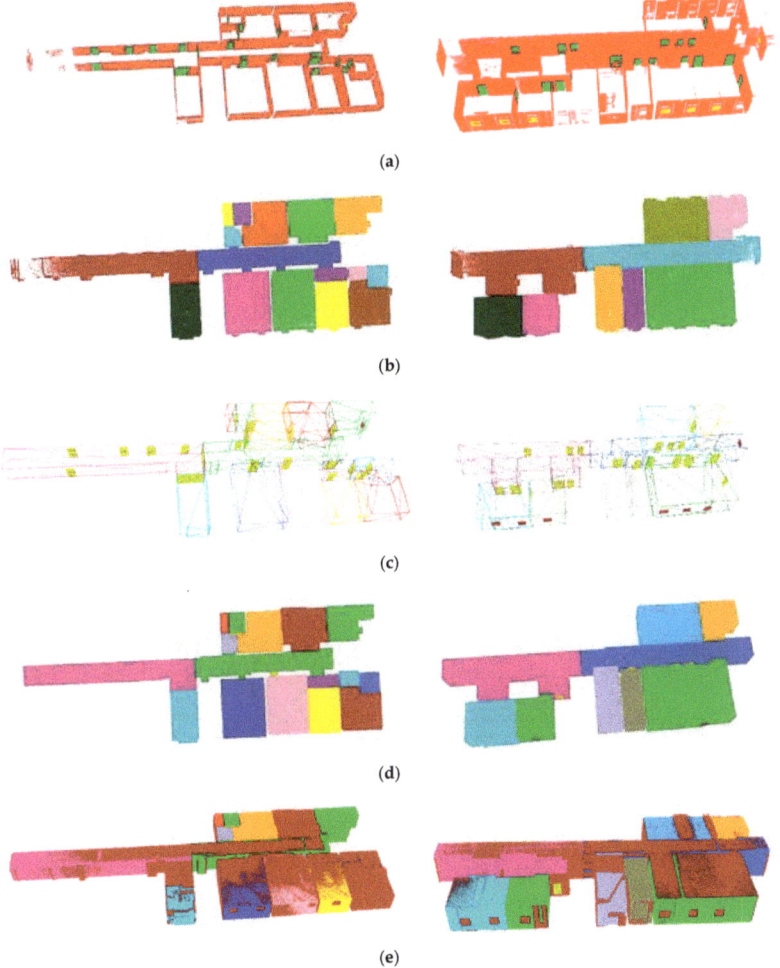

Figure 17. Opening extraction, room segmentation, structural model and wireframe model results with the benchmark point clouds. (**a**) Doors (green) and windows (yellow) of the first and the second floors. (**b**) Segmented rooms of the first and second floors. (**c**) The structural models with doors and windows of the first and the second floors. (**d**) The wireframe models with doors and windows of the first and second floors. (**e**) Matching between the point clouds and the structured models on the first and second floors.

For the corridor data of Shenzhen University, the acquired point clouds suffered from multiple reflections and refraction due to the glass walls, which resulted in the dramatic challenges during model reconstruction. Figure 18a shows the structured models, while Figure 18b shows that the detailed vector models of walls, pillars, the doors (green). The closed-loop polyhedron was created using the constrained Delaunay triangulation [51], which the detected closed polygons are as boundary rings. Figure 18c,d illustrate that the reconstructed models and point clouds are well matched. For the Xiamen University corridor, a high accuracy indoor structured model was obtained. Figure 19a,b show the structural models and the wall models, which are presented with detailed regularization information and accurate room representation with uneven ceiling heights. The doors (green) and windows (red) were correctly detected and completely embedded within the walls. The point clouds and the reconstructed model are well-matched, as presented in Figure 19c,d.

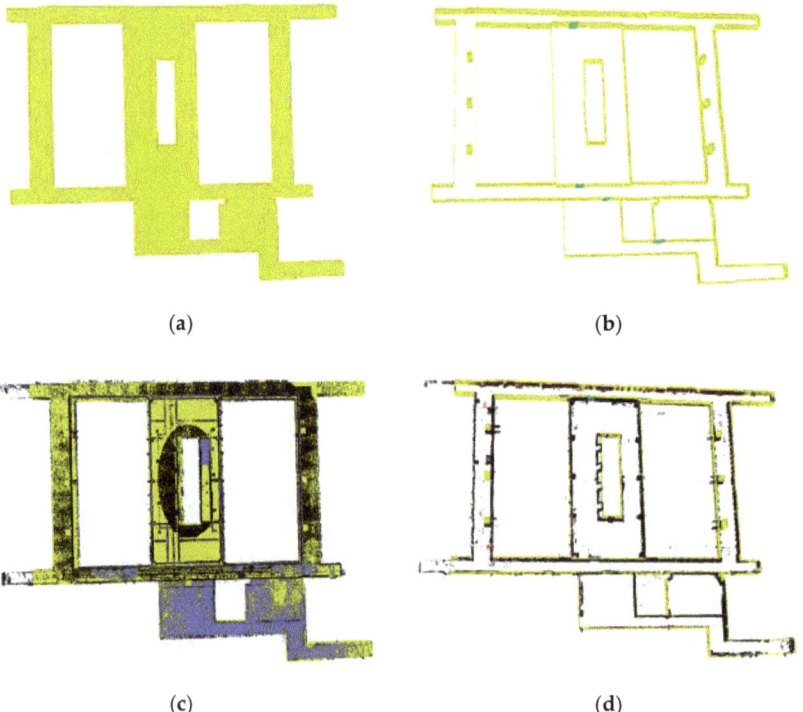

Figure 18. Structural model results of the corridor at the Shenzhen University. (**a**) The structural model. (**b**) Vector model of walls and pillars. (**c**) Matching between point clouds and the structural model. (**d**) Matching between point clouds and vector model of walls and pillars.

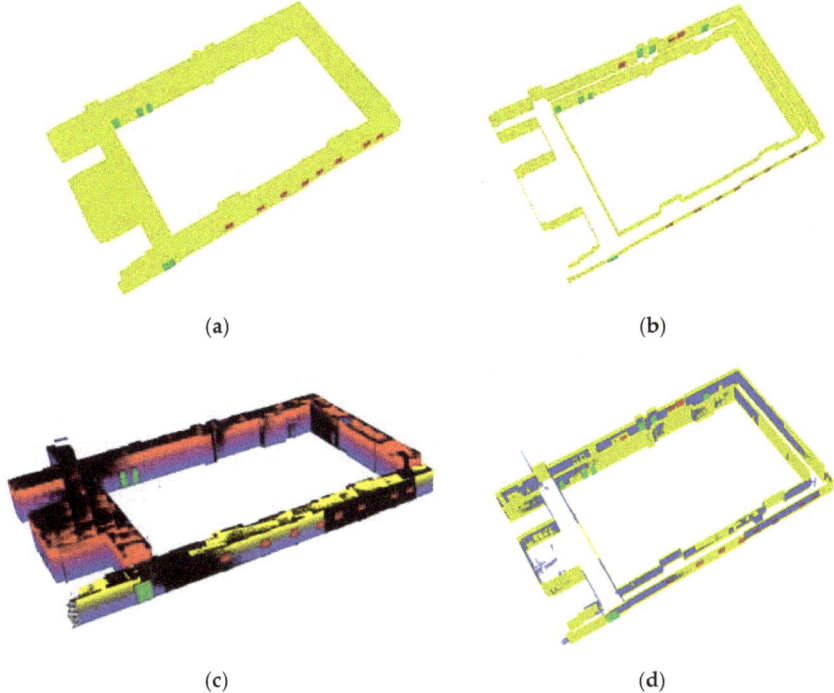

Figure 19. Structural model results of the corridor at the Xiamen University. (**a**) The structural model. (**b**) Vector model of walls. (**c**) Matching between point clouds and structured model. (**d**) Matching between.

For the parking lot in Xiamen University, the point clouds showed an excessively high level of noise. However, our proposed framework is still well-built even with incomplete data caused by severe occlusion (see Figure 20a,b). Our approach can auto-complete and generate closed-loop polyhedrons, and also correctly reconstruct the slant ceiling, floor, vertical walls, and regularized pillars. However, some curved walls are represented by many small polygons. The reconstructed models matched well with the original point clouds, as shown in Figure 20c,d.

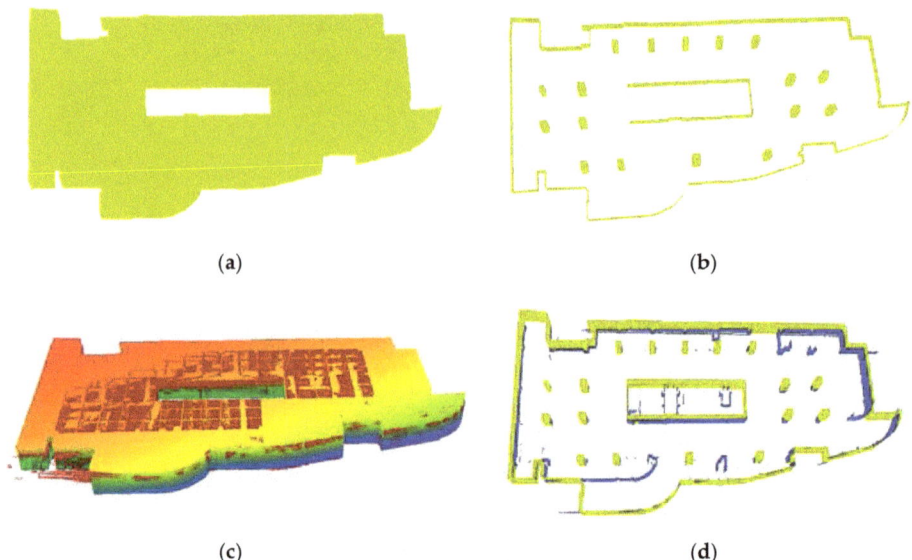

Figure 20. Structural model results of the parking lot at the Xiamen university (**a**) The structural model with slant floor and ceiling. (**b**) Vector model of walls and pillars. (**c**) Matching between point clouds and structured model. (**d**) Matching between point clouds and the vector model of walls and pillars.

More details of the results are displayed in Figure 21, despite the presence of noises and incomplete in the point clouds, our reconstructed models are of high correctness and well fit to the original point clouds.

For 5G signal intensity simulation, we tested our method on the structural model reconstructed from the benchmark data, as shown in Figure 22. The signal intensity is shown to drastically decrease from the base stations when three were mounted on the first floor, as illustrated by the changing colors of intensity (red to blue) in Figure 22a. In order to visualize the trend of signal intensity loss, a horizontal intensity profile was generated using the IDW interpolation method and is shown in Figure 22b. Similarly, the results from the multipath signal propagation and horizontal profile from the second floor are shown in Figure 22c,d. In Figure 22e, the received energy value is shown to significantly decrease with increasing distance under the path loss model.

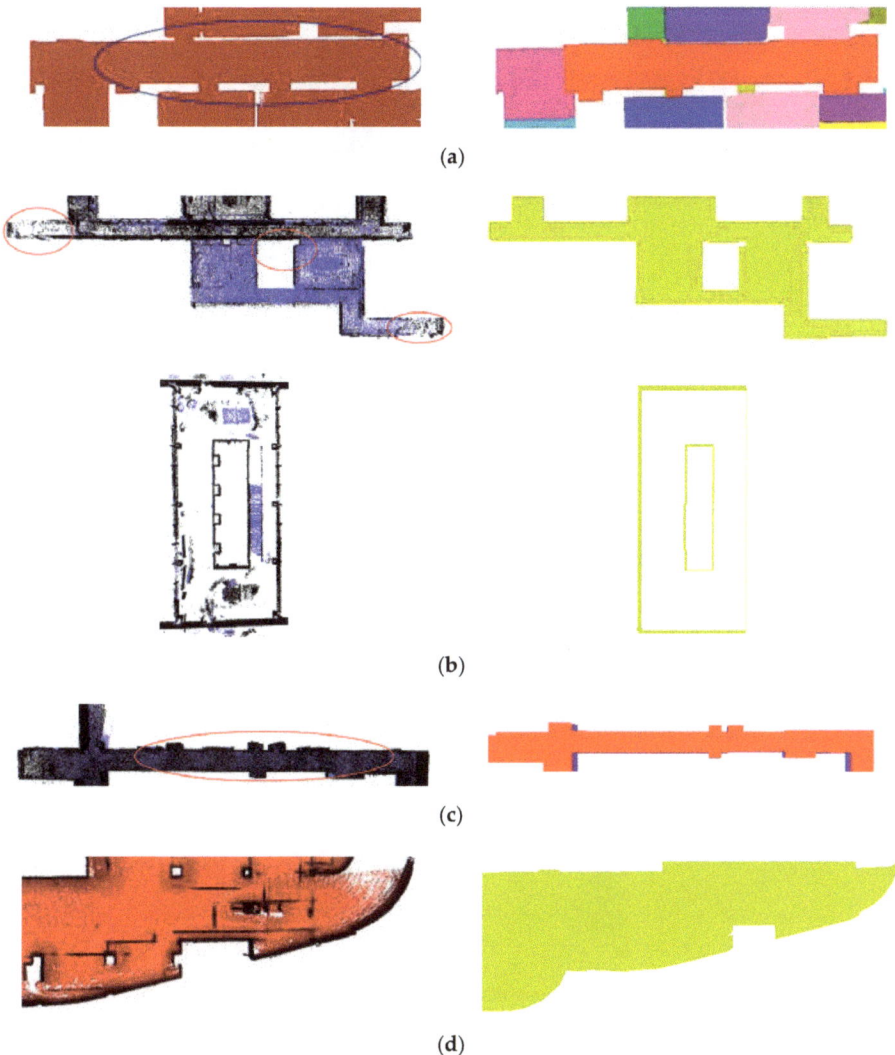

Figure 21. Close-up views of selected details. (**a**) Benchmark point clouds and reconstructed model. (**b**) The point clouds and reconstructed model of the corridor at the Shenzhen University. (**c**) The point clouds and reconstructed model of the corridor at Xiamen University. (**d**) The point clouds and reconstructed model of the parking lot at Xiamen University.

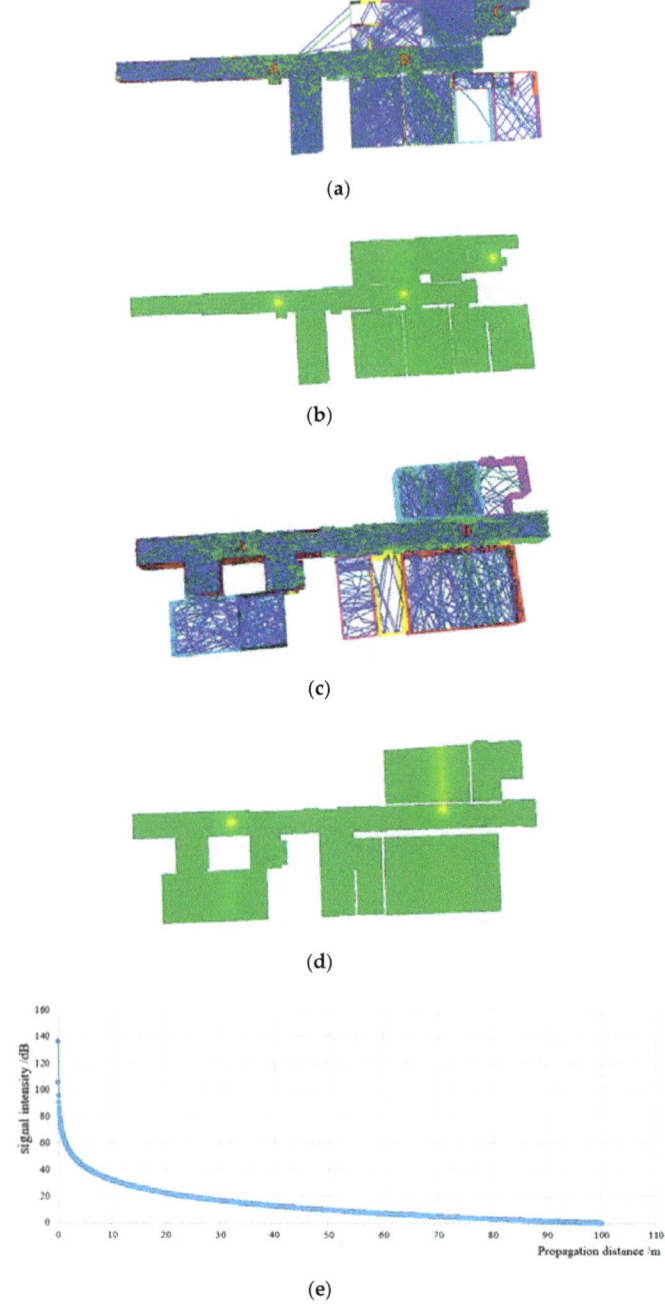

Figure 22. 5G signal intensity simulation based on the structural model by benchmark data. (**a**) The multipath signal propagation on the first floor. (**b**) Horizontal profile of signal intensity on the first floor. (**c**) The multipath signal propagation on the second floor. (**d**) Horizontal profile of signal intensity on the second floor. (**e**) Received energy value.

5. Evaluation and Discussion

Four real-world datasets captured using MLS were used to test our proposed methodology. Field experiments were used to analyze the visualization results and correctness of the semantic information and the spatial and topological relations of reconstructed models, as shown in Figures 17–21. The quantitative evaluation of the model included basic element extraction, running time, and geometric errors, as shown in Tables 4–6 and Figure 23.

Table 4. Results of basic element extraction.

Description	Number of Points	Actual/Detected Doors	Actual/Detected Windows	Actual/Detected Rooms	Actual/Detected Pillars
Benchmark data	11,628,186	51/42	21/8	25/25	0/0
Corridor (Shenzhen University)	1,980,911	4/4	0/0	1/1	6/6
Corridor (Xiamen University)	7,683,766	8/8	11/11	1/1	0/0
Parking Lot (Xiamen University)	2,098,634	0/0	0/0	1/1	23/18

Table 5. Running time for different scenes.

Description	Surface Extraction (s)	Opening Detection (s)	Room Segmentation (s)	Line Regularization and Model Reconstruction (s)	Total Time (s)
Benchmark data	80	19	287	49	435
Corridor (Shenzhen University)	9	4	0	24	37
Corridor (Xiamen University)	7	6	0	20	33
Parking Lot (Xiamen University)	28	0	0	32	60

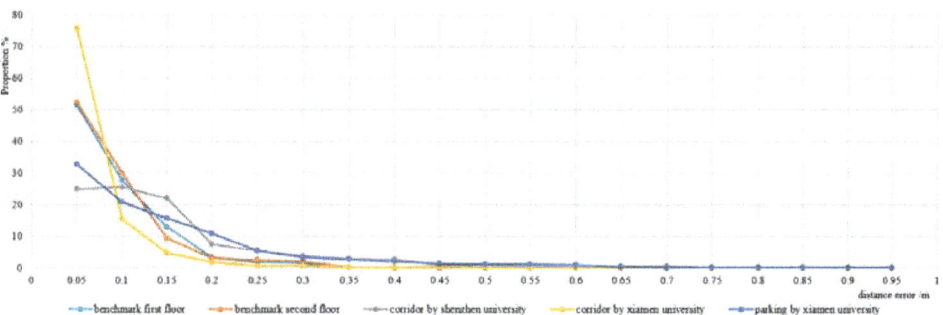

Figure 23. Euclidean distance deviation distribution map.

Table 6. Euclidean distance deviation for different scenes.

Error/m	0.05	0.10	0.15	0.20	0.25	0.30	0.35	0.40	0.45	0.50	0.55	0.60	0.65	0.70	0.75	0.80	0.85	0.90	0.95
Benchmark first floor (%)	51.50	27.68	12.92	3.26	1.73	1.61	0.28	0.21	0.20	0.11	0.10	0.09	0.07	0.07	0.08	0.05	0.02	0.01	0.01
Benchmark second floor (%)	52.31	30.09	9.36	3.20	2.41	2.11	0.31	0.07	0.01	0.02	0.02	0.01	0.01	0.02	0.01	0.01	0.01	0.01	0.01
Corridor, Shenzhen University (%)	25.10	25.81	22.02	7.45	5.51	3.81	3.02	2.55	1.10	0.81	0.82	0.40	0.51	0.50	0.14	0.12	0.21	0.01	0.11
Corridor, Xiamen University (%)	75.83	15.49	4.81	1.75	0.62	0.60	0.11	0.11	0.41	0.10	0.02	0.01	0.02	0.05	0.02	0.02	0.01	0.01	0.01
Parking lot, Xiamen University (%)	32.82	20.87	15.71	10.92	5.38	3.30	2.62	2.01	1.37	1.23	1.06	0.91	0.44	0.26	0.27	0.23	0.20	0.25	0.15

5.1. Quantitative Evaluation

Tables 4 and 5 enumerate key properties of the reconstructed model, including the number of points, actual and extracted basic elements, and runtime. For the statistical analysis of geometric errors, the reconstructed results with real data did not contain ground-truth data, so we used the distance from each original point cloud to its corresponding model plane as geometric error. The summary is presented in Table 6 and Figure 23.

In Table 4, for the benchmark data, 42 doors and 8 windows were correctly detected: 20 doors and 1 window on the first floor and 22 doors and 7 windows on the second floor. The closed doors cannot be detected. The detection failure for the other windows was due to high sparsity and noise in the wall point clouds. However, all rooms and corridors were correctly segmented and had no under-segmentation or over-segmentation. The Shenzhen corridor dataset had high levels of noise and sparsity due to multiple reflections of moving objects and refraction from the glass wall. However, the structured information of pillars and doors were accurately extracted. For the Xiamen corridor dataset with high-quality point clouds, all openings were correctly detected. For the Xiamen parking lot dataset, 18 pillars were correctly detected. In terms of time efficiency, the reconstruction of the four real-world datasets required little runtime (see Table 5), with only the room segmentation taking relatively more processing time. The results indicate our proposed methodology has high modelling efficiency with wide-ranging applications in different indoor scenes.

The summary of Euclidean distance deviation and diagram are shown in Table 6 and Figure 23. The reconstruction accuracy from the Xiamen corridor was highest, having 75.83% of point distance deviation within the 0.05 m range. The two floors from the benchmark showed comparable results with 51.50% (1st floor) and 52.31% (2nd floor) of deviations coming from the 0.05 m range. For the Shenzhen Corridor and Xiamen parking lot, the percentage of deviation within 0.05 m reached 25.10% and 32.82% respectively. This indicates that when using our approach, the reconstruction quality is heavily dependent on the quality of point clouds. Nevertheless, our method shows it can provide reliable and accurate reconstruction of indoor scenes within the 0.10 m range without the need for manual intervention.

5.2. Limitations

A major technical limitation of our method is that the detection of openings are highly dependent on the geometric quality of point clouds, which for indoor scenes with high amount of noise, could be very problematic. Also, the curved walls are represented with many small polygons, indicating that irregular structures could not be expressed as meshes. Then, the output results in this study are surface models; however, BIM standard models are often represented as volumetric building entities with walls, floors, ceilings, and topological information, thus, the surface models will lead to limit the expression of model thickness in practice. Lastly, in 5G signal simulation, the type of wall materials, which could create varying degrees of signal loss, was ignored for simplification, which results in some errors with actual situation.

6. Conclusions and Outlook

The current bottleneck in 3D indoor reconstruction is the low level of automation and accuracy in the reconstruction of the complex indoor environment. To address this problem, we proposed a novel method that combines the rich structures of lines and 3D geometric information of surfaces to automatically build a three-dimensional structured model from MLS point clouds. First, a fully automatic room segmentation is performed on the unstructured point clouds via multi-label graph cuts to overcome over-segmentation problems. The floorplan lines are then extracted and regularized from the image to obtain detail structural information. Finally, the segmented room, line, and surface elements are used as semantic information, and the 3D structured models are reconstructed by multi-label graph cuts. We showed how our proposed approach is able to accurately reconstruct real-world datasets without requiring manual operation. Also, the signal intensity simulation for 5G small base station was conducted using the results of our 3D model, which showed how our proposed technique can be very useful in such an application.

We tested our method on four real-world datasets acquired using the MLS. In analyzing the results, we included the assessment of the geometric elements, time-efficiency, and geometric errors in the evaluation. Experimental results show that the reconstructed structured models, including ceilings, floors, walls, doors, windows, and pillars, etc. The Combination of linear structures with 3D geometric surfaces to reconstruct structured models, which improve the computational efficiency and structural accuracy. The resulting models show that the geometric error of is within 0.1m for different indoor scenes. The detection of geometric elements is highly dependent on the geometric quality of point clouds. For our future endeavors, we will try to combine image and point clouds to further enrich the model results, which could help improve opening detection and compensate for poor point cloud quality. We will also reconstruct the full volumetric models using the extracted geometric elements, and further close to the requirements for Building Information Modeling. Finally, we will be investigating further the use of our approach in optimal location for 5G small base stations and other similar technologies, as well as considering other applications that may benefit from our approach.

Author Contributions: Conceptualization, Y.C., Q.L.; methodology, Y.C., Q.L. and Z.D.; software, Y.C. and Z.D.; validation, Y.C. and Q.L.; formal analysis, Y.C. and Q.L.; investigation, Y.C. and Z.D.; resources, Q.L.; writing—original draft preparation, Y.C., Q.L. and Z.D.; writing—review and editing, Y.C., Q.L. and Z.D.; visualization, Y.C. and Z.D.; supervision, Q.L. and Z.D.; project administration, Q.L. and Z.D.

Funding: This research was funded by the Key Program of the National Natural Science Foundation of China (No. 41531177), the National Science Fund for Distinguished Young Scholars of China (No. 41725005), the Technical Cooperation Agreement between Wuhan University and Huawei Space Information Technology Innovation Laboratory (No. YBN2018095106), the National Natural Science Foundation of China (No. 41901403), the National Key Research and Development Program of China (No. 2016YFB0502203).

Acknowledgments: We would like to thank the ISPRS Commission WG IV/5 for provision of the data. In addition, we also thank to Key Laboratory of Sensing and Computing for Smart City and the School of Information Science and Engineering, Xiamen University for providing the corridor and parking lot datasets, and special thanks to the professional English editing service from EditX to improve the language.

Conflicts of Interest: The authors declare no conflict of interest.

References

1. Becker, S.; Peter, M.; Fritsch, D. Grammar-Supported 3d Indoor Reconstruction from Point Clouds for As-Built Bim. *ISPRS Int. Arch. Photogramm. Remote Sens. Spat. Inf. Sci.* **2015**, *II-3/W4*, 17–24. [CrossRef]
2. Vilariño, L.D.; Boguslawski, P.; Khoshelham, K.; Lorenzo, H.; Mahdjoubi, L. Indoor Navigation from Point Clouds: 3d Modelling and Obstacle Detection. *ISPRS Int. Arch. Photogramm. Remote Sens. Spatial Inf. Sci.* **2016**, *XLI-B4*, 275–281.
3. Vilariño, L.D.; Frias, E.; Balado, J.; Gonzalezjorge, H. Scan planning and route optimization for control of execution of as-designed BIM. *ISPRS Int. Arch. Photogramm. Remote Sens. Spatial Inf. Sci.* **2018**, *XLII-4*, 143–148.
4. Boyes, G.; Ellul, C.; Irwin, D. Exploring bim for operational integrated asset management-a preliminary study utilising real-world infrastructure data. *ISPRS Int. Arch. Photogramm. Remote Sens. Spatial Inf. Sci.* **2017**, *IV-4/W5*, 49–56. [CrossRef]
5. Tomasi, R.; Sottile, F.; Pastrone, C.; Mozumdar, M.M.R.; Osello, A.; Lavagno, L. Leveraging bim interoperability for uwb-based wsn planning. *IEEE Sens. J.* **2015**, *15*, 5988–5996. [CrossRef]
6. Rafiee, A.; Dias, E.; Fruijtier, S.; Rafiee, A.; Dias, E.; Fruijtier, S.; Scholten, H. From bim to geo-analysis: View coverage and shadow analysis by bim/gis integration. *Procedia Environ. Sci.* **2014**, *22*, 397–402. [CrossRef]
7. Tang, D.; Kim, J. Simulation support for sustainable design of buildings. In Proceedings of the CTBUH International Conference, Shanghai, China, 16–19 September 2014.
8. Boguslawski, P.; Mahdjoubi, L.; Zverovich, V.E.; Fadli, F. Two-graph building interior representation for emergency response applications. *ISPRS Int. Arch. Photogramm. Remote Sens. Spatial Inf. Sci.* **2016**, *III-2*, 9–14.

9. Bassier, M.; Vergauwen, M. Clustering of Wall Geometry from Unstructured Point Clouds Using Conditional Random Fields. *Remote Sens.* **2019**, *11*, 1586. [CrossRef]
10. Ikehata, S.; Yang, H.; Furukawa, Y. Structured Indoor Modeling. In Proceedings of the IEEE International Conference on Computer Vision, Santiago, Chile, 7–13 December 2015.
11. Wang, J.; Xu, K.; Liu, L.; Cao, J.; Liu, S.; Yu, Z.; Gu, X. Consolidation of low-quality point clouds from outdoor scenes. *Comput. Graph. Forum.* **2013**, *32*, 207–216. [CrossRef]
12. Mura, C.; Mattausch, O.; Villanueva, A.J.; Gobbetti, E.; Pajarola, R. Automatic room detection and reconstruction in cluttered indoor environments with complex room layouts. *Comput. Graph.* **2014**, *44*, 20–32. [CrossRef]
13. Ochmann, S.; Vock, R.; Wessel, R.; Tamke, M.; Klein, R. Automatic generation of structural building descriptions from 3D point cloud scans. In Proceedings of the International Conference on Computer Graphics Theory and Applications, Lisbon, Portugal, 5–8 January 2014.
14. Turner, E.; Cheng, P.; Zakhor, A. Fast, Automated, Scalable Generation of Textured 3D Models of Indoor Environments. *IEEE J. Sel. Top. Signal. Process.* **2015**, *9*, 409–421. [CrossRef]
15. Ochmann, S.; Vock, R.; Wessel, R.; Klein, R. Automatic reconstruction of parametric building models from indoor point clouds. *Comput. Graph.* **2016**, *54*, 94–103. [CrossRef]
16. Ambrus, R.; Claici, S.; Wendt, A. Automatic Room Segmentation from Unstructured 3-D Data of Indoor Environments. In Proceedings of the International Conference on Robotics and Automation, Singapore, 29 May–3 June 2017.
17. Wang, R.; Xie, L.; Chen, D. Modeling Indoor Spaces Using Decomposition and Reconstruction of Structural Elements. *Photogramm. Eng. Remote Sens.* **2017**, *83*, 827–841. [CrossRef]
18. Vilariño, L.D.; Verbree, E.; Zlatanova, S.; Diakité, A. Indoor Modelling from Slam-Based Laser Scanner: Door Detection to Envelope Reconstruction. *ISPRS Int. Arch. Photogramm. Remote Sens.* **2017**, *XLII-2/W7*, 345–352.
19. Li, L.; Su, F.; Yang, F.; Zhu, H.; Li, D.; Zuo, X.; Li, F.; Liu, Y.; Ying, S. Reconstruction of Three—Dimensional (3D) Indoor Interiors with Multiple Floors via Comprehensive Segmentation. *Remote Sens.* **2018**, *10*, 1281. [CrossRef]
20. Yang, F.; Li, L.; Su, F.; Li, D.L.; Zhu, H.H.; Ying, S.; Zuo, X.K.; Tang, L. Semantic decomposition and recognition of indoor spaces with structural constraints for 3D indoor modelling. *Automat. Constrn.* **2019**, *106*, 102913. [CrossRef]
21. Stichting, C.; Centrum, M.; Dongen, S.V. *A Cluster Algorithm for Graphs*; CWI: Amsterdam, The Netherlands, 2000; pp. 1–40.
22. Ochmann, S.; Vock, R.; Klein, R. Automatic reconstruction of fully volumetric 3D building models from oriented point clouds. *ISPRS J. Photogramm. Remote Sens.* **2019**, *151*, 251–262. [CrossRef]
23. Sanchez, V.; Zakhor, A. Planar 3D modeling of building interiors from point cloud data. In Proceedings of the IEEE International Conference on Image Processing, Orlando, FL, USA, 30 September–3 October 2012.
24. Lafarge, F.; Alliez, P. Surface Reconstruction through Point Set Structuring. *Comput. Graph. Forum.* **2013**, *32*, 225–234. [CrossRef]
25. Monszpart, A.; Mellado, N.; Brostow, G.; Mitra, N. RAPter: Rebuilding man-made scenes with regular arrangements of planes. *Acm Trans. Graph.* **2015**, *34*, 103. [CrossRef]
26. Awrangjeb, M.; Gilani, S.A.; Siddiqui, F.U. An Effective Data-Driven Method for 3-D Building Roof Reconstruction and Robust Change Detection. *Remote Sens.* **2018**, *10*, 1512. [CrossRef]
27. Xiao, J.; Furukawa, Y. Reconstructing the world's museums. In Proceedings of the European Conference on Computer Vision, Florence, Italy, 7–13 October 2012.
28. Schnabel, R.; Wahl, R.; Klein, R. Efficient RANSAC for point-cloud shape detection. *Comput. Graph. Forum* **2007**, *26*, 214–226. [CrossRef]
29. Mura, C.; Mattausch, O.; Pajarola, R. Piecewise-planar Reconstruction of Multi-room Interiors with Arbitrary Wall Arrangements. *Comput. Graph. Forum* **2016**, *35*, 179–188. [CrossRef]
30. Boulch, A.; Gorce, M.D.L.; Marlet, R. Piecewise-Planar 3D Reconstruction with Edge and Corner Regularization. *Comput. Graph. Forum.* **2014**, *33*, 55–64. [CrossRef]
31. Cui, Y.; Li, Q.; Yang, B.; Xiao, W.; Chen, C.; Dong, Z. Automatic 3-D Reconstruction of Indoor Environment with Mobile Laser Scanning Point Clouds. *IEEE J. Sel. Top. Appl. Earth Observ. Remote Sens.* **2019**, *99*, 1–14. [CrossRef]

32. Lin, Y.; Wang, C.; Chen, B.L.; Zai, D.W.; Li, J. Facet Segmentation-Based Line Segment Extraction for Large-Scale Point Clouds. *IEEE Trans. Geosci. Remote Sens.* **2017**, *55*, 4839–4854. [CrossRef]
33. Xia, S.; Wang, R. Façade Separation in Ground-Based LiDAR Point Clouds Based on Edges and Windows. *IEEE J. Sel. Top. Appl. Earth Observ. Remote Sens.* **2019**, *12*, 1041–1052. [CrossRef]
34. Lu, X.; Liu, Y.; Li, K. Fast 3D Line Segment Detection from Unorganized Point Cloud. In Proceedings of the IEEE Conference on Computer Vision Pattern Recognition, Long Beach UA, CA, USA, 15–21 June 2019.
35. Liu, C.; Wu, J.; Furukawa, Y. FloorNet: A Unified Framework for Floorplan Reconstruction from 3D Scans. In Proceedings of the European Conference on Computer Vision, Munich, Germany, 8–14 September 2018.
36. Oesau, S.; Lafarge, F.; Alliez, P. Indoor scene reconstruction using feature sensitive primitive extraction and graph-cut. *ISPRS J. Photogramm. Remote Sens.* **2014**, *90*, 68–82. [CrossRef]
37. Wang, C.; Hou, S.; Wen, C.; Gong, Z.; Li, Q.; Sun, X.; Li, J. Semantic line framework-based indoor building modeling using backpacked laser scanning point cloud. *ISPRS J. Photogramm. Remote Sens.* **2018**, *143*, 150–166. [CrossRef]
38. Bauchet, J.; Lafarge, F. KIPPI: KInetic Polygonal Partitioning of Images. In Proceedings of the IEEE Conference on Computer Vision Pattern Recognition, Salt Lake City, UT, USA, 18–22 June 2018.
39. Sui, W.; Wang, L.; Fan, B.; Xiao, H.; Wu, H.; Pan, C. Layer-Wise Floorplan Extraction for Automatic Urban Building Reconstruction. *IEEE Trans. Vis. Comput. Graph.* **2016**, *22*, 1261–1277. [CrossRef]
40. Novakovic, G.; Lazar, A.; Kovacic, S.; Vulic, M. The Usability of Terrestrial 3D Laser Scanning Technology for Tunnel Clearance Analysis Application. *Appl. Mech. Mater.* **2014**, *683*, 219–224. [CrossRef]
41. Suzuki, S.; Be, K. Topological structural analysis of digitized binary images by border following. *Comput. Vis. Graph. Image Process.* **1985**, *30*, 32–46. [CrossRef]
42. OpenCV. Available online: https://opencv.org/ (accessed on 2 April 2019).
43. Boykov, Y.; Kolmogorov, V. An experimental comparison of min-cut/max-flow algorithms for energy minimization in vision. *IEEE Trans. Pattern Anal. Mach. Intell.* **2004**, *26*, 1124–1137. [CrossRef] [PubMed]
44. Boykov, Y.; Veksler, O.; Zabih, R. Fast approximate energy minimization via graph cuts. *IEEE Trans. Pattern Anal. Mach. Intell.* **2001**, *23*, 1222–1239. [CrossRef]
45. Kolmogorov, V.; Zabin, R. What energy functions can be minimized via graphcuts? *IEEE Trans. Pattern Anal. Mach. Intell.* **2002**, *26*, 147–159. [CrossRef] [PubMed]
46. Von Gioi, R.G.V.; Jakubowicz, J.; Morel, J.M.; Randall, G. LSD: A Fast Line Segment Detector with a False Detection Control. *IEEE Trans. Pattern Anal. Mach. Intell.* **2010**, *32*, 722–732. [CrossRef] [PubMed]
47. Kümmerle, R.; Grisetti, G.; Strasdat, H.; Konolige, K.; Burgard, W. G 2 o: A general framework for graph Optimization. In Proceedings of the IEEE International Conference on Robotics and Automation, Shanghai, China, 9–13 May 2011.
48. Yang, G.; Chen, J. Research on Propagation Model for 5G Mobile Communication Systems. *Mob. Commun.* **2018**, *42*, 28–33.
49. 5G. Study on Channel Model for Frequencies from 0.5 to 100 GHZ (3GPP TR 38.901 version 14.0.0 release 14). Available online: http://www.etsi.org/standards-search (accessed on 5 May 2017).
50. Khoshelham, K.; Vilariño, L.D.; Peter, M.; Kang, Z.; Acharya, D. The ISPRS benchmark on indoor modelling. *Int. Arch. Photogramme. Remote Sens. Spat. Inf. Sci.* **2017**, *XLII-2/W7*, 367–372. [CrossRef]
51. Chew, L.P. Constrained Delaunay triangulations. *Algorithmica* **1989**, *4*, 97–108. [CrossRef]

© 2019 by the authors. Licensee MDPI, Basel, Switzerland. This article is an open access article distributed under the terms and conditions of the Creative Commons Attribution (CC BY) license (http://creativecommons.org/licenses/by/4.0/).

Article

Web-Net: A Novel Nest Networks with Ultra-Hierarchical Sampling for Building Extraction from Aerial Imageries

Yan Zhang, Weiguo Gong *, Jingxi Sun and Weihong Li

Key Lab of Optoelectronic Technology & Systems of Education Ministry, Chongqing University, Chongqing 400044, China
* Correspondence: wggong@cqu.edu.cn; Tel.: +86-138-830-13563

Received: 13 June 2019; Accepted: 9 August 2019; Published: 14 August 2019

Abstract: How to efficiently utilize vast amounts of easily accessed aerial imageries is a critical challenge for researchers with the proliferation of high-resolution remote sensing sensors and platforms. Recently, the rapid development of deep neural networks (DNN) has been a focus in remote sensing, and the networks have achieved remarkable progress in image classification and segmentation tasks. However, the current DNN models inevitably lose the local cues during the downsampling operation. Additionally, even with skip connections, the upsampling methods cannot properly recover the structural information, such as the edge intersections, parallelism, and symmetry. In this paper, we propose the Web-Net, which is a nested network architecture with hierarchical dense connections, to handle these issues. We design the Ultra-Hierarchical Sampling (UHS) block to absorb and fuse the inter-level feature maps to propagate the feature maps among different levels. The position-wise downsampling/upsampling methods in the UHS iteratively change the shape of the inputs while preserving the number of their parameters, so that the low-level local cues and high-level semantic cues are properly preserved. We verify the effectiveness of the proposed Web-Net in the Inria Aerial Dataset and WHU Dataset. The results of the proposed Web-Net achieve an overall accuracy of 96.97% and an IoU (Intersection over Union) of 80.10% on the Inria Aerial Dataset, which surpasses the state-of-the-art SegNet 1.8% and 9.96%, respectively; the results on the WHU Dataset also support the effectiveness of the proposed Web-Net. Additionally, benefitting from the nested network architecture and the UHS block, the extracted buildings on the prediction maps are obviously sharper and more accurately identified, and even the building areas that are covered by shadows can also be correctly extracted. The verified results indicate that the proposed Web-Net is both effective and efficient for building extraction from high-resolution remote sensing images.

Keywords: remote sensing; deep learning; building extraction; web-net; ultra-hierarchical sampling

1. Introduction

Large numbers of satellites and drones have been launched alongside the rapid development of aerospace technology. Hence, high-resolution remote sensing images are getting easier to acquire. An important use for remote sensing images is extracting and mapping artificial objects, such as buildings [1], roads [2], and vehicles [3] at the pixel-level. Among them, building extraction is the most critical task, and it is commonly applied to monitor the subtle changes in urban areas, urban planning, and estimating the population. However, different from roads and vehicles, building areas always contain complex scenic backgrounds. Meanwhile, in some areas, the visual features (shapes and colours) of buildings and that of other natural objects (hills and lakes) are highly similar, which makes the building extraction task greatly challenging, not only for designing auto-detection models, but also for the artificial labelling tasks in the remote sensing field. In general, a high-quality image provides more cues for identifying the building areas, whereas the abundant local information that is provided by the remote sensing images with higher resolution also raises higher requirements for the models' denoising and feature extraction abilities.

1.1. Building Extraction with Machine Learning

The building extraction task has drawn the attention of researchers over recent years. Before the common application of deep learning, there were massive machine learning models that tried to handle this task. In general, the pixel-wise labelling model consists of two sub-modules: the feature descriptor for extracting the semantic features from the original images and the pixel-wise classifier for determining the classes of the pixels. Some carefully designed feature descriptors were widely used in early approaches. Tuermer et al. [4] firstly used the histogram of gradient (HOG) feature descriptor in remote sensing for detecting vehicles. The Haar feature is applied in [5] for detecting buildings' outlines and determining the location of buildings' corners. Additionally, Yang et al. [6] applied the Scale-invariant feature transform [7] (SIFT) for classifying objects in remote sensing images. Unlike the artificially designed feature descriptors, the trainable models are the mainstream for the choices of classifiers. In [8], Mountrakis et al. reviewed the early applications of Support Vector Machines (SVMs) on remote sensing images. They stated that there are hundreds of relevant papers that apply SVMs to remote sensing images for various tasks. Except for SVMs, [9] researched the Bayes classifier and demonstrated that the naive Bayes can achieve comparable performance under most conditions. In [10], an assembly model, called the Fuzzy Stacked Generalization (FSG), which combined the detection results of multiple classifiers under a hierarchical architecture, was designed such that the building extraction performance can be further boosted. Although models that were based on classical machine learning methods achieved remarkable results in building extraction, how to properly and automatically extract the building areas are still challenging and expensive due to the time consuming artificial feature selections and the poor generalization abilities of the aforementioned classifiers.

1.2. Building Extraction with Deep Learning

Recently, with the rapid improvement of GPU computing, deep convolutional neural networks have become cornerstone in computer vision and remote sensing areas due to their great capability of extracting hierarchical features in an end-to-end fashion. Fully Convolutional Networks (FCNs) [11] are the common choice for most current deep learning models for the pixel-level labelling task. Within the framework of FCN, there are two keypoints affecting the performance with respect to the segmentation accuracy. The first one is the feature extraction backbone network, and the other is the upsampling design that preserves the features' structural consistency. VGG [12], ResNet [13], Inception [14], and their mutation models [15,16] are the most popular backbones because of their high structural flexibilities and great generalization abilities. Recently, DenseNet [17] and its dense connection patterns have become the mainstream backbones due to the efficiency of their feature reuse. By extending the FCN architectures, U-Net [18] and SegNet [19] propose an encoder-decoder structure to compensate the semantic features with local cues and enhance the structural consistency of the prediction map. In addition, Deeplab [20] proposed Atrous Spatial Pyramid Pooling (ASPP) to encode the context and scene information via a pyramid scene parsing (PSP) [21] structure and atrous convolution [22]. Deeplab made great progress on semantic segmentation tasks by embedding the ASPP into the encoder-decoder architectures. In the remote sensing area, according to the mentioned properties of high-resolution remote sensing images, some carefully designed models have been proposed and optimized for building extraction tasks that are based on these above semantic segmentation approaches. In early research, [23,24] used naive FCN architectures with deconvolutional layers to extract buildings or roads, and these works demonstrated the effectiveness and efficiency of the FCN architecture. [25,26] trained FCNs to extract the buildings using the patch-wise method. In [27], Wu et al. built a multi-constraint network to sharpen the boundaries of artificial object predictions. A trainable block, called the field-of-view (FoV), is proposed in [28] to boost the performance of the FCN. With the successful applications of U-Net in the pixel-wise area labellings, most current models [28–33] use encoder-decoder architectures. The mutation models enhance the buildings' semantic boundaries by introducing a new loss or fusing features in more effective ways. Moreover, Yang et al. [29] proposed an encoder-decoder network that was based on DenseNet and an attention mechanism, which is called the dense-attention network (DAN), which achieves remarkable improvements in building extraction. Meanwhile, Mou et al. analyzed and encoded the long-range relationships in remote sensing images over sequences of time. Furthermore, [30,31] applied the recurrent neural networks to fuse the hierarchical features from the different levels of the FCN. Audebert et al. [32] proposed an efficient multi-scale approach to leverage both a large spatial context and the high-resolution data and investigated the early and late fusion of Lidar and multispectral data to cover the scale variance of buildings from different areas. In [33,34], the extra geographical information (DSM, DEM, and Lidar images) are fed into a carefully designed FCN, together with high-resolution RGB images, and the results indicate that abundant features always lead to sharper predicted building boundaries. Moreover, post-processing methods, such as Guider Filter [1] and Conditional Random Field (CRF) methods [35,36], have been heavily researched and attempted to preserve the structure consistency between the building predictions and the original images.

1.3. The Motivation and Our Contribution

As mentioned in 1.2, the models that are based on the encoder-decoder framework have achieved the best performance on building extraction tasks; however, there are three main dilemmas that remain for the current building extraction tasks. (1) Early approaches easily classify non-buildings as buildings. This is caused by the semantic feature maps that still contain noises and the long-range reliabilities not being properly extracted. Generally, it is an inevitable problem for Convolutional Neural Network (CNN)-based models, since its denoising operation, such as Max-pooling and Average-pooling, is always accompanied by local cue losses. (2) The contours of the extracted building maps are blurred and irregular. (3) The generalization abilities of the current building extraction models are weak, as described in [37], since FCN-based networks only get high-quality predictions for areas where the landforms are highly similar to that of the training areas; meanwhile, we found that the building areas that are covered by shadows are likely to be labelled as non-buildings. To some extent, these three dilemmas are partly conflictive. (1) requires less noises in the feature maps, while (2) needs more local information to obtain the regular contours and retain the structural consistency. Although (3) could be relieved by applying deeper networks, the deeper network that apparently needs many more parameters also faces training difficulties and overfitting issues. In this paper, we propose a novel nested encoder-decoder deep network, named Web-Net, to simultaneously overcome the above conflicting obstacles that exist in the building extraction task. The main contributions of this paper can be listed, as follows.

1. We first propose a cobweb-like fully nested and symmetric network architecture, named Web-Net. Following the dense connection patterns, the output of every node layer is fed into all the subsequent node layers in both the horizontal and vertical directions. The harmony nested and dense-connected fashion leads to better features reuse abilities and generalization abilities.

2. We build a novel feature sampling and feature fusing block, named Ultra-Hierarchical Sampling (UHS), which is applied to every node layer in the proposed Web-Net. The UHS block consists of a pair of position-wise downsampling and upsampling sub-layers: an Ultra-Hierarchical Downsampling (UHDS) sub-layer and an Ultra-Hierarchical Upsampling (UHUS) sub-layer. By iteratively feeding the feature maps from different levels into the UHDS and the UHUS, they can be reshaped to a fixed size and then embedded together. Benefitting from the fully position-wise operation in the down/upsampling, the number of the feature map parameters and their spatial structure are preserved. Therefore, UHS achieves a better balance between the preservation of local cues, the structural consistency, and feature denoising as compared with normal downsampling and upsampling methods, which results in more accurate building extraction contours and better classification accuracies.

3. We analyze the effects of the deep supervision methods on the nested Web-Net. Based on the pruning of Web-Net, we propose the efficient mode, the balance mode, and the high-performance mode for the proposed Web-Net to make it more flexible and easier to adopt in either time-sensitive tasks or accuracy sensitive tasks.

This paper is organized, as follows. After Section 1 introduced the building segmentation in remote sensing image processing, Section 2 details the semantic segmentation frameworks that are related to our work. Subsequently, Section 3 gives the proposed method, describes the architecture of Web-Net, and lists the implementation details of the UHS block and the deep supervision method. The experimental results and discussions are illustrated in Section 4. Finally, we provide a conclusion in Section 5.

2. Related Work

In this part, we review the early classical network architectures and the state-of-the-art models for object segmentation tasks. These architectures are widely applied to remote sensing object extraction and other similar binary semantic segmentation tasks.

2.1. Fully Convolution Methods

In early research, the patch-based CNN was commonly used and it was the mainstream method in the remote sensing building extraction field. In it, the images are firstly divided into several mini patches and then fed into CNN networks to extract the semantic features. Afterwards, fully connected (FC) layers are used to classify each pixel. The patch-based CNN is strictly restrained by the number of parameters, and the extremely small patches (always less than 25 pixels) would consume a large amount of memory. Therefore, the final prediction usually lacks structural integrity, especially in the large scale building areas. The FCN [11] replaces all of the FC layers with convolution layers, and this procedure is mainly based on the assumption that every patch in an image follows the same probability distribution; hence, applying convolutional layers whose parameters are locally shared can achieve comparable performance with FC layers with several orders of magnitude fewer parameters. Therefore, when encountering fixed GPU memory, an FCN can achieve a larger image patch as its input and better long-range reliabilities can be obtained, which significantly improves the prediction quality with fewer structural errors. Meanwhile, fewer parameters also benefit the model's robustness and ease the difficulties of training.

2.2. Encoder-Decoder Architectures

The encoder-decoder structure is widely applied on pixelwise labelling tasks, such as semantic segmentation, object segmentation, etc. [18], first built a highly symmetric architecture, called U-Net, in which the structures and dimensions of the decoders mirror the encoders. The outputs of each level from the encoder are directly linked to the corresponding level of the decoder as inputs through the jump connection. When compared with FCN-based networks, U-Net built a more sophisticated decoder to gradually upsample the semantic feature maps to the original image size, and the local cues from the encoder are compensated at the corresponding decoder level, which enhances the predicted contours. SegNet [19], which was further extended from U-Net, implemented a memorized Max-pooling operation in the encoder model that stores the indices of the maximum pixel, and the decoders in SegNet upsample its input feature maps while using the memorized max-pooling indices. Rather than ordinary max-pooling, the memorized max-pooling preserves the location information of the maximum pixel in an adjacent area, which allows for the upsampling in the decoder blocks to better recover the lost local cues.

2.3. Nested Connected Architectures

Motivated by the idea of densely connected networks, the nested connected architectures are designed to reuse more features. Nested architectures always have sophisticated and carefully designed adjacent/jump connections, and different bundles of the inner layer can be explicitly assigned to corresponding sub-networks architectures. To the best of our knowledge, GridNet [38] is the first approach towards implementing a nested connected architecture in the semantic segmentation area. The feature propagation paths in GridNet can be separately divided into the U-Net, the FCN, the Fully Resolution residual Network [39], and other symmetric or asymmetric encoder-decoder architectures, which allow for the model to contain more complicated feature paths and extract deeper semantic features. Furthermore, Unet++ [40] introduces the idea of nested architecture into U-Net; there are various levels of U-Nets that are stacked in Unet++, and hence the entire structure of Unet++ looks similar to an equilateral triangle (the same number of layers on every edge). One of the most critical contributions of Unet++ is introducing the Deep Supervision [41] method (DS) into the nested architecture networks. Unet++ applies the DS method on every sub-U-Net. Benefitting from the DS method, Unet++ can be easily trained and it achieves better performance on the segmentation task, rather than early nested networks.

3. Proposed Method

3.1. Overview of the Proposed Networks

Figure 1 shows the high-level structure of the proposed Web-Net and its skip connection patterns. Different from the encoder-decoder architectures, such as U-Net [18] and SegNet [19], the proposed network consists of a backbone encoder and nests of node layers (decoders). These node layers absorb the feature maps from the adjacent node layers (Figure 1a) and the long-range node layers (Figure 1b) in the horizontal direction (red dotted line) and the vertical direction (green and blue dotted lines). Benefitting from the nested connection pattern, the node layers can simultaneously work as parts of the encoder and decoder in the proposed framework. Apparently, in the vertical direction, the input hierarchical features from different levels of Web-Net need to be resized to the same size for further processing for every node layer; therefore, we propose the carefully designed Ultra-Hierarchical Sampling (UHS) block to accomplish this. Web-Net can be seen as the densest version of the nested encoder-decoder networks by applying the UHS block in every node layer. The implementation of the UHS blocks and then the abundance of message paths and the deep supervision method for Web-Net will be described in Sections 3.2 and 3.3, respectively.

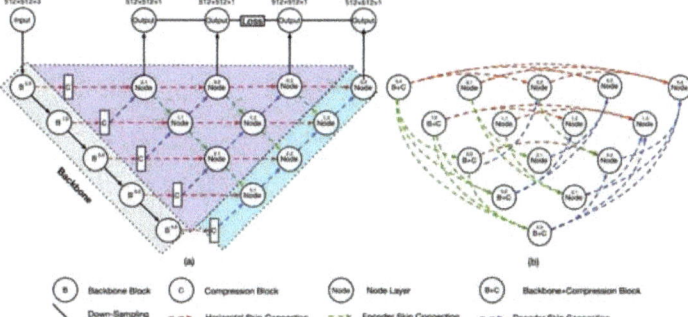

Figure 1. Architecture of the proposed Web-Nets. (**a**) shows the structure of Web-Net with adjacent connections. The five blocks in the grey areas represent the encoder backbone. The compression blocks are linked behind each level of encoder blocks to reduce the dimensions, and the solid arrow shows the normal downsampling (Max Pooling) operation. The light blue area is the decoder part of Web-Net, and the outputs of every node layers in the decoder are only fed into the node layers of upper levels in the vertical direction. The purple triangular area located at the corner of the Web-Net architecture is the node layers, which work as both encoders and decoders. The node layers obtain the features from their neighbouring layers, and then we simultaneously fuse the features and feed them into the adjacent node layers. The green, blue and red dotted lines indicate the feature transfers in the top-bottom, bottom-top and horizontal directions, respectively. (**b**) details the skip connections in Web-Net. The red dotted line represents the horizontal skip connections among the same levels, the green dotted lines denote the hierarchical top-bottom skip connections, and the blue dotted lines show the hierarchical bottom-top jump connections. Apparently, each regular triangle with different coloured edges constructs a mini encoder-decoder architecture.

3.2. Node Layer and Ultra-Hierarchical Sampling Block

As the corner components of the proposed network structure, Figure 2 details the workflow of Ultra Hierarchical Sampling (UHS) block in the node layer, which down-samples/up-samples the dimensions of inputs by iteratively applying the position-wise reshape operation. At first, as shown in Figure 1, the input features from both adjacent and long-range node layers are fed into the Feature Gather block. Depending on which levels the inputs coming from, the inputs are divided into three groups, which are top-bottom, bottom-top, and horizontal groups, and they are represented by green, blue, and red dotted lines, respectively. Subsequently, these hierarchical features with the different shapes are fed into Ultra-Hierarchical Sample (UHS) block to reshape them into the same size and concatenate them together for further processing. Finally, the outputs of the UHS block are delivered into the Feature Fusing sub-block, which contains two 3 × 3 convolution layers with an Relu Activation function and Batch Normalization Layer and a Squeeze and Excitation (SE) Block [42]. Here, the SE block adaptively recalibrates features with channel dimensions through a simple gate mechanism. For further convenient analysis, we define the necessary symbolic representations for the node layer in priority. Assuming that the scale factor between two neighbored levels in Web-Net is d, in general d is set to 2. The specified node layer is represented as $N_{(i,j)}$, where $i, j \in [0, n-1]$, i indicates which level the node layer belongs to and j is the index of the node layers in the ith level. After the compression block, the shapes of the feature maps in level i are (C_i, H_i, W_i). Moreover, the relationship of the feature map shapes between the level a and the level b can be computed as in Equation (1):

$$(C_b, H_b, W_b) = \left(d^{b-a}C_a, d^{a-b}H_a, d^{a-b}W_a\right) \qquad (1)$$

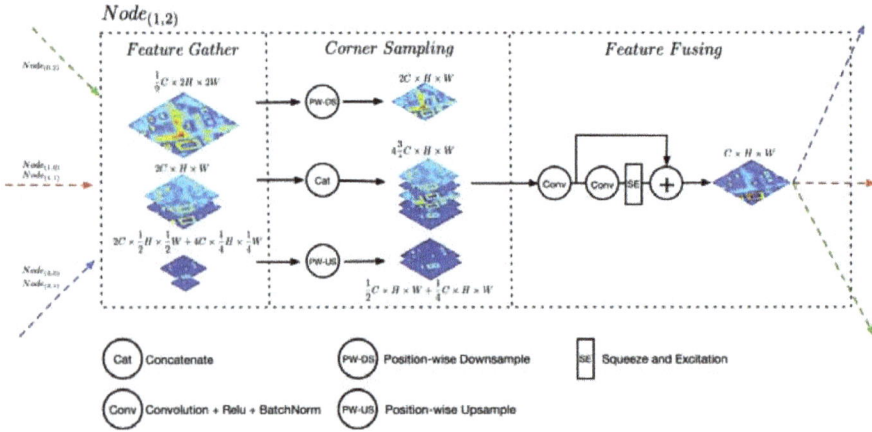

Figure 2. The workflow of Node Layer for $Node_{(1,2)}$. C, H, W represent channel numbers, heights, and widths, respectively, of the feature maps in level 1. For clear visualization, the intermediate feature maps coming from intra-level and inter-level node layers are summarized with the dimension channels.

As mentioned above, the hierarchical input features need to be reshaped to the same shape. The early methods usually apply the classical down-sampling methods (Max Pool, Mean Pool, et.al), which simply select the maximum value or the averaged value from each pooling grid and the classical up-sampling methods (Bilinear, Nearest, et.al), which complement the missing values in each pooling grid through the Bilinear or Nearest interpolation methods. As analyzed in Section 1.3, the local cues lose during pooling operation cannot be recovered in the up-sampling processing. There are two keypoints in order to preserve the local cues: one is that every value in the feature map cannot be directly dropped out, in another word, the total amount of the feature map parameters needs to be unchanged. Another is that the structural consistency of the feature map must be kept. For this, we design the position-wise operations to change the shape of the feature maps in the UHS block. As shown in Figure 3, the position-wise here means the operation just acts on the positions of each pixel in the feature map, the parameter amount and their corresponding values are unchanged. Assuming that the feature maps of three dimensions (channel, height, width) of A (B) is the result of applying the position-wise downsample (position-wise upsample) on B (A), c, i, and j are the indices of the produced feature maps, the output of the position-wise downsample (position-wise upsample) can be calculated as Equations (2) and (3), where | and % indicate the exact division and remainder operations.

$$A_{(c,\ i,\ j)} = B_{(c|s^2,\ s\ i+c\%s^2|s,\ sj+c\%s^2\%s)} \tag{2}$$

$$B_{(c,\ i,\ j)} = B_{(cs^2+(i+j)\%s^2,\ i|s,\ j|s)} \tag{3}$$

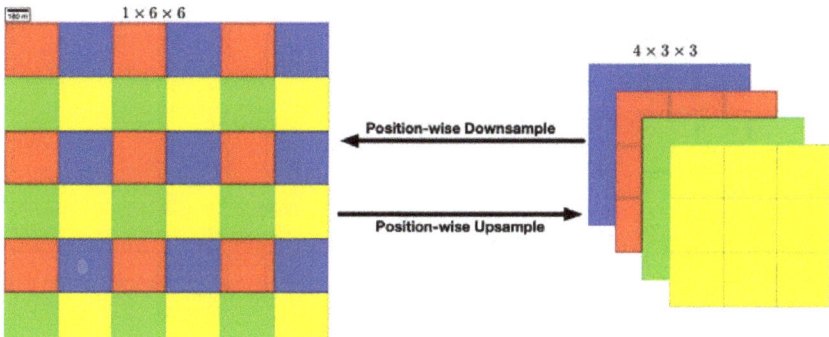

Figure 3. Diagram of the position-wise operations where the red, blue, green, and yellow cubes indicate pixels of different positions in the feature map respectively. The position-wise Downsample operation halves the size of the original feature map while the dimensions are stretched for four times. In contrast, the position-wise Upsample operation doubles the size of the feature map and reduces the dimension channel according to the position of every pixel.

As shown in Figure 4, the proposed UHS block involves a position-wise Downsample (PW-DS) flow and a position-wise Upsample (PW-US) flow. In the PW-DS flow, we simultaneously use four ordinary pooling layers with different hyperparameter initializations on the input features to simulate the position-wise down-sampling operation. For the kernel size s, when $s = 1$, every pixel in the feature map would be properly preserved. With the increasing of s, the larger pooling kernel size can filter out noises but blur the local cues. The pooling stride is set to 2, while the padding is $(0, 0, 0, 0)$, $(0, s-1, 0, 0)$, $(s-1, 0, 0, 0)$, and $(s-1, s-1, 0, 0)$ individually, and each pooling layer would reduce the size of the feature maps by half, the results of four pooling layers would be concatenated together as the final output. The PW-DS operation is iteratively applied on the input feature map until the output is reshaped to the target size. Assuming that the scale factor between the input feature map and the output target is f, the number of iterations is equal to $log_2 f$. In simple terms, the PW-DS squeezes the input feature maps into the target size, the local cues and structural information are encoded into the dimension channels. In the PW-US flow, similar to Dense Upsampling Convolution (DUC) [43], the position-wise up-sampling operation doubles the size of the input feature map and it reduces the input channel number to a quarter in each iteration, the PW-US operation is looped on the input feature map until the size of output is enlarged to the target. For example, if the input shape and the target shape are (d^4c, h, w) and (c, d^2h, d^2w), respectively, the PW-US would be applied twice. In the first iteration, every feature strip with the shape $(d^4c, 1, 1)$ is reshaped to (d^2c, d, d), therefore the feature map after the first iteration of PW-US has a shape of $(d^2c, dh,\ dw)$. Similarly, the output shape of the second iteration of PW-US is changed to (c, d^2h, d^2w). Corresponding to the PW-DS stream, the PW-US stream can be seen as flattening the squeezed feature map back to a specific shape, it decodes the local cues and structural information into the high-resolution feature maps.

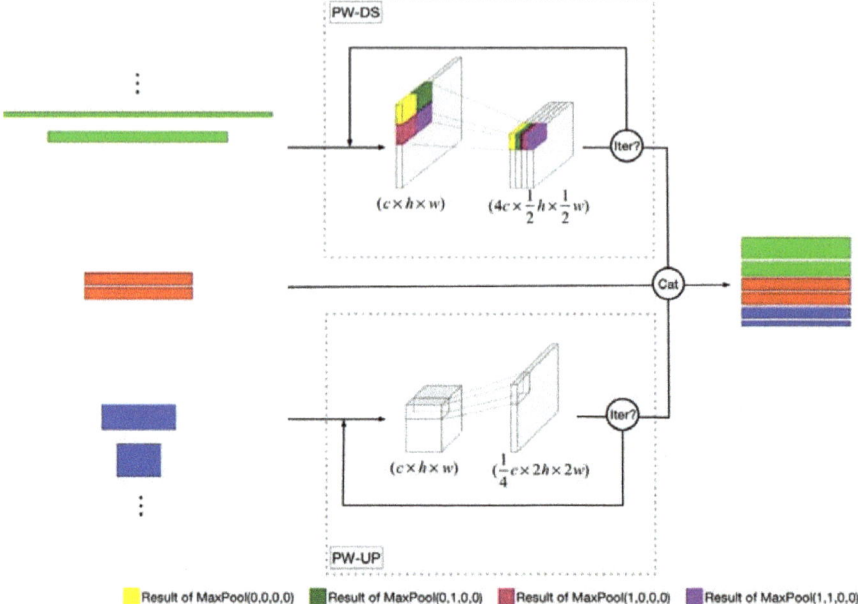

Figure 4. The workflow of the Ultra-Hierarchical Sampling (UHS) blocks. Similar to Figure 1, the green, red, and blue cubes indicate the feature maps from the top, horizontal, and bottom levels respectively. c, h, w represent the dimensions of the input for the position-wise Downsample (PW-DS) and the position-wise Upsample (PW-US). The downsample and upsample rates for position-wise downsample and position-wise upsample are initialized to 2. In the PW-DS flow, the yellow, dark green, bright red, and purple cubes indicate the results of four individual pooling layers with different padding initializations.

3.3. Dense Hierarchical Pathways and Deep Supervision

As described in Section 3.1, the proposed Web-Net contains the dense skip pathways, both in the horizontal and in the vertical directions. The horizontal connections just work like DenseNets, where all the preceding feature maps in the same level would pass directly to the layers behind them as part of the input feature maps. In the vertical direction, profiting by the proposed UHS block, the encoder and decoder node layers can also gather the preceding hierarchical feature maps as the inputs, the dense connection patterns can greatly shorten the message paths in both directions. Suppose that the feature fuse function and outputs of the node layer (i, j) are $H_{(i,j)}$ and $X_{(i,j)}$, respectively, the transform functions for the up stream and the down stream are defined as $PWUS$ and $PWDS$, respectively, and w, m, n are the indexes of feature maps from corresponding levels, the transform of $Node_{(i,j)}$ is shown as Equation (4).

$$X_{(i,j)} = H_{(i,j)}(Cat([X_{i,w}|w \in [0, j-1]], PWUS(X_{m,i+j-m}|m \in [j+1, i+j]), PWDS(X_{n,j}|n \in [0, i-1]))) \quad (4)$$

It can be seen that the Web-Net is a densest connected, symmetric, and elegant architecture, where the features can efficiently propagate to each node in every level within the shortest path. Additionally, the nested architecture makes the Web-Net contain numbers of Web-Nets with smaller levels in it.

In Figure 5, there are eight different encoder-decoder feature propagation paths in a basic 3 level Web-Net, and each graph describes a special encoder-decoder structure. Specifically, the input features of every node layer are coming from other small Web-Net architecture, therefore, the semantic feature can be not only extracted in the nested pattern, but also compensated with the local cues by jump connections, this results in a sharper and more accurate prediction. Moreover, in Web-Net, there are just a few extra parameters when compared with the U-Nets architecture with the same encoder backbone, because we share and reuse the feature maps rather than create new ones. Hence, as compared with other complicated network structures, the proposed Web-Net can partly avoid the over-fittings that are caused by the large parameter amounts of deeper encoder or wider decoder benefitting from the elegant feature reuse manners.

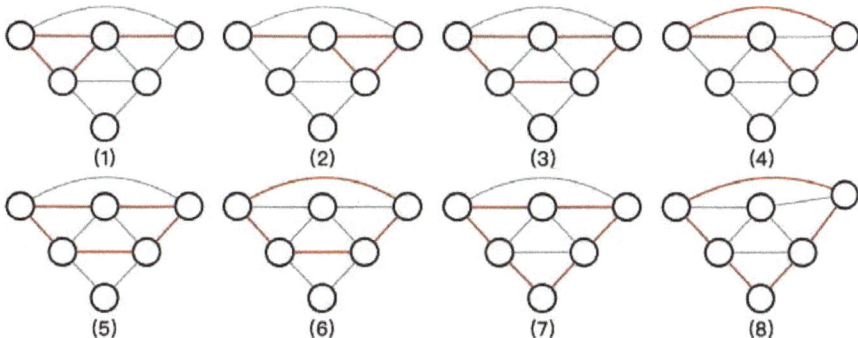

Figure 5. The 8 encoder-decoder structures in a basic 3-level Web-Net. The red lines in each graph constitute an independent encoder-decoder structure.

Profiting from the nested design of Web-Net, all of the outputs of the node layers in level 0 have full resolutions as Ground Truths; therefore, we can apply the deep supervision method on them. For the prediction layer $Node_{(i,j)}$, we use l_n to represent its loss function. l_n is the fusion loss, which is a linear weighted summary of the binary cross-entropy (BCE) and the Dice coefficient. The final loss L is simply a combination of l_n, as shown in Equations (5) and (6):

$$l_i = -(\omega_1 \frac{1}{N} Y log \hat{Y}_i + \omega_2 \frac{2Y * \hat{Y}_i + \varepsilon}{Y + \hat{Y}_i + \varepsilon}) \tag{5}$$

$$L = \sum_{i=1}^{4} l_i(Y, \hat{Y}_i) \tag{6}$$

where Y and $\hat{Y}$ denote the ground truth and prediction probabilities, respectively, and ε is set as 0.01 to prevent the value of the denominator from being 0. ω_1 and ω_2 are the coefficients that balance the Binary Cross-Entropy and Dice loss. As depicted in [13], the identity mapping that is constructed by residual connections in the UHS blocks ensures that the optimization loss L is equal to the optimizing series of encoder-decoder sub-networks; this indicates that the performance of Web-Net would not be worse than anyone of sub-networks even in the worst case. Section 4, discusses pruning and ablation studies that are applied to exploit the benefits of deep supervision methods.

4. Experiments and Discussions

In this section, to demonstrate the efficiency and effectiveness of the proposed Web-net, we have evaluated it for the building extraction task on very high-resolution remote-sensing images among different areas.

4.1. Training Details

4.1.1. Datasets

We conduct all experimental evaluations on the challenging Inria Aerial Image Labelling Dataset [37] and WHU Dataset [44]. The Inria dataset mainly contains five open-access land-cover types from Austin, Chicago, Kitsap County, Vienna, and West Tyrol. There are 36 ortho-rectified images that cover 81 km^2 for each region. Additionally, the five areas cover abundant landscapes ranging from highly dense metropolitan financial districts to alpine resorts, as shown in Figure 6.

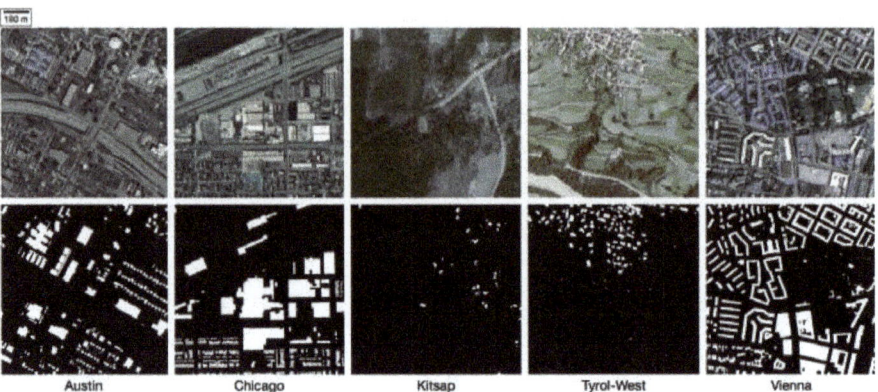

Figure 6. Visual close-ups of the Inria dataset images from five different regions and their corresponding reference data.

The images in this dataset contain three bands (RGB) with very high spatial resolution (0.3 m). There are just two semantic labels of building and nonbuilding, and the target area in the dataset is the footprint of the roof. Therefore, it is completely suitable for our research purposes and convenient for validating model performances. In our experiments, we split each image into 100 sub-images, with a resolution of 500 × 500. In total, there are 18000 split images. Because the test set reference data is not publicly released, we choose the first five unsplit images from each area as the test set (images 1–5 for the testing and images 6–36 for the training) following the official validation suggestions [37] to achieve fair results and comparisons.

The WHU Dataset contains 8189 tiles of 512 × 512 pixels with more than 187,000 well-labelled buildings in New Zealand as compared to the Inria Dataset. The dataset covers approximately 450 km^2 and it has the same spatial resolution of 0.3 m as that of the Inria Dataset. This dataset was officially divided into a train set, a validation set, and a test set, consisting of 4736 images, 1036 images, and 2416 images respectively.

4.1.2. Metrics

The intersection over union (IoU) of the positive class (building) and the overall accuracy are applied as the evaluation criteria to evaluate the performance of the different building extraction methods on the remote sensing images, which are also following the official guidance of the Inria Dataset [37]. The Overall Accuracy can actually evaluate the percent of the correctly predicted pixels. For the balanced dataset, the overall accuracy can objectively represent the model's classification ability. However, the buildings always cover small areas on the aerial imageries and they are easy to be ignored. In the extreme situation, only one small building is located in a large area. Regardless of whether the model can correctly extract the building or not, there are few differences in the overall accuracy metric. The Intersection over Union (IoU), which is a widely used non-linear measure that robustly evaluates how close two distributions are, is introduced in the segmentation task to overcome the effect of the unbalanced phenomenon.

4.1.3. Implement Details

We build the proposed Web-Net based on the PyTorch library [45]. We train the models both from scratch and fine-tune the encoder backbones using the pretrained parameters from ImageNet [46]. We apply the Adam algorithm [47] with the default settings ($\beta_1 = 0.9, \beta_2 = 0.999$, and the weight decay is 0) to optimize the model parameters during training. We follow the popular polylearning rate schedule that is computed as Equation (7) to adjust the learning rate:

$$lr = lr_{init}(1 - \frac{iter}{max\_iter})^{power} \tag{7}$$

where the initial learning rate lr_{init} is 0.001, $power = 8$, and the max iterations is set to 30. In addition, in each iteration, the whole training set is sequentially fed into the model. It takes approximately 27 hours to train our model with the Inria dataset on one NVIDIA GTX1080Ti.

4.2. Ablation Evaluation

In this section, we aim to study how the proposed Web-Net works with the different backbones and the sampling methods. For convenient analysis, we build all the ablation experiments on the Inria Aerial Dataset.

4.2.1. Backbone Encoder Evaluation

In this section, we evaluate the performance of the proposed Web-Net with different backbone encoders that are trained from scratch. VGG [12], ResNet [13], [17], ResNext [15], Xception [16], and DenseNet [17] are applied as the encoders; in addition, the pooling size in the UHS is set to 2 and the batch size is fixed as 4 in each model in order to obtain fair comparisons. The other hyperparameter settings follow the description of Section 4.1.3. Table 1 lists the results.

Table 1. The Intersection over Union (IoU) and Acc.. of various backbone encoders for the validation set.

Backbones	IoU (%)	Acc. (%)	[1] TT(Min)	[2] MS(GB)
VGG-16	75.10	96.10	50	6.86
Res-50	75.33	96.06	62	5.88
Res-101	75.58	96.17	70	7.07
ResNext-50	76.23	96.25	52	5.91
ResNext-101	76.39	96.30	87	7.68
Dense-121	75.93	96.20	-	-
Dense-161	76.58	96.38	-	-
Xception	75.58	96.14	58	7.72

[1] Training time per epoch, [2] Memory space cost on a GPU.

From Table 1, it can be seen that even the encoder with a very basic VGG-16 can acquire a quite good result on the validation dataset, which proves the effectiveness of the proposed Web-Net architecture. Furthermore, ResNet achieves comparable metric scores as the VGG network, but it requires considerably fewer parameters, since it benefits from the residual learning method. A significant performance boost comes from the ResNxet network that replaces the convolution layers in it with aggregated sets of sub-convolution layers, which is also known as group convolution. Similar to ResNext, Xception also applies group convolution operations, but it obtains lower metrics than ResNext due to the lake of residual transform. Unexpectedly, there is little improvement when we apply the widely used DenseNet as the encoder backbone in Web-Net. We believe that it is because the proposed nested hierarchical structure has applied the dense connection patterns among the node layers, and so the dense connections in the encoder backbone blocks are not as critical and necessary. The original DenseNet that is implemented in deep learning platforms is computationally expensive, since the high frequency concatenating operations exponentially expand the memory costs, and the Efficient-DenseNet [48] may save memory, but it decreases the training efficiency. Therefore, we do not test their training time and memory costs in Table 1. Furthermore, we apply deeper backbones to evaluate time and memory consumptions. The deeper networks, such as Xception, Dense-161, and ResNext-101 obtain less than a 1% improvement with respect to the IoU, while they take much more training time and consume more GPU memories. Therefore, we choose ResNext-50 as the backbone in further experiments to retain the best balance between the model's performance, time and memory costs. At the same time, an oracle model (best performance) is proposed in Section 4.3.

4.2.2. Ultra-Hierarchical Samplings Evaluation

As mentioned in Section 3.2, the size of the pooling kernels in the PWDS flow is vital to the performance of the UHS block, since the pooling size determines the capability to balance the denoising and information preservation. Meanwhile, we create four comparable down-up sampling blocks by replacing the PWDS and the PWUS with max pooling/average pooling and a bilinear interpolation/Deconv layer, respectively, in order to evaluate the effectiveness of the proposed UHS block. These four blocks are named the Max-Bilinear, Avg-Bilinear, Max-Deconv, and Avg-Deconv. The best results for each model are given in Table 2, and Figure 7 shows how the IoU scores vary with the kernel sizes of the downsampling operations.

Table 2. Evaluation of Web-Net for the validation set.

Models	IoU (%)	Acc. (%)
Max-Bilinear	75.96	96.19
Avg-Bilinear	75.82	96.16
Max-Deconv	76.20	96.25
Avg-Deconv	76.23	96.25
UHS	76.50	96.33

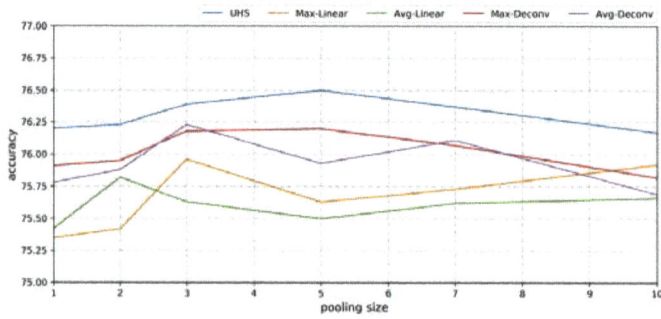

Figure 7. IoU line chart of Web-Net with the UHS block, Max-Linear, Avg-Linear, Max-Deconv, and Avg-Deconv operations.

From Table 2, the proposed Web-Net with the UHS blocks achieves the best results with an IoU of 76.50% and an Acc. of 96.33%, which are observably higher than those the other four comparable models. Additionally, the learnable upsampling method deconvolution reaches nearly 76.20% with respect to the IoU. The Web-Nets with the naive Max-Bilinear and Avg-Bilinear acquire the worst performance with respect to both the IoU and Acc. These results in Table 2 verify that the structure of the UHS block, as well as the position-wise down/upsample operations in Web-Net, play pivotal roles on boosting the model's performance. In Figure 7, it can be seen that the IoU of the UHS increases as the kernel size increases from 1 to 5 and then slowly decreases as the kernel size further increases. In addition, with the increase of the pooling size, the IoU curves of the other four models fluctuate more and are more chaotic. Meanwhile, the optimal kernel size of the UHS is 5, which is nearly twice as large as those of other models. These observations support the assumptions that the highly symmetric structures of the down/upsampling methods in the UHS blocks generate better and more stable denoising and local cue preservation abilities. Figure 8 lists some representative predictions from the Bilinear, Deconv, and the UHS-based Web-Net. It can be seen that the Web-Net with the UHS acquires sharper boundaries for larger buildings, and buildings with surrounding vegetation, which are easy to misclassify, are correctly extracted. All of these observations prove that the UHS block has better denoising and feature preservation abilities.

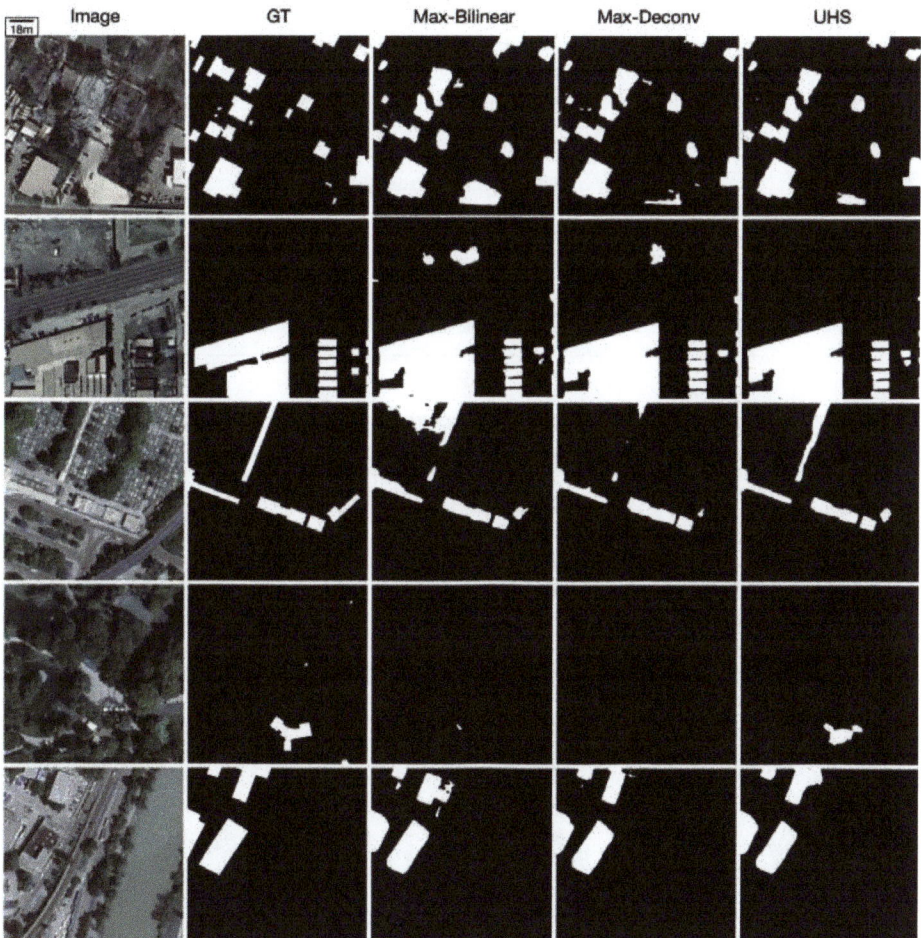

Figure 8. Examples (500 × 500 patches) of Web-Net with UHS, Max-Bilinear, and Max-Deconv blocks.

4.2.3. Pruning and Deep Supervision Evaluation

In this section, we prune the Web-Net into four scales according to the depth level to prove the efficiency and adaptability of the proposed Web-Net for both time-sensitive task and performance-sensitive task.

Due to the highly symmetric structure of Web-Net, we can partly supervise the outputs of $Node_{(0,1)}$, $Node_{(0,2)}$, $Node_{(0,3)}$, and $Node_{(0,4)}$ to individually simulate the Web-Net with different depths, which are represented as Web-Net-L_n, where n is the depth.

Table 3 reports the metric scores of each sub-Web-Net on the Inria Dataset. Web-Net-L_0 is much worse than the others due to the different numbers of parameters and network depth. Web-Net-L_1 achieves a 6.18% higher IoU than Web-Net-L_0, and the further improvements of the depth gradually increase the IoU to 76.50%. From Figure 9, it can be seen that the feature map of Web-Net-L_0 obtains significant low-level information, and the object maps are incomplete while lots of non-building areas are detected. The deeper and more complicated structures of Web-Net can efficiently extract the semantic information and involve fewer local features and details. It should be noted that the contours of the feature map from Web-Net-L_3 are not blurred, which proves that the proposed Ultra-Hierarchical Upsampling sub-block is effective for completely eliminating the local cues from the features that are encoded in the channel dimension. Meanwhile, Table 3 lists the time costs of each pruned model. Apparently, except for L_0, every five extra seconds of inference time can increase the IoU by at least 1.2%. Therefore, there are three modes that are involved in Web-Net to make inferences balanced with different accuracies and time costs, which are efficient (L_1), balanced (L_2), and effective (L_3) modes.

Table 3. Model Pruning.

Models	IoU (%)	Acc. (%)	Time(s)
Web-Net-L_0	67.90	94.73	15.4
Web-Net-L_1	74.02	95.94	18.7
Web-Net-L_2	75.20	96.14	23.2
Web-Net-L_3	76.50	96.33	28.8

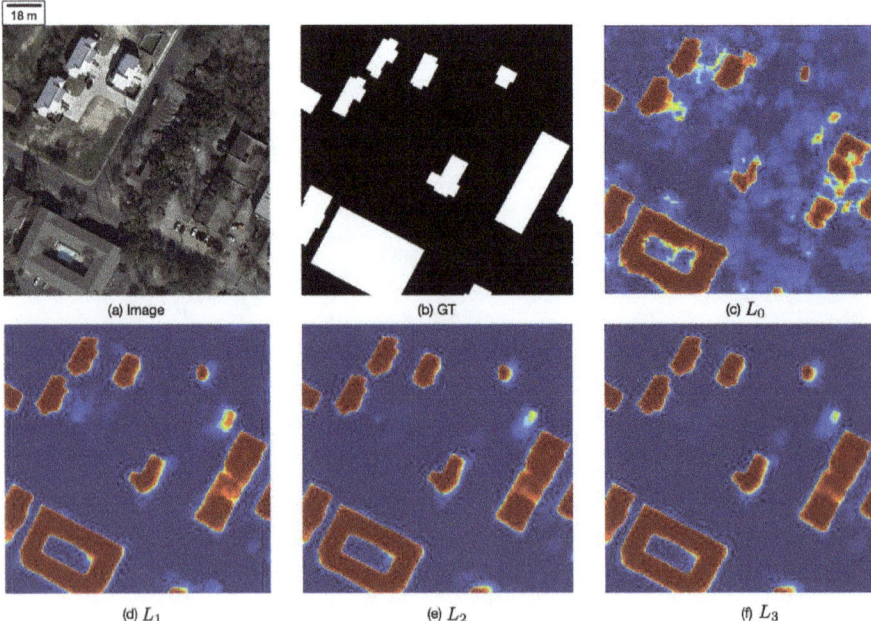

Figure 9. Comparison of the heat maps from different nodes in level 0. (a) is the original image, (b) is the ground truth, and (c–f) are the heat maps from different output nodes.

4.3. Best Performance and Comparisons with Related Networks

In this part, we investigate the best performance of the proposed Web-Net using four NVIDIA GTX1080Tis, and we then compare the Web-Net and the related state-of-the-art models to verify the effectiveness and efficiency of the proposed Web-Net.

4.3.1. Best Performance Model

A series of ablation experiments are built to determine the best performance of the proposed model. We just start with the basic encoder-decoder model (ResNext-50) and iteratively improve the performance by applying the proposed UHS blocks, nested structures, and learning strategies. Moreover, we apply the parameters that are pretrained on ImageNet to initiate the model. Bigger training sample sizes and deeper basic encoder structures are also used to obtain a better performing model with respect to the metrics. Table 4 shows the results.

Table 4. Various Design Results.

	Web-Net					
Unet++ (ResNext50 × 32 × 4d)	✓	✓	✓	✓	✓	✓
Web-Net (ResNext50 × 32 × 4d)		✓	✓	✓	✓	✓
Pretrained			✓	✓	✓	✓
DS [1]				✓	✓	✓
Web-Net (ResNext101 × 64 × 4d)					✓	✓
Batch = 16						✓
Acc. (%)	95.79	96.33	96.65	96.72	96.86	96.97
IoU (%)	73.32	76.50	78.37	78.69	79.52	80.10

[1] Deep supervision.

Rather than the basic U-Net architectures, the Web-Net architecture improves the IoU by 3.54% and the Acc. by 0.58%. The pretrained encoder backbone (ResNext-50) also results in a 1.87% improvement compared with training from scratch. The deep supervision method obtains small but consistent improvements of 0.32% and 0.07% for the IoU and Acc., respectively. Applying the deeper ResNext-101 as the encoder backbone could further obtain additional 0.83% and 0.14% improvements in the IoU and Acc, respectively, over ResNext-101. Finally, a large batch size (16) results in the best performance of the proposed Web-Net with an IoU of 80.10% and an Acc. of 96.97%.

4.3.2. Comparison Experiments on the Inria Aerial Dataset

Next, we provide the performance comparisons of the proposed Web-Net and other aforementioned state-of-the-art models on the Inria Aerial Dataset, and the results are listed in Table 5. Compared with the FCN-based baselines in [37], Web-Net outperforms the FCN and multi-layer perceptron (MLP) by 26.28% and 15.43%, respectively, for the IoU. Moreover, the result of Web-Net is 6.10% higher than that of the recurrent network in fully convolutional network (RiFCN), which applies a time consuming recurrent backward stream to fuse the hierarchical features in the time sequence. Web-Net achieves a 20.57% higher IoU when comparing the proposed Web-Net with the Mask R-CNN, which is a popular framework that simultaneously conducts instance detection and semantic segmentation tasks. The naive SegNet acquire a 70.14% IoU and a 95.17% overall accuracy, which indicates that the mainstream encoder-decoder architecture can work well on the extracted building areas. Including the latest nested Unet++, we can observe an improvement of at least 7.1% among all Unet and SegNet models with respect to the IoU. By combining the encoder-decoder architecture with dense connection patterns into the Dual-Resolution U-Net, the two-level U-Net acquires remarkable performance for the building extraction task. When compared with them, Web-Net acquires 5.88% and 5.55% higher IoUs, respectively. The recent GAN [49]-based approaches, Building-A-Net, acquire the state-of-the-art results on the Inria Aerial Dataset. Benefiting from the great generalization abilities of the GAN, the original Building-A-Net with 52 layers achieves a 74.75% IoU, and the deeper version further acquires an impressive IoU and Acc. of 78.73% and 96.71%, respectively. Although it is not fair to compare Web-Net with Building-A-Net, the Web-Net architecture can be easily embedded into the GAN framework. Thus, we build the corresponding Web-Net-ResNext50 and Web-Net-ResNext101, where the numbers of parameters are similar to the generated Building-A-Net with 52 and 152 dense layers, respectively, to verify the effectiveness of Web-Net. Compared with the Building-A-Net-52, The 50-layer Web-Net obtains 1.75% better performance, while the 101-layer Web-Net with the pretrained parameter initialization achieves an 80.10% IoU, which is 1.37% higher than the performance of the 152-layer pretrained Building-A-Net. Figure 10 lists some randomly chosen prediction maps from the MLP, SegNet, Unet++, and Web-Net, in order to provide a more intuitive view. It can be seen that there is a vital performance improvement from Web-Net on the large-sized building areas. The MLP and SegNet frequently misclassify the building pixels that are located in shadows into non-buildings, and therefore there are many "holes" in their prediction maps due to their weak abilities for extracting long-range correlations. Although the nested connections in Unet++ can partly relieve this phenomenon, the shadows in the building areas still have negative effects on the accuracy of the building extraction. The proposed Web-Net achieves a surprising prediction quality for large-sized building areas and shadow areas (red circle areas). Additionally, the false extractions and missed extractions of Web-Net are significantly reduced (yellow circle areas), and the boundaries of the extracted building maps are sharper than those of the other models. In Table 5, we also compare the efficiency of Web-Net with the other models.

Table 5. Numerical Results of the State-of-the-art models on the Inria Dataset.

Methods	Acc. (%)	IoU (%)	Time (s)
FCN [50]	92.79	53.82	
Mask R-CNN [51]	92.49	59.53	-
MLP [50]	94.42	64.67	20.4
SegNet (Single-Loss) [52]	95.17	70.14	26.0
SegNet (Multi-Task Loss) [52]	95.73	73.00	-
Unet++ (ResNext-50) [40]	95.79	73.32	26.5
RiFCN [30]	95.82	74.00	-
Dual-Resolution U-Nets [53]	-	74.22	-
2-levels U-Nets [54]	96.05	74.55	208.8
Building-A-Net (Dense 52 layers) [55]	96.01	74.75	-
Proposed (ResNext-50)	96.33	76.50	28.8
Building-A-Net (Dense 152 layers pretrained) [55]	96.71	78.73	150.5
Proposed (ResNext-101 Pretrained)	96.97	80.10	56.5

Although we apply the overlapping-tile strategy [18], the proposed Web-Net only takes 56.5 s to process one 5000 × 5000 image, which is three times faster than the state-of-the-art building-A-Net method. Meanwhile, the lighter version of Web-Net that applies ResNext50 as the encoder just takes 28.8 s and it also achieves a satisfactory extraction result. The efficiency of Web-Net mainly arises from the efficient backbone encoders (ResNext) structure and a smaller number of layers in the decoder that are built by the parameter-efficient UHS block. The run time of Web-Net is even similar to the FCN with the same encoder structure.

We test the performance of the Web-Net and other models [30,37,40,52] on five areas with the different landforms from the Inria Dataset to verify the performance of the Web-Net for buildings of various styles. Table 6 shows the results.

When compared with the basic SegNet, the proposed Web-Net gains +7.68%, +21.07%, +2.65%, +18.04%, and +10.59% better IoUs for Austin, Chicago, Kitsap Country, Western Tyrol, and Vienna, respectively. Additionally, Web-Net outperforms Unet++ by +7.8%, +6.73%, +6.61%, +8.66%, and +5.45% with respect to the IoU, respectively. From Figure 10, we can observe that the performance boost of Web-Net mostly comes from the sharper building contours and the areas that are covered by vegetation and shadows. Furthermore, we find some inaccurate labels in Chicago and Vienna according to the abnormally low IoUs and Accs., and some examples are shown in Figure 11.

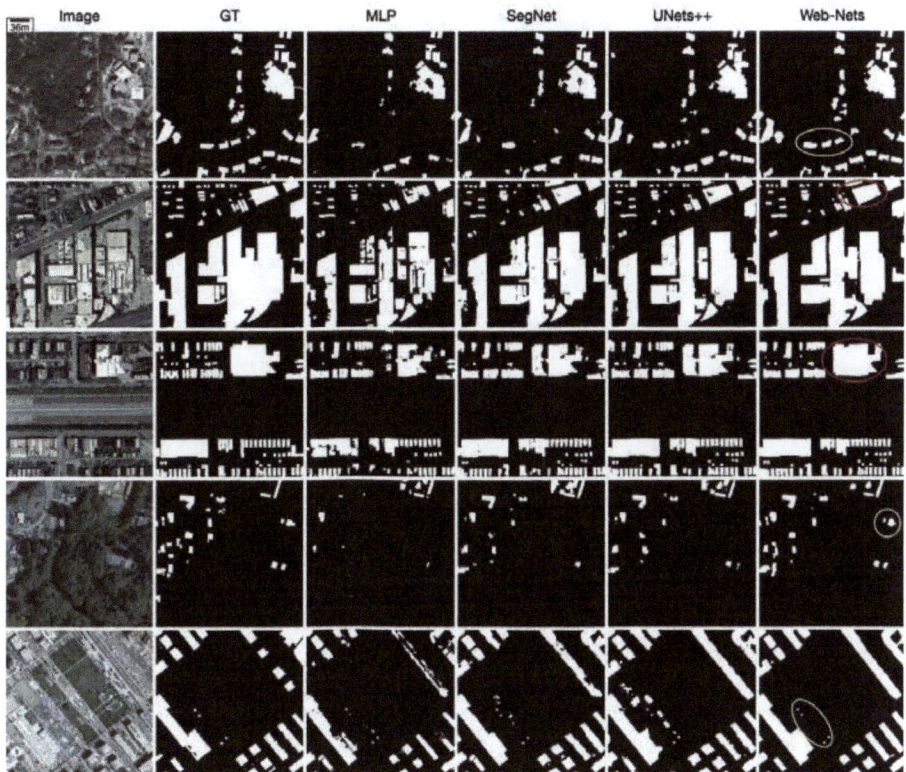

Figure 10. The images in each row are randomly chosen from Austin, Chicago, Kitsap, Tyrol, and Vienna, and the patch size is 1000 × 1000. Columns 2–6 are the ground truth and prediction maps from MLP, SegNet, Unet++, and Web-Net, respectively. The areas in red circles are correct predictions in shadowed areas, and the yellow circles are correct classifications where other models fail.

Table 6. Numerical Results among Cities.

Methods		Austin	Chicago	Kitsap Country	Western Tyrol	Vienna	Overall
SegNet (Single-Loss) [52]	IoU	74.81	52.83	68.06	65.68	72.90	70.14
	Acc.	92.52	98.65	97.28	91.36	96.04	95.17
SegNet (Multi-Task Loss [52]	IoU	76.76	67.06	73.30	66.91	76.68	73.00
	Acc.	93.21	99.25	97.84	91.71	96.61	95.73
Unet++ (ResNext-50) [40]	IoU	74.69	67.17	64.10	75.06	78.04	73.32
	Acc.	96.28	91.88	99.21	97.99	93.61	95.79
RiFCN [30]	IoU	76.84	67.45	63.95	73.19	79.18	74.00
	Acc.	96.50	91.76	99.14	97.75	93.95	95.82
Proposed (ResNext-101 Pretrained)	IoU	82.49	73.90	70.71	83.72	83.49	80.10
	Acc.	97.47	93.90	99.35	98.73	95.35	96.97

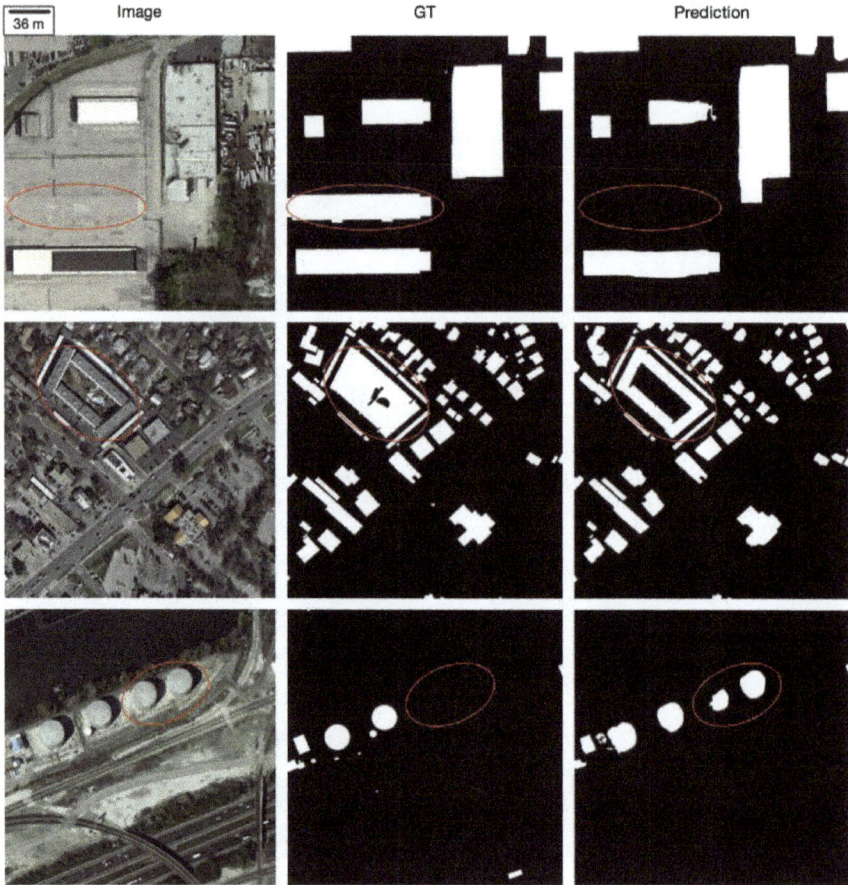

Figure 11. Examples of some mismatched ground truth labels on Inria Aerial Image Labelling Dataset, mislabelled areas are annotated by the red circles.

4.3.3. Comparison Experiments on WHU Dataset

We construct a comparison between the Web-Nets and a number of state-of-the-art encoder-decoder architectures on the WHU Dataset, where the distribution is different from that of the Inria Aerial Dataset, in order to test the generalization abilities and robustness of the proposed Web-Net, which has shown great performance on the Inria Aerial Dataset. All the models are trained from scratch and Table 7 lists the results.

Table 7. Numerical Results of the State-of-the-art models on WHU Dataset

Methods	Acc. (%)	IoU (%)
SegNet [51]	98.12	84.47
U-Net [18]	98.45	86.80
Unet++ [41]	98.48	87.30
Web-Net(Proposed)	98.54	88.76

It can be seen from Table 7 that the very simple encoder-decoder architecture such as SegNet, can achieve a satisfying result (98.12% and 84.47% on Acc and IoU, respectively) on WHU Dataset. With the more complicated encoder-decoder architecture, Unet achieves 2.33% higher scores than SegNet on the IoU metric. The naive nested encoder-decoder architecture Unet++ also works well on the WHU Dataset and gains an improvement of 2.83% on the IoU against SegNet. When compared with the aforementioned architectures, the proposed Web-Net shows great building extraction ability where the Acc. and IoU of the Web-Net is 98.54% and 88.76%, respectively, which is even 0.06% and 1.46% higher than the Acc. and IoU of the state-of-the-art Unet++. In addition to the quantitative analysis, we also perform a visual analysis on WHU Dataset, illustrating some randomly chosen prediction maps that are listed in Figure 12.

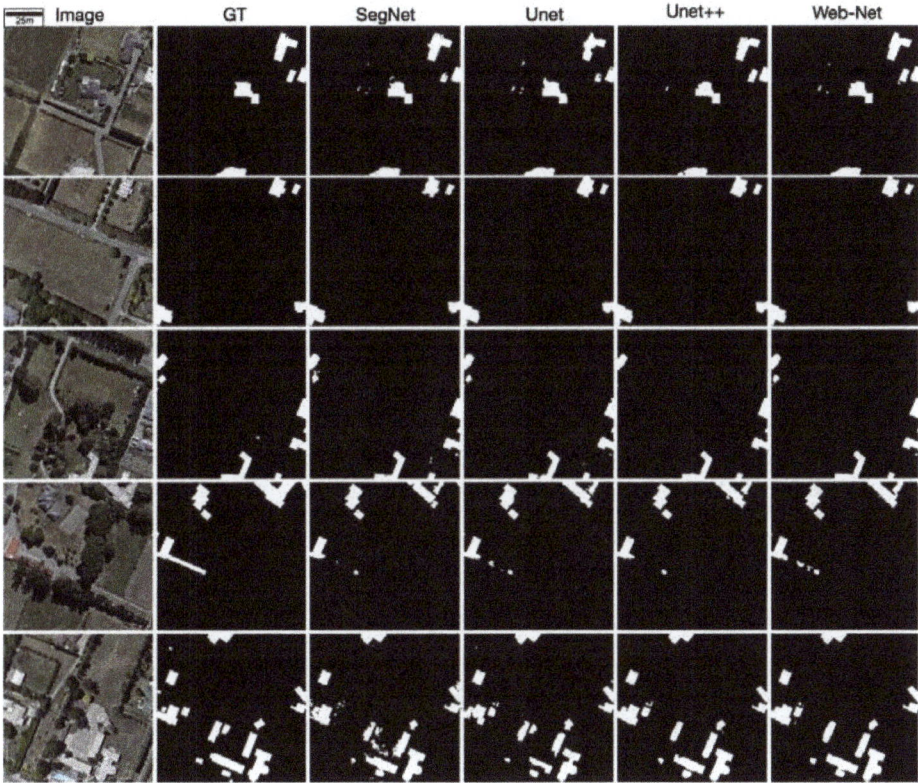

Figure 12. Samples of binary predictions of SegNet, Unet, Unet++, and Web-Nets with corresponding aerial imageries and ground truths employing the WHU Dataset.

Apparently, the prediction maps of naive encoder-decoder architectures such as SegNet and Unet, implemented on the WHU Dataset achieves better performance than that employing the Inria Labelling Dataset benefitting from the lower image complexities and higher labelling accuracies. However, the nested encoder-decoder architecture Unet++ still outperforms the naive encoder-decoder architecture on the visual effect such that the consistency of the prediction maps of Unet++ is much better than that of SegNet and Unet; in other words, there are fewer holes and discrete small misclassified areas on the prediction maps. When compared with Unet++, the proposed Web-Net obviously enhances the visual result of building extraction with much sharper and more accurate contours and higher accuracy in extracting the small scale buildings. The quality and visual analysis both prove the generalization abilities of the proposed Web-Net, which can robustly achieve state-of-the-art building extraction results on imageries from different areas.

5. Conclusions

In this paper, we proposed a novel fully convolutional network, called the Web-Net, which uses the UHS block to perform the building extraction on high-resolution remote sensing images. In particular, the architecture of the proposed model looks similar to a spider web, and except for the encoder nodes, every node is connected to its neighbours, highlighting our reason for naming it Web-Net. Inspired by DenseNet, we designed the dense jump connections in both the vertical direction and in the horizontal direction to efficiently extract and utilize more abundant features. To fuse the hierarchical features from the different levels, we also designed the UHS block to iteratively change the shape of the feature maps while using position-wise upsampling/downsampling operations, and the UHS block is applied on every node of the Web-Net. The key benefit of the UHS block is that the local cues can be preserved and encoded into the channel dimension during the downsampling, while no extra parameters are added during upsampling. This is helpful for retaining the consistency of the semantic cues and the structural information. Within the highly symmetric and harmonious architecture of Web-Net and the UHS block, the proposed method can extract and propagate the low-level and high-level features throughout the network in an efficient way. With the benefits that are outlined above, the proposed Web-Net can significantly improve the ability to fuse the high-level semantic features and the boundary-aware low-level features and achieve a higher quality building extraction result. Moreover, by replacing the encoder with other backbones, further improvements of the deep neural networks can be easily embedded into the proposed Web-Net architecture to further boost the performance. The experiments that were executed on the Inria Aerial Image Labelling Dataset have demonstrated that the proposed Web-Net outperforms other encoder-decoder-based models on both the IoU and Acc metrics and it extracts sharper and more accurate building predictions. In addition, the time costs of the proposed Web-Net are significantly shorter than those of other state-of-the-art models. Moreover, the proposed Web-Net performed well in the extraction of buildings that were mixed with vegetation or shadows. Nevertheless, the buildings in high-resolution aerial imageries have extremely complex morphological characteristics, such as straight lines, curves, and orientations. These characteristics cannot be directly extracted by the FCN-based networks, and determining how to embed morphological characteristics into CNN structure is an open and urgent problem.

Author Contributions: Conceptualization, Y.Z.; Formal analysis, Y.Z. and J.S.; Funding acquisition, W.G. and W.L.; Methodology, Y.Z.; Project administration, W.G.; Software, Y.Z.; Supervision, W.G. and W.L.; Validation, J.S.; Writing – original draft, Y.Z. and J.S.; Writing – review & editing, W.G. and W.L.

Funding: This work was funded by the Key Projects of Science and Technology Agency of Guangxi province, China (Guike AA 17129002); National Science and Technology Key Program of China (2013GS500303); and the Municipal Science and Technology Project of CQMMC, China (2017030502).

Acknowledgments: We thank Inria for providing the Inria Aerial Image Labelling Dataset in their website (https://project.inria.fr/aerialimagelabelling/). We are also very grateful for the valuable suggestions and comments of peer reviewers.

Conflicts of Interest: The authors declare no conflict of interest.

References

1. Xu, Y.; Wu, L.; Xie, Z.; Chen, Z. Building Extraction in Very High Resolution Remote Sensing Imagery Using Deep Learning and Guided Filters. *Remote Sens.* **2018**, *10*, 144. [CrossRef]
2. Gao, L.; Shi, W.; Miao, Z.; Lv, Z. Method based on edge constraint and fast marching for road centerline extraction from very high-resolution remote sensing images. *Remote Sens.* **2018**, *10*, 900. [CrossRef]
3. Audebert, N.; Le Saux, B.; Lefèvre, S. Segment-before-detect: Vehicle detection and classification through semantic segmentation of aerial images. *Remote Sens.* **2017**, *9*, 368. [CrossRef]

4. Tuermer, S.; Kurz, F.; Reinartz, P.; Stilla, U. Airborne vehicle detection in dense urban areas using HoG features and disparity maps. *IEEE J. Sel. Top. Appl. Earth Observ. Remote Sens.* **2013**, *6*, 2327–2337. [CrossRef]
5. Cote, M.; Saeedi, P. Automatic rooftop extraction in nadir aerial imagery of suburban regions using corners and variational level set evolution. *IEEE Trans. Geosci. Remote Sens.* **2013**, *51*, 313–328. [CrossRef]
6. Yang, Y.; Newsam, S. Comparing SIFT descriptors and Gabor texture features for classification of remote sensed imagery. In Proceedings of the 2008 15th IEEE International Conference on Image Processing, San Diego, CA, USA, 12–15 October 2008; pp. 1852–1855.
7. Lowe, D.G. Object recognition from local scale-invariant features. In Proceedings of the 7th IEEE International Conference on Computer Vision, Kerkyra, Greece, 20–27 September 1999; pp. 1150–1157.
8. Mountrakis, G.; Im, J.; Ogole, C. Support vector machines in remote sensing: A review. *ISPRS J. Photogramm.* **2011**, *66*, 247–259. [CrossRef]
9. Maloof, M.A.; Langley, P.; Binford, T.O.; Nevatia, R.; Sage, S. Improved rooftop detection in aerial images with machine learning. *Mach. Learn.* **2003**, *53*, 157–191. [CrossRef]
10. Senaras, C.; Ozay, M.; Vural, F.T.Y. Building detection with decision fusion. *IEEE J. Sel. Top. Appl. Earth Observ. Remote Sens.* **2013**, *6*, 1295–1304. [CrossRef]
11. Long, J.; Shelhamer, E.; Darrell, T. Fully convolutional networks for semantic segmentation. In Proceedings of the IEEE Conference on Computer Vision and Pattern Recognition, Boston, MA, USA, 8–10 June 2015; pp. 3431–3440.
12. Simonyan, K.; Zisserman, A. Very deep convolutional networks for large-scale image recognition. *arXiv* **2014**, arXiv:1409.1556.
13. He, K.; Zhang, X.; Ren, S.; Sun, J. Deep residual learning for image recognition. In Proceedings of the IEEE Conference on Computer Vision and Pattern Recognition, Las Vegas, NV, USA, 27–30 June 2016; pp. 770–778.
14. Szegedy, C.; Ioffe, S.; Vanhoucke, V.; Alemi, A.A. Inception-v4, inception-resnet and the impact of residual connections on learning. In Proceedings of the Thirty-First AAAI Conference on Artificial Intelligence, San Francisco, CA, USA, 4–9 February 2017.
15. Xie, S.; Girshick, R.; Dollár, P.; Tu, Z.; He, K. Aggregated residual transformations for deep neural networks. In Proceedings of the IEEE Conference on Computer Vision and Pattern Recognition, Honolulu, HI, USA, 22–25 July 2017; pp. 1492–1500.
16. Chollet, F. Xception: Deep learning with depthwise separable convolutions. In Proceedings of the IEEE Conference on Computer Vision and Pattern Recognition, Honolulu, HI, USA, 22–25 July 2017; pp. 1251–1258.
17. Huang, G.; Liu, Z.; Van Der Maaten, L.; Weinberger, K.Q. Densely connected convolutional networks. In Proceedings of the IEEE conference on computer vision and pattern recognition, Honolulu, HI, USA, 22–25 July 2017; pp. 4700–4708.
18. Ronneberger, O.; Fischer, P.; Brox, T. U-net: Convolutional networks for biomedical image segmentation. In Proceedings of the International Conference on Medical Image Computing and Computer-Assisted Intervention, Munich, Germany, 5–9 October 2015; pp. 234–241.
19. Badrinarayanan, V.; Kendall, A.; Cipolla, R. Segnet: A deep convolutional encoder-decoder architecture for image segmentation. *IEEE Trans. Pattern Anal. Mach. Intell.* **2017**, *39*, 2481–2495. [CrossRef]
20. Chen, L.-C.; Papandreou, G.; Kokkinos, I.; Murphy, K.; Yuille, A.L. Deeplab: Semantic image segmentation with deep convolutional nets, atrous convolution, and fully connected crfs. *IEEE Trans. Pattern Anal. Mach. Intell.* **2018**, *40*, 834–848. [CrossRef] [PubMed]
21. Zhao, H.; Shi, J.; Qi, X.; Wang, X.; Jia, J. Pyramid scene parsing network. In Proceedings of the IEEE conference on computer vision and pattern recognition, Honolulu, HI, USA, 22–25 July 2017; pp. 2881–2890.
22. Chen, L.-C.; Zhu, Y.; Papandreou, G.; Schroff, F.; Adam, H. Encoder-decoder with atrous separable convolution for semantic image segmentation. In Proceedings of the European Conference on Computer Vision (ECCV), Munich, Germany, 8–14 September 2018; pp. 801–818.
23. Huang, Z.; Cheng, G.; Wang, H.; Li, H.; Shi, L.; Pan, C. Building extraction from multi-source remote sensing images via deep deconvolution neural networks. In Proceedings of the 2016 IEEE International Geoscience and Remote Sensing Symposium (IGARSS), Beijing, China, 10–15 July 2016; pp. 1835–1838.
24. Zhong, Z.; Li, J.; Cui, W.; Jiang, H. Fully convolutional networks for building and road extraction: Preliminary results. In Proceedings of the 2016 IEEE International Geoscience and Remote Sensing Symposium (IGARSS), Beijing, China, 10–15 July 2016; pp. 1591–1594.

25. Bittner, K.; Cui, S.; Reinartz, P. Building Extraction from Remote Sensing Data Using fully Convolutional networks. In Proceedings of the International Archives of the Photogrammetry, Remote Sensing & Spatial Information Sciences, Hannover, Germany, 6–9 June 2017.
26. Alshehhi, R.; Marpu, P.R.; Woon, W.L.; Dalla Mura, M. Simultaneous extraction of roads and buildings in remote sensing imagery with convolutional neural networks. *ISPRS J. Photogramm.* **2017**, *130*, 139–149. [CrossRef]
27. Wu, G.; Shao, X.; Guo, Z.; Chen, Q.; Yuan, W.; Shi, X.; Xu, Y.; Shibasaki, R. Automatic building segmentation of aerial imagery using multi-constraint fully convolutional networks. *Remote Sens.* **2018**, *10*, 407. [CrossRef]
28. Chen, K.; Fu, K.; Yan, M.; Gao, X.; Sun, X.; Wei, X. Semantic segmentation of aerial images with shuffling convolutional neural networks. *IEEE Geosci. Remote Sens. Lett.* **2018**, *15*, 173–177. [CrossRef]
29. Yang, H.; Wu, P.; Yao, X.; Wu, Y.; Wang, B.; Xu, Y. Building Extraction in Very High Resolution Imagery by Dense-Attention Networks. *Remote Sens.* **2018**, *10*, 1768. [CrossRef]
30. Mou, L.; Zhu, X.X. RiFCN: Recurrent network in fully convolutional network for semantic segmentation of high resolution remote sensing images. *arXiv* **2018**, arXiv:1805.02091.
31. Mou, L.; Ghamisi, P.; Zhu, X.X. Deep recurrent neural networks for hyperspectral image classification. *IEEE Trans. Geosci. Remote Sens.* **2017**, *55*, 3639–3655. [CrossRef]
32. Audebert, N.; Le Saux, B.; Lefèvre, S. Beyond RGB: Very high resolution urban remote sensing with multimodal deep networks. *ISPRS J. Photogramm.* **2018**, *140*, 20–32. [CrossRef]
33. Pan, B.; Shi, Z.; Xu, X. MugNet: Deep learning for hyperspectral image classification using limited samples. *ISPRS J. Photogramm.* **2018**, *145*, 108–119. [CrossRef]
34. Bittner, K.; Adam, F.; Cui, S.; Körner, M.; Reinartz, P. Building Footprint Extraction From VHR Remote Sensing Images Combined with Normalized DSMs Using Fused Fully Convolutional Networks. *IEEE J. Sel. Top. Appl. Earth Observ. Remote Sens.* **2018**, *11*, 2615–2629. [CrossRef]
35. Shrestha, S.; Vanneschi, L. Improved Fully Convolutional Network with Conditional Random Fields for Building Extraction. *Remote Sens.* **2018**, *10*, 1135. [CrossRef]
36. Wang, Y.; Liang, B.; Ding, M.; Li, J. Dense Semantic Labelling with Atrous Spatial Pyramid Pooling and Decoder for High-Resolution Remote Sensing Imagery. *Remote Sens.* **2019**, *11*, 20. [CrossRef]
37. Maggiori, E.; Tarabalka, Y.; Charpiat, G.; Alliez, P. Can semantic labelling methods generalize to any city? the inria aerial image labelling benchmark. In Proceedings of the 2017 IEEE International Geoscience and Remote Sensing Symposium (IGARSS), Fort Worth, TX, USA, 23–28 July 2017; pp. 3226–3229.
38. Fourure, D.; Emonet, R.; Fromont, E.; Muselet, D.; Tremeau, A.; Wolf, C. Residual conv-deconv grid network for semantic segmentation. *arXiv* **2017**, arXiv:1707.07958.
39. Pohlen, T.; Hermans, A.; Mathias, M.; Leibe, B. Full-resolution residual networks for semantic segmentation in street scenes. In Proceedings of the IEEE Conference on Computer Vision and Pattern Recognition, Honolulu, HI, USA, 22–25 July 2017; pp. 4151–4160.
40. Zhou, Z.; Siddiquee, M.M.R.; Tajbakhsh, N.; Liang, J. Unet++: A nested u-net architecture for medical image segmentation. In *Deep Learning in Medical Image Analysis and Multimodal Learning for Clinical Decision Support*; Springer: Berlin/Heidelberg, Germany, 2018; pp. 3–11.
41. Wang, L.; Lee, C.-Y.; Tu, Z.; Lazebnik, S. Training deeper convolutional networks with deep supervision. *arXiv* **2015**, arXiv:1505.02496.
42. Hu, J.; Shen, L.; Sun, G. Squeeze-and-excitation networks. In Proceedings of the IEEE Conference on Computer Vision and Pattern Recognition, Salt Lake City, UT, USA, 18–22 June 2018; pp. 7132–7141.
43. Wang, P.; Chen, P.; Yuan, Y.; Liu, D.; Huang, Z.; Hou, X.; Cottrell, G. Understanding convolution for semantic segmentation. In Proceedings of the 2018 IEEE Winter Conference on Applications of Computer Vision (WACV), Lake Tahoe, CA, USA, 12–15 March 2018; pp. 1451–1460.
44. Ji, S.; Wei, S.; Lu, M. Fully Convolutional Networks for Multisource Building Extraction From an Open Aerial and Satellite Imagery Data Set. *IEEE Trans. Geosci. Remote Sens.* **2018**, *574*, 574–586. [CrossRef]
45. Paszke, A.; Gross, S.; Chintala, S.; Chanan, G.; Yang, E.; DeVito, Z.; Lin, Z.; Desmaison, A.; Antiga, L.; Lerer, A. Automatic differentiation in pytorch. In Proceedings of the NIPS 2017 Autodiff Workshop: The Future of Gradient-basedMachine Learning Software and Techniques, Long Beach, CA, USA, 9 December 2017.
46. Deng, J.; Dong, W.; Socher, R.; Li, L.-J.; Li, K.; Li, F.-F. Imagenet: A large-scale hierarchical image database. In Proceedings of the IEEE Conference on Computer Vision and Pattern Recognition, Miami, FL, USA, 20–25 June 2009.

47. Kingma, D.P.; Ba, J. Adam: A method for stochastic optimization. *arXiv* **2014**, arXiv:1412.6980.
48. Pleiss, G.; Chen, D.; Huang, G.; Li, T.; van der Maaten, L.; Weinberger, K.Q. Memory-efficient implementation of densenets. *arXiv* **2017**, arXiv:1707.06990.
49. Goodfellow, I.; Pouget-Abadie, J.; Mirza, M.; Xu, B.; Warde-Farley, D.; Ozair, S.; Courville, A.; Bengio, Y. Generative adversarial nets. In Proceedings of the Advances in Neural Information Processing Systems, Montréal, Canada, 8–13 December 2014; pp. 2672–2680.
50. Chen, Q.; Wang, L.; Wu, Y.F.; Wu, G.M.; Guo, Z.L.; Waslander, S.L. Aerial imagery for roof segmentation: A large-scale dataset towards automatic mapping of buildings. *ISPRS J. Photogramm.* **2019**, *147*, 42–55. [CrossRef]
51. He, K.; Gkioxari, G.; Dollár, P.; Girshick, R. Mask R-CNN. In Proceedings of the IEEE International Conference on Computer Vision, Honolulu, HI, USA, 22–25 July 2017; pp. 2961–2969.
52. Bischke, B.; Helber, P.; Folz, J.; Borth, D.; Dengel, A. Multi-task learning for segmentation of building footprints with deep neural networks. *arXiv* **2017**, arXiv:1709.05932.
53. Lu, K.; Sun, Y.; Ong, S.-H. Dual-Resolution U-Net: Building Extraction from Aerial Images. In Proceedings of the 2018 24th International Conference on Pattern Recognition (ICPR), Beijing, China, 20–24 August 2018; pp. 489–494.
54. Khalel, A.; El-Saban, M. Automatic pixelwise object labelling for aerial imagery using stacked u-nets. *arXiv* **2018**, arXiv:1803.04953.
55. Li, X.; Yao, X.; Fang, Y. Building-A-Nets: Robust Building Extraction From High-Resolution Remote Sensing Images With Adversarial Networks. *IEEE J. Sel. Top. Appl. Earth Observ. Remote Sens.* **2018**, *11*, 3680–3687. [CrossRef]

© 2019 by the authors. Licensee MDPI, Basel, Switzerland. This article is an open access article distributed under the terms and conditions of the Creative Commons Attribution (CC BY) license (http://creativecommons.org/licenses/by/4.0/).

Article

Semantic Segmentation of Urban Buildings from VHR Remote Sensing Imagery Using a Deep Convolutional Neural Network

Yaning Yi [1,2,†], Zhijie Zhang [3,†], Wanchang Zhang [1,*], Chuanrong Zhang [3], Weidong Li [3] and Tian Zhao [4]

1. Key Laboratory of Digital Earth Science, Institute of Remote Sensing and Digital Earth, Chinese Academy of Sciences, Beijing 100094, China
2. University of Chinese Academy of Sciences, Beijing 100049, China
3. Department of Geography, University of Connecticut, Storrs, CT 06269, USA
4. Department of Computer Science, University of Wisconsin, Milwaukee, WI 53211, USA
* Correspondence: zhangwc@radi.ac.cn; Tel.: +86-10-8217-8131
† The first two authors are contributed equally to the work presented and are considered as equal first authors of this manuscript.

Received: 22 June 2019; Accepted: 26 July 2019; Published: 28 July 2019

Abstract: Urban building segmentation is a prevalent research domain for very high resolution (VHR) remote sensing; however, various appearances and complicated background of VHR remote sensing imagery make accurate semantic segmentation of urban buildings a challenge in relevant applications. Following the basic architecture of U-Net, an end-to-end deep convolutional neural network (denoted as DeepResUnet) was proposed, which can effectively perform urban building segmentation at pixel scale from VHR imagery and generate accurate segmentation results. The method contains two sub-networks: One is a cascade down-sampling network for extracting feature maps of buildings from the VHR image, and the other is an up-sampling network for reconstructing those extracted feature maps back to the same size of the input VHR image. The deep residual learning approach was adopted to facilitate training in order to alleviate the degradation problem that often occurred in the model training process. The proposed DeepResUnet was tested with aerial images with a spatial resolution of 0.075 m and was compared in performance under the exact same conditions with six other state-of-the-art networks—FCN-8s, SegNet, DeconvNet, U-Net, ResUNet and DeepUNet. Results of extensive experiments indicated that the proposed DeepResUnet outperformed the other six existing networks in semantic segmentation of urban buildings in terms of visual and quantitative evaluation, especially in labeling irregular-shape and small-size buildings with higher accuracy and entirety. Compared with the U-Net, the F1 score, Kappa coefficient and overall accuracy of DeepResUnet were improved by 3.52%, 4.67% and 1.72%, respectively. Moreover, the proposed DeepResUnet required much fewer parameters than the U-Net, highlighting its significant improvement among U-Net applications. Nevertheless, the inference time of DeepResUnet is slightly longer than that of the U-Net, which is subject to further improvement.

Keywords: semantic segmentation; urban building extraction; deep convolutional neural network; VHR remote sensing imagery; U-Net

1. Introduction

One of the fundamental tasks in remote sensing is building extraction from remote sensing imagery. It plays a key role in applications such as urban construction and planning, natural disaster and crisis management [1–3]. In recent years, owing to the rapid development of sensor technology, very high resolution (VHR) images with spatial resolution from 5 to 30 cm have become available [4], making small-scale objects (e.g., cars, buildings and roads) distinguishable and identifiable via semantic segmentation methods. Semantic segmentation as an effective technique aims to assign each pixel in the target image into a given category [5]; therefore, it was quickly developed and extensively applied to urban planning and relevant studies including building/road detection [6–8], land use/cover mapping [9–12], and forest management [13,14] with the emergence of a large number of publicly available VHR images.

In previous research, some machine learning methods were adopted to enhance the performance of VHR semantic segmentation with focus on the feature learning methods [15–18]. Song and Civco [19] adopted the support vector machine (SVM) with the shape index as a feature to detect roads in urban areas. Tian et al. [20] applied the random forest classifier to classify wetland land covers from multi-sensor data. Wang et al. [21] used the SVM-based joint bilateral filter to classify hyperspectral images. Das et al. [22] presented a probabilistic SVM to detect roads from VHR multispectral images with the aid of two salient features of roads and the design of a leveled structure. As pointed by Ball et al. [15], traditional feature learning approaches can work quite well, but several issues remain in the applications of these techniques and constrain their wide applicability.

The last few years witnessed the progress of deep learning, which has become one of the most cutting-edge and trending technologies thanks to hardware development of graphics processing unit (GPU). Owing to the successful application of deep convolutional neural network (DCNN) in object detection [23–25], image classification [26,27] and semantic segmentation [28–31], deep learning was introduced to remote sensing field for resolving the classic problems in a new and efficient way [32]. DCNN was adopted in many traditional remote sensing tasks, such as data fusion [33], vehicle detection [34,35] and hyperspectral classification [36,37]. As for building extraction, many DCNN-based methods have been proposed by many researchers [38–40]. For example, Saito et al. [41] directly extracted roads and buildings from raw VHR remote sensing image by applying a single convolutional neural network, and an efficient method to train the network for detecting multiple types of objects simultaneously was proposed. Marmanis et al. [4] proposed a trainable DCNN for image classification by combining semantic segmentation and edge detection, which significantly improved the classification accuracy. Bittner et al. [42] proposed the Fused-FCN4s model consisting of three parallel FCN4s networks to learn the spatial and spectral building features from three-band (red, green, blue), panchromatic and normalized digital surface model (nDSM) images. Vakalopoulou et al. [43] combined the SVM classifier and the Markov random field (MRF) model as a deep framework for building segmentation with Red-Green-Blue and near-infrared multi spectral images in high resolution. In contrast to feature learning approaches, deep learning approaches took advantage of several significant characteristics as summarized in Ball et al. [15]. However, adopting very successful deep networks to fit remote sensing imagery analysis can be challenging [15].

Very recent studies indicated that a deeper network would have a better performance when it came to object detection, visual recognition and semantic segmentation tasks. However, the deeper the network, the more significant the issues such as vanishing gradients. In order to account for this, He et al. [44] presented a deep residual learning approach, which reformulated the layers as learning residual functions with reference to the layer inputs, instead of learning unreferenced functions and achieved training of residual nets of 152 layers. This is eight times deeper than the VGG network while still maintaining lower complexity. Ronneberger et al. [30] presented a network and training strategy named U-Net, which performed data augmentation to make efficient use of annotated samples. High-level semantic information and low-level detailed information were combined by using the concatenate operation, and such a network can be trained in an end-to-end fashion from very few training images and still outperform the previous best approach.

In this paper, an end-to-end deep convolutional neural network (denoted as DeepResUnet) was proposed to complement semantic segmentation at pixel scale on urban buildings from VHR remote sensing imagery. Since according to the literature [15,26], a deeper network would have better performance for semantic segmentation, we decided to follow network structure that enables the existence of larger number of layers in the network without running into training problems, thus the idea of residual learning is adopted in our network. Following the basic structure of U-Net, the proposed DeepResUnet contains two sub-networks: a cascade down-sampling network which extracts feature maps of buildings from the VHR image; and an up-sampling network which reconstructs the extracted feature maps of buildings back to the same size of the input VHR image. The deep residual learning approach was adopted to facilitate training in order to alleviate the degradation problem that often occurred in the model training process, and finally a softmax classifier was added at the end of the proposed network to obtain the final segmentation results.

To summarize, the main contributions of this paper are as follows. First, an end-to-end deep convolutional neural network, i.e., DeepResUnet, was proposed for complex urban building segmentation at pixel scale with three-band (red, green, blue) VHR remote sensing imagery. No additional data or any post-processing methods were adopted in this study. Second, in the DeepResUnet, the residual block (ResBlock) was designed as the basic processing unit to optimize model training and deep residual learning approach was applied to alleviate gradient-related issues. Third, in addition to comparing the performance of different deep models, the applicability of deep models was also explored by testing the trained models in a new urban area. Results indicated that DeepResUnet has the ability to identify urban buildings and it can be applied to dense urban areas such as big cities and even megacities. The purpose of this paper is not just to come up with a novel approach with better performance, it means more than just a higher accuracy. Our work can generate raw data (e.g., building boundaries) for geographical analysis such as urban planning and urban geography study. Only with more accurate raw data to begin with can those geographical analysis be accurate and instructive. We also used another totally new dataset to test and show that our proposed method is transferable with other datasets and can still maintain high performance, which means this method can have a very wide range of application. Last but not least, we hope that our proposed network structure can inspire scholars to build an even greater network.

Following the introduction, the remainder of this paper is arranged as follows: Section 2 introduces the architecture of the proposed DeepResUnet, with a focus on ResBlock, the down-sampling network, and up-sampling network. Detailed implementations of the DeepResUnet, extensive experimental results, and comparisons with other existing networks are presented in Section 3. The discussion is provided in Section 4, followed by Section 5 with the conclusions.

2. Methodology

DeepResUnet is an end-to-end DCNN that follows the basic structure of U-Net. DeepResUnet contains two sub-networks: A cascade down-sampling network for extracting building feature maps from the VHR image, and an up-sampling network for reconstructing the extracted building feature maps back to the same size of input VHR image. To reduce gradient degradation in model training, the deep-residual-learning approach was adopted in model training and a softmax classifier was used at the end of the network to obtain the final segmentation results. Figure 1 shows the detailed architecture of the proposed DeepResUnet network.

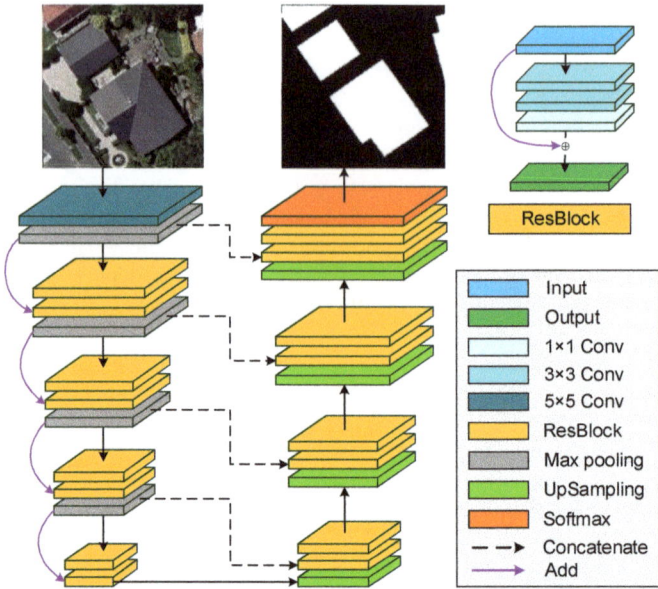

Figure 1. The architecture of the proposed DeepResUnet network. The input layer is an aerial image with three channels (red, green, blue) and the output is a binary segmentation map. The left part represents a down-sampling sub-network and the right part represents an up-sampling sub-network.

Table 1 presents the detailed information of DeepResUnet network. The input layer is a VHR aerial image with three channels (red, green, blue), and the output image is a binary segmentation map in which the pixel in white denotes the building and the pixel in black denotes the background. Similar to the U-Net, the architecture of DeepResUnet is mostly symmetrical but much deeper than that of U-Net. To accelerate the training, the batch normalization [45] layer was used after each convolutional layer. Note that no fully connected layers were used in DeepResUnet.

In the following subsections, we first give a brief description of the ResBlock architecture used in DeepResUnet and then provide the down-sampling and up-sampling sub-networks in detail.

Table 1. The architecture of the proposed DeepResUnet network.

Name	Kernel Size	Stride	Pad	Output Size
Down-sampling network				
Input	–	–	–	$256 \times 256 \times 3$
Conv_1	5×5	1	2	$256 \times 256 \times 128$
Pooling_1	2×2	2	0	$128 \times 128 \times 128$
ResBlock_1	$3 \times 3/3 \times 3/1 \times 1$	1	1	$128 \times 128 \times 128$
ResBlock_2	$3 \times 3/3 \times 3/1 \times 1$	1	1	$128 \times 128 \times 128$
Add_1	–	–	–	$128 \times 128 \times 128$
Pooling_2	2×2	2	0	$64 \times 64 \times 128$
ResBlock_3	$3 \times 3/3 \times 3/1 \times 1$	1	1	$64 \times 64 \times 128$
ResBlock_4	$3 \times 3/3 \times 3/1 \times 1$	1	1	$64 \times 64 \times 128$
Add_2	–	–	–	$64 \times 64 \times 128$
Pooling_3	2×2	2	0	$32 \times 32 \times 128$
ResBlock_5	$3 \times 3/3 \times 3/1 \times 1$	1	1	$32 \times 32 \times 128$
ResBlock_6	$3 \times 3/3 \times 3/1 \times 1$	1	1	$32 \times 32 \times 128$
Add_3	–	–	–	$32 \times 32 \times 128$
Pooling_4	2×2	2	0	$16 \times 16 \times 128$
ResBlock_7	$3 \times 3/3 \times 3/1 \times 1$	1	1	$16 \times 16 \times 128$
ResBlock_8	$3 \times 3/3 \times 3/1 \times 1$	1	1	$16 \times 16 \times 128$
Add_4	–	–	–	$16 \times 16 \times 128$

Table 1. Cont.

Name	Kernel Size	Stride	Pad	Output Size
\multicolumn{5}{c}{Up-sampling network}				
UpSampling_1	2 × 2	2	0	32 × 32 × 128
Concat_1	–	–	–	32 × 32 × 256
Conv_1U	1 × 1	1	0	32 × 32 × 128
ResBlock_1U	3 × 3/3 × 3/1 × 1	1	1	32 × 32 × 128
ResBlock_2U	3 × 3/3 × 3/1 × 1	1	1	32 × 32 × 128
UpSampling_2	2 × 2	2	0	64 × 64 × 128
Concat_2	–	–	–	64 × 64 × 256
Conv_2U	1 × 1	1	0	64 × 64 × 128
ResBlock_3U	3 × 3/3 × 3/1 × 1	1	1	64 × 64 × 128
ResBlock_4U	3 × 3/3 × 3/1 × 1	1	1	64 × 64 × 128
UpSampling_3	2 × 2	2	0	128 × 128 × 128
Concat_3	–	–	–	128 × 128 × 256
Conv_3U	1 × 1	1	0	128 × 128 × 128
ResBlock_5U	3 × 3/3 × 3/1 × 1	1	1	128 × 128 × 128
ResBlock_6U	3 × 3/3 × 3/1 × 1	1	1	128 × 128 × 128
UpSampling_4	2 × 2	2	0	256 × 256 × 128
Concat_4	–	–	–	256 × 256 × 256
Conv_4U	1 × 1	1	0	256 × 256 × 128
ResBlock_7U	3 × 3/3 × 3/1 × 1	1	1	256 × 256 × 128
ResBlock_8U	3 × 3/3 × 3/1 × 1	1	1	256 × 256 × 128
Conv_5U	1 × 1	1	0	256 × 256 × 2
Output	–	–	–	256 × 256 × 2

2.1. ResBlock

With increasing depth of deep neural networks, problems like vanishing gradients start to emerge. To resolve this issue, the deep residual framework (ResNet) was proposed to ensure the gradient be directly propagated from top to bottom of the network during the backward propagation [44]. Previous studies suggested that the residual framework can improve accuracy considerably with increased layer depth and is easier to optimize [46].

Formally, by denoting the input as x_l, the output of a residual unit as x_{l+1} and the residual function as $F(\cdot)$, the residual unit can be expressed as:

$$x_{l+1} = F(x_l) + x_l \qquad (1)$$

Inspired by the ResNet, the residual block (ResBlock) was designed as the basic processing unit in DeepResUnet to optimize model training. Figure 2 illustrates the structure of different types of Resblocks. In general, they are all shaped like a bottle neck. A 1 × 1 convolutional layer following two successive 3 × 3 convolutional layers in the ResBlock with ReLU as activation function between successive layers was designed to account for gradient degradation of image in the training process. For better performance, two successive 3 × 3 convolution kernels were adopted in DeepResUnet by following the suggestions from other studies [47,48]. It is worthwhile mentioning that the number of channels of the first 3 × 3 convolution layer was twice than that of the latter. A small number of channels of the first 3 × 3 convolutional layer can reduce model parameters without losing too much image information. Additionally, a 1 × 1 convolution layer was added to the ResBlock.

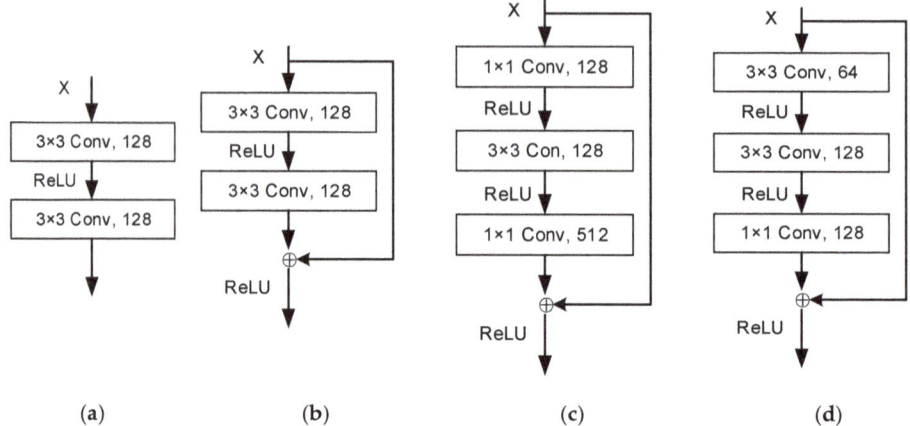

Figure 2. Illustration of the ResBlock structure. (**a**) plain neural unit used in U-Net; (**b**) basic residual unit used in ResNet-34; (**c**) "bottleneck" shaped residual unit used in ResNet-50/101/152; (**d**) the designed ResBlock used in DeepResUnet.

2.2. Down-Sampling Network

Inspired by ENet [49], early downscale-sampling method was employed after the input layer in the proposed DeepResUnet network. The assumption behind this is that the feature maps from the initial image layer contain adverse noise that would directly contribute to segmentation, which should be filtered. Therefore, max-pooling layer with the size of 2 × 2 was added to reduce the input size. It is worthwhile mentioning that although the pooling operation is capable of reducing the learning parameters while keeping scaling invariant, spatial information essential for pixelwise segmentation was indeed partially lost in this process [15]. To keep as much spatial information as possible, a 5 × 5 convolutional layer with 128 channels was added before the first max-pooling layer to gain a larger receptive field.

As exhibited in Figure 1, two successive ResBlock modules were set after the first max-pooling layer to obtain image features, and then a 2 × 2 max-pooling layer was added to reduce the learning parameters while keeping scaling invariant for enlarging the receptive field. To make better use of the previous features and propagate the gradient, the feature maps from the pooling layer were added to the output of two successive ResBlock modules in a residual manner. The detailed description of the down-sampling process is exhibited in Figure 3a. The input of the latter pooling layer was computed by:

$$y = f(f(P(x))) + P(x) \qquad (2)$$

where $P(\cdot)$ represents the pooling function, $f(\cdot)$ represents the ResBlock operation, x represents the input, and y is the output which is used as the input for the subsequent pooling and up-sampling operations.

To effectively exploit image information, a down-sampling network with two successive ResBlock layers and one pooling layer repeated for three times was developed in DeepResUnet. Consequently, the size of the input image was reduced from 256 × 256 to 16 × 16. In addition, two successive ResBlock layers without pooling layer were employed at the end of down-sampling network to serve as a bridge connecting the down-sampling and up-sampling networks.

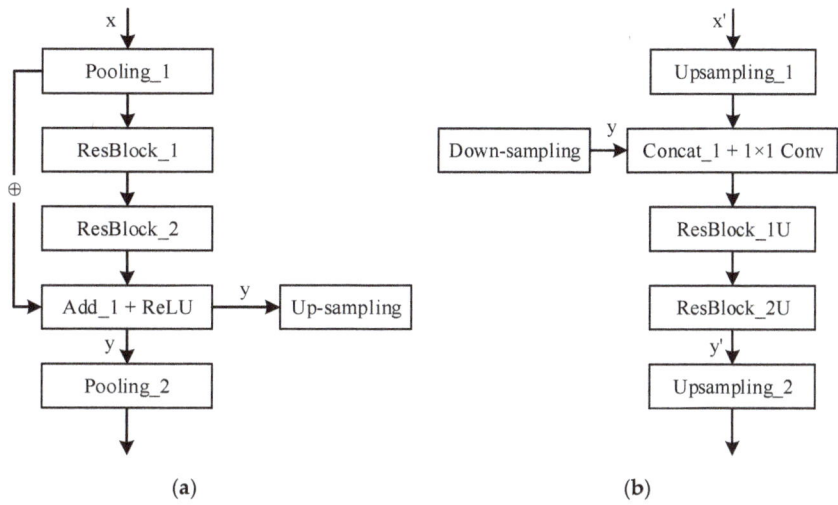

Figure 3. The details of the down-sampling (a) and up-sampling structures (b).

2.3. Up-sampling Network

Similar to the U-Net, our up-sampling network was symmetric to the down-sampling network. As illustrated in Figure 1, the up-sampling operation functions in recovering the details of feature maps. For pixelwise segmentation, aggregating multi-context information can effectively improve the performance of deep learning models [50]. The intuition is that low-level features contain finer details which can compensate for high-level semantic features [51]. In the up-sampling network, the low-level features at a pixel scale were propagated to the corresponding high levels in a concatenation manner, and then a 1 × 1 convolutional layer was employed to change the output channels. Subsequently, two successive ResBlock modules were added after the concatenation operation. As illustrated in Figure 3b, the output of ResBlock can be formulated as:

$$y' = f(f(f_{1\times1}(U(x') \otimes y))) \quad (3)$$

where $f_{1\times1}(\cdot)$ is the 1 × 1 convolutional operation, $U(\cdot)$ is the up-sampling operation, x' is the previous feature maps from the down-sampling network, and the symbol of $\otimes$ denotes the concatenation operation.

Different from the down-sampling network, the up-sampling network adopted the up-sampling layer instead of the max-pooling layer. Four up-sampling layers were used in the up-sampling network to facilitate reconstructing the feature maps to the same size as the input image. Following the last ResBlock layer in the up-sampling network, a softmax layer was used to derive the final segmentation maps.

3. Experiments and Results

In this section, we first describe the dataset used in the experiments and experimental setting. We then provide qualitative and quantitative comparisons of performances between DeepResUnet and other state-of-the-art approaches in semantic segmentation of urban buildings from the same data source (VHR remote sensing imagery).

3.1. Dataset

Large-scale training samples are required for deep learning models to learn various features. Aerial images with a spatial resolution of 0.075 m and the corresponding building outlines were collected from a public source (https://data.linz.govt.nz) as our experiment dataset. The aerial images covered an urban area of Christchurch City, New Zealand, and were taken during the flying season (summer period) 2015 and 2016. Note that the aerial images had been converted into orthophotos (three-band, red-green-blue) by the provider and were provided in format of tiles. The pixel value of the aerial image varies from 0 to 255. The image we selected was mosaiced with 335 tiles covering 25,454 building objects, and the mosaic image is 38,656 × 19,463 pixels, as shown in Figure 4. To train and test the DeepResUnet, the mosaic image was split into two parts for training and testing, with almost equivalent areas including 12,665 and 12,789 building objects, respectively (Figure 4). Meanwhile, the building outlines corresponding to the aerial image, which were stored as polygon shapefiles, were similarly converted into a raster image and further sub-divided into two parts as was done to the aerial image. In the experiment, the building outlines were regarded as the ground truth to train and evaluate the methods. Note that here the elevation data like the digital surface model (DSM) were not used in the dataset, and data augmentation algorithms were not used in this study.

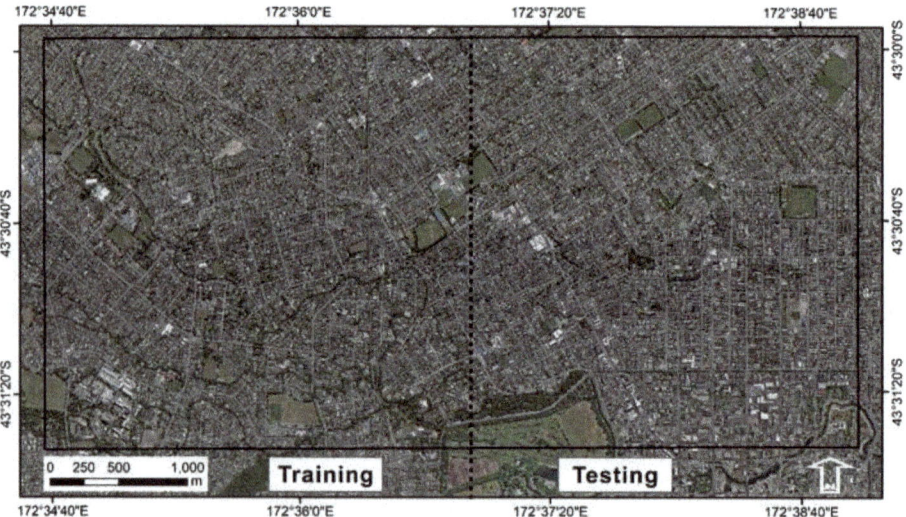

Figure 4. Overview of the dataset (an urban area of Christchurch City, New Zealand) used in the present study. The black dotted line separates the image into training and testing datasets for evaluating the performance of the proposed DeepResUnet by comparing it with other state-of-the-art deep learning approaches.

3.2. Experimental Setup

DeepResUnet was implemented in Keras using the Tensorflow framework as the backend. In the experiment, the aerial imagery of the training area was split into patches with the size of 256 × 256 by using a sliding window algorithm with a stride of 128. Thus, a total of 17,961 training patches of the same size (256 × 256 image block) were prepared. During training, 80% of training patches were used to train the models while the remaining 20% was used for cross-validation. The proposed network and other comparison ones were trained on a NVIDIA GeForce GTX 1080Ti GPU (11 GB RAM). The glorot normal initializer [52] was used to initialize weights and parameters of networks, and the cross-entropy loss was employed in training process. The Adam optimizer [53] was adopted to optimize the training loss. Due to the limited memory of GPU, the batch size of 6 was chosen in the experiment. The learning rate was set at 0.001 initially, but it was gradually reduced by a factor of 10 in every 11,970 iterations. In this experiment, the proposed DeepResUnet converged after 35,910 iterations.

During the inference phase, the aerial imagery of the test area was also split into patches with the size of 256 × 256 given the limitation of GPU memory. To reduce the impact of boundaries, a sliding window algorithm with a stride of 64 was applied to generate the overlap images and the predictions of overlap images were averaged as final segmentation results. For the purpose of clearly reflecting the performance of the DeepResUnet, no post-processing methods such as the filters [54] and conditional random field [55] were used in the study in order to guarantee that a fair comparison in terms of the pure network performance.

3.3. Results

To evaluate the performance of DeepResUnet, six existing state-of-the-art deep learning approaches (i.e., FCN-8s, SegNet, DeconvNet, U-Net, ResUNet and DeepUNet) were selected for comparison in the exact same experimental environment. Each network was trained from scratch, without using pretrained models, and all networks converged during training. The inference procedure of six existing deep learning approaches was the same as that of DeepResUnet. The overall results from different networks of a randomly selected test area are shown together in Figure 5.

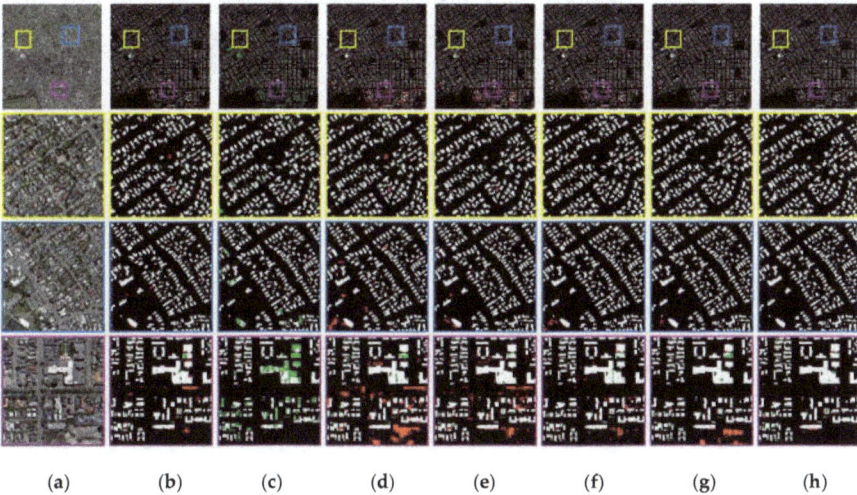

Figure 5. Visual comparison of segmentation results (an urban area of Christchurch City, New Zealand) obtained by seven approaches. The first row shows the overall results of a randomly picked test area generated by the seven networks, while the last three rows exhibit the zoomed-in results of the corresponding regions that were also randomly picked from the test area. In the colored images of the last three rows, the white, red and green colors represent true positive, false positive and false negative predictions, respectively. (**a**) Image. (**b**) FCN-8s. (**c**) SegNet. (**d**) DeconvNet. (**e**) U-Net. (**f**) ResUNet. (**g**) DeepUNet. (**h**) DeepResUnet.

By visual inspection, it appears that DeepResUnet outperformed the other approaches. As illustrated in Figure 5, the major parts of buildings were accurately extracted by DeepResUnet, while more false positives (red) and more false negatives (green) were found in the semantic segmentation of urban buildings by the other approaches, especially SegNet, DeconvNet and U-Net. Visually, FCN-8s, ResUNet and DeepUNet models performed better than other three approaches selected for comparison, but they still did not accurately identify small-size buildings in dense building area and FCN-8s model often misclassified roads as buildings. It is worthwhile mentioning that ResUNet and DeepUNet models as the improved versions of U-Net, outperformed the original U-Net model, similar to the proposed DeepResUnet, but more false positives (red) occurred in their segmentation results. There were few misclassifications (false positive) in the segmentation results of SegNet model, but many false negatives (green) appeared in the segmentation results, indicating that many buildings were not accurately identified by SegNet. The DeconvNet and U-Net models frequently misclassified roads and bare surfaces as buildings, and many false positives (red) were obtained in the corresponding segmentation results, implying that these models did not make full use of the image features. Overall, DeepResUnet outperformed the other six approaches with less false negatives and false positives in the semantic segmentation image of urban buildings.

For further comparison, the partial results of the test area for semantic segmentation of urban buildings are presented in Figure 6. It is also clear that all the seven tested approaches can identify the regular shaped buildings in general with acceptable accuracy, such as rectangle and square shaped buildings. However, for small-size and irregularly shaped buildings (as shown in Figure 6), DeepResUnet had very competitive performance on better preservation of patch edges, followed by DeepUNet, while more false positives (red) and false negatives (green) were obtained by the other approaches. Among these approaches, SegNet, DeconvNet and U-Net generated considerably more incomplete and inaccurate labelings than FCN-8s and ResUNet did. Although the proposed DeepResUnet was not yet perfect in the semantic segmentation of urban buildings, relatively more accurate extraction of building boundaries and relatively coherent building object labeling with fewer false positive returns made it rank in a high position among all the seven compared state-of-the-art deep networks.

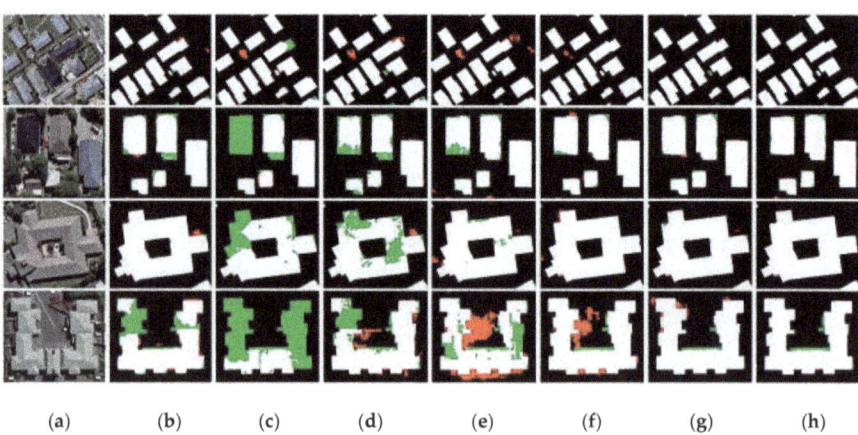

Figure 6. Visual comparison of segmentation results by seven approaches. In the colored images, the white, red and green colors represent true positive, false positive and false negative predictions, respectively. (a) Image. (b) FCN-8s. (c) SegNet. (d) DeconvNet. (e) U-Net. (f) ResUNet. (g) DeepUNet. (h) DeepResUnet.

To quantitatively evaluate the performance of DeepResUnet, five conventionally used criteria, including precision, recall, F1 score (F1), Kappa coefficient and overall accuracy (OA), were employed. The Kappa coefficient and OA are the global measures of segmentation accuracy [56]. Precision measures the percentage of matched building pixels in the segmentation map, while Recall represents the proportion of matched building pixels in the ground truth. F1 is the geometric mean between precision and recall, which is formulated as:

$$F1 = 2 \times \frac{Precision \times Recall}{Precision + Recall} \qquad (4)$$

As illustrated in Table 2, DeepResUnet had the best performance among all networks in terms of all the five criteria, followed by DeepUNet. All evaluation criteria were improved a considerable amount in the test area by the proposed DeepResUnet in this study. Among these deep models, SegNet and DeconvNet had the worst performance, followed by U-Net, ResUNet and FCN-8s. It is worthwhile mentioning that both ResUNet and DeepUNet models had a better performance than the original U-Net model, indicating the effectiveness of the combined methods. Although most models achieved relatively high values in one of the evaluations metrics, none of them have good performance in all metrics. For instance, DeepResUnet outperformed SegNet in term of Precision index only by about 0.7%. However, in terms of Recall index, DeepResUnet outperformed SegNet more than by 12%. The proposed DeepResUnet had the best performance in terms of the Recall index compared with other six networks, indicating that DeepResUnet was more effective in suppressing false negatives in semantic segmentation of urban buildings. With respect to the F1 score, Kappa coefficient and OA, the proposed DeepResUnet in this study was still the best among all deep models. This further demonstrated the superiority of the proposed network in semantic segmentation of urban buildings from VHR remotely sensed imagery.

Table 2. Quantitative comparison of five conventionally used metrics obtained from the segmentation results (for an urban area of Christchurch City, New Zealand) by FCN-8s, SegNet, DeconvNet, U-Net, ResUNet, DeepUNet and the proposed DeepResUnet, where the values in bold format are the highest numbers for corresponding metrics.

Models	Precision	Recall	F1	Kappa	OA
FCN-8s [28]	0.9163	0.9102	0.9132	0.8875	0.9602
SegNet [29]	0.9338	0.8098	0.8674	0.8314	0.9431
DeconvNet [31]	0.8529	0.9001	0.8758	0.8375	0.9413
U-Net [30]	0.8840	0.9190	0.9012	0.8709	0.9537
ResUNet [51]	0.9074	0.9315	0.9193	0.8948	0.9624
DeepUNet [57]	0.9269	0.9245	0.9257	0.9035	0.9659
DeepResUnet	**0.9401**	**0.9328**	**0.9364**	**0.9176**	**0.9709**

4. Discussion

4.1. About the DeepResUnet

DeepResUnet adopted the U-Net as its basic structure and meanwhile took advantages of deep residual learning by replacing the plain neural units of the U-Net with the residual learning units to facilitate the training process of the network. Although some combined approaches of the U-Net with the deep residual network have been reported in recent studies [51,57,58], significant differences can be found between their network architectures and that of ours.

First, in the network architecture, a new residual block, namely, Resblock, was designed as the basic processing unit to learn various representations of remote sensing images. ResBlock consisted of two successive 3 × 3 convolutional layers and a single 1 × 1 convolutional layer, which was designed to replace the basic residual unit that was used by other combined approaches of combining U-Net with the residual learning approach. Although a single 1 × 1 convolutional layer has been proved effective in dimension reduction [26], some important information for pixelwise segmentation might be compressed under the limited number of network layers and consequently would cause the loss of the transferred information, thus affecting the final performance of the deep network. For resolving this knotty problem, we made a tradeoff between the number of parameters and network layers by using two successive 3 × 3 convolutional layers and a single 1 × 1 convolutional layer in the ResBlock and the number of channels of the latter 3 × 3 convolution layer doubles that of the former. Hence, the structure of DeepResUnet is much deeper than that of U-Net.

Second, the main purpose of this study was to perform semantic segmentation of urban buildings from VHR remotely sensed imagery. To some extent, the proposed network was designed for pixel-level urban building semantic segmentation, and it may be applied to dense urban areas such as big cities and even megacities. For clearly and fairly reflecting the performance of different networks, any post-processing operations, such as the filters or conditional random field, were not applied in the proposed DeepResUnet. Additionally, the infrared band data and DSM data, which are the two data sources that have much potential in improving the final performance if combined with aforementioned post-processing operations, were also not used in this study. In addition, the performance of these combined approaches (i.e., ResUNet [51] and DeepUNet [57]) was compared in the Section 3.3. Among all combined models, the proposed DeepResUnet model achieved the best performance, indicating the effectiveness of DeepResUnet. Note that those combined models obtained very high values compared with the original U-Net model; therefore, the combination of different methods will be a new trend to improve the performance in the future.

4.2. Effects of Resblock

To further confirm the effectiveness of Resblock, the performances of DeepResUnet (Baseline + Resblock) and its variants were compared and are summarized in Table 3. Here, baseline refers to the basic structure of our network which is "U" shaped. The basic residual unit refers to the structure used in ResNet-34. Bottleneck refers to the "bottleneck" shaped residual unit used in ResNet-50/101/152. Plain neural units are just convolution layers that do not use the residual structure. As shown in Table 3, the performance of Baseline + bottleneck was similar to that of Baseline + plain neural unit, close to that of Baseline + basic residual unit, indicating that the bottleneck cannot improve the performance of the baseline model. However, the Baseline + basic residual unit outperformed Baseline + plain neural unit, implying that the residual learning is effective in building semantic segmentation task. The poor performance of Baseline + bottleneck indicated that the 1×1 convolutional layer in the bottleneck may have a negative impact on information transmission, and more research is needed to confirm this. A comparison between Baseline + bottleneck and Baseline + basic residual unit, Baseline + Resblock (i.e., DeepResUnet) achieved better segmentation results. This implies the effectiveness of replacing the 1×1 convolutional layer in bottleneck with a 3×3 convolutional layer and also indicates that reducing the number of channels by half in the first 3×3 convolution layer has little to almost no effect on the accuracy of the segmentation result, but this procedure greatly reduces the number of parameters which would benefit the training process and create a more robust model.

Table 3. Comparisons of building segmentation results and model complexity among the different variants of DeepResUnet.

Metrics	Baseline + Plain Neural Unit	Baseline + Basic Residual Unit	Baseline + Bottleneck	Baseline + Resblock (DeepResUnet)
Precision	0.9234	0.9329	0.9277	0.9401
Recall	0.9334	0.9330	0.9321	0.9328
F1	0.9283	0.9329	0.9299	0.9364
Kappa	0.9068	0.9129	0.9089	0.9176
OA	0.9669	0.9691	0.9677	0.9709
Parameters (m)	4.89	4.89	3.06	2.79
Training time (second/epoch)	1485	1487	1615	1516
Inference time (ms/image)	63.5	63.8	72.5	69.3

In terms of complexity, Baseline + Resblock requires fewer parameters than the other models because a small number of channels was applied in the first convolutional layer of Resblock (as exhibited in Figure 2). Although the structure of Resblock is similar to that of the bottleneck, Baseline + bottleneck requires a longer training time and inference time than Baseline + Resblock. In addition, Baseline + Resblock needs a slightly longer training time and inference time than Baseline + plain neural unit and Baseline + basic residual unit, but generated better segmentation results. Overall, the structure of Resblock has an obvious effect in improving the performance of deep models.

4.3. Complexity Comparison of Deep Learning Models

In recent years, with the rapid advancement of the computer hardware, high-end GPU or GPU clusters have made network training easier, and some deeper networks have been proposed, such as DenseNets [59] and its extended network [60]. However, for experimental research and practical application, the trade-offs among the layer depth, number of channels, kernel sizes, and other attributes of the network must still be considered when designing the architectures of networks [61], given the concern of cost-effectiveness in training time and commercial cost.

To evaluate the complexity of DeepResUnet, the number of parameters, training time and inference time were compared with six existing state-of-the-art deep learning approaches (i.e., FCN-8s, SegNet, DeconvNet, U-Net, ResUNet and DeepUNet). It is worthwhile mentioning that the running time of deep models including training and testing time can be affected by many factors [62], such as the parameters and the model structure. Here, we simply compared the complexity of deep models. As shown in Table 4, DeconvNet has the largest number of parameters, and the longest training time and inference time among all models. The numbers of parameters of DeepResUnet are much fewer than most networks except the DeepUNet, because DeepUNet adopted a very small convolutional channel (each convolutional layer with 32 channels). Even though the proposed DeepResUnet followed the basic structure of U-Net, the number parameters of U-Net are nearly eleven times higher than that of DeepResUnet. However, DeepResUnet requires a longer training time and inference time than U-Net and its combined networks. The main reason may be that the deep residual learning may have a negative effect on model operations. The training time and inference time of ResUNet are longer than those of U-Net, also indicating the negative impact of deep residual learning. Additionally, the structure of DeepResUnet is much deeper than those of deep networks, which may also increase the operation time of DeepResUnet. Compared with FCN-8s and SegNet, they require less training time than DeepResUnet, but the inference time of DeepResUnet is shorter than that of FCN-8s, close to that of SegNet. From the viewpoint of accuracy improvement and reducing computing resources, such a minor time increase should be acceptable. Overall, DeepResUnet achieves a relative trade-off between the model performance and complexity.

Table 4. Complexity comparison of FCN-8s, SegNet, DeconvNet, U-Net, ResUNet, DeepUNet and the proposed DeepResUnet.

Model	Parameters (m)	Training Time (Second/Epoch)	Inference Time (ms/image)
FCN-8s [28]	134.27	979	86.1
SegNet [29]	29.46	1192	60.7
DeconvNet [31]	251.84	2497	214.3
U-Net [30]	31.03	718	47.2
ResUNet [51]	8.10	1229	55.8
DeepUNet [57]	0.62	505	41.5
DeepResUnet	2.79	1516	69.3

4.4. Applicability Analysis of DeepResUnet

To further explore the applicability of DeepResUnet, the urban area of Waimakariri, New Zealand was used to test the effectiveness of DeepResUnet. The aerial images of Waimakariri were taken during 2015 and 2016, with a spatial resolution of 0.075 m. The corresponding building outlines were also provided by the website (https://data.linz.govt.nz). The aerial image of Waimakariri is 13,526 × 12,418 pixels. Note that DeepResUnet was trained using the aerial images of Christchurch City, New Zealand,

and we did not train DeepResUnet again on other images. Additionally, any pre- and post-processing methods were not applied when testing the aerial images of Waimakariri.

The results of the Waimakariri area are presented in Figure 7. Many false negatives (green) appeared in the segmentation results of SegNet model and many misclassifications (false positive) in the segmentation results of DeconvNet model. It is obvious that FCN-8s and DeepResUnet outperformed the other approaches. Neither ResUNet nor DeepUNet performed well in this new urban area, indicating the relatively poor applicability of these combined models. As exhibited in last three rows of Figure 7, few false positives (red) and false negatives (green) were obtained in the segmentation results of DeepResUnet model. DeepResUnet accurately identified small-size and irregularly shaped buildings. Quantitative results are provided in Table 5. The overall performance of DeepResUnet in this study was the best among deep models, followed by FCN-8s and ResUNet. The performance of these deep models was basically consistent with the testing results of Christchurch City, indicating that DeepResUnet has the ability to identify urban buildings and it can be applied to dense urban areas such as big cities and even megacities.

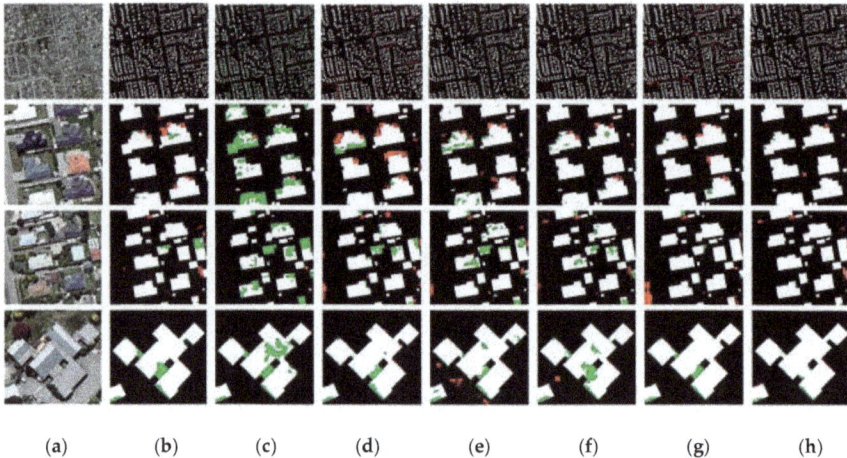

Figure 7. Visual comparison of segmentation results by seven approaches (for an urban area of Waimakariri, New Zealand). The first row shows the overall results, and the last three rows exhibit the zoomed-in results. In the colored figures, the white, red and green colors represent true positive, false positive and false negative predictions, respectively. (**a**) Image. (**b**) FCN-8s. (**c**) SegNet. (**d**) DeconvNet. (**e**) U-Net. (**f**) ResUNet. (**g**) DeepUNet. (**h**) DeepResUnet.

Table 5. Quantitative comparison of five conventionally used metrics (for an urban area of Waimakariri, New Zealand) obtained from the segmentation results by FCN-8s, SegNet, DeconvNet, U-Net, ResUNet, DeepUNet and the proposed DeepResUnet, where the values in bold format are the highest numbers for corresponding metrics.

Models	Precision	Recall	F1	Kappa	OA
FCN-8s [28]	0.8831	**0.9339**	0.9078	0.8807	0.9581
SegNet [29]	**0.9475**	0.6174	0.7477	0.6944	0.9079
DeconvNet [31]	0.8004	0.9135	0.8532	0.8080	0.9306
U-Net [30]	0.8671	0.8621	0.8646	0.8263	0.9403
ResUNet [51]	0.9049	0.8895	0.8972	0.8683	0.9549
DeepUNet [57]	0.8305	0.9219	0.8738	0.8356	0.9412
DeepResUnet	0.9101	0.9280	**0.9190**	**0.8957**	**0.9638**

The bolded numbers indicate the largest number in the column, easier to find out which one performs better in this format.

4.5. Limitations of Deep Learning Models in This Study

Although deep learning models achieved impressive results in semantic segmentation, some limitations exist in those models due to the complexity of remotely sensed images. As can be seen in Figure 8, in some particular areas, many false positives appeared in the result of building segmentations of all networks used in this paper. This problem may be caused by the phenomenon of "different objects with the same spectral reflectance" or "same objects with the different spectral reflectance". Actually, this phenomenon extensively exists in remotely sensed images. For DCNN-based methods, it is difficult to learn robust and discriminative representations from insufficient training samples and to distinguish subtle spectral differences [32]. In addition, many buildings were not fully identified by deep learning models because of roadside trees or shadows. This is also a challenge for DCNN-based methods.

Recently, some studies [63–66] found that the effective fusion of color imagery with elevation (such as DSM) might be helpful to resolving these problems. The elevation data containing the height information of the ground surface make it easy to discriminate the building roofs from impervious surfaces. Additionally, only three-band (red, green, blue) images were used to extract buildings. The near-infrared band was not used in the study which might be helpful to identify vegetation. In the future work, the use of elevation data and multispectral images (including the near-infrared band) will be considered for alleviating these issues.

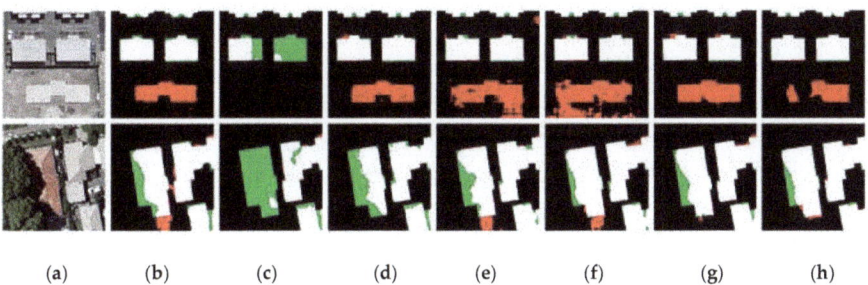

Figure 8. Visual comparison of urban building segmentation results by using different networks. In the colored figures, the white, red and green colors represent true positive, false positive and false negative predictions, respectively. (**a**) Image. (**b**) FCN-8s. (**c**) SegNet. (**d**) DeconvNet. (**e**) U-Net. (**f**) ResUNet. (**g**) DeepUNet. (**h**) DeepResUnet.

5. Conclusions

An end-to-end DCNN, denoted as DeepResUnet, for VHR image semantic segmentation of urban buildings at pixel scale, was proposed by adopting the architecture of U-Net as the basic structure. Specifically, the proposed DeepResUnet contains two sub-networks, that is, a cascade down-sampling network that is used to extract building feature maps from the VHR image, and an up-sampling network that is used to reconstruct the extracted feature maps of buildings back to the same size of input VHR image. To reduce gradient degradation, deep residual learning was incorporated in the proposed network.

To evaluate the performance of DeepResUnet, six existing state-of-the-art deep networks, including FCN-8s, SegNet, DeconvNet, U-Net, ResUNet and DeepUNet were selected for comparison in the exact same experiment environment, both visually and quantitatively. Each network was trained from scratch rather than being pretrained before the experiments. One of the advantages of DeepResUnet was that it requires far less parameters than most methods except DeepUNet. However, it does require slightly longer inference time than some other networks. For visual comparison, it was clear that all the seven tested networks were capable of extracting the regular-shape buildings, such as rectangle and square shaped buildings. However, other networks were less capable in accurate extraction of the irregularly shaped buildings, demonstrating that the proposed DeepResUnet outperformed the other six networks in a way that fewer false negatives and false positives appeared in the semantic segmentation image of urban buildings. Five conventionally used criteria, that is, precision, recall, F1 score (F1), Kappa coefficient and overall accuracy (OA), were used to evaluate the performance of the networks quantitatively, where DeepResUnet outperformed all the others because it suppressed false negatives, especially in semantic segmentation of irregular-shape and small-size buildings with higher accuracy and shape entirety. DeepResUnet is relatively better at suppressing false negatives as shown by its superior recall. Compared with the U-Net, DeepResUnet increased the F1 score, Kappa coefficient and overall accuracy by 3.52%, 4.67% and 1.72%, respectively. Additionally, DeepResUnet was further tested using the aerial images of an urban area of Waimakariri, New Zealand, further indicating the effectiveness and applicability of DeepResUnet.

More research is needed to improve DeepResUnet to better discriminate different objects with similar spectral characteristics or the same objects with different spectral characteristics. To some extent, the proposed network was designed for pixel-level urban building semantic segmentation and it may be applied to dense urban areas such as big cities and even megacities. As a continuation of this work, the fusion of image with elevation data (such as DSM) may be considered in the future to refine the performance of the proposed method.

Author Contributions: W.Z., Y.Y. and Z.Z. conceived this research. Y.Y. and Z.Z. performed the experiments, analyzed the results and wrote the paper. W.Z., C.Z., W.L. and T.Z. gave comments and modified the manuscript.

Funding: This research was funded by the National Key Research and Development Program of China, grant number 2016YFA0602302 and 2016YFB0502502.

Acknowledgments: The aerial images are provided by National Topographic Office of New Zealand (https://data.linz.govt.nz).

Conflicts of Interest: The authors declare no conflict of interest.

References

1. Grinias, I.; Panagiotakis, C.; Tziritas, G. MRF-based segmentation and unsupervised classification for building and road detection in peri-urban areas of high-resolution satellite images. *ISPRS J. Photogramm. Remote Sens.* **2016**, *122*, 145–166. [CrossRef]
2. Montoya-Zegarra, J.A.; Wegner, J.D.; Ladicky, L.; Schindler, K. Semantic segmentation of aerial images in urban areas with class-specific higher-order cliques. In Proceedings of the Joint ISPRS workshops on Photogrammetric Image Analysis (PIA) and High Resolution Earth Imaging for Geospatial Information (HRIGI), Munich, Germany, 25–27 March 2015; pp. 127–133.
3. Erener, A. Classification method, spectral diversity, band combination and accuracy assessment evaluation for urban feature detection. *Int. J. Appl. Earth Obs. Geoinf.* **2013**, *21*, 397–408. [CrossRef]
4. Marmanis, D.; Schindler, K.; Wegner, J.D.; Galliani, S.; Datcu, M.; Stilla, U. Classification with an edge: Improving semantic with boundary detection. *ISPRS J. Photogramm. Remote Sens.* **2018**, *135*, 158–172. [CrossRef]
5. Li, J.; Ding, W.; Li, H.; Liu, C. Semantic segmentation for high-resolution aerial imagery using multi-skip network and Markov random fields. In Proceedings of the IEEE International Conference on Unmanned Systems (ICUS), Beijing, China, 27–29 October 2017; pp. 12–17.
6. Zhou, H.; Kong, H.; Wei, L.; Creighton, D.; Nahavandi, S. On Detecting Road Regions in a Single UAV Image. *IEEE Trans. Intell. Transp. Syst.* **2017**, *18*, 1713–1722. [CrossRef]
7. Yi, Y.; Zhang, Z.; Zhang, W. Building Segmentation of Aerial Images in Urban Areas with Deep Convolutional Neural Networks. In Proceedings of the Advances in Remote Sensing and Geo Informatics Applications, Tunisia, 12–15 November 2018; pp. 61–64.
8. Shu, Z.; Hu, X.; Sun, J. Center-Point-Guided Proposal Generation for Detection of Small and Dense Buildings in Aerial Imagery. *IEEE Geosci. Remote Sens. Lett.* **2018**, *15*, 1100–1104. [CrossRef]
9. Moser, G.; Serpico, S.B.; Benediktsson, J.A. Land-Cover Mapping by Markov Modeling of Spatial-Contextual Information in Very-High-Resolution Remote Sensing Images. *Proc. IEEE* **2013**, *101*, 631–651. [CrossRef]
10. Huang, B.; Zhao, B.; Song, Y. Urban land-use mapping using a deep convolutional neural network with high spatial resolution multispectral remote sensing imagery. *Remote Sens. Environ.* **2018**, *214*, 73–86. [CrossRef]

11. Matikainen, L.; Karila, K. Segment-Based Land Cover Mapping of a Suburban Area—Comparison of High-Resolution Remotely Sensed Datasets Using Classification Trees and Test Field Points. *Remote Sens.* **2011**, *3*, 1777–1804. [CrossRef]
12. Zhang, W.; Li, W.; Zhang, C.; Hanink, D.M.; Li, X.; Wang, W. Parcel-based urban land use classification in megacity using airborne LiDAR, high resolution orthoimagery, and Google Street View. *Comput. Environ. Urban Syst.* **2017**, *64*, 215–228. [CrossRef]
13. Dalponte, M.; Bruzzone, L.; Gianelle, D. Tree species classification in the Southern Alps based on the fusion of very high geometrical resolution multispectral/hyperspectral images and LiDAR data. *Remote Sens. Environ.* **2012**, *123*, 258–270. [CrossRef]
14. Solórzano, J.V.; Meave, J.A.; Gallardo-Cruz, J.A.; González, E.J.; Hernández-Stefanoni, J.L. Predicting old-growth tropical forest attributes from very high resolution (VHR)-derived surface metrics. *Int. J. Remote Sens.* **2017**, *38*, 492–513. [CrossRef]
15. Ball, J.E.; Anderson, D.T.; Chan, C.S. Comprehensive survey of deep learning in remote sensing: Theories, tools, and challenges for the community. *J. Appl. Remote Sens.* **2017**, *11*, 54. [CrossRef]
16. Turker, M.; Koc-San, D. Building extraction from high-resolution optical spaceborne images using the integration of support vector machine (SVM) classification, Hough transformation and perceptual grouping. *Int. J. Appl. Earth Obs. Geoinf.* **2015**, *34*, 58–69. [CrossRef]
17. Yousefi, B.; Mirhassani, S.M.; AhmadiFard, A.; Hosseini, M. Hierarchical segmentation of urban satellite imagery. *Int. J. Appl. Earth Obs. Geoinf.* **2014**, *30*, 158–166. [CrossRef]
18. Gilani, A.S.; Awrangjeb, M.; Lu, G. An Automatic Building Extraction and Regularisation Technique Using LiDAR Point Cloud Data and Orthoimage. *Remote Sens.* **2016**, *8*, 258. [CrossRef]
19. Song, M.J.; Civco, D. Road extraction using SVM and image segmentation. *Photogramm. Eng. Remote Sens.* **2004**, *70*, 1365–1371. [CrossRef]
20. Tian, S.H.; Zhang, X.F.; Tian, J.; Sun, Q. Random Forest Classification of Wetland Landcovers from Multi-Sensor Data in the Arid Region of Xinjiang, China. *Remote Sens.* **2016**, *8*, 954. [CrossRef]
21. Wang, Y.; Song, H.W.; Zhang, Y. Spectral-Spatial Classification of Hyperspectral Images Using Joint Bilateral Filter and Graph Cut Based Model. *Remote Sens.* **2016**, *8*, 748. [CrossRef]
22. Das, S.; Mirnalinee, T.T.; Varghese, K. Use of Salient Features for the Design of a Multistage Framework to Extract Roads From High-Resolution Multispectral Satellite Images. *IEEE Trans. Geosci. Remote Sens.* **2011**, *49*, 3906–3931. [CrossRef]
23. Girshick, R. Fast R-CNN. In Proceedings of the IEEE International Conference on Computer Vision (ICCV), Santiago, Chile, 11–18 December 2015; pp. 1440–1448.
24. Szegedy, C.; Toshev, A.; Erhan, D. Deep Neural Networks for object detection. In Proceedings of the 27th Annual Conference on Neural Information Processing Systems (NIPS), Montreal, QC, Canada, 8–13 December 2013; pp. 2553–2561.
25. Liu, W.; Anguelov, D.; Erhan, D.; Szegedy, C.; Reed, S.; Fu, C.-Y.; Berg, A.C. SSD: Single shot multibox detector. In Proceedings of the 14th European Conference on Computer Vision, ECCV 2016, Amsterdam, The Netherlands, 11–14 October 2016; pp. 21–37.
26. Szegedy, C.; Liu, W.; Jia, Y.Q.; Sermanet, P.; Reed, S.; Anguelov, D.; Erhan, D.; Vanhoucke, V.; Rabinovich, A. Going Deeper with Convolutions. In Proceedings of the IEEE Conference on Computer Vision and Pattern Recognition (CVPR), Boston, MA, USA, 7–12 June 2015; pp. 1–9.
27. Krizhevsky, A.; Sutskever, I.; Hinton, G.E. ImageNet Classification with Deep Convolutional Neural Networks. In Proceedings of the 26th Annual Conference on Neural Information Processing Systems (NIPS), Lake Tahoe, NV, USA, 3–6 December 2012; pp. 1097–1105.
28. Long, J.; Shelhamer, E.; Darrell, T. Fully convolutional networks for semantic segmentation. In Proceedings of the IEEE Conference on Computer Vision and Pattern Recognition (CVPR), Boston, MA, USA, 7–12 June 2015; pp. 3431–3440.
29. Badrinarayanan, V.; Kendall, A.; Cipolla, R. SegNet: A Deep Convolutional Encoder-Decoder Architecture for Image Segmentation. *IEEE Trans. Pattern Anal. Mach. Intell.* **2017**, *39*, 2481–2495. [CrossRef]
30. Ronneberger, O.; Fischer, P.; Brox, T. U-Net: Convolutional Networks for Biomedical Image Segmentation. In Proceedings of the 18th International Conference on Medical Image Computing and Computer-Assisted Intervention (MICCAI), Munich, Germany, 5–9 October 2015; pp. 234–241.

31. Noh, H.; Hong, S.; Han, B. Learning Deconvolution Network for Semantic Segmentation. In Proceedings of the 15th IEEE International Conference on Computer Vision, ICCV 2015, Santiago, Chile, 11–18 December 2015; pp. 1520–1528.
32. Zhang, L.; Zhang, L.; Du, B. Deep Learning for Remote Sensing Data: A Technical Tutorial on the State of the Art. *IEEE Geosci. Remote Sens. Mag.* **2016**, *4*, 22–40. [CrossRef]
33. Scarpa, G.; Gargiulo, M.; Mazza, A.; Gaetano, R. A CNN-Based Fusion Method for Feature Extraction from Sentinel Data. *Remote Sens.* **2018**, *10*, 236. [CrossRef]
34. Schilling, H.; Bulatov, D.; Niessner, R.; Middelmann, W.; Soergel, U. Detection of Vehicles in Multisensor Data via Multibranch Convolutional Neural Networks. *IEEE J. Sel. Top. Appl. Earth Obs. Remote Sens.* **2018**, *11*, 4299–4316. [CrossRef]
35. Chen, X.; Xiang, S.; Liu, C.L.; Pan, C.H. Vehicle Detection in Satellite Images by Hybrid Deep Convolutional Neural Networks. *IEEE Geosci. Remote Sens. Lett.* **2014**, *11*, 1797–1801. [CrossRef]
36. Zhong, Z.; Fan, B.; Ding, K.; Li, H.; Xiang, S.; Pan, C. Efficient Multiple Feature Fusion With Hashing for Hyperspectral Imagery Classification: A Comparative Study. *IEEE Trans. Geosci. Remote Sens.* **2016**, *54*, 4461–4478. [CrossRef]
37. Ma, X.R.; Fu, A.Y.; Wang, J.; Wang, H.Y.; Yin, B.C. Hyperspectral Image Classification Based on Deep Deconvolution Network With Skip Architecture. *IEEE Trans. Geosci. Remote Sens.* **2018**, *56*, 4781–4791. [CrossRef]
38. Pan, X.; Yang, F.; Gao, L.; Chen, Z.; Zhang, B.; Fan, H.; Ren, J. Building Extraction from High-Resolution Aerial Imagery Using a Generative Adversarial Network with Spatial and Channel Attention Mechanisms. *Remote Sens.* **2019**, *11*, 917. [CrossRef]
39. Yuan, J. Learning Building Extraction in Aerial Scenes with Convolutional Networks. *IEEE Trans. Pattern Anal. Mach. Intell.* **2018**, *40*, 2793–2798. [CrossRef]
40. Ji, S.; Wei, S.; Lu, M. Fully Convolutional Networks for Multisource Building Extraction From an Open Aerial and Satellite Imagery Data Set. *IEEE Trans. Geosci. Remote Sens.* **2019**, *57*, 574–586. [CrossRef]
41. Saito, S.; Yamashita, T.; Aoki, Y. Multiple Object Extraction from Aerial Imagery with Convolutional Neural Networks. *J. Imaging Sci. Technol.* **2016**, *60*, 104021–104029. [CrossRef]
42. Bittner, K.; Adam, F.; Cui, S.; Körner, M.; Reinartz, P. Building Footprint Extraction From VHR Remote Sensing Images Combined With Normalized DSMs Using Fused Fully Convolutional Networks. *IEEE J. Sel. Top. Appl. Earth Obs. Remote Sens.* **2018**, *11*, 2615–2629. [CrossRef]
43. Vakalopoulou, M.; Karantzalos, K.; Komodakis, N.; Paragios, N. Building detection in very high resolution multispectral data with deep learning features. In Proceedings of the IEEE International Geoscience and Remote Sensing Symposium (IGARSS), Milan, Italy, 26–31 July 2015; pp. 1873–1876.
44. He, K.M.; Zhang, X.Y.; Ren, S.Q.; Sun, J. Deep Residual Learning for Image Recognition. In Proceedings of the IEEE Conference on Computer Vision and Pattern Recognition (CVPR), Las Vegas, NV, USA, 27–30 June 2016; pp. 770–778.
45. Ioffe, S.; Szegedy, C. Batch normalization: Accelerating deep network training by reducing internal covariate shift. In Proceedings of the 32nd International Conference on Machine Learning (ICML), Lile, France, 6–11 July 2015; pp. 448–456.
46. Wang, H.; Wang, Y.; Zhang, Q.; Xiang, S.; Pan, C. Gated Convolutional Neural Network for Semantic Segmentation in High-Resolution Images. *Remote Sens.* **2017**, *9*, 446. [CrossRef]
47. Cheng, D.; Meng, G.; Xiang, S.; Pan, C. FusionNet: Edge Aware Deep Convolutional Networks for Semantic Segmentation of Remote Sensing Harbor Images. *IEEE J. Sel. Top. Appl. Earth Obs. Remote Sens.* **2017**, *10*, 5769–5783. [CrossRef]
48. Simonyan, K.; Zisserman, A. Very Deep Convolutional Networks for Large-Scale Image Recognition. *arXiv* **2014**, arXiv:1409.1556.
49. Paszke, A.; Chaurasia, A.; Kim, S.; Culurciello, E. Enet: A deep neural network architecture for real-time semantic segmentation. *arXiv* **2016**, arXiv:1606.02147.
50. Liu, Y.; Fan, B.; Wang, L.; Bai, J.; Xiang, S.; Pan, C. Semantic labeling in very high resolution images via a self-cascaded convolutional neural network. *ISPRS J. Photogramm. Remote Sens.* **2017**, *145*, 78–95. [CrossRef]
51. Zhang, Z.; Liu, Q.; Wang, Y. Road Extraction by Deep Residual U-Net. *IEEE Geosci. Remote Sens. Lett.* **2018**, *15*, 749–753. [CrossRef]

52. Glorot, X.; Bengio, Y. Understanding the difficulty of training deep feedforward neural networks. In Proceedings of the 13th International Conference on Artificial Intelligence and Statistics (AISTATS), Sardinia, Italy, 13–15 May 2010; pp. 249–256.
53. Kingma, D.P.; Ba, J. Adam: A Method for Stochastic Optimization. In Proceedings of the 3rd International Conference on Learning Representations (ICLR), San Diego, CA, USA, 7–9 May 2015; p. 13.
54. Xu, Y.; Wu, L.; Xie, Z.; Chen, Z. Building Extraction in Very High Resolution Remote Sensing Imagery Using Deep Learning and Guided Filters. *Remote Sens.* **2018**, *10*, 144. [CrossRef]
55. Chen, L.; Papandreou, G.; Kokkinos, I.; Murphy, K.; Yuille, A.L. DeepLab: Semantic Image Segmentation with Deep Convolutional Nets, Atrous Convolution, and Fully Connected CRFs. *IEEE Trans. Pattern Anal. Mach. Intell.* **2018**, *40*, 834–848. [CrossRef]
56. Volpi, M.; Tuia, D. Dense Semantic Labeling of Subdecimeter Resolution Images With Convolutional Neural Networks. *IEEE Trans. Geosci. Remote Sens.* **2017**, *55*, 881–893. [CrossRef]
57. Li, R.; Liu, W.; Yang, L.; Sun, S.; Hu, W.; Zhang, F.; Li, W. DeepUNet: A Deep Fully Convolutional Network for Pixel-Level Sea-Land Segmentation. *IEEE J. Sel. Top. Appl. Earth Obs. Remote Sens.* **2018**, *11*, 3954–3962. [CrossRef]
58. Wu, G.; Shao, X.; Guo, Z.; Chen, Q.; Yuan, W.; Shi, X.; Xu, Y.; Shibasaki, R. Automatic Building Segmentation of Aerial Imagery Using Multi-Constraint Fully Convolutional Networks. *Remote Sens.* **2018**, *10*, 407. [CrossRef]
59. Huang, G.; Liu, Z.; van der Maaten, L.; Weinberger, K.Q. Densely Connected Convolutional Networks. In Proceedings of the 30th IEEE Conference on Computer Vision and Pattern Recognition (CVPR), Honolulu, HI, USA, 21–26 July 2017; pp. 2261–2269.
60. Jégou, S.; Drozdzal, M.; Vazquez, D.; Romero, A.; Bengio, Y. The One Hundred Layers Tiramisu: Fully Convolutional DenseNets for Semantic Segmentation. In Proceedings of the IEEE Conference on Computer Vision and Pattern Recognition Workshops (CVPRW), Honolulu, HI, USA, 21–26 July 2017; pp. 1175–1183.
61. He, K.; Sun, J. Convolutional neural networks at constrained time cost. In Proceedings of the IEEE Conference on Computer Vision and Pattern Recognition (CVPR), Boston, MA, USA, 7–12 June 2015; pp. 5353–5360.
62. Canziani, A.; Paszke, A.; Culurciello, E. An Analysis of Deep Neural Network Models for Practical Applications. *arXiv* **2016**, arXiv:1605.07678.
63. Sun, W.; Wang, R. Fully Convolutional Networks for Semantic Segmentation of Very High Resolution Remotely Sensed Images Combined With DSM. *IEEE Geosci. Remote Sens. Lett.* **2018**, *15*, 474–478. [CrossRef]
64. Audebert, N.; Le Saux, B.; Lefèvre, S. Beyond RGB: Very high resolution urban remote sensing with multimodal deep networks. *ISPRS J. Photogramm. Remote Sens.* **2018**, *140*, 20–32. [CrossRef]
65. Zhang, W.; Huang, H.; Schmitz, M.; Sun, X.; Wang, H.; Mayer, H. Effective Fusion of Multi-Modal Remote Sensing Data in a Fully Convolutional Network for Semantic Labeling. *Remote Sens.* **2018**, *10*, 52. [CrossRef]
66. Huang, J.; Zhang, X.; Xin, Q.; Sun, Y.; Zhang, P. Automatic building extraction from high-resolution aerial images and LiDAR data using gated residual refinement network. *ISPRS J. Photogramm. Remote Sens.* **2019**, *151*, 91–105. [CrossRef]

© 2019 by the authors. Licensee MDPI, Basel, Switzerland. This article is an open access article distributed under the terms and conditions of the Creative Commons Attribution (CC BY) license (http://creativecommons.org/licenses/by/4.0/).

Article

A Building Extraction Approach Based on the Fusion of LiDAR Point Cloud and Elevation Map Texture Features

Xudong Lai [1,2], Jingru Yang [1], Yongxu Li [1,*] and Mingwei Wang [2,3]

1. School of Remote Sensing and Information Engineering, Wuhan University, Wuhan 430079, China
2. Key Laboratory for National Geographic Census and Monitoring, National Administration of Surveying, Mapping and Geoinformation, Wuhan 430079, China
3. Institute of Geological Survey, China University of Geosciences, Wuhan 430074, China
* Correspondence: Liyongxu@whu.edu.cn; Tel.: +86-131-6462-5398

Received: 15 May 2019; Accepted: 3 July 2019; Published: 10 July 2019

Abstract: Building extraction is an important way to obtain information in urban planning, land management, and other fields. As remote sensing has various advantages such as large coverage and real-time capability, it becomes an essential approach for building extraction. Among various remote sensing technologies, the capability of providing 3D features makes the LiDAR point cloud become a crucial means for building extraction. However, the LiDAR point cloud has difficulty distinguishing objects with similar heights, in which case texture features are able to extract different objects in a 2D image. In this paper, a building extraction method based on the fusion of point cloud and texture features is proposed, and the texture features are extracted by using an elevation map that expresses the height of each point. The experimental results show that the proposed method obtains better extraction results than that of other texture feature extraction methods and ENVI software in all experimental areas, and the extraction accuracy is always higher than 87%, which is satisfactory for some practical work.

Keywords: LiDAR point cloud; building extraction; elevation map; Gabor filter; feature fusion

1. Introduction

Remote sensing is the acquisition of information about objects or phenomena without physical contact [1]. A large amount of remote sensing data has been generated and applied, and improvements in the spatial and temporal resolution of remote sensing images have made them become the main data source for object extraction [2], such as tree crown extraction [3], coastal zone detection [4], road recognition [5], etc. Buildings constitute the main component of urban areas, and building extraction using remote sensing images has become a hot research topic as remote sensing technology has the advantage of being fast, large-scale, and economical. Some researchers provided information of the spectral, geometrical, contextual, and rooftop segment patch via the morphological building index (MBI) and saliency cue to extract building information, which had good performance and versatility under different image conditions. However, the image-based building extraction technique is limited by large intra-class differences and small inter-class differences in spectral features [6,7]. 3D information is valid for buildings, especially elevation information, while for the image. it is complicated to realize, and it is mainly reflected by the change of elevation, which is important information of buildings.

As one of the active remote sensing data sources, LiDAR uses laser pulses to measure the distance between the sensor and different objects. It is widely used in geodesy [8], geo-statistics [9], archeology [10], geography [11], the control and navigation of autonomous vehicles [12], etc. Compared with 2D images, which only provide position and shape information, LiDAR can conveniently acquire

3D information on objects in terrain. Therefore, many studies have applied the LiDAR point cloud to conduct building extraction [13]. Wang et al. adopted a building extraction technique based on the point voxel group by using the class-oriented fusion method and "horizontal hollow ratio", which was effective for large-scale and complex urban environments [14]. Qin et al. demonstrated the use of geometric and radiation features of the waveform and the point cloud with parametric and non-parametric classification methods. The experimental results suggested that it was efficiently used for urban land cover mapping [15]. Zhao et al. utilized connected operators to extract building regions from LiDAR data, neither producing new contours, nor changing positions, which was effective, and the average offset values of simple and complex building boundaries were 0.2–0.4 m [16]. Huang et al. proposed a novel object and region-based top-down strategy to extract buildings, and the experimental result proved that the proposed method achieved good performance and was robust when parameters were within reasonable ranges [17]. Yi et al. detailed a method for reconstructing the volume structure of urban buildings directly from the original LiDAR point cloud. The experimental results demonstrated the advantage of the approach in terms of effectiveness on large-scale and raw LiDAR point data [18].

However, the discreteness of the point cloud may lead to the loss of some features, and it is difficult to distinguish objects with similar heights, while it is able to extract different objects with texture features in 2D images. As elevation map is a kind of 2D image obtained by projecting the point cloud onto 2D planes, and it can provide abundant texture features and has been utilized in the field of building extraction. Fasahat et al. realized building extraction by transforming the point cloud into an elevation map and analyzing gradient information from the elevation map. Experimental results showed the effectiveness in eliminating trees, extracting buildings of all sizes, and extracting buildings with and without a transparent roof [19]. Liu et al. combined remote sensing data of multiple sources to draw height maps of different object types for land cover and land use mapping, which coincided well with the ground survey data with an accuracy of 5.7 m by root mean squared error (RMSE) [20]. Kang et al. achieved the rendering of barren terrain by enhancing the geometric features of elevation maps and increased the number of landscape features, which was most suitable for rendering barren terrain or planet surfaces [21]. He et al. proposed to organize LiDAR point data as three different maps: dense depth map, height map, and surface normal map. It was proven to recover successfully object hierarchies, boundary sharpness, and global integrity regardless of point cloud sparsity, large loss, and 3D to 2D degradation uncertainty [22]. In addition, the texture feature extraction method can be used to obtain features to extract objects on the basis of the elevation map, which can robustly detect buildings from satellite images and outperforms state-of-the-art building detection method [23]. Cao et al. constructed a unified multilevel channel characteristic framework and realized target detection based on histograms of oriented gradient (HoG) features. The experimental results showed that this method could reduce the missed detection rate and improve the detection speed [24]. Du et al. used the gray level co-occurrence matrix (GLCM) features to obtain textures from an elevation map and combined them with point cloud information to achieve area and object-level building extraction, and the results suggested a good potential for large-sized LiDAR data [25]. Niemi et al. inventoried soil damage from forwarding trails and fitted a logistic regression model for predicting the event of soil damage, which showed that DTM-derived local binary patterns (LBP) were useful in terrain trafficability mapping [26].

Point cloud information can reflect the spatial structure of ground objects, but its discrete type may lead to the lack of correlation information of each part. Texture features can reflect the correlation of each part and help to distinguish different objects. Therefore, point cloud and texture features can be fused to achieve complementarity, as well as reflect the features of objects from multiple dimension so as to obtain better results. However, the increased data dimension may lead to an increase in time complexity, and feature selection is always utilized to solve the problem. As the essence of feature selection is a combinatorial optimization problem, which means selecting a satisfactory feature subset to conduct building extraction, it is usually solved by swarm intelligence algorithms [27]. In this paper, by fusing the point cloud and texture features, as well as conducting feature selection, a building extraction technique is realized. Point cloud features are extracted based on the eigenvalue, density, and elevation, and the point cloud is also transformed into an elevation map to extract texture features. After that, the fusion of the point cloud and texture features is used to extract buildings from different experimental areas. Among various swarm intelligence algorithms, particle swarm optimization (PSO) is easy to implement, and stably converges to the optimal solution. Therefore, it is adopted to obtain the superior feature subset for building extraction in the paper.

This paper is structured as follows. In Section 2, the core method and basic principles of this paper are elaborated in detail. The steps of the method are described in Section 3. Section 4 describes the experiments that are carried out according to the method, the experimental data, the final results, and the evaluation of the accuracy. Section 5 summarizes the work of this paper and research prospects.

2. Basic Theory of Gabor Filters

As for 2D images, the Gabor filter is one of the efficient filtering techniques and is based on a sinusoidal plane wave. Its use has been explored in many applications [28,29]. The Gabor filter can not only characterize the spatial frequency structure of an image, but also retain spatial relationship information, and the spatial frequency positioning ability is essential to extract orientation-dependent frequency content from the pattern [30]. Furthermore, as the Gabor filter is invariant to zoom, rotation, and translation, it is suitable for texture representation and recognition [31]. In the spatial domain, a 2D Gabor filter is a Gaussian kernel function modulated by a sinusoidal plane wave, which consists of a real part and an imaginary part representing the orthogonal direction. These two parts can either form a plurality or be used separately [32,33].

The formula for the Gabor filter is expressed as below:

$$g(x,y) = \left(\frac{1}{2\pi\sigma_x\sigma_y}\right) exp\left(-\frac{1}{2}\left(\frac{\tilde{x}^2 + \tilde{y}^2}{\sigma_x^2 + \sigma_y^2}\right) + 2\pi j W \tilde{x}\right) \quad (1)$$

$$\tilde{x} = xcos\theta + ysin\theta \quad \tilde{y} = -xsin\theta + ycos\theta \quad (2)$$

where σ_x and σ_y are parameters that describe the spread of the current pixel in the neighborhood in which weighted summation occurs, W is the central frequency of the complex sinusoid, $\theta \in [0, \pi)$ is the orientation of the horizontal to vertical stripes in the equation above, and æ represents the imaginary unit.

The extraction of texture features using the Gabor filter includes two main processes: filter design and the effective extraction of texture feature sets from the filter's output. The process of acquiring texture features from an image using the Gabor filter is as follows. Firstly, the input image is divided into blocks. Secondly, the Gabor filter banks are established, and thirdly, we convolve the Gabor filter templates with each image block in the spatial domain; each image block obtains the filter outputs. These outputs are of the image blocks' size. Fourthly, each image block is passed through the outputs of the Gabor filter templates and is "condensed" into the texture feature of the image block [34].

3. Building Extraction Based on the Fusion of Point Cloud and Texture Features

3.1. Point Cloud Features

At first, as the LiDAR system generates a number of noise points when acquiring data, which is usually manifested as elevation anomaly points and will affect the accuracy of building extraction, the point cloud is denoised, and elevation anomalies are filtered out. After that, the features of the point cloud, which include various eigenvalues, are obtained. Unlike eigenvectors, eigenvalues have good rotationally-invariant properties [35], and therefore, feature extraction based on the point cloud's eigenvalues was used for building extraction. Besides, density and elevation are both critical attributes of point cloud. Thus, features based on eigenvalues, density, and elevation were extracted as the reference data of building extraction. The specific meanings and formulas used in the calculations are shown in Table 1.

Table 1. Point cloud features.

Category	Name	Abbreviation	Meaning	Formula
Eigenvalue-based features	Sum	SU	Sum of eigenvalues	$\lambda_1 + \lambda_2 + \lambda_3$
	Total variance	TV	Total variance	$(\lambda_1 \lambda_2 \lambda_3)^{1/3}$
	Eigen entropy	EI	Characteristic entropy	$-\sum_{i=-3}^{3} \lambda_i \cdot In(\lambda_i)$
	Anisotropy	AN	Anisotropy	$(\lambda_1 - \lambda_3)/\lambda_1$
	Planarity	PL	Planarity	$(\lambda_2 - \lambda_3)/\lambda_1$
	Linearity	LI	Linearity	$(\lambda_1 - \lambda_2)/\lambda_1$
	Surface roughness	SR	Surface roughness	$\lambda_3/(\lambda_1 + \lambda_2 + \lambda_3)$
	Sphericity	SP	Sphericity	λ_3/λ_1
Density-based feature	Point Density	PD	Point Density	$0.75 * \frac{N_{3D}}{\pi r^3}$
Elevation-based features	Height above	HA	The height difference between the current point and the lowest point	$Z - Z_{min}$
	Height below	HB	The height difference between the highest point and the current point	$Z_{max} - Z$
	Sphere Variance	SPV	Standard deviation of the height difference in the spherical neighborhood	$-\sqrt{\frac{\sum_{i=1}^{n}(Z_i - Z_{ave})^2}{n-1}}$

λ_1, λ_2, and λ_3 are eigenvalues of the point cloud, where $\lambda_1 < \lambda_2 < \lambda_3$. An analysis of the eigenvalues and eigenvectors can often provide important information for extraction decisions. According to the points in the neighborhood, the covariance matrix of the center point was calculated, and then the eigenvalues of the point were obtained. Based on these eigenvalues, 12 kinds of features can be calculated, including sum of eigenvalues (SU), total variance (TV), eigenvalues (EI), anisotropy (AN), planarity (PL), linearity (LI), surface roughness (SR), and sphericity (SP). AN refers to the uniformity of the point distribution on three arbitrary vertical axes, which helps to separate anisotropic structures, such as power lines and buildings, from vegetation. PL is a measurement of planar characteristics of the point cloud, and planar structures have high PL values. As the surface of a building's roof reflects laser directly, this feature is remarkable. LI is a measurement of the linear attributes of a point cloud. The power lines and edges of buildings have obvious linear structures, and the linearity of these points is characterized by high values. SR is the average number of points allocated by the point cloud in three directions. The distribution of vegetation points in all directions has no tendency, so the SR values of vegetation are high. The density of the point cloud in the neighborhood of penetrable targets, such as vegetation, reflects the distribution of the point cloud and is usually higher than that of buildings. In the vicinity of the cylinder at the center point, the height differences, including that between the current point and the lowest point (height above (HA)) and that between the highest point and the current point (height below (HB)), were calculated. The standard deviation of the elevation included the elevation of each point in the spherical and cylindrical neighborhoods. Z_{ave} is the average of the current neighborhood's interior point elevation; n is the number of points in the current neighborhood's interior point cloud; and Z_i is the i-th point in the neighborhood. The sphere variance (SPV) value is high for objects with few changes in elevation [36]. For high-rise building facades and roofs, the differences between the current and the lowest elevation are usually much larger than that of other points. Therefore, building facades can be distinguished effectively, while the standard deviations of elevations in spherical neighborhoods can be used to identify ground and other horizontal planes. All of the mentioned features above show the properties of point clouds from the point of view of eigenvalues, elevations, and densities. They can provide more effective information for building extraction than single-scale features [37].

3.2. Texture Feature Extraction Based on the Elevation Map

In this study, the point cloud was transformed into an elevation map for texture feature extraction. The process of transformation was based on the elevation distribution of the point cloud, and it was easy to operate. Firstly, the grid size was set as 1 m, and then, the point cloud was rasterized according to its x and y coordinates, while each grid corresponded to one pixel in the elevation map [38]. After that, the height threshold was set, and the elevation variance of all points in the corresponding grid of each pixel was calculated. If the variance was below the threshold, the average elevation of points in the gird was selected as the gray reference value of the corresponding pixel. Otherwise, the height distribution curve was interpolated based on triangulation in the natural neighborhood, and half of the peak value was taken as the gray reference value. For a grid with few or even no points, the median elevation value of the points in the K-nearest-neighbor was taken. After all the above, the gray reference values were normalized to 0–255. In this way, the elevation map corresponding to the point cloud can be obtained as shown in Figure 1.

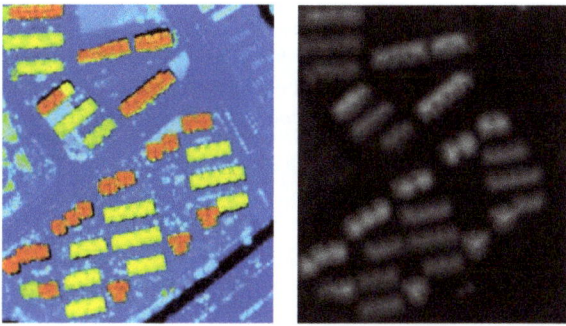

Figure 1. Point cloud rendering image (left) and elevation map (right).

After the elevation map was obtained, the corresponding texture features were extracted for further building extraction. Compared with other methods, the Gabor filter can capture those features that correspond to different spatial frequencies (scales) and orientations, so it can be used to discriminate features of images. In this study, a 2D Gabor filter was used to extract texture features. The texture features of elevation maps in different orientations and scales were obtained by changing the values of the orientation and frequency parameters. The orientation and frequency values were updated as follows:

$$\theta(i) = \frac{(i-1)\pi}{O}, \text{ where } i = 1, 2, ..., O \tag{3}$$

$$f(i) = \frac{f_{max}}{\sqrt{(2)^{i-1}}}, \text{ where } i = 1, 2, ..., S \tag{4}$$

where $\theta(i)$ is the orientation parameter, O is the number of orientation parameters, $f(i)$ is the frequency variable, and S is the number of frequency variables. In this study, four frequency values and six orientation values were combined to obtain 24 texture features. The frequency values changed gradually with 0.2, 1.414, 0.1, and 0.0707. The Gabor filter convolution kernel functions were in six different orientations: 0, $\pi/6$, $\pi/3$, $\pi/2$, $2\pi/3$, and $5\pi/6$ with the same frequency value.

3.3. Feature Selection for Reducing the Number of Features

In this paper, PSO was used for feature selection to decrease the data dimension, which is a kind of swarm intelligence algorithm using a group of particles. It has been noted that members of a group seem to share information among themselves, which is a fact that leads to increased efficiency of the group. A particle moves toward the optimum based on its present velocity, its previous experience, and the experience of its neighbors. In an n-Dimensional search space, the position and velocity of the i-th particle are represented as vectors $X_i = x_{i1}, ..., x_{in}$ and $V_i = v_{i1}, ..., v_{in}$. Let $Pbest_i$ and $Gbest$ be the best position of the i-th particle and the group's best position so far, respectively. The velocity and position of each particle are updated as follows [39]:

$$V_i^{k+1} = \omega \cdot V_i^k + r_1 \cdot c_1 \cdot (Pbest_i^k - X_i^k) + r_2 \cdot c_2 \cdot (Gbest^k - X_i^k) \tag{5}$$

$$X_i^{k+1} = X_i^k + V_i^{k+1} \tag{6}$$

where V_i^k is the velocity of the i-th particle at iteration k, ω is the inertia weight factor, c_1 and c_2 are the acceleration coefficients, r_1 and r_2 are random numbers between zero and one, and X_i^k is the position of the i-th particle at iteration k. In the velocity updating process, the values of the parameters such as ω, c_1, and c_2 should be determined in advance, which makes it cumbersome to solve large-scale optimization problems.

However, decimal coding may not be suitable for discrete optimization such as feature selection; thus, the position vector of a particle should be coded as a binary form. The velocity of the i-th element in the i-th particle is related to the possibility that the position of the particle takes a value of one or zero. It is implemented by defining an intermediate variable $S(v_{ij}^{k+1})$, called a sigmoid limiting transformation, as follows [40,41]:

$$S(v_{ij}^{k+1}) = \frac{1}{1 + exp(-v_{ij}^{k+1})} \tag{7}$$

The value of $S(v_{ij}^{k+1})$ can be interpreted as a probability threshold. If a random number selected from a uniform distribution in [0,1] is less than the threshold, the value of the position of the j-th element in the i-th particle at iteration $k+1$ (i.e., x_{ij}^{k+1}) is set to one, and otherwise to zero, and the position vector is replaced as follows:

$$x_{ij}^{k+1} = \begin{cases} 1 & if \quad rand < S(v_{ij}^{k+1}) \\ 0 & otherwise \end{cases} \tag{8}$$

where *rand* denotes random numbers uniformly distributed between zero and one; $S(v_{ij}^{k+1})$ is a sigmoid limiting transformation.

In this paper, PSO was used to extract as high an accuracy as possible with few features. To improve the training process, the feature combination was adjusted by PSO, and the optimized results could be obtained by choosing the feature combination with the minimum error as the most suitable one [42]. Finally, a reasonable combination of point cloud and texture features was obtained for building extraction, and the whole process is shown in Figure 2.

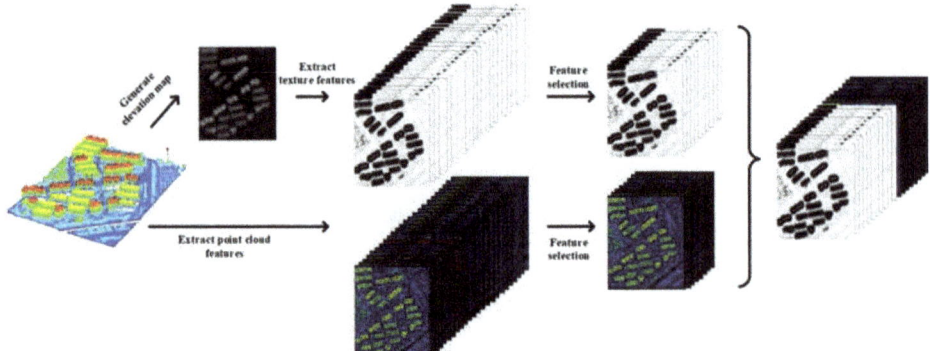

Figure 2. The process to obtain the optimal combination of features.

3.4. Definition of the Objective Function

To obtain the results of high extraction accuracy and reduce the number of features, an objective function was defined as an auxiliary in this paper. As the Fisher discriminant criterion has been shown to have good performance in building extraction and other extraction problems that include two categories, maximizing the differences between classes and minimizing the differences within classes, and accurately identifying the target category from other classes, it was used to define the objective function for feature selection [43]. The formula of the objective function is expressed as follows:

$$fit = \frac{(\mu_1 - \mu_2)^2}{(\sigma_1^2 + \sigma_2^2) \cdot n} \qquad (9)$$

where fit represents the value of the objective function, μ_1 and μ_2 are the eigen mean vectors of two types of objects, σ_1 and σ_2 are the eigen variance vectors of two types of objects, respectively, and n is the number of points. The output of point cloud features was the vectors. The texture features were in the form of a 2D image. They can be converted into vectors, and finally, these two kinds of features can be merged into a vector, while a higher feature vector dimension of each point can be obtained by combining the two vectors. Besides, the larger value of the objective function demonstrated better quality of classification.

3.5. Implementation of the Proposed Method

The proposed method was easy to implement, and the key issues of building extraction were the fusion of point cloud and texture features, as well as feature selection. The process of the proposed method is shown as follows:

- Step 1: Input the testing images, and compute the feature vectors of the point cloud. Generate elevation maps, and extract texture features via the Gabor filter from them.
- Step 2: Build the training and testing samples based on the fusion of point cloud and texture features;
- Step 3: Randomly generate the initial population of PSO in the range of −10–10 via decimal coding, and transform it into binary coding;
- Step 4: Conduct building extraction, and compute the fitness value of each particle by Equation (9);
- Step 5: Operation of PSO:

    Step 5-1: Update the velocity of each particle by using Equation (5);

    Step 5-2: Switch the population into the form of binary coding by Equation (8);

- Step 6: Conduct building extraction, and compute the fitness value of each particle by Equation (9);
- Step 7: If the solution is better, replace the current particle; otherwise, the particle does not change, and then, find the current global best solution;
- Step 8: Judge whether the maximum number of iterations is reached, and if it is, go to Step 9; otherwise, go to Step 5;
- Step 9: Output the optimal feature combination, and compare it with other building extraction methods via the extraction accuracy.

4. Experimental Results and Discussion

The experimental environment in this study was a computer with a 2.30-GHz CPU and 8 G of RAM. The data-processing operation was realized using MATLAB 2016a and VS2017 software. The manual extraction process was accomplished using LiDAR software and visual interpretation by researchers with relevant working experience.

4.1. Experimental Platform and Data Information

The data used in this study were point cloud data obtained from a Riegl LMS-Q780 laser scanner in Fuzhou, China. The experimental data included five non-overlapping urban areas, which contained buildings, vegetation, and other types of objects. Since the high density of the experimental point cloud may result in a large amount of calculation, it was necessary to down-sample the data in order to reduce the amount of calculation. According to the density of the point cloud after down-sampling, the data areas were divided into Low-Density Region 1 (LDR 1), LDR 2, the medium-density region (MDR), High-Density Region 1 (HDR 1), and HDR 2. Details on the experimental data are shown in Table 2.

Table 2. Experimental data information. LDR, low-density region; MDR, medium-density region; HDR, high-density region.

Experimental Data	Data Area (m^2)	Number of Points		Point Cloud Density	
		Original Data	After Dilution	Original Data	After Dilution
LDR 1	174,080	4,486,763	19,320	25.799339	0.111040
LDR 2	155,595	3,989,310	21,926	25.683631	0.140958
MDR	186,147	585,024	23,675	26.261592	0.183575
HDR 1	99,470	2,283,275	29,127	23.062170	0.294197
HDR 2	68,040	1,897,760	20,663	27.936171	0.303810

The experimental data were colored according to the elevation rendering, and the results of the manual extraction are shown in Figures 3–7.

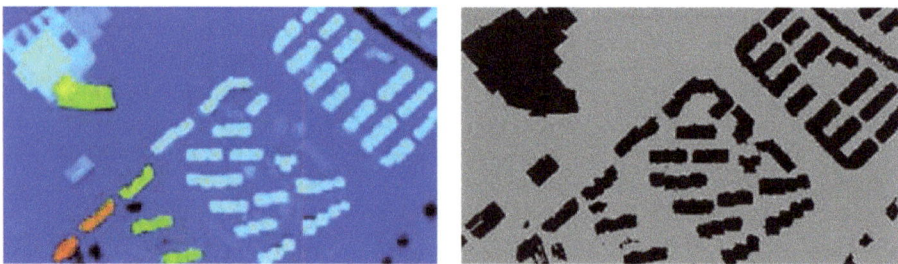

Figure 3. LDR 1 elevation coloration and manual extraction results.

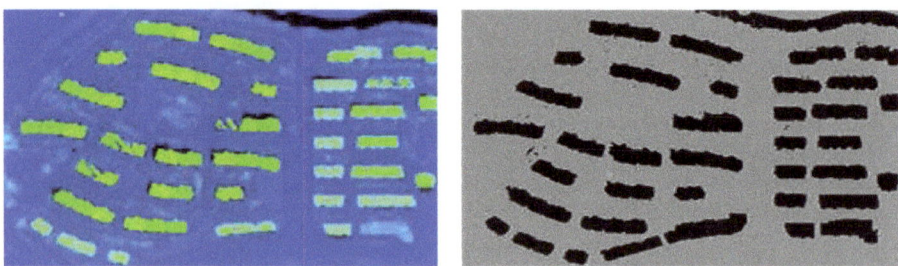

Figure 4. LDR 2 elevation coloration and manual extraction results.

Figure 5. MDR elevation coloration and manual extraction results.

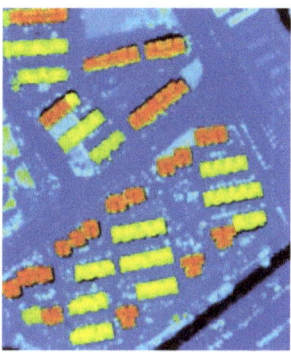 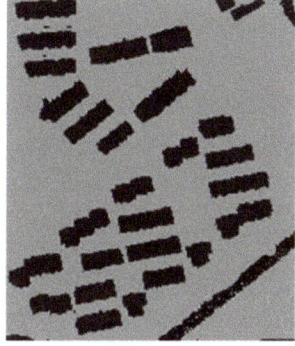

Figure 6. HDR 1 elevation coloration and manual extraction results.

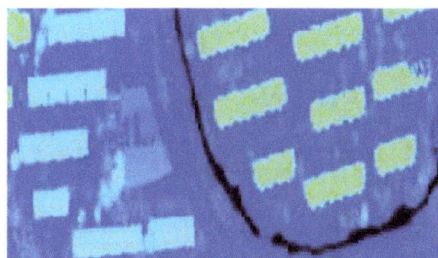 

Figure 7. HDR 2 elevation coloration and manual extraction results.

4.2. Extraction of Texture Features

The process of extracting texture features using the Gabor filter in this study is shown below:

Figure 8 shows the process of the Gabor filter, where it is formed on the basis of different values of the orientation and frequency parameters. Different texture features can be yielded after elevation map convoluting with templates. A group of parameter combination results is shown in Figure 8 with the same frequency value of 0.2, and the orientation varied from $0-5\pi/6$ via steps of $\pi/6$, while the local display of the common part is also shown on the right side. It can be concluded that variation of the parameter combination caused the change of the convolution module and resulted in differences in texture features, especially on the edge and corner of the buildings.

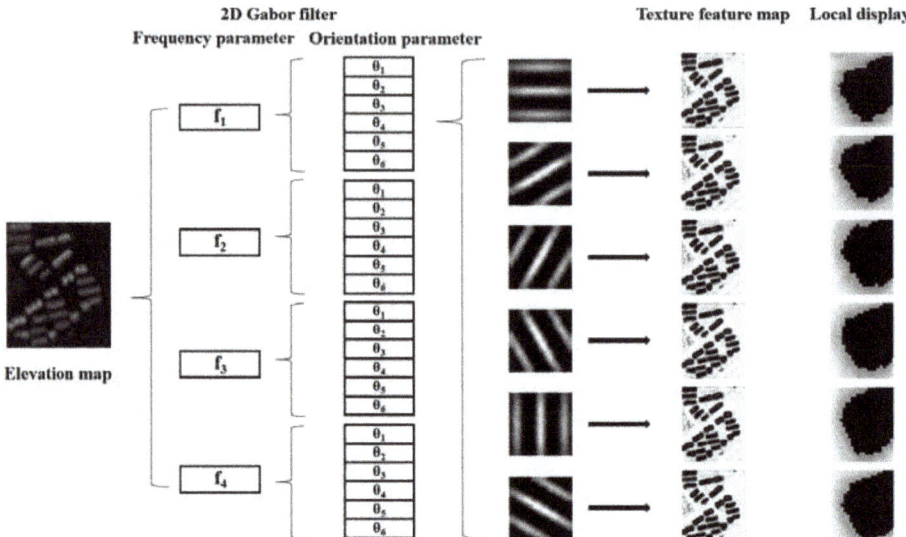

Figure 8. Texture features extraction using the Gabor filter.

4.3. Comparative Analysis and Accuracy Evaluation of Building Extraction

In order to prove the effectiveness of the proposed method, the experimental results were compared with those obtained using GLCM, LBP, and HoG for texture feature extraction. Those of building extraction based only on point cloud features (OPCF), building extraction with no feature selection (NFS), and building extraction using ENVI software were also compared in this paper. The extraction accuracy of different methods is shown in Table 3, and the building extraction locations of different experimental areas are shown in Figures 9–13.

Table 3. Comparison of the experimental results with other methods for building extraction (%). OPCF, only point cloud features; NFS, no feature selection.

Experimental Data	GLCM	HoG	LBP	OPCF	NFS	ENVI	Proposed
LDR 1	86.9984	75.9503	88.3870	80.4586	78.7330	87.4203	90.4238
LDR 2	65.5523	85.1865	74.5297	85.5651	89.6949	91.3310	92.2558
MDR	75.8902	78.9356	73.3347	81.7022	82.3527	83.5180	87.1679
HDR 1	87.5064	90.8264	90.0470	87.4961	81.6047	90.2660	92.1138
HDR 2	62.3917	76.6975	75.2795	79.4367	84.2762	86.2752	89.1207

From Table 3, it can be seen that the extraction accuracy with the proposed method was superior to other texture feature extraction methods and ENVI software. For HDR 2, the extraction accuracy of the proposed method was over 10% higher than that of GLCM HoG and LBP. Especially for LDR 2 and HDR 2, the extraction accuracy of GLCM was only around 60%, while the proposed method could still achieve an extraction accuracy higher than 87%. Although the extraction accuracy by using ENVI software exceeded 80%, and even reached 90% for LDR 2 and HDR 1, the extraction accuracy of the proposed method could still be 1.2874% higher than ENVI software. Comparing with the result of NFS, this suggested that feature selection benefited the building extraction by improving its extraction accuracy and efficiency. In all, when the follow-up operations were the same, the final results obtained by the Gabor filter applied in this paper were more accurate than those of other texture feature extraction methods. After feature selection, not only the extraction accuracy was higher, but also the computational time was shorter, as the data dimension was decreased. Besides, using about 10 features, we were able to achieve such satisfactory results.

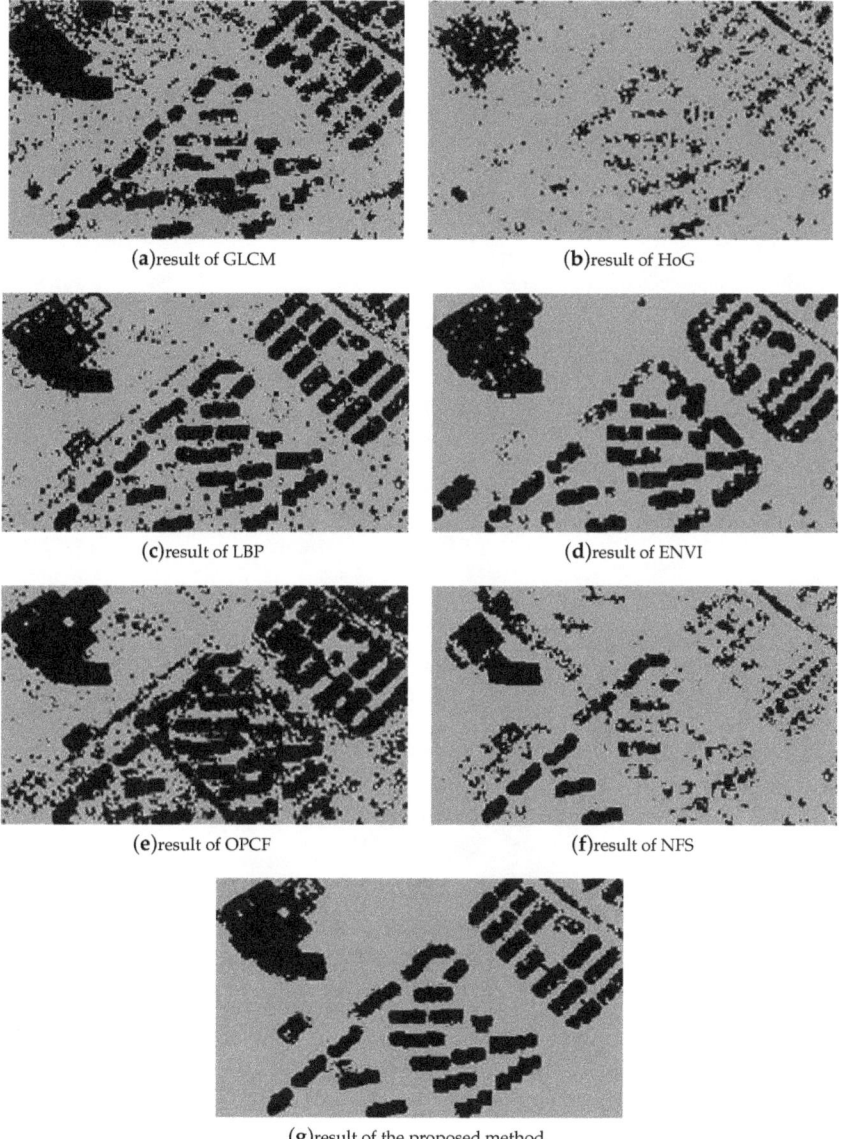

(a) result of GLCM

(b) result of HoG

(c) result of LBP

(d) result of ENVI

(e) result of OPCF

(f) result of NFS

(g) result of the proposed method

Figure 9. Building extraction results of LDR 1.

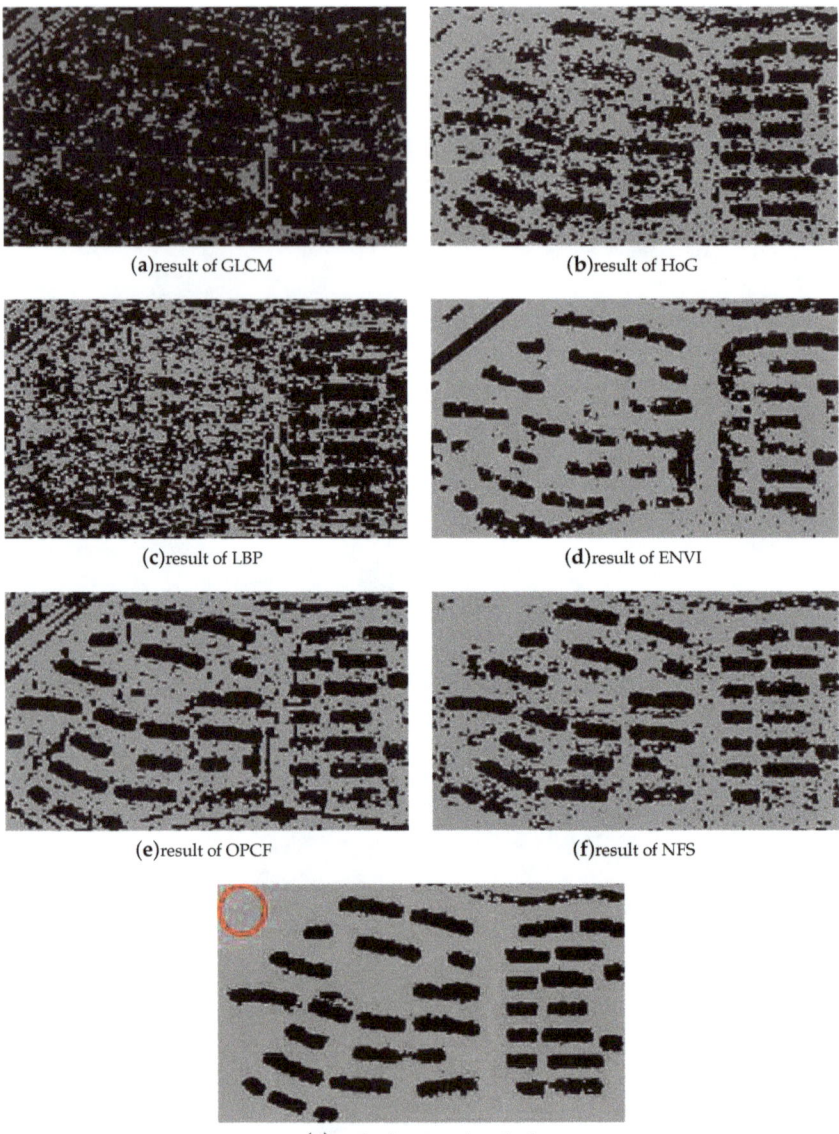

(a)result of GLCM (b)result of HoG
(c)result of LBP (d)result of ENVI
(e)result of OPCF (f)result of NFS
(g)result of proposed method

Figure 10. Building extraction results of LDR 2.

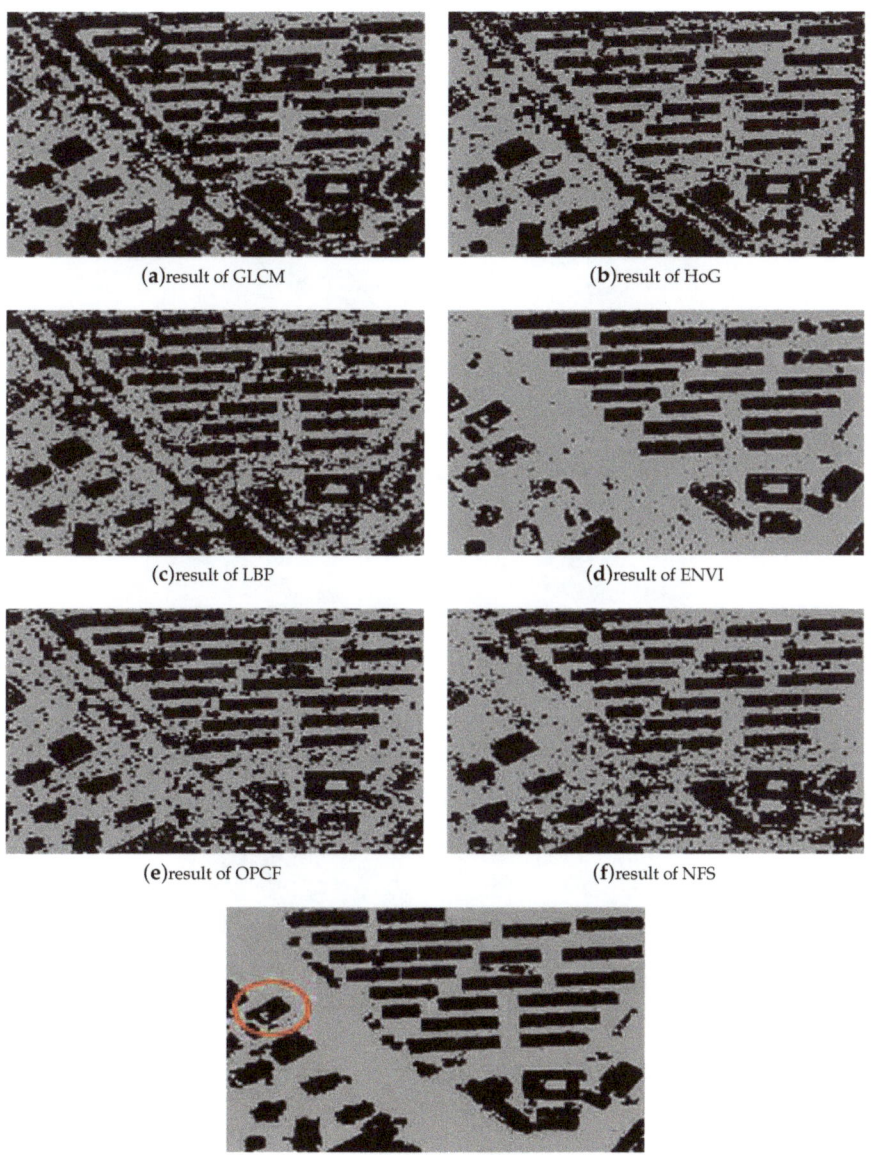

(a) result of GLCM
(b) result of HoG
(c) result of LBP
(d) result of ENVI
(e) result of OPCF
(f) result of NFS
(g) result of proposed method

Figure 11. Building extraction results of MDR.

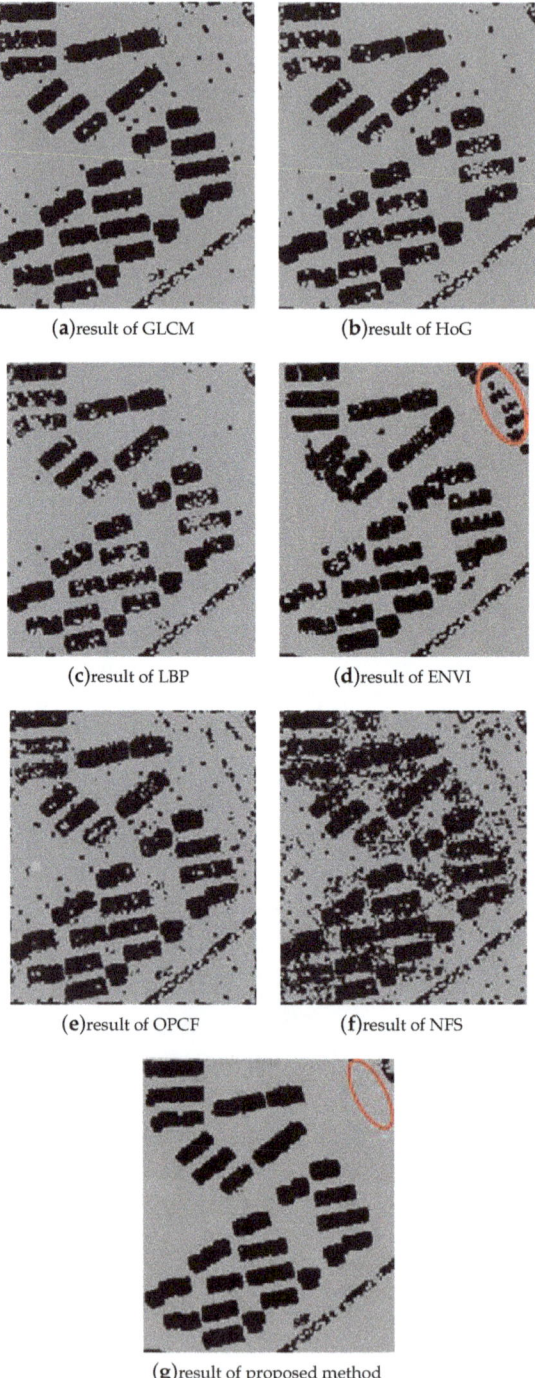

Figure 12. Building extraction results of HDR 1.

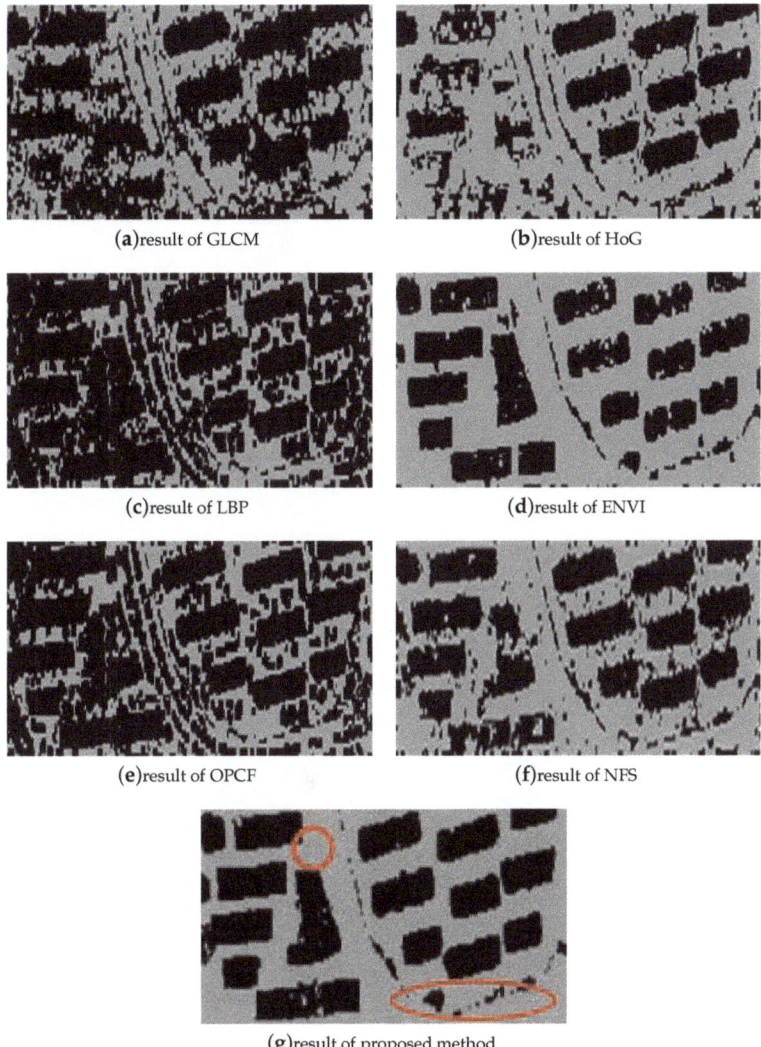

Figure 13. Building extraction results of HDR 2.

As shown in Figures 9–13, the experimental results of the proposed method were superior to other texture extraction methods, such as GLCM, HoG, and LBP, as well as NFS, OPCF, and ENVI software in the five experimental areas, as it generated a lower number of errors in the building's interior area. In addition, the proposed method preserved shape better and the interior integrity of the building. HoG and NFS were incapable of extracting complete buildings in LDR1, and the proposed method was better at preserving the integrity of the large complex building on the top left corner than LBP. For LDR2, GLCM and LBP were unable to be applied for building extraction, and the proposed method produced more correct results obviously, especially in the red circle in Figure 10g, than other methods. For MDR, only ENVI software and the proposed method extracted the complete building in the red circle in Figure 11g. However, less points were extracted as building points by the proposed method in other non-building areas than ENVI software. For HDR1, all of the methods, except NFS, obtained good results in most of the test area. However, more non-building points were obviously extracted as building points in the area of the red circle, which is shown in Figure 12d, and there were also some discrete errors in the non-building areas for other methods, while the proposed method obtained better extraction results, as Figure 12g shows. Furthermore, more buildings were extracted correctly by GLCM, LBP, and OPCF than other methods in HDR2, and for the proposed method, this was less in the red circle areas in Figure 13g.

5. Conclusions

This paper presented a building extraction method based on the fusion of point cloud and texture features, by calculating the feature values, elevation, and density of the point cloud and transforming the point cloud into an elevation map. The Gabor filter was used to extract texture features based on the elevation map, and the features could be assigned to the point cloud again. Then, point cloud and texture features were fused, and feature selection was done to realize more accurate and efficient building extraction. The experiments showed that the fusion of point cloud and texture features was able to obtain higher extraction accuracy than other methods. Besides, because of the large number of features, PSO was used to select a better feature combination to realize building extraction from the point cloud. Compared with the results from other building extraction methods, as well as NFS, OPCF, and ENVI software, the extraction accuracy by using the proposed method could satisfy practical applications preferably. In summary, the proposed method was proven to be efficient and valid for building extraction, with satisfactory extraction accuracy, which always exceeded 87%. It could provide a convenient and effective way to extract buildings in urban areas. On the basis of this work, future work will be performed on the optimization of the texture feature extraction method in the entire data-processing process.

Author Contributions: Conceptualization, X.L. and M.W.; Methodology, J.Y.; Software, J.Y. and M.W.; Validation, X.L., M.W., J.Y. and Y.L.; Formal analysis, X.L. and M.W.; Investigation, Y.L. and J.Y.; Resources, X.L. and M.W.; Data curation, X.L.; Writing—original draft preparation, X.L. and J.Y.; Writing—review and editing, X.L., M.W., J.Y. and Y.L.; Visualization, J.Y.; Supervision, X.L.; Project administration, X.L.; Funding acquisition, X.L.

Funding: This work was funded by the National Key Research & Development Program of China under Grant No. 41771368, the Key Laboratory for National Geographic Census and Monitoring, National Administration of Surveying, Mapping and Geoinformation under Grant No. 2018NGCM06, the Technical Research Service of Airborne LiDAR Data Acquisition and Digital Elevation Model Updating Project in Guangdong Province under Grant No. 0612-1841D0330175, and the Airborne LiDAR Data Acquisition and Digital Elevation Model Updating in Guangdong Province under Grant No. GPCGD173109FG317F.

Conflicts of Interest: The authors declare no conflict of interest.

References

1. Li, X.; Ling, F.; Foody, G.M.; Du, Y. A Superresolution Land-Cover Change Detection Method Using Remotely Sensed Images with Different Spatial Resolutions. *IEEE Trans. Geosci. Remote Sens.* **2016**, *54*, 3822–3841. [CrossRef]
2. Zhang, L.; Zhang, L.; Du, B. Deep Learning for Remote Sensing Data: A Technical Tutorial on the State of the Art. *IEEE Geosci. Remote Sens. Mag.* **2016**, *4*, 22–40. [CrossRef]
3. Panagiotidis, D.; Abdollahnejad, A.; Chiteculo, V. Determining Tree Height and Crown Diameter from High-resolution UAV Imagery. *Int. J. Remote Sens.* **2017**, *38*, 2392–2410. [CrossRef]
4. Marullo, S.; Patsaeva, S.; Fiorani, L. Remote sensing of the coastal zone of the European seas. *Int. J. Remote Sens.* **2018**, *39*, 9313–9316. [CrossRef]
5. Zhang, J.; Chen, L.; Wang, C.; Zhuo, L.; Tian, Q.; Liang, X. Road Recognition From Remote Sensing Imagery Using Incremental Learning. *IEEE Trans. Intell. Transp. Syst.* **2017**, *99*, 1–13. [CrossRef]
6. Huang, X.; Yuan, W.; Li, J.; Zhang, L. A New Building Extraction Postprocessing Framework for High-Spatial-Resolution Remote-Sensing Imagery. *IEEE J. Sel. Top. Appl. Earth Obs. Remote Sens.* **2017**, *10*, 654–668. [CrossRef]
7. Li, E.; Xu, S.; Meng, W.; Zhang, X. Building Extraction from Remotely Sensed Images by Integrating Saliency Cue. *IEEE J. Sel. Top. Appl. Earth Obs. Remote Sens.* **2017**, *10*, 906–919. [CrossRef]

8. Magnússon, E.; Belart, J.; Pálsson, F.; Ágústsson, H.; Crochet, P. Geodetic Mass Balance Record with Rigorous Uncertainty Estimates Deducedfrom Aerial Photographs and LiDAR Data–Case Study from Drangajökull Icecap, NW Iceland. *Cryosphere* **2016**, *9*, 4733–4785. [CrossRef]
9. Höfler, V.; Wessollek, C.; Karrasch, P. Knowledge-Based Modelling of Historical Surfaces Using LiDAR Data. *Earth Resour. Environ. Remote Sens./GIS Appl. VII* **2016**, 1–11. [CrossRef]
10. Harmon, J.M.; Leone, M.P.; Prince, S.D.; Snyder, M. LiDAR for Archaeological Landscape Analysis: A Case Study of Two Eighteenth-Century Maryland Plantation Sites. *Am. Antiq.* **2017**, *71*, 649–670. [CrossRef]
11. Baek, N.; Shin, W.S.; Kim, K.J. Geometric primitive extraction from LiDAR-scanned point clouds. *Clust. Comput.* **2017**, *20*, 741–748. [CrossRef]
12. Rozsa, Z.; Sziranyi, T. Obstacle Prediction for Automated Guided Vehicles Based on Point Clouds Measured by a Tilted LiDAR Sensor. *IEEE Trans. Intell. Transp. Syst.* **2018**, *99*, 1–13. [CrossRef]
13. Zheng, Y.; Weng, Q.; Zheng, Y. A Hybrid Approach for Three-Dimensional Building Reconstruction in Indianapolis from LiDAR Data. *Remote Sens.* **2017**, *9*, 310. [CrossRef]
14. Wang, Y.; Cheng, L.; Chen, Y.; Wu, Y.; Li, M. Building Point Detection from Vehicle-Borne LiDAR Data Based on Voxel Group and Horizontal Hollow Analysis. *Remote Sens.* **2016**, *8*, 419. [CrossRef]
15. Qin, Y.; Li, S.; Vu, T.T.; Niu, Z.; Ban, Y. Synergistic Application of Geometric and Radiometric Features of LiDAR Data for Urban Land Cover Mapping. *Opt. Express* **2015**, *23*, 13761–13775. [CrossRef] [PubMed]
16. Zhao, Z.; Duan, Y.; Zhang, Y.; Cao, R. Extracting Buildings from and Regularizing Boundaries in Airborne liDAR Data Using Connected Operators. *Int. J. Remote Sens.* **2016**, *37*, 889–912. [CrossRef]
17. Huang, R.; Yang, B.; Liang, F.; Dai, W.; Li, J.; Tian, M.; Xu, W. A top-down Strategy for Buildings Extraction from Complex Urban Scenes Using Airborne LiDAR Point Clouds. *Infrared Phys. Technol.* **2018**, *92*, 203–218. [CrossRef]
18. Yi, C.; Zhang, Y.; Wu, Q.; Xu, Y.; Remil, O.; Wei, M.; Wang, J. Urban Building Reconstruction from Raw LiDAR Point Data. *Comput.-Aided Des.* **2017**, *93*, 1–14. [CrossRef]
19. Siddiqui, F.; Teng, S.; Awrangjeb, M.; Lu, G. A Robust Gradient Based Method for Building Extraction from LiDAR and Photogrammetric Imagery. *Sensors* **2016**, *16*, 1110. [CrossRef]
20. Liu, C.; Wang, X.; Huang, H.; Gong, P.; Wu, D.; Jiang, J. The Importance of Data Type, Laser Spot Density and Modelling Method for Vegetation Height Mapping in Continental China. *Int. J. Remote Sens.* **2016**, *37*, 6127–6148. [CrossRef]
21. Kang, H.; Sim, Y.; Han, J. Terrain Rendering with Unlimited Detail and Resolution. *Graph. Models* **2018**, *97*, 64–79. [CrossRef]
22. He, Y.; Chen, L.; Chen, J.; Li, M. A Novel Way to Organize 3D LiDAR Point Cloud as 2D Depth Map Height Map and Surface Normal Map. In Proceedings of the IEEE International Conference on Robotics and Biomimetics (ROBIO), Zhuhai, China, 6–9 December 2015; pp. 1383–1388.
23. Konstantinidis, D.; Stathaki, T.; Argyriou, V.; Grammalidis, N. Building Detection Using Enhanced HoG–LBP Features and Region Refinement Processes. *IEEE J. Sel. Top. Appl. Earth Obs. Remote Sens.* **2017**, *10*, 888–905. [CrossRef]
24. Cao, J.; Pang, Y.; Li, X. Learning Multilayer Channel Features for Pedestrian Detection. *IEEE Trans. Image Process.* **2017**, *26*, 3210–3220. [CrossRef] [PubMed]
25. Du, S.; Zhang, Y.; Zou, Z.; Xu, S.; He, X.; Chen, S. Automatic Building Extraction from LiDAR Data Fusion of Point and Grid-based Features. *ISPRS J. Photogramm. Remote Sens.* **2017**, *130*, 294–307. [CrossRef]
26. Niemi, M.T.; Vastaranta, M.; Vauhkonen, J.; Melkas, T.; Holopainen, M. Airborne LiDAR-derived Eelevation Data in Terrain Trafficability Mapping. *Scand. J. For. Res.* **2017**, *32*, 761–773. [CrossRef]
27. Alatas, B. Sports Inspired Computational Intelligence Algorithms for Global Optimization. *Artif. Intell. Rev.* **2017**, *12*, 1–49. [CrossRef]
28. Li, C.; Wei, W.; Li, J.; Song, W. A Cloud-based Monitoring System via Face Recognition Using Gabor and CS-LBP Features. *J. Supercomput.* **2017**, *73*, 1532–1546. [CrossRef]
29. Kaggwa, F.; Ngubiri, J.; Tushabe, F. Combined Feature Level and Score Level Fusion Gabor Filter-Based Multiple Enrollment Fingerprint Recognition. *Int. Conf. Signal Process.* **2017**, 159–165. [CrossRef]
30. Kim, J.; Um, S.; Min, D. Fast 2D Complex Gabor Filter with Kernel Decomposition. *IEEE Trans. Image Process.* **2018**, *27*, 1713–1722. [CrossRef]
31. Luan, S.; Chen, C.; Zhang, B.; Han, J.; Liu, J. Gabor Convolutional Networks. *IEEE Trans. Image Process.* **2017**, *99*, 4357–4366.

32. Karanam, S.; Gou, M.; Wu, Z.; Rates-Borras, A.; Camps, O.; Radke, R.J. A Systematic Evaluation and Benchmark for Person Re-Identification: Features, Metrics, and Datasets. *IEEE Trans. Pattern Anal. Mach. Intell.* **2019**, *31*, 523–536. [CrossRef] [PubMed]
33. Thanou, D.; Chou, P.; Frossard, P. Graph-Based Compression of Dynamic 3D Point Cloud Sequences. *IEEE Trans. Image Process.* **2016**, *25*, 1765–1778. [CrossRef] [PubMed]
34. Song, W.; Lei, Y.; Chen, S.; Pan, Z.; Yang, J.J.; Pan, H.; Du, X.; Cai, W.; Wang, Q. Multiple Facial Image Features-based Recognition for The Automatic Diagnosis of Turner Syndrome. *Comput. Ind.* **2018**, *100*, 85–95. [CrossRef]
35. Meng, F.; Wang, X.; Shao, F.; Wang, D.; Hua, X. Energy-Efficient Gabor Kernels in Neural Networks with Genetic Algorithm Training Method. *Electronics* **2019**, *8*, 105. [CrossRef]
36. Lei, H.; Jiang, G.; Quan, L. Fast Descriptors and Correspondence Propagation for Robust Global Point Cloud Registration. *IEEE Trans. Image Process.* **2017**, *26*, 3614–3623. [CrossRef]
37. Fu, Y.; Chiang, H.D. Toward Optimal Multiperiod Network Reconfiguration for Increasing the Hosting Capacity of Distribution Networks. *IEEE Trans. Power Deliv.* **2018**, *33*, 2294–2304. [CrossRef]
38. Yang, J.; Cao, Z.; Qian, Z. A Fast and Robust Local Fescriptor for 3D Point Cloud Registration. *Inf. Sci.* **2016**, *346*, 163–179. [CrossRef]
39. Hasanipanah, M.; Amnieh, H.B.; Arab, H.; Zamzam, M.S. Feasibility of PSO–ANFIS model to Estimate Rock Fragmentation Produced by Mine Blasting. *Neural Comput. Appl.* **2018**, *30*, 1015–1024. [CrossRef]
40. Lin, J.C.W.; Yang, L.; Fournier-Viger, P.; Hong, T.P.; Voznak, M. A Binary PSO Approach to Mine High-utility Itemsets. *Soft Comput.* **2017**, *21*, 5103–5121. [CrossRef]
41. Wang, M.; Wu, C.; Wang, L.; Xiang, D.; Huang, X. A feature selection approach for hyperspectral image based on modified ant lion optimizer. *Knowl.-Based Syst.* **2019**, *168*, 39–48. [CrossRef]
42. Phan, A.; Nguyen, M.; Bui, L. Feature weighting and SVM parameters optimization based on genetic algorithms for classification problems. *Appl. Intell.* **2016**, *46*, 455–469. [CrossRef]
43. Wan, Y.; Wang, M.; Ye, Z.; Lai, X. A "Tuned" Mask Learnt Approach Based on Gravitational Search Algorithm. *Comput. Intell. Neurosci.* **2016**, *2016*, 1–16. [CrossRef] [PubMed]

© 2019 by the authors. Licensee MDPI, Basel, Switzerland. This article is an open access article distributed under the terms and conditions of the Creative Commons Attribution (CC BY) license (http://creativecommons.org/licenses/by/4.0/).

Article

The Comparison of Fusion Methods for HSRRSI Considering the Effectiveness of Land Cover (Features) Object Recognition Based on Deep Learning

Shiran Song [1], Jianhua Liu [1,2,*], Heng Pu [1], Yuan Liu [1] and Jingyan Luo [1]

1. School of Geomatics and Urban Spatial Information, Beijing University of Civil Engineering and Architecture, Beijing 100044, China; songshiran@stu.bucea.edu.cn (S.S.); 2108521516018@stu.bucea.edu.cn (H.P.); 2108160218007@stu.bucea.edu.cn (Y.L.); 2108521518006@stu.bucea.edu.cn (J.L.)
2. Key Laboratory for Urban Geomatics of National Administration of Surveying, Mapping and Geoinformation, Beijing 100044, China
* Correspondence: liujianhua@bucea.edu.cn; Tel.: +86-010-6832-2377

Received: 21 May 2019; Accepted: 13 June 2019; Published: 17 June 2019

Abstract: The efficient and accurate application of deep learning in the remote sensing field largely depends on the pre-processing technology of remote sensing images. Particularly, image fusion is the essential way to achieve the complementarity of the panchromatic band and multispectral bands in high spatial resolution remote sensing images. In this paper, we not only pay attention to the visual effect of fused images, but also focus on the subsequent application effectiveness of information extraction and feature recognition based on fused images. Based on the WorldView-3 images of Tongzhou District of Beijing, we apply the fusion results to conduct the experiments of object recognition of typical urban features based on deep learning. Furthermore, we perform a quantitative analysis for the existing pixel-based mainstream fusion methods of IHS (Intensity-Hue Saturation), PCS (Principal Component Substitution), GS (Gram Schmidt), ELS (Ehlers), HPF (High-Pass Filtering), and HCS (Hyper spherical Color Space) from the perspectives of spectrum, geometric features, and recognition accuracy. The results show that there are apparent differences in visual effect and quantitative index among different fusion methods, and the PCS fusion method has the most satisfying comprehensive effectiveness in the object recognition of land cover (features) based on deep learning.

Keywords: image fusion; high spatial resolution remotely sensed imagery; object recognition; deep learning; method comparison

1. Introduction

With the development of earth observation technology, a large number of remote sensing satellites have been launched, which further improves the acquisition ability of high spatial resolution and high spectral resolution imagery, and provides extensive data sources for applications [1]. Object recognition of urban typical land features from High Spatial Resolution Remote Sensing Imagery (HSRRSI) is an active and important research task driven by many practical applications. Traditional methods are based on hand-crafted or shallow-learning-based features with limited representation power [2].

In recent years, the application of deep learning in the field of remote sensing has become more and more extensive, and its progress has solved many problems, especially in target detection [3], target recognition [4], and semantic segmentation [5], which has taken the current research to a new height. High spatial resolution remotely sensed imagery often contains multiple types of land-cover with distinct spatial, spectral, and geometric characteristics, and the manual labeling sample is not enough, which limits the applications of deep learning in object recognition from HSRRSI [6].

Over the last decades, a number of relevant methods have been proposed by combining the spatial and the spectral information to extract spatial–spectral features [7–19]. In a recent study, Cheng propose a unified metric learning-based framework to alternately learn discriminative spectral-spatial features; they further designed a new objective function that explicitly embeds a metric learning regularization term into SVM (Support Vector Machine) training, which is used to learn a powerful spatial–spectral feature representation by fusing spectral features and deep spatial features, and achieved state-of-the-art results [20]. It is now commonly accepted that spatial–spectral-based methods can significantly improve the classification performance, which also reflects the importance of spatial and spectral features of image data in application-level in deep learning. However, the number of labeled samples in HSRRSI is quite limited because of the high expense of manually labeling, and even the available labels are not always reliable. Making full use of HSRRSI to produce high-quality training data will be a challenge.

The Worldview-3 images used in this study are composed of the panchromatic band and multispectral bands. The former has high spatial resolution and the latter has high spectral resolution. The questions remains of how to effectively utilize these remote sensing image data and take them as a whole to the greatest extent for comprehensive analysis and application. Spatial–spectral fusion can solve the constraints between spatial resolution and spectral resolution. In the processing stage, remote sensing images with different spatial and spectral resolutions in the same region are fused to obtain images with both high spatial resolution and high spectral resolution. Panchromatic-multispectral fusion is the most classical method of spatial–spectral fusion and the first choice for various applications. The fusion technology originated in the 1980s [21–23]. Since the SPOT-1 satellite system first provided panchromatic and multispectral images simultaneously in 1986, panchromatic-multispectral fusion technology has developed rapidly; a lot of methods have been proposed [24–26].

In general, the existing mainstream panchromatic and multispectral fusion methods can be divided into four categories: component substitution-based fusion [27,28], multi-resolution analysis-based fusion [29,30], model optimization-based fusion [31,32], and sparse expression-based fusion methods [33]. Although there are many existing fusion methods, it is still challenging to find a suitable image fusion method for specific data sources and specific application scenarios.

In this study, six traditional spatial–spectral fusion methods are selected for panchromatic and multispectral bands in the study area to generate remote sensing images with both high spatial resolution and high spectral resolution. Then, we apply image fusion results to conduct the experiments of land cover (features) object recognition for remote sensing images based on Mask R-CNN [34]. The experimental results demonstrate the effectiveness of the proposed method and reveal the potential application of image fusion technology in target recognition and feature-oriented primitive processing, analysis, and understanding. By comparing the recognition results of different fusion methods, we obtain a fusion image that is more suitable for network generalization ability. It also verifies that a fusion image with high spatial resolution and high spectral resolution achieves better recognition effect.

2. Methodology

2.1. Image Fusion Methods

In order to improve the quality of remote sensing image data, such as resolution, contrast, integrity, and other indicators, various fusion methods have been developed. The common methods are IHS (Intensity-Hue Saturation), PCS (Principal Component Substitution), ELS (Ehlers) [35,36], GS (Gram Schmidt), HPF (High-Pass Filtering) [21], and HCS (Hyper spherical Color Space). In this study, we use these six methods to evaluate the adaptability of six fusion methods to high-resolution imagery and their effectiveness of land cover (features) object recognition based on deep learning.

IHS transformation can effectively separate spatial (intensity) and spectral (hue and saturation) information from a standard Red-Green-Blue (RGB) image [37]. First, the IHS method transforms an RGB image into the IHS image space. The IHS color space is represented by Intensity (I), Hue (H), and Saturation (S). The effect of this representation of remote sensing images aligns better with human visual habits, making the image objects look more similar to the color changes of real objects, and closer to the human perception mechanism of color. Next, the intensity component (I) is replaced by the panchromatic image. Then, an inverse IHS transformation is performed to obtain a fused image that has high spatial resolution and hyperspectral resolution. The specific process is shown in Figure 1. The fused image obtained by transformation, substitution, and inverse transformation not only has the advantage of high resolution of panchromatic image, but also maintains the hue and saturation of the multispectral image. These characteristics will be beneficial to the subsequent deep learning models to capture the fine features of the complex land-use images used for generalization.

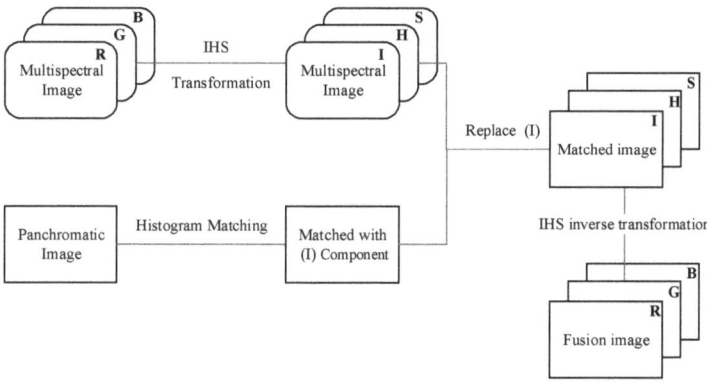

Figure 1. IHS (Intensity-Hue Saturation) fusion flow chart.

Compared with IHS transformation, the Principal Component Substitution (PCS) technique uses Principal Component Transformation (PCT), with which number of input bands is not limited to three. PCS is a multi-dimensional linear transformation fusion based on image statistical features, which can concentrate variance information and compress data. When using the PCT in image fusion, the first component of the low spatial resolution images is replaced by the high spatial resolution images. The fused images are obtained by applying an inverse PCT on the new set of components [38,39]. This fusion method has a wide range of applications, and the image after inverse transformation of principal components is clearer and richer. It can more accurately reveal the internal structure of multi band remote sensing information data, thereby reducing the difficulty and complexity of the subsequent deep learning feature extraction model. The specific process is shown in Figure 2.

GS fusion method mainly uses Gram–Schmidt Transformation in mathematics, which can effectively eliminate the correlation between multi-spectral bands. It is similar to the Principal Component Transform (PCA) method and is commonly used in mathematics. GS fusion firstly obtains a low-resolution panchromatic image from a multi-spectral image and uses the image as the first band of multi-spectral image to recombine with the original multi-spectral image. Then, GS transformation is applied to the reconstructed multi-band image. Equation (1) is the concrete equation of the GS transformation. The panchromatic image is used to replace the first band of the image after GS transformation. Lastly, the fused image is obtained by GS inverse transformation. GS transformation gives the image higher contrast, and can better maintain the spectral information of the original image with less information distortion. This algorithm can weaken the correlation between multispectral bands, thus reducing information redundancy, highlighting more useful or discriminative information in the data itself, thus increasing the effectiveness of land cover (features) object recognition based on deep learning.

$$\left. \begin{array}{rcl} GS_T(i,j) &=& (B_T(i,j) - \mu_T) - \sum_{i=1}^{T-1} (\varnothing(B_T, GS_i) \times GS_i(i,j)) \\ \mu_T &=& \frac{\sum_{j=1}^{N} \sum_{i=1}^{M} B_T(i,j)}{M \times N} \\ \varnothing(B_T, GS_i) &=& \left[\frac{\delta(B_T, GS_i)}{\delta(GS_i, GS_i)^2} \right] \end{array} \right\} \quad (1)$$

In Equation (1), GS_T denotes the T orthogonal component after GS transformation; B_T represents the T band of the original low spatial resolution remote sensing image; M and N represent the total number of rows and columns of the image; i and j represent the rows and columns of the original low spatial resolution image, respectively; μ_T is the mean of the gray value of pixels in the T band of the original low spatial resolution remote sensing image; $\varnothing(B_T, GS_i)$ is the covariance between the T band of the original low spatial resolution image and GS_i.

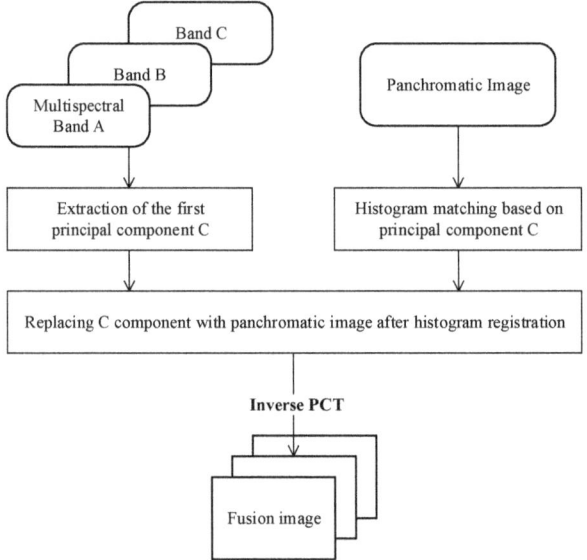

Figure 2. Principal Component Transformation (PCT) fusion flow chart.

The difference between GS and PCA is that the information contained in the PCA is mainly in the first component with the most information, and its information decreases in turn in the remaining color components, while the components transformed by Gram-Schmidt are only orthogonal, and the amount of information contained in each component is not significantly different [40]. Therefore, the GS transform can preserve the spectral information of original multispectral images and the spatial texture features of panchromatic images to the greatest extent, so as to solve the problem of excessive concentration of the first component in PCT. However, the GS transformation is relatively complex and unsuitable for large-scale image fusion.

Both IHS and PCS fusion methods pertain to component substitution methods, which are remarkable image fusion techniques that are able to meet user's needs in most application scenarios. Insufficiently, the computational complexity of these methods is too high to merge massive volumes of data from new satellite images quickly and effectively. For that reason, many research studies have been carried out to develop an advanced image fusion method with a fast computing capability and to preserve the high spatial and spectral quality [41]. One of the improved standard data fusion techniques is the Ehlers fusion method. The Ehlers fusion algorithm was founded by Professor

Manfred Ehlers of the University of Osnabluk. The basic principle of the Ehlers fusion algorithm is to sharpen the panchromatic band by using Fast Fourier Transform (FFT filtering), and then use the IHS transform for image fusion. The advantage of this algorithm is that it provides three preset filtering models for different regional images, which are urban, suburban, and suburban mixed areas, respectively. This feature preserves the same spatial characteristics of the fused image as the original image. The specific process of the ELS fusion method are shown in Figure 3.

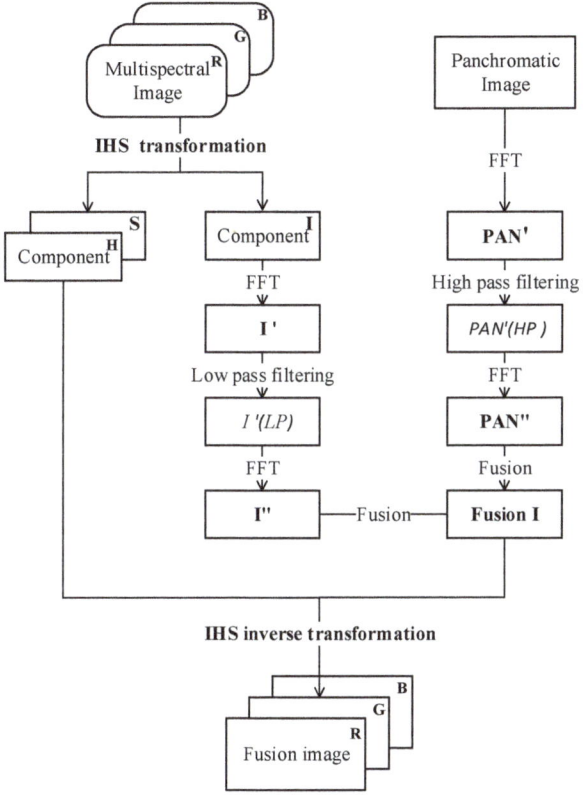

Figure 3. The Ehler transformation flow chart.

High-Pass Filtering is often used in image texture and detailed processing to improve the high-frequency details of images, and highlight the linear features and edge information of images. For a remote sensing image, the spectral information of the image is included in the low-frequency part, and the details, edges, and textures of the image are included in the high-frequency part. The basic principle of High-Pass Filtering (HPF) fusion is to extract the high-frequency part of the image and apply it to the low-resolution image (multi-spectral image) to form the high-frequency feature prominent fusion image [42], which can improve the application accuracy of image target recognition. Equation (1) is the concrete equation of the GS transformation.

$$HP_i = (W_a \times MSI_{iLP}) + (W_b \times PAN_{HP}), \qquad (2)$$

In Equation (2), W_a and W_b are weighted, respectively, and added to 1; MSI_{iLP} represents the result of low-pass filtering of low-resolution image (multi-spectral image) in band i, PAN_{HP} represents the result of high-pass filtering of high-resolution image (panchromatic image), and HP_i represents the fusion image of original multi-spectral image in band *i* after the above processing. This method not only effectively reduces the low-frequency noise in high-resolution images, but also can be used in all multi-spectral bands after filtering.

Hyper spherical Color Sharpening (HCS) is a fusion method suitable for multi-band images. For a remote sensing image with N bands, it is shown as a strength I component and N-1 angle component on the hypersphere. The mathematical function of the fusion of HCS method is shown in Equation (3), in which x_i is the *i* component of the original color space.

$$\left.\begin{array}{rl} I &= \sqrt{x_1^2 + x_2^2 + x_3^2 + \ldots x_n^2} \\ \varphi_1 &= arctan(\frac{\sqrt{x_n^2 + x_{n-1}^2 + x_{n-2}^2 + \ldots x_2^2}}{x_1}) \\ \varphi_{n-2} &= arctan(\frac{\sqrt{x_n^2 + x_{n-1}^2}}{x_{n-2}}) \\ \varphi_n &= arctan(\frac{x_n}{x_{n-1}}) \end{array}\right\} \quad (3)$$

This method converts multi-band remote sensing data into the hyper spherical color space by constructing models and simulating the panchromatic band I component, so as to obtain a sharpened panchromatic band I component, and finally to reverse the data to obtain the fusion image. The fusion image of this method highlights the edge contour of the object, improves the utilization rate of image information, improves the accuracy and reliability of subsequent computer interpretation, and is conducive to feature extraction and classification recognition of a subsequent deep learning model.

The aforementioned six methods extract the most valuable information from the original image according to their own characteristics and fuse it into high-quality images, which can improve the spatial resolution and spectral resolution of the original image. In the subsequent experiments, for the optimal fusion results, we demonstrate the effectiveness and practicability of the image in land cover (features) object recognition based on deep learning.

2.2. Object Recognition Method based Mask R-CNN

2.2.1. Mask R-CNN Network Architecture

In recent years, deep learning has been applied to remote sensing measurements in replacement of the empirical feature design process by automatically learning multi-level representations [43]. Since 2012, Convolutional Neural Networks (CNN) have been widely used in image classification. Novel CNN structures, such as AlexNet [4], VGGNet [44], GoogLeNet [45], and ResNet [46], have been shown to be remarkable.

Since 2015, a special CNN structure, known as region-based models, which detects objects by predicting a bounding box of each object, have been developed for pixel-wise semantic segmentation and object detection [43]. For example, these region-based models include R-CNN [47], Fast R-CNN [48], Faster R-CNN [49], and Mask R-CNN [34]. In this paper, the most representative Mask R-CNN in the current field is used as the basic model.

Mask R-CNN is a two-stage framework (Figure 4). The first stage scans the image and generates proposals. The second stage classifies proposals and generates bounding boxes and masks. The backbone network of Mask R-CNN is ResNet101 and FPN (Feature Pyramid Networks). ResNet is the champion of the classification task of the ImageNet competition in 2015, which can increase the network depth to hundreds of layers and has excellent performance. FPN utilizes the feature pyramid generated by ResNet to fully fuse the high-resolution and high-semantic information of low-level features of the image and generates feature maps of different scales into RPN (Region Proposal Networks) and ROI (Region of Interest) Align layers by top-down up sampling and horizontal connection processes. RPN is a lightweight neural network that scans images with sliding windows and searches for areas where objects exist. ROI Align is a regional feature aggregation method proposed by Mask-RCNN, which solves the problem of misalignment caused by two quantifications in ROI Pooling operation. After using

RoIAlign, the accuracy of the mask is improved from 10% to 50%. Another breakthrough of the model is the introduction of semantic segmentation branch, which realizes the decoupling of the relationship between mask and class prediction. The mask branch only does semantic segmentation, and the task of type prediction is assigned to another branch.

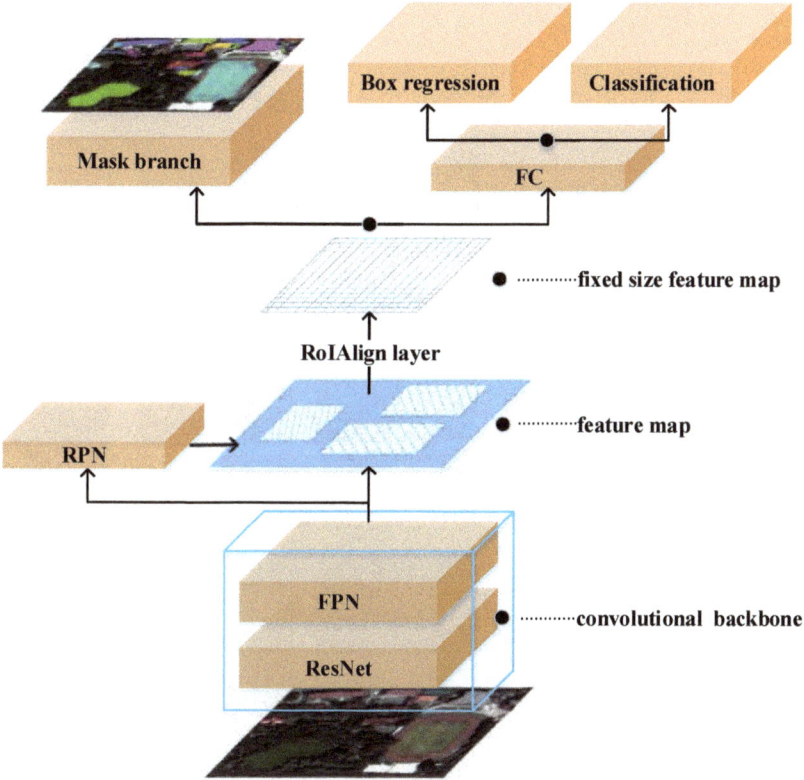

Figure 4. Mask R-CNN network architecture.

2.2.2. Network Training

As an initial experiment, we trained a Mask R-CNN model on the RSSCN7 Dataset [50] and RSDataset [51]. The pixel resolution of the above dataset is 0.3m. We selected 996 images as training data and 200 images as testing data (Table 1). Several typical landmark, including buildings and water bodies, a=were selected as marking samples (Figure 5). All target objects in training and testing data sets were labeled manually, including attributes and masking information of objects. Figure 6 is the examples of training sample masks of Figure 5.

Table 1. The strategy of training and testing division for different datasets.

Class	Dataset	Training	Samples	Testing	Samples
Building	RSSCN7	708	19,116	100	2174
	RSDataset	98	2641	25	676
Total		806	21,757	125	2850
Water	RSSCN7	100	117	50	66
	RSDataset	90	99	25	30
Total		190	216	75	96

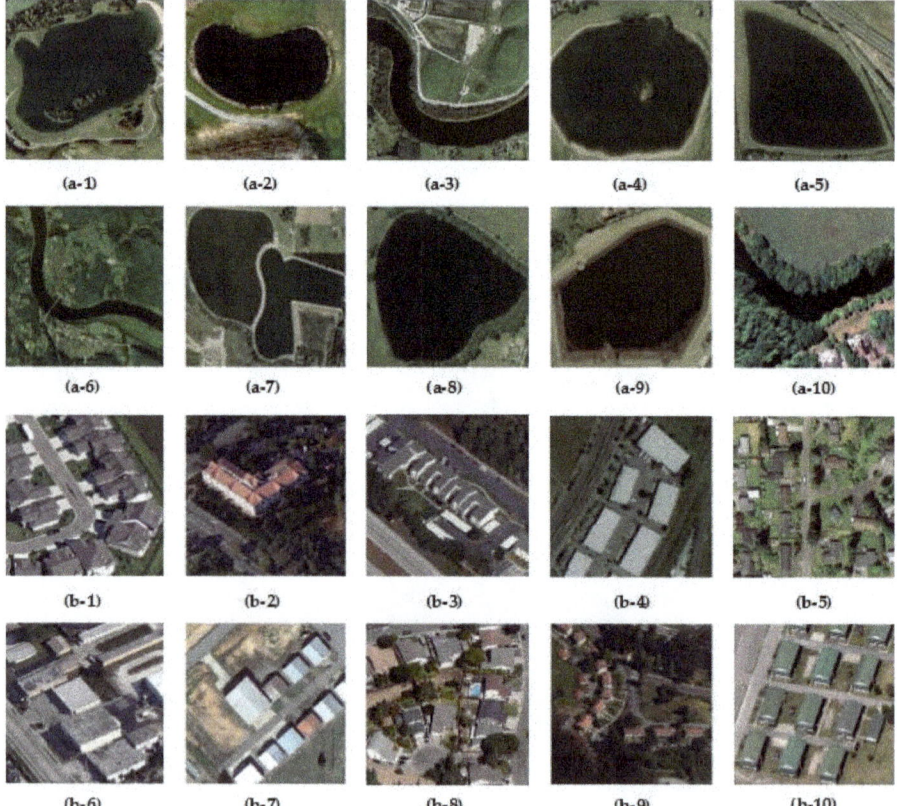

Figure 5. Examples of training samples covering water bodies (**a1–a10**) and buildings (**b1–b10**). The pixel resolution of each sample is 0.3 m.

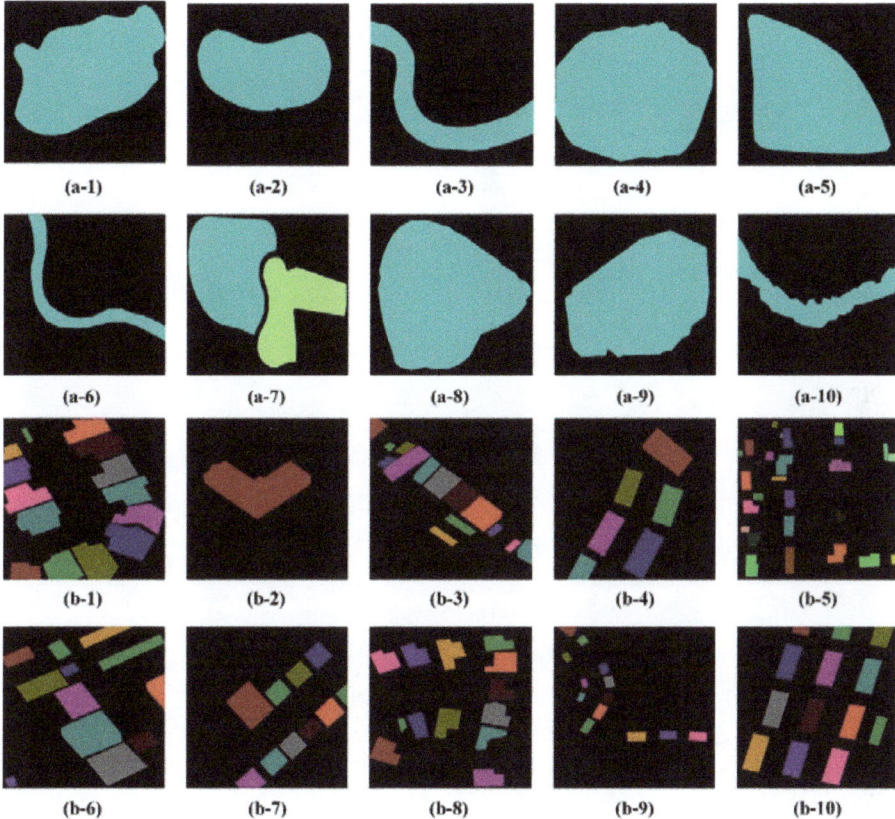

Figure 6. Examples of training sample masks from Figure 5 (color = positive; black = negative).

Based on Tensorflow, keras, and Anaconda deep learning libraries, we use NVIDIA GeForce GTX 1060 with a single GPU to train 996 training data samples (21,757 buildings, 216 water bodies) and generate a recognition model. The total training time is 58 hours. The main parameters of the Mask R-CNN model are set in Table 2. We use the trained model to recognize buildings and water bodies in 200 test data sets, including 3254 buildings and 185 water bodies. Figures 7 and 8 are examples of recognition results of buildings and water bodies, respectively. Table 3 reflects the recognition accuracy. The precision and recall rate of building recognition reached 0.8275 and 0.7828, respectively, and the precision and recall rate of water body recognition reached 0.8529 and 0.9062, respectively, which confirm the availability of the model.

Table 2. Main parameters information of model.

Parameter	Values	Parameter	Values
GPU_COUNT	1	TRAIN_ROIS_PER_IMAG	200
IMAGES_PER_GPU	1	MAX_GT_INSTANCES	200
BACKBONE	ResNet	DETECTION_MAX_INSTANCES	200
BACKBONE_STRIDES	[4, 8, 16, 32, 64]	Batch Size	1
NUM_CLASSES	3	Epochs	30
RPN_ANCHOR_SCALES	(32, 64, 128, 256, 512)	LEARNING_RATE	0.0001
RPN_ANCHOR_RATIOS	[0.5, 1, 2]	LEARNING_MOMENTUM	0.9
RPN_NMS_THRESHOLD	0.7	WEIGHT_DECAY	0.0001

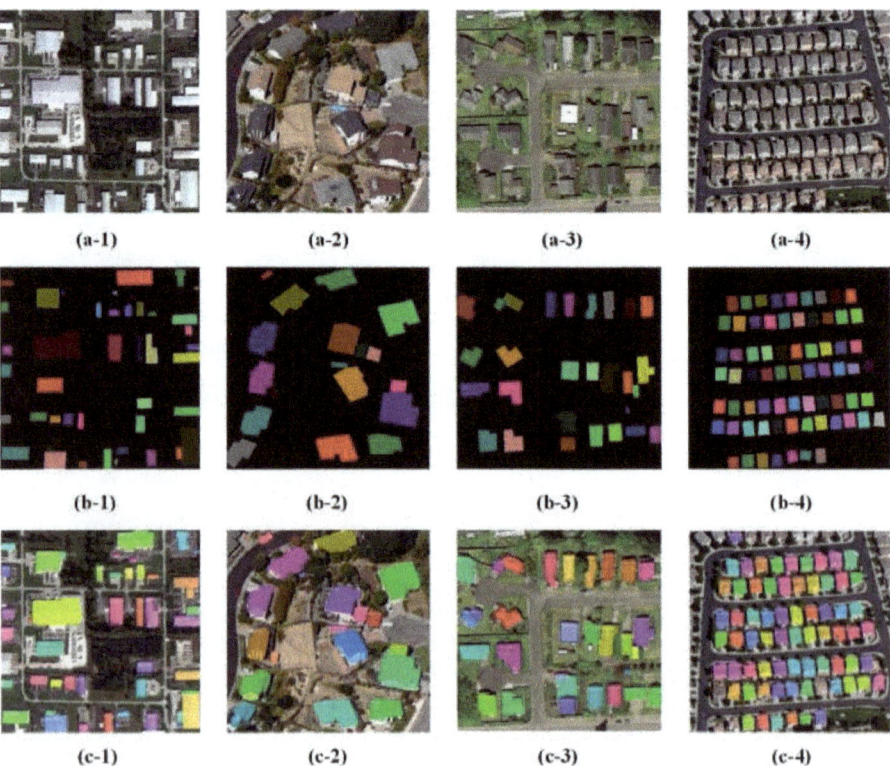

Figure 7. Examples of building recognition result for (**a1–a4**) original image. (**b1–b4**) Examples of test sample masks (color = positive; black = negative). (**c1–c4**) Building recognition result (color mask is recognition mark).

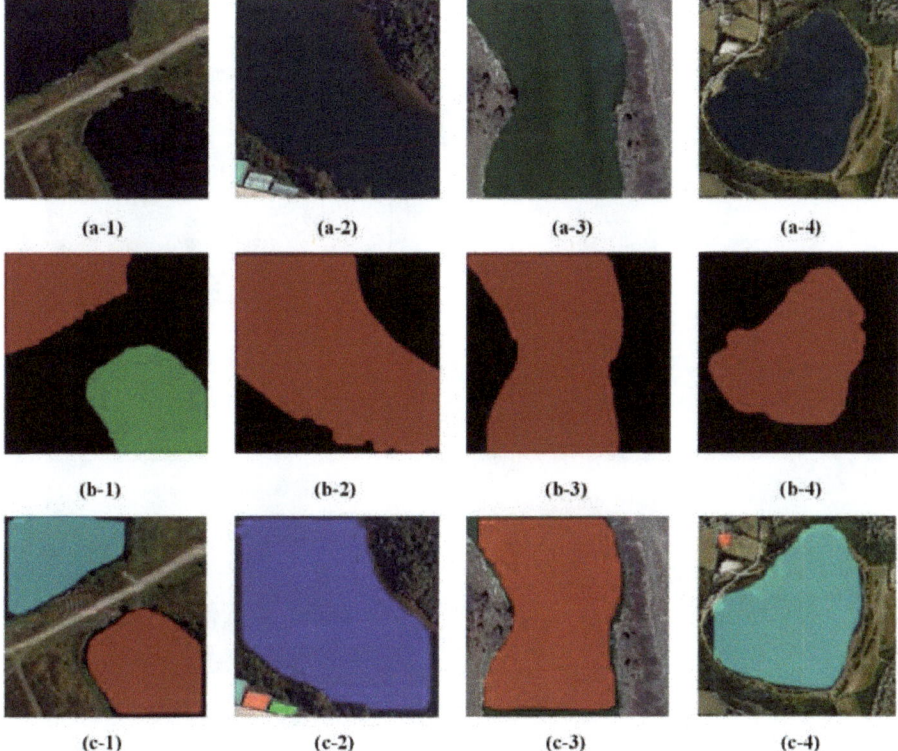

Figure 8. Examples of water recognition result for (**a1**–**a4**) original image. (**b1**–**b4**) Examples of test sample masks (color = positive; black = negative). (**c1**–**c4**) Water recognition result (color mask is recognition mark).

Table 3. Accuracy evaluation of test dataset recognition results.

Class	Actual	Detection	Matching	Precision	Recall
Buildings	2850	2696	2231	0.8275	0.7828
Water	96	102	87	0.8529	0.9062

3. Experiment

3.1. Experimental Area

This study chooses the high-resolution remote sensing image data of Tongzhou New Town in Beijing City acquired by Digital globe's WorldView-3 satellite on September 19, 2017, as the research area, including panchromatic and multispectral images (Figure 9). The longitude and latitude ranges are nwLat = 39°96′, nwLong = 116°63′, seLat = 39°84′, seLong = 116°78′. The image is geometrically corrected to ensure the effect of data fusion. Because cloud cover is only 0.004 when the image of the study area is acquired, there is no atmospheric correction operation. This area contains countryside, residential, cultural, and industrial areas. Various and versatile architecture types of Surface Coverage Elements with different color, size, and usage make it an ideal study area to evaluate the potential of a building extraction algorithm.

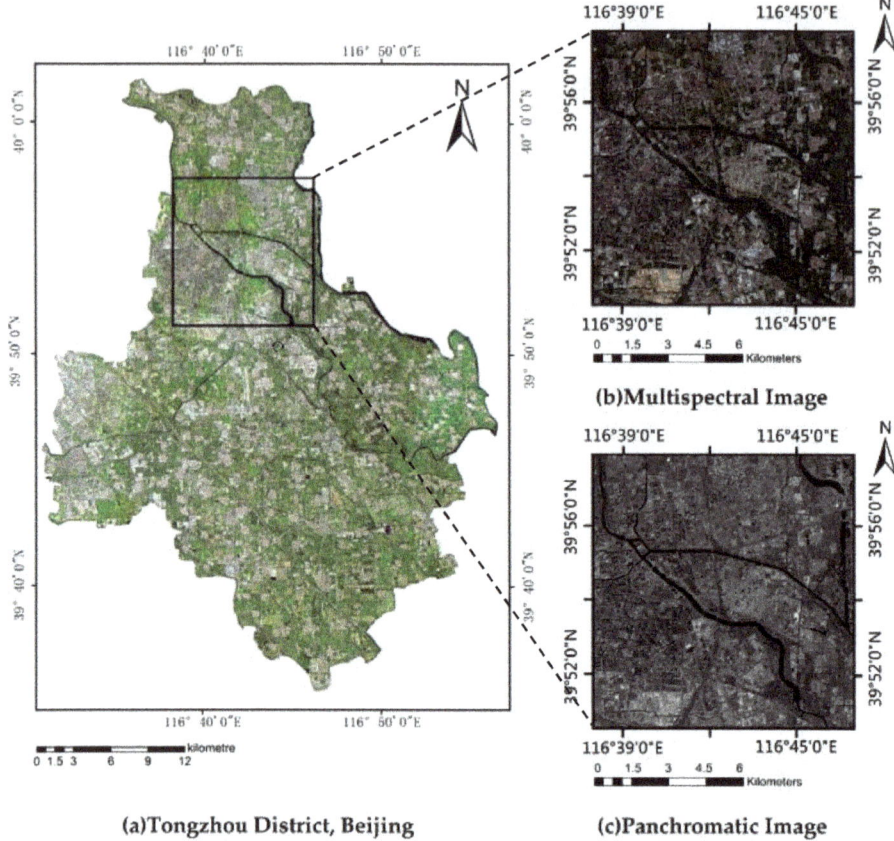

Figure 9. Study areas (**a**) and image data. (**b**) Multispectral image. (**c**) Panchromatic image, including the WorldView-3 images of Tongzhou new town.

In this study, 800 pixels × 800 pixels were selected as the experimental area (Figure 10). The Mask R-CNN algorithm used in this experiment supports three-band images. Generally, the information content of the three bands is sufficient to support the research of land features (cover) recognition. In this study, a 0.3-m panchromatic image and the R, G, and B bands of a 1.24-m multispectral image are fused by six methods. Finally, a 0.3-m true color high resolution remote sensing image is obtained, which is conducive to subsequent artificial visual interpretation and qualitative and quantitative analysis. This experimental area contains trees, houses, roads, grasslands, and water bodies. The types of land cover are diverse to ensure the effectiveness of the experiment.

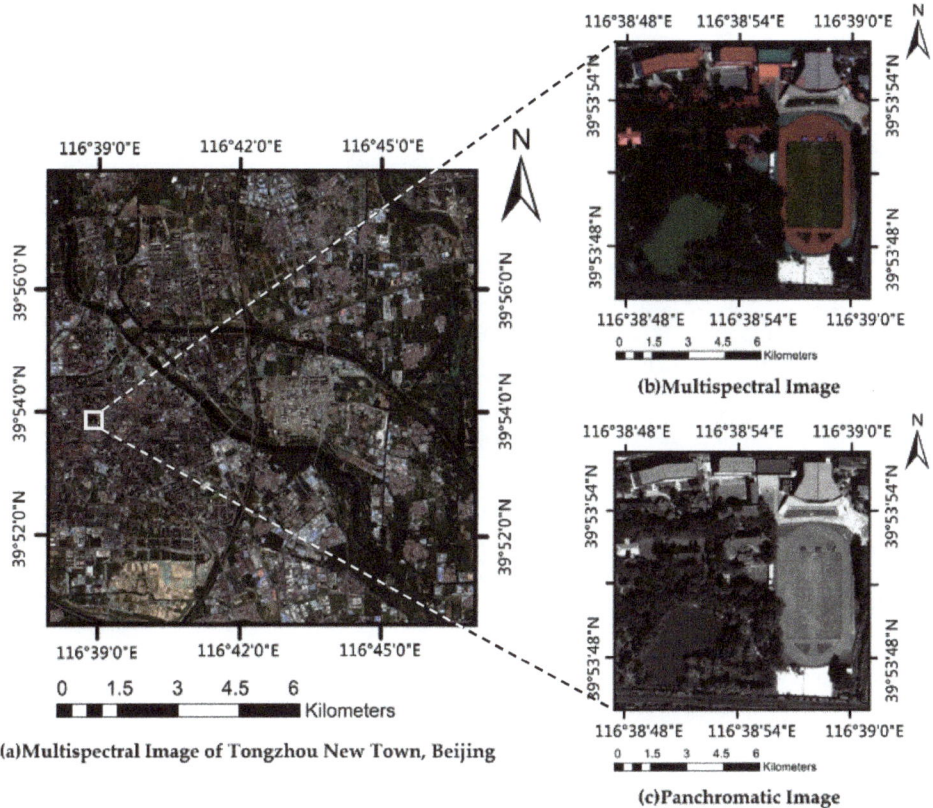

Figure 10. Image sample of experimental area-1.

3.2. Experimental Method

In this study, the model based on Mask R-CNN is used to recognize the typical urban features from the fused high-resolution remote sensing images. This paper compares and evaluates the adaptability of IHS, PCS, GS, ELS, HPF, and HCS fusion methods to the object recognition of surface coverage (elements) based on deep learning from three aspects: visual effect, quantitative analysis, and object recognition accuracy.

3.3. Result and Discussion

3.3.1. Visual Assessment

The results of six fusion methods are shown in Figure 11. By checking whether the spectral resolution of the fused image is maintained and whether the spatial resolution is enhanced, the quality of the fused image and the adaptability of the method can be evaluated on the whole.

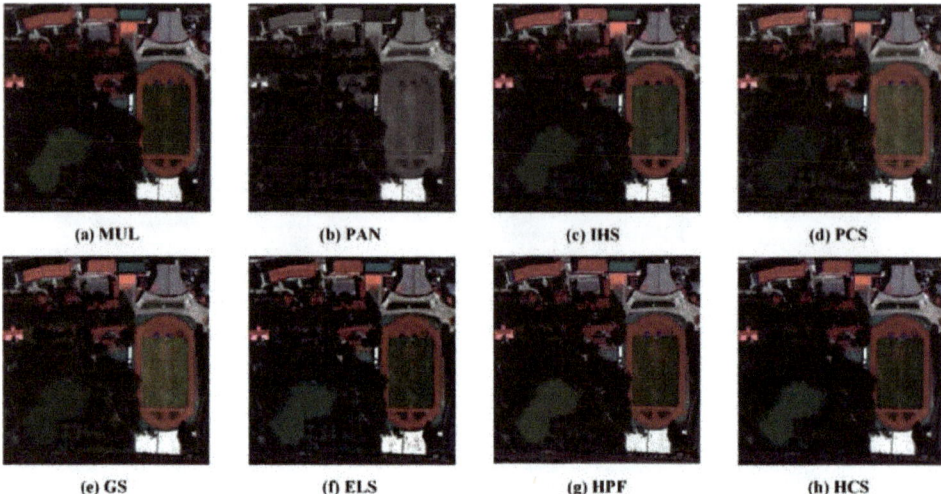

Figure 11. Comparison between the original images. (**a**) Multispectral image and (**b**) Panchromatic image and six methods (**c–h**) for the WorldView-3 multispectral and pan images. Fused images from (**c**) Intensity-Hue Saturation (IHS), (**d**) Principal Component Substitution (PCS), (**e**) Gram-Schmidt Transformation (GST), (**f**) Ehlers Fusion (ELS), (**g**) High Pass Filter Resolution Merge (HPF), and (**h**) Hyper spherical Color Space (HCS) methods.

In order to achieve a more detailed visual interpretation effect, this study fuses the images of water, vegetation, roads, playgrounds, and bare land to establish the corresponding atlas (Table 4).

Firstly, we analyze the spectral feature preservation ability of fused images. Compared with the original multispectral image, the fused image basically maintains the spectral characteristics, but there are also significant differences in local color.

The overall color of IHS, PCS, and HCS fusion images is consistent with the original multi-spectral remote sensing image, and the color contrast is moderate. The color of construction land in the GS fusion image is brighter than that of original image combination, and the color of river water changes from grass green to dark green compared with original image combination. After ELS fusion, the brightness of the fused image is darker and the contrast of the image is reduced. After HPF fusion, the edge of the fused image is obvious and the sharpening effect is outstanding.

In terms of spatial information enhancement ability, by comparing the results of the six methods with panchromatic images, we can find that the linear objects, such as roads, water bodies, playgrounds, and residential contours, can be better distinguished and the texture structure of fused images becomes clearer.

PCS, GS, and HPF fusion images have better visual effects, with prominent edges and clear textures of residential areas, roads, playgrounds, water, and woodlands, which are easy to visually interpret. Compared with the three aforementioned methods, IHS and HCS fusion images have some deficiencies in texture clarity of woodland and water bodies, and the other features are very similar in both spectral and spatial characteristics. The ELS transform has serious duplication and blurring phenomena on the boundaries of roads, playgrounds, water, and woodland. In the interlaced area of woodland and architecture, the land features appear inconsistent.

By subjective evaluation of the color, texture structure, clarity, and spatial resolution of the fused image, the quality of the fused image and the superiority of the fusion method can be evaluated as a whole. However, the subtle differences of spectral and spatial characteristics of many fusion images cannot be well distinguished by visual assessment alone. It is necessary to use objective evaluation methods to quantitatively describe the differences between different algorithms.

Table 4. Fusion effect of six fusion methods and multispectral image.

Image \ Methods	Water	Vegetation	Road	Roof	Playground Greenspace	Playground Runway
Pan						
Mul						
IHS						
PCS						
GS						
ELS						
HPF						
HCS						

3.3.2. Quantitative Assessment

This Quantitative evaluation of fusion effect is a complex problem. The objective evaluation method is to determine the statistical parameters of fused images. This method greatly improves the certainty and stability of evaluation.

In this paper, the mean value of each band of the fused image is used to reflect the overall brightness of the image. The main expression is as follows:

$$\mu = \frac{1}{M \times N} \sum_{i=1}^{M} \sum_{j=1}^{N} F(i,j), \qquad (4)$$

$F(i, j)$ is the gray value of the fused image F at the pixels (i, j), M and N are the size of image F. The higher the mean value is, the brighter the overall brightness of the image is.

Standard deviation and information entropy are used as quantitative indicators to evaluate image information richness. The standard deviation is obtained indirectly from the mean value, which indicates the degree of dispersion between the gray value and the mean value of the image pixel. The expression of the standard deviation is as follows:

$$Std = \sqrt{\frac{1}{M \times N} \sum_{i=1}^{M} \sum_{j=1}^{N} (F(i,j) - \mu)^2}, \qquad (5)$$

The information entropy of fusion image can reflect the amount of image information. Generally, the larger the entropy of the fused image is, the better the fusion quality will be. The calculation equation is as follows:

$$Ce = -\sum_{i=0}^{L} P_i \log_2 P_i, \qquad (6)$$

Ce is the information entropy; P_i is the occurrence probability of the gray value i of the pixel in the image; L is the maximum gray level of the image.

The definition of the average gradient response image is used to reflect the contrast of minute details and texture transformation features in the image simultaneously. The main expression is as follows:

$$G = \frac{1}{M \times N} \sum_{i=1}^{M} \sum_{j=1}^{N} \sqrt{((\frac{\partial F(i,j)}{\partial_i})^2 + (\frac{\partial F(i,j)}{\partial_j})^2)/2}, \quad (7)$$

Spectral distortion degree is used to evaluate the spectral distortion degree of the fused image relative to the original image. The spectral distortion expression is defined as:

$$\text{Warp} = \frac{1}{W} \sum_i \sum_j |V'_{i,j} - V_{i,j}|, \quad (8)$$

W is the total number of pixels in the image, $V'_{i,j}$ and $V_{i,j}$ are the gray values of the fused image and the original image (i, j).

The correlation coefficient with each band of the multispectral image and coefficient of correlation between each band of the fused image and panchromatic image are used as quantitative indicators to measure the spectral fidelity (i.e., the degree of preservation of the advantages of the original panchromatic band and the multispectral band in both geometric and spectral information). The related expression is as follows:

$$Cc = \frac{\sum_{i=1}^{M} \sum_{j=1}^{N} (F(i,j) - \mu_F)(A(i,j) - \mu_A)}{\sqrt{\sum_{i=1}^{M} \sum_{j=1}^{N} (F(i,j) - \mu_F)^2 (A(i,j) - \mu_A)^2}}, \quad (9)$$

where μ_F and μ_A represent the average gray level of fusion image and source image, respectively. The larger the correlation coefficient, the more information the fusion image gets from the source image, the better the fusion effect.

Table 5 shows the statistical analysis between different fusion methods. Comparing the brightness information of fused images, we find that PCS has the largest mean value in R and G bands. The values of the two bands are 322.45 and 396.63, respectively. ELS has the largest mean value in the B band with 397.25 values. The brightness information of GS(R band: 313.25; G band: 380.74; B band: 243.82), HPF(R band: 312.58; G band: 379.63; B band: 242.70,) and HCS(R band: 313.43; G band: 381.05; B band: 243.88) is not significantly different from that of the original multispectral image (R band: 313.08; G band: 380.13; B band: 243.20). Compared with other fusion methods, these three methods have more scene reductively.

Firstly, we can evaluate the spatial information of fused images by standard deviation. From the comparative analysis of standard deviation information, the values of PCS, HCS, HPF, and HCS are stable, which are close to or higher than the original multispectral images. It shows that the gray level distribution of the images is more discrete than the original ones, and the amount of image information has increased, among which PCS (R band: 95.694; G band: 167.71; B band: 173.94) has achieved the best results. The standard deviation of IHS (B band: 88.860) and ELS (B band: 82.816) in the B band is small, which indicates that the gray level distribution of the image is convergent and the amount of image information is reduced.

From the information entropy index, the information entropy of each fusion image is close to or higher than that of the original multi-spectral image. The increase of information entropy indicates that the information of each fusion image is richer than that of the pre-fusion image. In the R band, IHS and ELS reached 1.558 and 1.583 above average, respectively. PCS and HCS fusion methods achieve the overall optimal effect in the G (PCS: 1.328; IHS: 1.324) and B bands (PCS: 1.557; IHS: 1.558), indicating that the information richness of PCS and HCS fusion images are higher than that of other fusion images.

Table 5. Statistical analysis between different fusion methods.

Method / Band	μ	Std	Ce	G	Warp	Spectral	Spatial
Panchromatic	287.77	124.60	0.788	13.096	—	—	1.0
Multispectral (R)	313.08	90.577	0.649	15.019	—	1.0	0.903
IHS	241.94	159.00	1.558	9.5800	105.40	0.872	0.939
PCS	322.45	95.694	0.679	8.7229	27.949	0.941	0.981
GS	313.25	88.027	0.596	8.0980	24.963	0.942	0.976
ELS	229.04	159.71	1.583	10.839	112.68	0.887	0.929
HPF	312.58	90.577	0.644	11.954	18.465	0.946	0.894
HCS	313.43	96.929	0.682	12.597	16.723	0.946	0.895
Multispectral (G)	380.13	155.30	1.307	27.238	—	1.0	0.929
IHS	377.14	152.98	1.291	12.883	27.602	0.966	0.954
PCS	396.63	167.71	1.328	15.227	47.888	0.944	0.983
GS	380.74	154.66	1.244	13.999	42.345	0.944	0.976
ELS	373.31	157.02	1.288	15.443	31.553	0.954	0.917
HPF	379.63	155.30	1.301	20.634	31.649	0.946	0.915
HCS	381.05	160.66	1.324	15.787	19.539	0.973	0.929
Multispectral (B)	243.20	158.38	1.554	25.665	—	1.0	0.941
IHS	310.06	88.860	0.572	9.7820	96.955	0.914	0.951
PCS	259.94	173.94	1.557	15.321	48.605	0.946	0.978
GS	243.82	160.02	1.497	14.266	43.372	0.944	0.978
ELS	397.25	82.816	1.131	10.366	165.51	0.852	0.881
HPF	242.70	158.37	1.511	20.892	32.541	0.946	0.927
HCS	243.88	160.53	1.558	10.907	12.685	0.988	0.944

Symbolic Meaning: μ = mean value; **Stud** = standard deviation; **Ce** = Information entropy; **G** = Mean gradient; **Warp** = Distortion degree; **Spectral** = Spectral correlation coefficient; **Spatial** = Spatial correlation coefficient.

From the analysis of the definition (average gradient) of the fused images, the average gradient of the six fusion images in the G and B bands is close to or higher than that of the original multi-spectral images, except for the R band. This shows that the six fusion methods have been enhanced in spatial information, and the expression effect of detailed information is higher. The HPF fusion method performs best in image clarity.

Comparing the spatial correlation coefficients of six fused images with the original color images, we find that PCS(R band: 0.981; G band: 0.983; B band: 0.973) and GS(R band: 0.976; G band: 0.976; B band: 0.978) achieve better results, indicating that the geometric details of fused images are more abundant, and the spatial correlation between the PCS fusion image and the original image is the highest.

The correlation coefficients of images need to consider not only the ability of the processed image to retain the spatial texture details of the original high spatial resolution image, but also the ability of the image to retain spectral characteristics. Comparing the spectral correlation coefficients between the fused image and the original multi-spectral image, we find that the fused image of PCS, GS, HPF, and HCS has a high correlation with the original image, among which the HCS fusion method has the lowest change in spectral information and the strongest spectral fidelity.

From the analysis of spectral distortion degree, the distortion of IHS in the R band and B band is 105.40 and 96.95, respectively, and ELS (R band: 112.68; B band: 165.51) is much higher than other methods. However, in the G band, the distortion of IHS and ELS is 27.602 and 31.553, respectively, while PCS obtained the maximum distortion of all methods of 47.888, which shows that the performance of different methods in different bands is greatly different. Overall, the distortion degree of his (R band: 105.40; G band: 27.602; B band: 96.955), ELS (R band: 112.68; G band: 31.553; B band: 165.51), and PCS (R band: 27.949; G bands: 0.976; B band: 0.978) is too large, which indicates that the distortion degree of image spectra is greater. The spectral distortion of GS, HPF, and HCS fusion methods is small, which indicates that the spectral distortion of fusion images is low, among which HCS (R band: 0.946; G band: 0.973; B band: 0.988) achieves the best effect.

3.3.3. Accuracy Assessment of Objectification Recognition

Figure 12 is the recognition result of six fusion methods using the Mask-RCNN model. In Figure 12c–h, the fusion images generated by different fusion methods have different recognition results for buildings, water bodies, and playgrounds. Among them, the overall recognition results obtained by GS and PCS are better, followed by HPF and HCS. The results for IHS and ELS have less satisfying effect. Figure 13 is the recognition result of typical buildings in the fusion image of six methods. The building recognition effect of the left-upper region (Building A, B, C, and D) and right-upper region (Building F), with obvious edge differences, achieves good performance, and almost all of them can be recognized correctly.

The confidence level of object recognition is shown in Table 6. The confidence of PCS in five buildings with clear outlines is the only one of the six methods where all values reach above 0.95. In building A, B, and F, the confidence of PCS is the highest, at 0.933, 0.974, and 0.983, respectively. In building C and D, the confidence of IHS is the highest, at 0.984 and 0.996, respectively. Building E in the lower left area interwoven with vegetation or shadows is shown to be partially unrecognizable. Among the recognition results of building E, the masks based on six fusion methods cannot clearly depict the edges of buildings. The confidence of the recognition results is the lowest among all building individuals. IHS achieves the highest value of 0.970, and PCS achieves confidence of 0.950.

The buildings with small building areas in the middle region (Area M) are easy to be missed due to the interference of spectral characteristics and shadows. Occasionally, buildings with prominent roof structures are misidentified as multiple buildings, resulting in repeated detection. Area N in Figure 13 is composed of stairs and lawns. Except for ELS and HPF, there is no misidentification in other methods. Area G is actually empty space. The HPF method misidentified area N as a building.

Considering the overall effect of building recognition, PCS is the best, achieving more accurate segmentation of building edges, followed by IHS, GS, and HCS, and lastly ELS and HPF, which have different degrees of omission and misidentification.

Table 6. Comparison of building confidence level of six fusion methods and multispectral images.

Building \ Method	MUL	IHS	PCS	GS	ELS	HPF	HCS
A	0.870	0.993	0.993	0.993	0.993	0.982	0.983
B	0.914	0.964	0.974	0.972	0.965	0.973	0.968
C	0.874	0.984	0.958	0.951	0.982	0.969	0.942
D	0.963	0.996	0.971	0.973	0.991	0.959	0.962
E	0.830	0.970	0.950	0.932	0.941	0.900	0.911
F	0.930	0.918	0.983	0.978	0.982	0.972	0.969
AVERAGE	0.897	0.971	0.972	0.967	0.977	0.959	0.956
Max Number	-	4	3	1	1	-	-

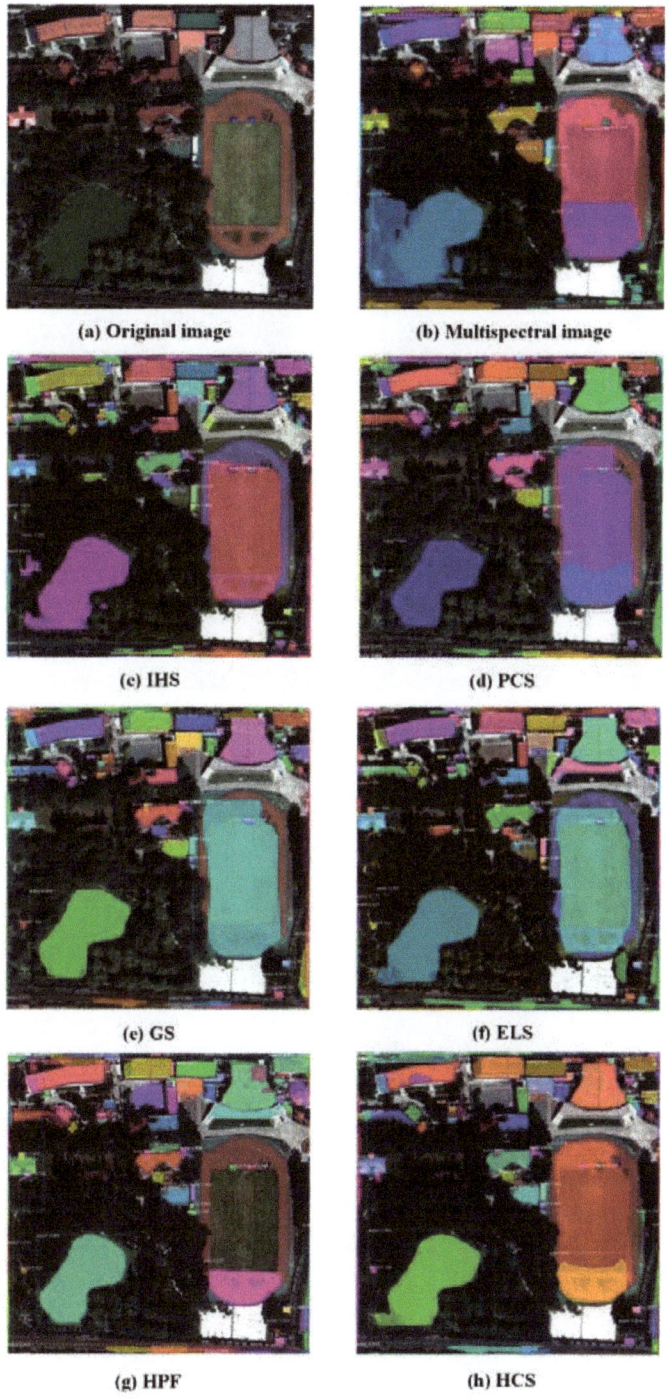

Figure 12. Comparison of recognition effect between the (**b**) Multispectral image and six fusion methods (**c–h**) for the WorldView-3 multispectral and pan images. Fused images from (c) Intensity-Hue

Saturation (IHS), (**d**) Principal Component Substitution (PCS), (**e**) Gram-Schmidt Transformation (GST), (**f**) Ehlers Fusion (ELS), (**g**) High Pass Filter Resolution Merge (HPF), and (**h**) Hyper spherical Color Space (HCS) methods.

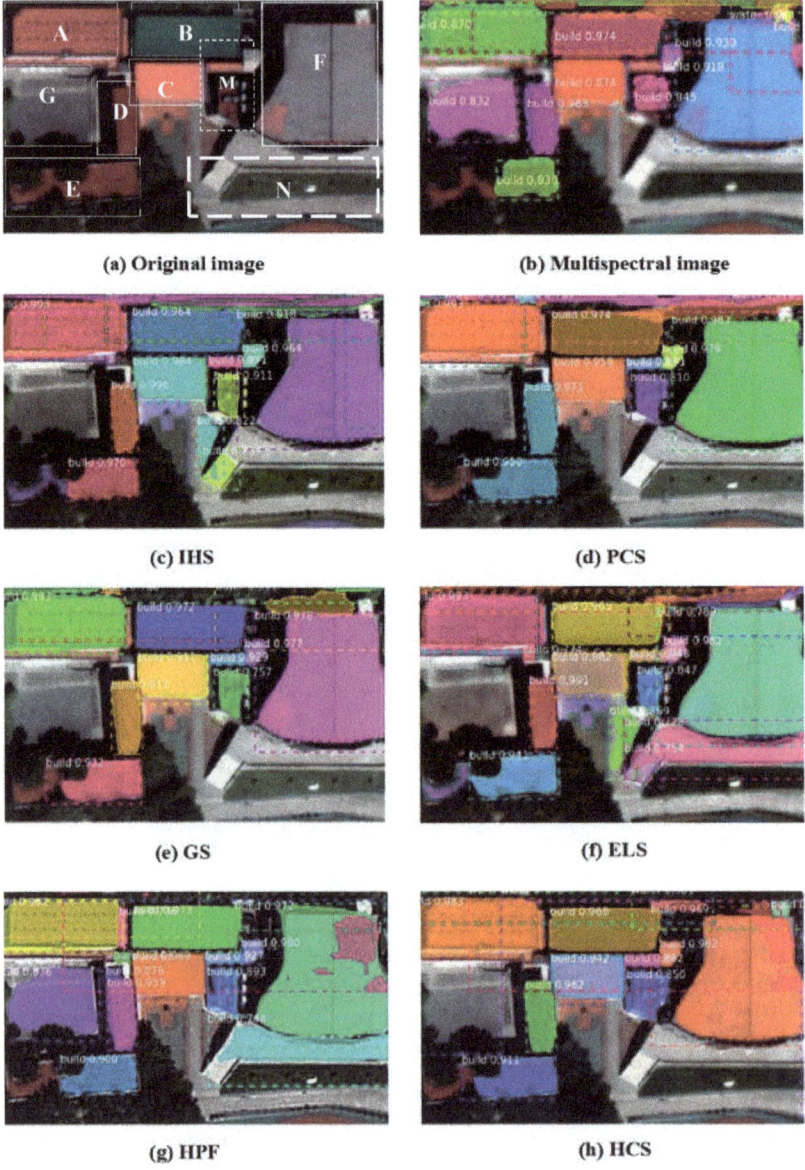

Figure 13. Comparison of accuracy of building recognition between the (**b**) Multispectral image and six fusion methods (**c**–**h**) for the WorldView-3 multispectral and pan images. Fused images from

(c) Intensity Hue Saturation (IHS), (d) Principal Component Substitution (PCS), (e) Gram-Schmidt Transformation (GST), (f) Ehlers Fusion (ELS), (g) High Pass Filter Resolution Merge (HPF), and (h) Hyper spherical Color Space (HCS) methods.

Figure 14 is the result of water body recognition in the fusion image of six methods. PCS, GS, and HPF have better segmentation effect. PCS and GS work better in segmentation of water body edges. The confidence of the two methods is 0.994 and 0.995 (Table 7), respectively. IHS, ELS, and HCS cannot distinguish the confusing parts of water bodies and vegetation shadows.

Table 7. Comparison of water confidence level of six fusion methods and a multispectral image.

Category \ Method	Mul	IHS	PCS	GS	ELS	HPF	HCS
Water	0.986	0.985	0.994	0.995	0.991	0.981	0.984

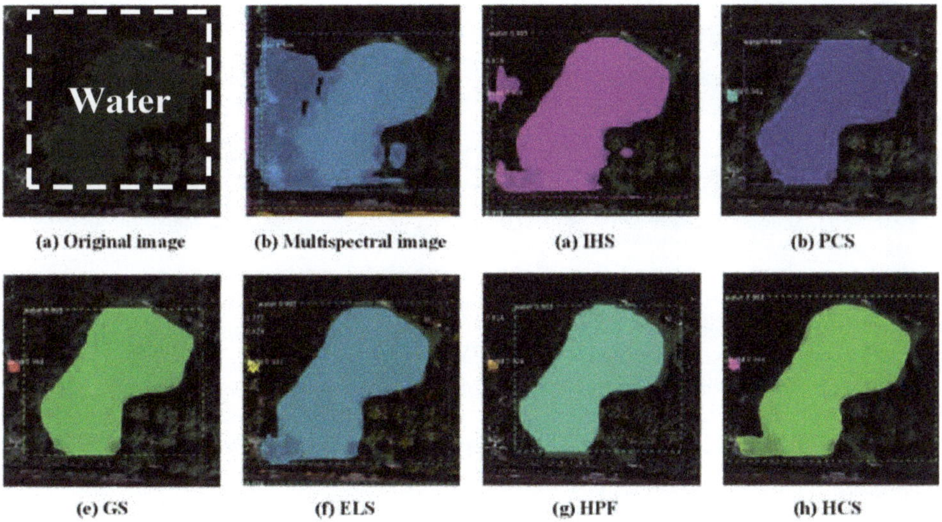

Figure 14. Comparison of accuracy of water recognition between the (b) Multispectral image and six fusion methods (c–h) for the WorldView-3 multispectral and pan images. Fused images from (c) Intensity Hue Saturation (IHS), (d) Principal Component Substitution (PCS), (e) Gram-Schmidt Transformation (GST), (f) Ehlers Fusion (ELS), (g) High Pass Filter Resolution Merge (HPF), and (h) Hyper spherical Color Space (HCS) methods.

In order to further verify the land cover (features) object recognition efficiency of the model for buildings by using six fusion methods, we selected an experimental area with denser buildings and more building types for building recognition (Figure 15). In total, 50 single buildings in the experimental area were identified by visual interpretation, and six complex building areas were divided for overall analysis.

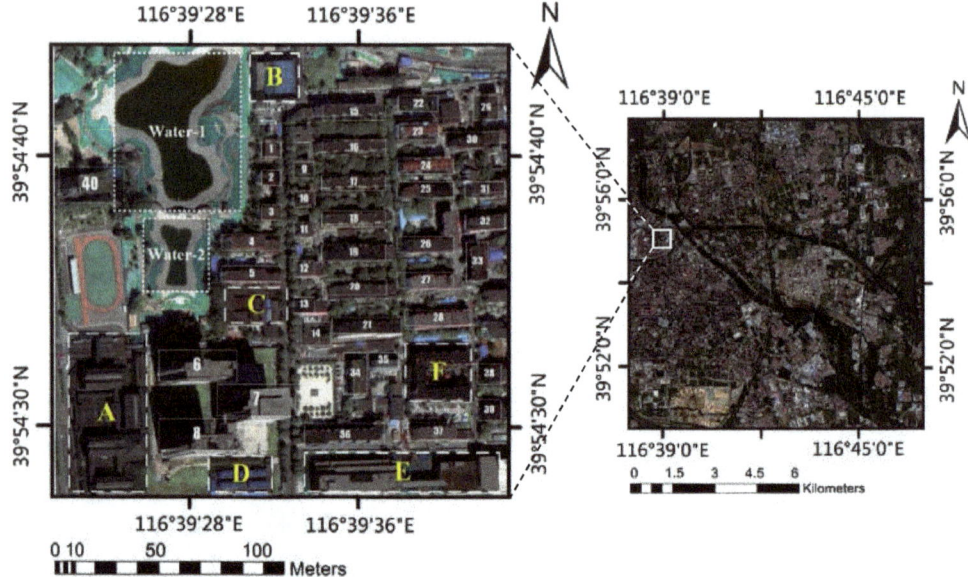

Figure 15. Experimental Area-2. (**1–50**) A total of 50 single buildings. (**A–F**) A total of six complex building areas.

Through the recognition results in Figure 16, we find that the image recognition effect after fusion has been greatly improved. The recognition results of multi-spectral remote sensing images have significant instances of misidentification and omission.

The IHS fusion method missed three buildings, ELS and HPF methods have miss one building. And PCS, GS, and HCS miss no detection. The six methods have achieved effective case segmentation for 40 single buildings. Combined with the data comparison results of Table 8 for confidence level, PCS (0.973) > HCS (0.965) > GS (0.956) > ELS (0.948) > HPF (0.930) > IHS (0.929), we can conclude that the PCS method achieves the best results. Through the recognition results of complex building areas, we find that multi-spectral remote sensing images are recognized as blurred edge masks, which cannot distinguish building units. A–F regions are irregular composite buildings. The recognition results of six fusion images show different degrees of fragmentation, which also provides a breakthrough point for our subsequent training of neural networks and the transformation of the network structure.

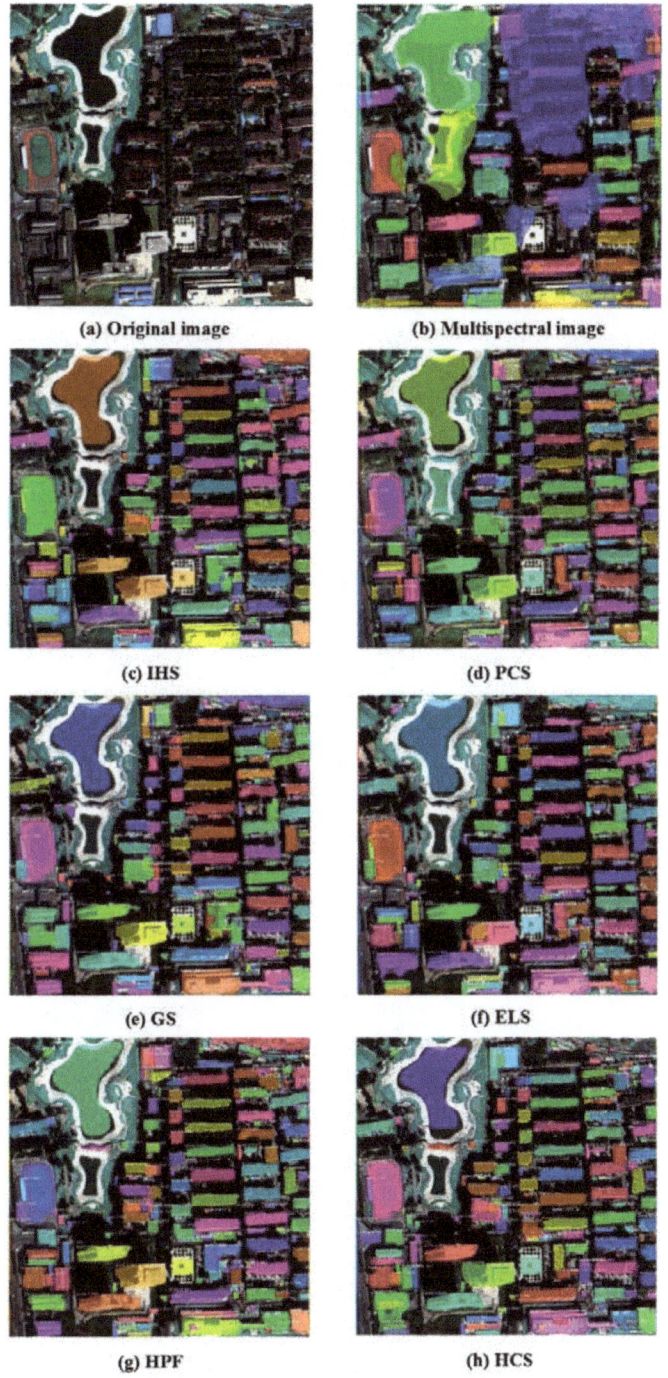

Figure 16. Comparison of accuracy of building recognition between the (**b**) Multispectral image and six fusion methods (**c**–**h**) for the WorldView-3 multispectral and pan images. Fused images from

(c) Intensity Hue Saturation (IHS), (d) Principal Component Substitution (PCS), (e) Gram-Schmidt Transformation (GST), (f) Ehlers Fusion (ELS), (g) High Pass Filter Resolution Merge (HPF), and (h) Hyper spherical Color Space (HCS) methods.

Table 8. Comparison of Building Confidence level of six fusion methods and the multispectral image.

Building \ Method	Mul	IHS	PCS	GS	ELS	HPF	HCS
1	0.851	0.994	0.994	0.990	0.993	0.983	0.993
2	-	0.993	0.990	0.994	0.990	0.979	0.996
3	-	0.991	0.989	0.990	0.990	0.985	0.991
4	0.965	0.990	0.988	0.981	0.993	0.974	0.993
5	-	0.864	0.829	0.869	0.932	0.866	0.836
6	0.906	0.995	0.994	0.987	0.995	0.982	0.994
7	-	-	0.907	0.828	0.779	0.904	0.875
8	-	0.978	0.977	0.978	0.982	0.982	0.982
9	0.807	0.995	0.996	0.994	0.996	0.985	0.995
10	0.846	0.978	0.977	0.979	0.982	0.978	0.987
11	0.788	0.983	0.986	0.978	0.985	0.979	0.985
12	-	-	0.965	0.735	0.964	0.904	0.725
13	-	0.988	0.993	0.991	0.992	0.984	0.992
14	-	0.992	0.986	0.990	0.990	0.977	0.989
15	-	0.994	0.990	0.992	0.996	0.987	0.996
16	-	0.986	0.991	0.992	0.994	0.987	0.992
17	-	0.993	0.994	0.994	0.995	0.988	0.994
18	-	0.991	0.989	0.983	0.995	0.991	0.991
19	-	0.861	0.891	0.932	0.879	0.783	0.870
20	-	0.988	0.992	0.984	0.985	0.977	0.99
21	-	0.997	0.997	0.990	0.995	0.985	0.996
22	-	0.994	0.997	0.994	0.996	0.992	0.996
23	-	0.988	0.994	0.983	0.992	0.950	0.991
24	-	0.996	0.997	0.993	0.994	0.990	0.996
25	0.896	0.960	0.995	0.991	0.994	0.982	0.994
26	0.901	0.993	0.993	0.993	0.991	0.967	0.993
27	-	0.980	0.989	0.974	0.978	0.805	0.976
28	-	0.985	0.978	0.976	0.982	0.915	0.984
29	-	0.995	0.996	0.997	0.996	0.994	0.997
30	-	0.987	0.993	0.994	0.994	0.990	0.995
31	-	0.979	0.979	0.968	0.987	0.976	0.985
32	0.916	0.988	0.990	0.991	0.990	0.976	0.992
33	0.951	0.995	0.993	0.994	0.993	0.989	0.995
34	0.951	0.994	0.996	0.995	0.996	0.988	0.996
35	0.953	0.979	0.972	0.952	0.918	-	0.939
36	-	0.987	0.952	0.744	0.976	0.935	0.939
37	-	0.890	0.908	0.866	0.776	0.888	0.902
38	-	0.986	0.987	0.983	0.987	0.972	0.984
39	-	0.986	0.794	0.738	-	0.767	0.792
40	-	0.985	0.986	0.982	0.991	0.973	0.991
AVERAGE	-	0.929	0.973	0.956	0.948	0.930	0.965
MaxNumber	-	11	17	3	13	1	12

In order to further quantitatively evaluate the recognition effect, we evaluate the detection accuracy of buildings in the experimental area by calculating precision and recall. Table 9 shows the results of six fusion methods and multispectral imagery. PCS achieves precision of 0.86 and recall of 0.80. These two indicators are the highest in the six methods. The precision of the multispectral image is only 0.48 and the recall is only 0.197, which also proves that image fusion can significantly improve the effectiveness of object recognition based on deep learning.

Table 9. Comparison of building detection results of six fusion methods and the multispectral image.

Index \ Method	Mul	IHS	PCS	GS	ELS	HPF	HCS
Detected objects	28	53	57	51	53	52	55
Matched objects	12	45	49	43	42	44	46
Precision	0.480	0.849	0.860	0.843	0.792	0.846	0.836
Recall	0.197	0.738	0.803	0.705	0.689	0.721	0.754

In addition, we selected two experimental areas (Figure 17) with similar and scattered water bodies to further verify the recognition efficiency of water bodies in the image generated by six fusion methods. The water bodies in the selected experimental area are different in size. Dense vegetation and shadows are interlaced with water bodies, which are liable to cause confusion.

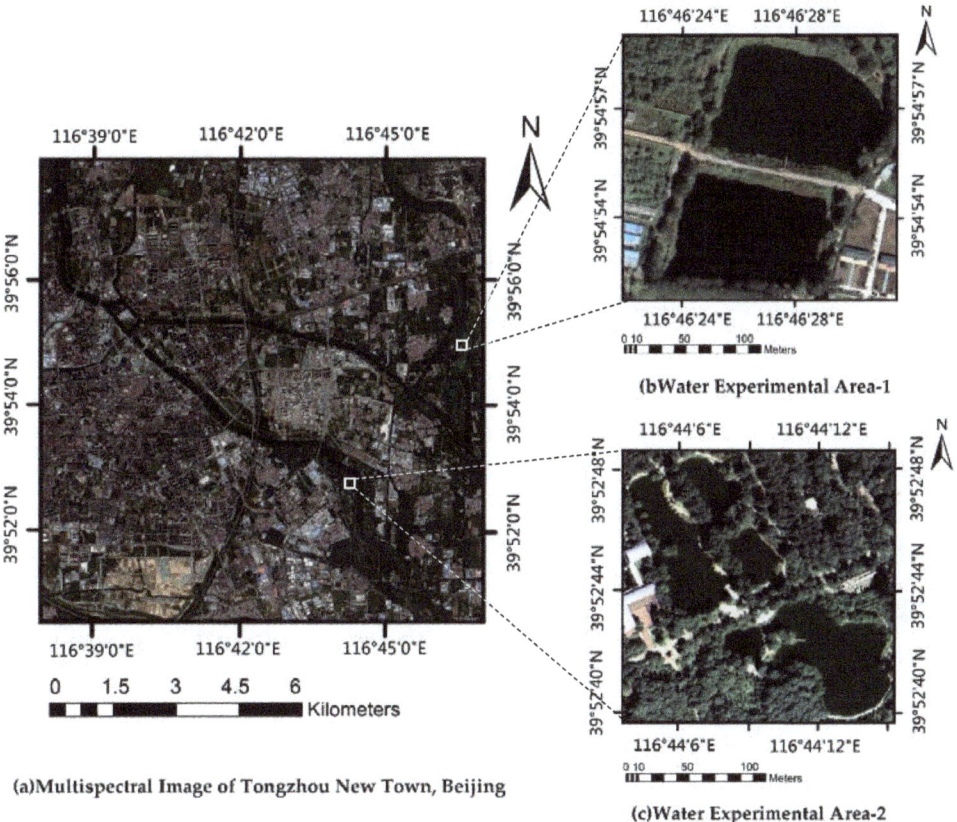

Figure 17. Water experimental area.

Through the recognition results of Figure 18, we find that in the multispectral remote sensing image, the edges of water bodies in regions A and B are significantly confused with vegetation, and the green space is mistaken for a water body. There is a serious missing detection phenomenon in regions C–G. Compared with this, the recognition results of six fusion methods have been greatly improved. PCS can better distinguish the confusing parts of water bodies and vegetation shadows. PCS has the best edge segmentation effect. There is no large area of water missing or overflowing in the segmentation mask. Table 10 shows the data comparison of confidence level, PCS (0.982) > GS

(0.944) > HCS (0.934) > IHS (0.931) > HPF (0.904) > ELS (0.767), combined with the water detection results of six fusion methods and multispectral imagery from Table 11, we conclude that PCS has the best effect.

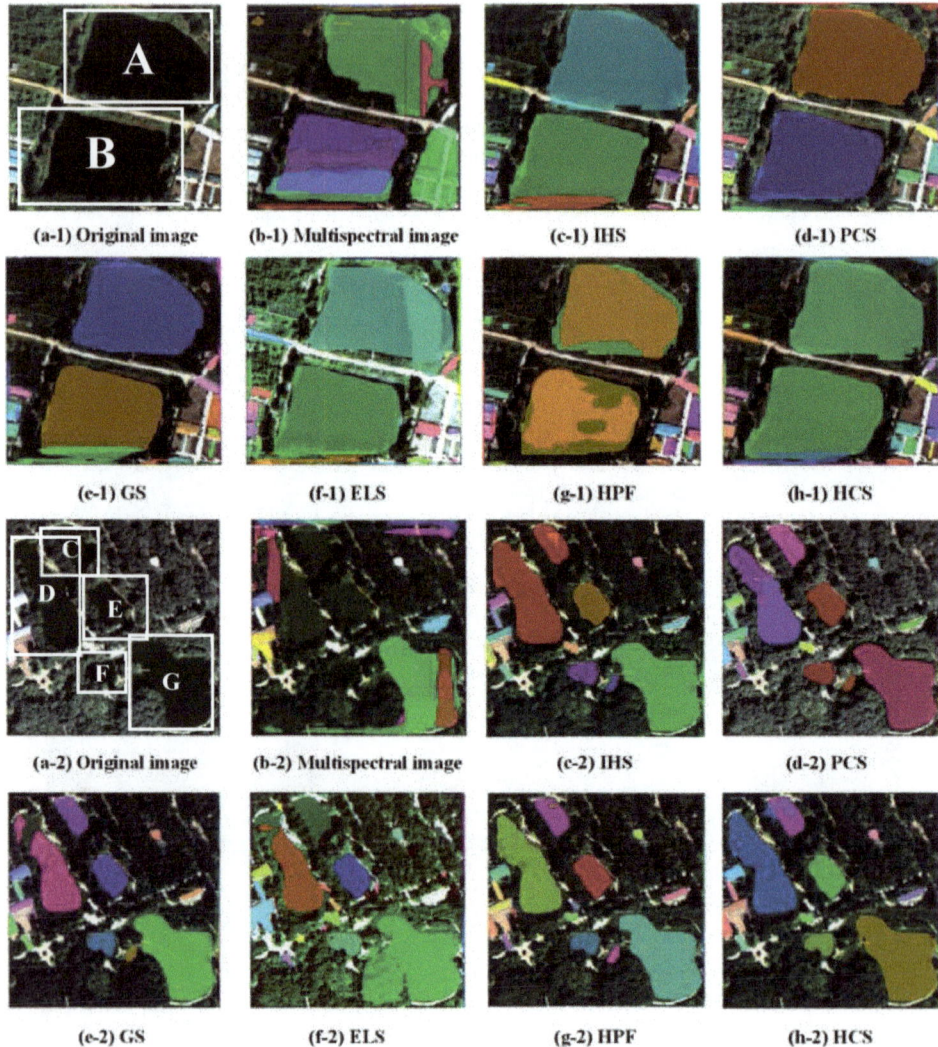

Figure 18. Comparison of Accuracy of Water Recognition between the (**b-1**, **b-2**) Multispectral image and six fusion methods (**c–h**) for the WorldView-3 multispectral and pan images. Fused images from (**c1**, **c2**) Intensity-Hue Saturation (IHS), (**d1**, **d2**) Principal Component Substitution (PCS), (**e1**, **e2**) Gram-Schmidt Transformation (GST), (**f1**, **f2**) Ehlers Fusion (ELS), (**g1**, **g2**) High Pass Filter Resolution Merge (HPF), and (**h1**, **h2**) Hyper spherical Color Space (HCS) methods.

Table 10. Comparison of water confidence level of the six fusion methods and the multispectral image.

Water \ Method	Mul	IHS	PCS	GS	ELS	HPF	HCS
A	0.770	0.995	0.994	0.969	0.984	0.965	0.989
B	0.774	0.986	0.985	0.970	0.893	0.936	0.985
C	-	0.732	0.982	0.898	-	0.810	0.827
D	-	0.941	0.943	0.891	0.777	0.773	0.861
E	-	0.946	0.993	0.977	0.837	0.953	0.964
F	-	0.931	0.996	0.924	0.898	0.919	0.941
G	0.946	0.984	0.984	0.982	0.984	0.974	0.972
AVERAGE	-	0.931	0.982	0.944	0.767	0.904	0.934
MaxNumber	-	2	5	-	1	-	-

Table 11. Comparison of water detection results of the six fusion methods and the multispectral image.

Index \ Method	Mul	IHS	PCS	GS	ELS	HPF	HCS
Detected objects	8	8	7	7	8	8	8
Matched objects	3	7	7	7	5	7	7
Precision	0.428	0.875	1	1	0.625	0.875	0.875
Recall	0.428	1	1	1	0.714	1	1

Table 12 reports the computation times of our proposed methods on three experimental areas (experimental area-1, water experimental area-1 and area-2). Combined with detection results in Tables 9 and 11, we can conclude that the computing time is proportional to detection of objects. How to achieve the balance of efficiency and effect through the improvement of the algorithm is the next important research direction.

Table 12. Computation time (second) comparison of the six different methods.

Running Time \ Method	Mul	IHS	PCS	GS	ELS	HPF	HCS	Average (6 Methods)
Building	5.430	16.32	16.95	15.35	16.46	15.58	16.65	16.218
Water-1	1.86	4.96	4.81	4.85	5.09	5.01	4.95	4.945
Water-2	4.51	5.13	4.95	4.98	5.01	5.15	5.08	5.050

From the above experiments, we can draw a conclusion that the object recognition method based on Mask-RCNN applied in this paper can recognize objects. The image recognition effect for two types of objects largely varies with different fusion methods, and the overall effect of PCS and GS is better. It is noteworthy that there is a common phenomenon in the results of object recognition. The segmentation effect of the edge of the mask is not satisfying, and the edge of the mask cannot be fully fitted to the object for segmentation. In the subsequent sample making and model improvement, the segmentation effect of the edge of the object needs to be further optimized.

4. Conclusions

Image fusion is the fundamental way to realize the complementary advantages between the high spatial resolution of the panchromatic band and the high spectral resolution of multispectral bands. In the pretreatment of model training data, we compared six fusion methods, IHS, PCS, GS, ELS, HPF, and HCS, by applying them to the same WorldView-3 satellite image. The results show that the fusion images obtained by different fusion methods are very different in visual effect and quantitative index.

Land cover (features) recognition effectiveness for buildings and water bodies using six fusion methods are notably different, and the recognition accuracy has been significantly improved compared with the original multi-spectral remote sensing images. Considering the subsequent segmentation and feature-oriented primitive processing, analysis, and understanding, PCS fusion method has the best comprehensive effect.

We use deep learning to perform typical object recognition of land cover (features) in remote sensing images, but there is still a long way to go to meet the standards of surveying and mapping products. Realizing the automation from remote sensing image to sematic vector map production is the ultimate goal. Compared with the object instance segmentation based on Mask-RCNN, the extraction of building vector contours based on instance segmentation will be more challenging in the current research field.

Author Contributions: The following contributions were given to this research effort: Conceptualization, S.S.; Methodology, S.S., and H.P.; Software, S.S.; Validation, S.S., Y.L. and H.P.; Formal analysis, S.S.; Investigation, H.P.; Resources, S.S.; Data curation, Y.L.; Writing—original draft preparation, S.S.; writing—review and editing, J.L.; Visualization, J.L.

Funding: This research is funded by National Natural Science Foundation of China (No. 41301489), Beijing Natural Science Foundation (No.4192018, No. 4142013), Outstanding Youth Teacher Program of Beijing Municipal Education Commission (No. YETP1647, No. 21147518608), Outstanding Youth Researcher Program of Beijing University of Civil Engineering and Architecture (No. 21082716012), and the Fundamental Research Funds for Beijing Universities (No. X18282).

Conflicts of Interest: The authors declare no conflict of interest.

References

1. Zhang, L.P.; Shen, H.F. Progress and future of remote sensing data fusion. *J. Remote Sens.* **2016**, *20*, 1050–1061. [CrossRef]
2. Deng, Z.; Sun, H.; Zhou, S.; Zhao, J.; Lei, L.; Zou, H. Multi-scale object detection in remote sensing imagery with convolutional neural networks. *ISPRS J. Photogramm. Remote Sens.* **2018**, *145*, 3–22. [CrossRef]
3. Girshick, R.; Donahue, J.; Darrelland, T.; Malik, J. Rich feature hierarchies for object detection and semantic segmentation. In Proceedings of the Conference on Computer Vision and Pattern Recognition (IEEE 2014), Columbus, OH, USA, 23–28 June 2014. [CrossRef]
4. Krizhevsky, A.; Sutskever, I.; Hinton, G. ImageNet Classification with Deep Convolutional Neural Networks. *Adv. Neural Inf. Process. Syst.* **2012**, *25*. [CrossRef]
5. Long, J.; Shelhamer, E.; Darrell, T. Fully Convolutional Networks for Semantic Segmentation. *IEEE Trans. Pattern Anal. Mach. Intell.* **2014**, *39*, 640–651. [CrossRef]
6. Bo, H.; Bei, Z.; Yimeng, S. Urban land-use mapping using a deep convolutional neural network with high spatial resolution multispectral remote sensing imagery. *Remote Sens. Environ.* **2018**, *214*, 73–86. [CrossRef]
7. Zhou, P.; Han, J.; Cheng, G.; Zhang, B. Learning Compact and Discriminative Stacked Autoencoder for Hyperspectral Image Classification. *IEEE Trans. Geosci. Remote Sens.* **2019**. [CrossRef]
8. Cheng, G.; Zhou, P.; Han, J. Learning Rotation-Invariant Convolutional Neural Networks for Object Detection in VHR Optical Remote Sensing Images. *IEEE Trans. Geosci. Remote Sens.* **2016**, *54*, 7405–7415. [CrossRef]
9. Zhong, Y.; Han, X.; Zhang, L. Multi-class geospatial object detection based on a position-sensitive balancing framework for high spatial resolution remote sensing imagery. *ISPRS J. Photogramm. Remote Sens.* **2018**, *138*, 281–294. [CrossRef]
10. Cheng, G.; Han, J.; Zhou, P.; Guo, L. Multi-class geospatial object detection and geographic image classification based on collection of part detectors. *ISPRS J. Photogramm. Remote Sens.* **2014**, *98*, 119–132. [CrossRef]
11. He, L.; Li, J.; Liu, C.; Li, S. Recent Advances on Spectral-Spatial Hyperspectral Image Classification: An Overview and New Guidelines. *IEEE Trans. Geosci. Remote Sens.* **2017**, *56*, 1579–1597. [CrossRef]
12. Zhao, J.; Zhong, Y.; Jia, T.; Wang, X.; Xu, Y.; Shu, H.; Zhang, L. Spectral-spatial classification of hyperspectral imagery with cooperative game. *ISPRS J. Photogramm. Remote Sens.* **2018**, *135*, 31–42. [CrossRef]
13. Peng, J.; Du, Q. Robust Joint Sparse Representation Based on Maximum Correntropy Criterion for Hyperspectral Image Classification. *IEEE Trans. Geosci. Remote Sens.* **2017**, *55*, 7152–7164. [CrossRef]
14. Dong, Y.; Du, B.; Zhang, L.; Zhang, L. Dimensionality Reduction and Classification of Hyperspectral Images Using Ensemble Discriminative Local Metric Learning. *IEEE Trans. Geosci. Remote Sens.* **2017**, *55*, 2509–2524. [CrossRef]

15. Liu, T.; Gu, Y.; Jia, X.; Benediktsson, J.A.; Chanussot, J. Class-Specific Sparse Multiple Kernel Learning for Spectral–Spatial Hyperspectral Image Classification. *IEEE Trans. Geosci. Remote Sens.* **2016**, *54*, 7351–7365. [CrossRef]
16. Xu, X.; Li, J.; Huang, X.; Dalla Mura, M.; Plaza, A. Multiple Morphological Component Analysis Based Decomposition for Remote Sensing Image Classification. *IEEE Trans. Geosci. Remote Sens.* **2016**, *54*, 3083–3102. [CrossRef]
17. Xiaoyong, B.; Chen, C.; Yan, X.; Du, Q. Robust Hyperspectral Image Classification by Multi-Layer Spatial-Spectral Sparse Representations. *Remote Sens.* **2016**, *8*, 985. [CrossRef]
18. Li, J.; Huang, X.; Gamba, P.; Bioucas-Dias, J.M.; Zhang, L.; Benediktsson, J.A.; Plaza, A. Multiple Feature Learning for Hyperspectral Image Classification. *IEEE Trans. Geosci. Remote Sens.* **2015**, *53*, 1592–1606. [CrossRef]
19. Chen, C.; Li, W.; Su, H.; Liu, K. Spectral-Spatial Classification of Hyperspectral Image Based on Kernel Extreme Learning Machine. *Remote Sens.* **2014**, *6*, 5795–5814. [CrossRef]
20. Cheng, G.; Li, Z.; Han, J.; Yao, X.; Guo, L. Ecploring Hierchical Convolutional Features for Hyperspectral Image Classification. *IEEE Trans. Geosci. Remote Sens.* **2018**, *56*, 6712–6722. [CrossRef]
21. Schowengerdt, R.A. Reconstruction of multispatial, multispectral image data using spatial frequency content. *Photogramm. Eng. Remote Sens.* **1980**, *46*, 1325–1334.
22. Hallada, W.A.; Cox, S. Image sharpening for mixed spatial and spectral resolution satellite systems. In Proceedings of the International Symposium on Remote Sensing of the Environment, Ann Arbor, MI, USA, 9–13 May 1983.
23. Cliche, G.; Bonn, F.; Teillet, P. Integration of the spot panchromatic channel into its multispectral mode for image sharpness enhancement. *Photogramm. Eng. Remote Sens.* **1985**, *51*, 311–316.
24. Wang, Z.; Ziou, D.; Armenakis, C.; Li, D.; Li, Q. A comparative analysis of image fusion methods. *IEEE Trans. Geosci. Remote Sens.* **2005**, *43*, 1391–1402. [CrossRef]
25. Thomas, C.; Ranchin, T.; Wald, L.; Chanussot, J. Synthesis of Multispectral Images to High Spatial Resolution: A Critical Review of Fusion Methods Based on Remote Sensing Physics. *IEEE Trans. Geosci. Remote Sens.* **2008**, *46*, 1301–1312. [CrossRef]
26. Aiazzi, B.; Alparone, L.; Baronti, S.; Garzelli, A.; Selva, M. 25 years of pansharpening: A critical review and new developments. In *Signal and Image Processing for Remote Sensing*, 2nd ed.; CRC Press: Boca Raton, FL, USA, 2012.
27. Gillespie, A.R.; Kahle, A.B.; Walker, R.E. Color enhancement of highly correlated images. II. Channel ratio and "chromaticity" transformation techniques. *Remote Sens. Environ.* **1987**, *22*, 343–365. [CrossRef]
28. Tu, T.M.; Su, S.C.; Shyu, H.C.; Huang, P.S. A new look at IHS-like image fusion methods. *Inf. Fusion* **2001**, *2*, 177–186. [CrossRef]
29. Aiazzi, B.; Alparone, L.; Baronti, S.; Garzelli, A. Context-driven fusion of high spatial and spectral resolution images based on oversampled multiresolution analysis. *IEEE Trans. Geosci. Remote Sens.* **2002**, *40*, 2300–2312. [CrossRef]
30. Otazu, X.; Gonzalez-Audicana, M.; Fors, O.; Núñez, J. Introduction of sensor spectral response into image fusion methods. Application to wavelet-based methods. *IEEE Trans. Geosci. Remote Sens.* **2005**, *43*, 2376–2385. [CrossRef]
31. Zhang, L.; Shen, H.; Gong, W.; Zhang, H. Adjustable Model-Based Fusion Method for Multispectral and Panchromatic Images. *IEEE Trans. Syst. Man Cybern. Part B Cybern.* **2012**, *42*, 1693–1704. [CrossRef]
32. Meng, X.C.; Shen, H.F.; Zhang, H.Y.; Zhang, L.; Li, H. Maximum a posteriori fusion method based on gradient consistency constraint for multispectral/panchromatic remote sensing images. *Spectrosc. Spectr. Anal.* **2014**, *34*, 1332–1337. [CrossRef]
33. Jiang, C.; Zhang, H.; Shen, H.; Zhang, L. A Practical Compressed Sensing-Based Pan-Sharpening Method. *IEEE Geosci. Remote Sens. Lett.* **2012**, *9*, 629–633. [CrossRef]
34. He, K.; Gkioxari, G.; Dollar, P.; Girshick, R. Mask R-CNN. In Proceedings of the 2017 IEEE International Conference on Computer Vision (ICCV), Venice, Italy, 22–29 October 2017. [CrossRef]
35. Ehlers, M.; Ehlers, M.; Posa, F.; Kaufmann, H.J.; Michel, U.; De Carolis, G. *Remote Sensing for Environmental Monitoring, GIS Applications, and Geology IV*; Society of Photo Optical: Bellingham, WA, USA, 2004; Volume 5574, pp. 1–13. [CrossRef]
36. Klonus, S.; Ehlers, M. Image Fusion Using the Ehlers Spectral Characteristics Preservation Algorithm. *Gisci. Remote Sens.* **2007**, *44*, 93–116. [CrossRef]

37. Cetin, M.; Musaoglu, N. Merging hyperspectral and panchromatic image data: Qualitative and quantitative analysis. *Int. J. Remote Sens.* **2009**, *30*, 1779–1804. [CrossRef]
38. Chavez, P.S.; Sides, S.C.; Anderson, J.A. Comparison of Three Different Methods to Merge Multiresolution and Multispectral Data: Landsat TM and SPOT Panchromatic. *Photogramm. Eng. Remote Sens.* **1991**, *57*, 265–303. [CrossRef]
39. Vrabel, N. Multispectral Imagery Band Sharpening Study. *Photogramm. Eng. Remote Sens.* **1996**, *62*, 1075–1084.
40. Jun, L.C.; Yun, L.L.; Hua, W.J.; Chao, W.R. Comparison of Two Methods of Fusing Remote Sensing Images with Fidelity of Spectral Information. *J. Image Graph.* **2004**, *9*, 1376–1385. [CrossRef]
41. Choi, M. A new intensity-hue-saturation fusion approach to image fusion with a tradeoff parameter. *IEEE Trans. Geosci. Remote Sens.* **2006**, *44*, 1672–1682. [CrossRef]
42. Yésou, H.; Besnus, Y.; Rolet, J. Extraction of spectral information from Landsat TM data and merger with SPOT panchromatic imagery—A contribution to the study of geological structures. *ISPRS J. Photogramm. Remote Sens.* **1993**, *48*, 23–36. [CrossRef]
43. Ji, S.; Wei, S.; Lu, M. A scale robust convolutional neural network for automatic building extraction from aerial and satellite imagery. *Int. J. Remote Sens.* **2018**. [CrossRef]
44. Simonyan, K.; Zisserman, A. Very Deep Convolutional Networks for Large-Scale Image Recognition. *arXiv* **2014**, arXiv:1409.1556.
45. Szegedy, C.; Liu, W.; Jia, Y.; Sermanet, P.; Reed, S.; Anguelov, D.; Erhan, D.; Vanhoucke, V.; Rabinovich, A. Going Deeper with Convolutions. In Proceedings of the IEEE Conference on Computer Vision and Pattern Recognition, Boston, MA, USA, 8–10 June 2015. [CrossRef]
46. He, K.; Zhang, X.; Ren, S.; Sun, J. Deep Residual Learning for Image Recognition. In Proceedings of the 2016 IEEE Conference on Computer Vision and Pattern Recognition (CVPR), Las Vegas, NV, USA, 27–30 June 2016. [CrossRef]
47. Girshick, R. Fast R-CNN. In Proceedings of the 2015 IEEE International Conference on Computer Vision (ICCV), Las Condes, Chile, 7–13 December 2015. [CrossRef]
48. Girshick, R.; Donahue, J.; Darrell, T.; Malik, J. Region-Based Convolutional Networks for Accurate Object Detection and Segmentation. *IEEE Trans. Pattern Anal. Mach. Intell.* **2016**, *38*. [CrossRef] [PubMed]
49. Ren, S.; He, K.; Girshick, R.; Sun, J. Faster R-CNN: Towards Real-Time Object Detection with Region Proposal Networks. *IEEE Trans. Pattern Anal. Mach. Intell* **2015**, *39*. [CrossRef] [PubMed]
50. Zou, Q.; Ni, L.; Zhang, T.; Wang, Q. Deep Learning Based Feature Selection for Remote Sensing Scene Classification. *IEEE Geosci. Remote Sens. Lett.* **2015**, *12*. [CrossRef]
51. Yang, Y.; Newsam, S. Bag-of-visual-words and spatial extensions for land-use classification. In Proceedings of the 18th SIGSPATIAL International Conference on Advances in Geographic Information Systems, San Jose, CA, USA, 2–5 November 2010; p. 270. [CrossRef]

© 2019 by the authors. Licensee MDPI, Basel, Switzerland. This article is an open access article distributed under the terms and conditions of the Creative Commons Attribution (CC BY) license (http://creativecommons.org/licenses/by/4.0/).

Article

Building Extraction from UAV Images Jointly Using 6D-SLIC and Multiscale Siamese Convolutional Networks

Haiqing He [1,2,*], Junchao Zhou [1], Min Chen [3], Ting Chen [4], Dajun Li [1] and Penggen Cheng [1]

1. School of Geomatics, East China University of Technology, Nanchang 330013, China; liuyuhui@ecit.cn (J.Z.); djli@ecit.cn (D.L.); pgcheng@ecit.cn (P.C.)
2. Key Laboratory of Watershed Ecology and Geographical Environment Monitoring, National Administration of Surveying, Mapping and Geoinformation, Nanchang 330013, China
3. Faculty of Geosciences and Environmental Engineering, Southwest Jiaotong University, Chengdu 611756, China; minchen@home.swjtu.edu.cn
4. School of Water Resources & Environmental Engineering, East China University of Technology, Nanchang 330013, China; ct_201607@ecit.cn
* Correspondence: hhq201360010@ecit.cn; Tel.: +86-181-4662-5391

Received: 28 February 2019; Accepted: 29 April 2019; Published: 1 May 2019

Abstract: Automatic building extraction using a single data type, either 2D remotely-sensed images or light detection and ranging 3D point clouds, remains insufficient to accurately delineate building outlines for automatic mapping, despite active research in this area and the significant progress which has been achieved in the past decade. This paper presents an effective approach to extracting buildings from Unmanned Aerial Vehicle (UAV) images through the incorporation of superpixel segmentation and semantic recognition. A framework for building extraction is constructed by jointly using an improved Simple Linear Iterative Clustering (SLIC) algorithm and Multiscale Siamese Convolutional Networks (MSCNs). The SLIC algorithm, improved by additionally imposing a digital surface model for superpixel segmentation, namely 6D-SLIC, is suited for building boundary detection under building and image backgrounds with similar radiometric signatures. The proposed MSCNs, including a feature learning network and a binary decision network, are used to automatically learn a multiscale hierarchical feature representation and detect building objects under various complex backgrounds. In addition, a gamma-transform green leaf index is proposed to truncate vegetation superpixels for further processing to improve the robustness and efficiency of building detection, the Douglas–Peucker algorithm and iterative optimization are used to eliminate jagged details generated from small structures as a result of superpixel segmentation. In the experiments, the UAV datasets, including many buildings in urban and rural areas with irregular shapes and different heights and that are obscured by trees, are collected to evaluate the proposed method. The experimental results based on the qualitative and quantitative measures confirm the effectiveness and high accuracy of the proposed framework relative to the digitized results. The proposed framework performs better than state-of-the-art building extraction methods, given its higher values of recall, precision, and intersection over Union (IoU).

Keywords: building extraction; simple linear iterative clustering (SLIC); multiscale Siamese convolutional networks (MSCNs); binary decision network; unmanned aerial vehicle (UAV)

1. Introduction

Building extraction based on remote sensing data is an effective technique to automatically delineate building outlines; it has been widely studied for decades in the fields of photogrammetry and remote sensing, and is extensively used in various applications, including urban planning, cartographic mapping, and land use analysis [1,2]. The significant progress in sensors and operating platforms has enabled us to acquire remote sensing images and 3D point clouds from cameras or Light Detection And Ranging (LiDAR) equipped in various platforms (e.g., satellite, aerial, and Unmanned Aerial Vehicle (UAV) platforms); thus, the methods based on images and point clouds are commonly used to extract buildings [3–5].

Building extraction can be broadly divided into three categories according to data source: 2D image-based methods, 3D point cloud-based methods, and 2D and 3D information hybrid methods. 2D image-based building extraction consists of two stages, namely, building segmentation and regularization. Many approaches have been proposed in recent years to extract buildings through very-high-resolution 2D imagery, including the active contour model-based method [6], multidirectional and multiscale morphological index-based method [7], combined binary filtering and region growing method [8], object-based method [9], dense attention network-based method [10], and boundary-regulated network-based method [2]. Although these methods have achieved important advancements, a single cue from 2D images remains insufficient to extract buildings under the complex backgrounds of images (e.g., illumination, shadow, occlusion, geometric deformation, and quality degradation), which cause inevitable obstacles in the identification and delineation of building outlines under different circumstances. Consequently, differentiating building and non-building objects that carry similar radiometric signatures is difficult by using spectral information alone. Existing methods focus more on building qualitative detection than accurate outline extraction, thus requiring further improvement in building contour extraction to satisfy various applications, such as automatic mapping and building change detection.

Unlike 2D remotely-sensed imagery, LiDAR data can provide the 3D information of ground objects, and are especially useful in distinguishing building and non-building objects by height variation. Various approaches based on LiDAR data, such as polyhedral building roof segmentation and reconstruction [11], building roof segmentation using the random sample consensus algorithm [12,13] and global optimization [14], and automatic building extraction using point- and grid-based features [15], have been proposed for building extraction. However, the utilization of height information alone may fail to distinguish building and non-building objects with similar heights, such as houses and surrounding trees with smooth canopies. The accuracy of building extraction often relies on the density of 3D point clouds, and the outline of poor-quality points at the edge of buildings is challenging to accurately delineate. Moreover, most LiDAR-based methods may only be applicable to urban building extraction and may be unsuitable for extracting rural buildings with topographic relief because of the difficulty in giving a certain height threshold to truncate non-building objects. Aside from these limitations, automatic building extraction is challenging in the contexts of complex shape, occlusion, and size. Therefore, automatically extracting buildings by using a single data type, either 2D remotely-sensed images or 3D LiDAR point clouds, remains insufficient.

Many approaches that combine spectral and height information have been proposed to overcome the shortcomings of building extraction using a single data type. In [16,17], Normalized Difference Vegetation Index (NDVI) and 3D LiDAR point clouds were used to eliminate vegetation and generate a building mask, and height and area thresholds were given to exclude other low-height objects and small buildings. A method based on LiDAR point clouds and orthoimage has been proposed to delineate the boundaries of buildings, which are then regulated by using image lines [1]. However, compared with satellite and aerial imagery, LiDAR data are actually difficult to access due to the high cost involved [5]. Tian et al. [18] proposed an approach to building detection based on 2D images and Digital Surface Model (DSM); unlike 3D LiDAR point clouds, height information is generated from stereo imagery by the dense matching algorithm. Moreover, the combination of 2D UAV orthoimages and image-derived 3D point clouds has been used for building extraction on the basis of low-cost and high-flexibility UAV photogrammetry and remote sensing [5,19]. Most civil UAVs only acquire remote sensing images with RGB channels and do not include multispectral bands (e.g., near-infrared bands), that is, eliminating vegetation by the NDVI is not feasible. As an alternative method, RGB-based

Multidimensional Feature Vector (MFV) and Support Vector Machine (SVM) classifiers were integrated by Dai et al. [5] to eliminate vegetation; in this method, buildings are extracted by using a certain height threshold (e.g., 2.5 m), and building outlines are regularized by jointly using a line-growing algorithm and a w-k-means clustering algorithm. However, this method is only useful for extracting buildings with linear and perpendicular edges and not applicable to extract buildings with irregular shapes.

On the basis of the advantages of UAV photogrammetry and remote sensing, this study concentrates on building segmentation and outline regularization based on UAV orthoimages and image-derived point clouds. First, image segmentation is implemented to cluster all pixels of UAV orthoimages; SLIC is a popular algorithm for segmenting superpixels and does not require much computational cost [20], but it easily confuses building and image backgrounds with similar radiometric signatures. We accordingly exploit a novel 6D simple linear iterative clustering (6D-SLIC) algorithm for superpixel segmentation by additionally imposing DSM that is generated from image-derived 3D point clouds; DSM helps to distinguish objects from different heights (e.g., building roof and road). Second, the vegetation superpixels are truncated by using a Gamma-transform Green Leaf Index (GGLI). Then, the boundaries of non-vegetation objects are shaped by merging the superpixels with approximately equal heights. Inspired by the progresses made in deep learning in recent years, the deep convolutional neural network is one of the most popular and successful deep networks for image processing because it can work efficiently under various complex backgrounds [21–26] and is suitable for identifying building objects under different circumstances. The Fully Convolutional Network (FCN) [27] is a specific type of deep network that is used for image segmentation and building extraction [28]. U-shaped convolutional Networks (U-Nets) are extended for image segmentation [29] and building extraction [30]. In this study, buildings are detected by Multiscale Siamese Convolutional Networks (MSCNs), including a feature learning network and a binary decision network, which are used to automatically learn a multiscale hierarchical feature representation and detect building objects. Finally, the building outlines are regulated by the Douglas–Peucker and iterative optimization algorithms.

The main contribution of this study is to propose a method for building extraction that is suitable for UAV orthoimage and image-derived point clouds. In this method, the improved SLIC algorithm for UAV image segmentation, which helps accurately delineate building boundaries under building and image backgrounds with similar radiometric signatures. MSCNs are used to improve the performance of building detection under various complex backgrounds, and the Douglas–Peucker algorithm and iterative optimization are coupled to eliminate jagged details generated from small structures as a result of superpixel segmentation.

The remainder of this paper is organized as follows. Section 2 describes the details of the proposed method for building extraction. Section 3 presents the comparative experimental results in combination with a detailed analysis and discussion. Section 4 concludes this paper and discusses possible future work.

2. Proposed Method

The proposed framework for building extraction consists of three stages, as presented in Figure 1. In the segmentation stage, 6D-SLIC is used to segment superpixels from UAV orthoimages and DSM (generated from image-derived point clouds), and the initial outlines of ground objects are shaped by merging the superpixels. In the building detection stage, a GGLI is used to eliminate vegetation, and the buildings are detected by using the proposed MSCNs (including a feature learning network for deep feature representation and a binary network for building detection). In the regularization stage, the building boundaries are decimated and simplified by removing insignificant vertices using the Douglas–Peucker algorithm. At the same time, the building outlines are regulated by using a proposed iterative optimization algorithm. Finally, the building outlines are validated and evaluated.

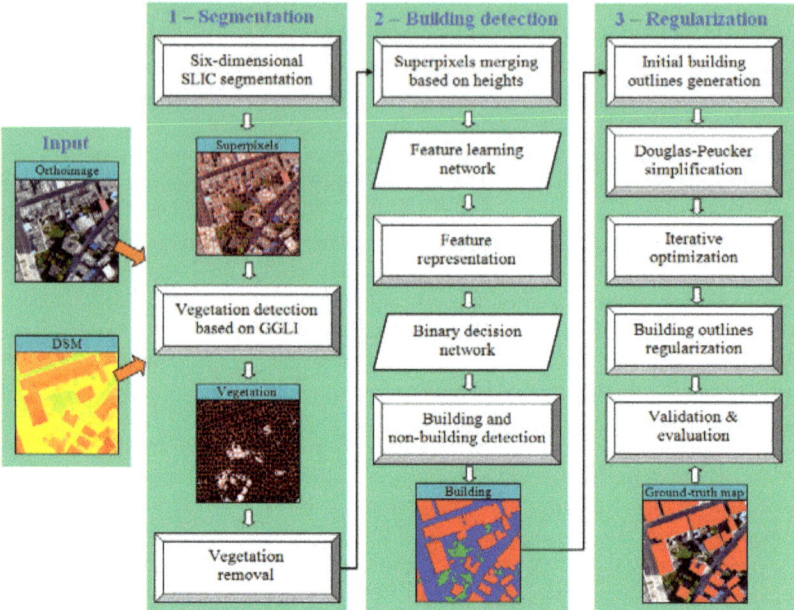

Figure 1. The proposed framework for building extraction.

2.1. D-SLIC-Based Superpixel Segmentation

Image segmentation is a commonly used and powerful technique for delineating the boundaries of ground objects. It is also a popular topic in the fields of computer vision and remote sensing. The classical segmentation algorithms for remotely-sensed imagery, such as quadtree-based segmentation [31], watershed segmentation [32], and Multi-Resolution Segmentation (MRS) [33], often partition an image into relatively homogeneous regions generally using spectral and spatial information while rarely introducing additional information to assist segmentation (e.g., height information) despite various improved methods for finding solutions to some image datasets [9,34–36]. Therefore, the commonly used segmentation methods that are highly dependent on spectral information cannot still break the bottleneck, i.e., sensitivity to illumination, occlusion, quality degradation, and various complex backgrounds. Especially for UAV remote sensing images, a centimeter-level ground resolution provides high-definition details and geometric structural information of ground objects but also generates disturbances, which pose a great challenge in accurately delineating boundaries.

Examples of four types of buildings are given in Figure 2a. The best results of segmentation obtained from classical methods are exhibited in Figure 2b,c; such results are achieved through multiple tests to find the optimal parameters (e.g., scale: 300, shape: 0.4, compactness: 0.8 in MRS). MRS performs better than quadtree-based methods do, but the building boundaries under MRS are still incomplete or confused with backgrounds relative to ground-truth outlines (Figure 2d) because the spectral difference is the insignificant gap at building edges. The accurate outlines of buildings are difficult to delineate from the spectral and spatial information of UAV images. Many strategies can be used to merge the segmented regions to the entities, but finding a generic rule to achieve a perfect solution in a single data source is actually difficult. Most classical algorithms (e.g., MRS) are time and memory consuming when used to segment large remotely-sensed imagery, because they use a pixel grid for the initial object representation [37].

Figure 2. Comparison of building extraction from UAV images using two classical segmentation methods. Column (**a**) includes four types of buildings in urban and rural areas. Columns (**b**,**c**) are the results of quadtree and MRS, respectively; the red lines are the outlines of ground objects. Column (**d**) is the ground-truth outlines corresponding to (**a**), with the red regions denoting the buildings.

Many deep learning-based algorithms, such as multiscale convolutional network [38], deep convolutional encoder–decoder [39], and FCN [40], have been proposed for the semantic segmentation of natural images or computer vision applications, and prominent progress has been made. However, deep learning-based methods dramatically increase computational time and memory and are thus inefficient for the fast segmentation of large UAV orthoimages. In the current study, a 6D-SLIC algorithm is used to extract initial building outlines by joining height information. SLIC is a state-of-the-art algorithm for segmenting superpixels that does not require much computational resource to achieve effective and efficient segmentation.

In the 6D-SLIC algorithm, superpixels are generated by clustering pixels according to their color similarity and proximity in the 2D image plane space; in this way, the proposed algorithm is similar to the SLIC algorithm [20]. Compared to the five-dimensional (5D) space $[l, a, b, x, y]$ in the SLIC algorithm, the height information obtained from image-derived 3D point clouds is then used to cluster pixels. Hence, a 6D space $[l, a, b, x, y, z]$ is used to generate compact, nearly uniform superpixels, where $[l, a, b]$ is defined by the pixel color vector of the CIELAB color space and $[x, y, z]$ is the 3D coordinate of a pixel. The pixels in the CIELAB color space are considered perceptually uniform for small color distances, and height information z is used to cluster the pixels into the building area with approximately equal heights.

Unlike that in the SLIC algorithm, the desired number of approximately equally sized superpixels K is indirectly given in the 6D-SLIC algorithm but is computed on the basis of the minimum area A_{min}, as follows:

$$K = \frac{N \cdot R^2}{A_{min}}, \qquad (1)$$

where N is the number of pixels in an image and R denotes the ground resolution (unit: m). A_{min} is commonly given as 10 m² with reference to the minimum area of buildings in The literature [5], whereas 5 m² is given to consider small buildings in the current study; each superpixel approximately contains N/K pixels, and a superpixel center would exist for roughly equally sized superpixels at every grid interval $S = \sqrt{N/K}$. K superpixel cluster centers $C_k = [l_k, a_k, b_k, x_k, y_k, z_k]$ with $k = [1, K]$ at regular grid intervals S are selected. Similar to the SLIC algorithm, the search area of the pixels associated with each cluster C_k is assumed to be within $2S \times 2S$ of the 2D image plane space. The Euclidean distance of the CIELAB color space and height are used to define pixel similarity, which is useful in clustering pixels for small distances. The distance measure D_S of the proposed 6D-SLIC algorithm is defined as follows:

$$D_S = \alpha \cdot d_{lab} + (1-\alpha) \cdot d_h + \frac{m}{S} d_{xy}, \qquad (2)$$

where D_S is the sum of the *lab* distance d_{lab}, height difference d_h, and the 2D image plane distance d_{xy} normalized by the grid interval S; α represents the weight to emphasize the contribution of d_{lab} and d_h, and it is the SLIC distance measure when α is set as 1, the weight α can be determined by selecting several building samples from the segmented data and performing multiple trials to obtain the optimal superpixel segmentation effect; and m is a variable that can be given to control the compactness of a superpixel. The distances of d_{lab}^{i,C_k}, d_h^{i,C_k}, and d_{xy}^{i,C_k} between a pixel i ($i \in \mathbb{R}^{2S \times 2S}$) and the cluster center C_k can be computed as follows:

$$\begin{aligned} d_{lab}^{i,C_k} &= \sqrt{(l_k - l_i)^2 + (a_k - a_i)^2 + (b_k - b_i)^2} \\ d_{xy}^{i,C_k} &= \sqrt{(x_k - x_i)^2 + (y_k - y_i)^2} \\ d_h^{i,C_k} &= |z_k - z_i|. \end{aligned} \qquad (3)$$

As a result of the high-definition details of UAV images, noisy pixels may be considerable and should be avoided in the selection of a cluster center. A 3D gradient is proposed to control the sampling of K cluster centers and move them to the lowest 3D gradient position in a 3×3 neighborhood to avoid placing a cluster center at the edge of buildings. The 3D gradients $G(x, y, z)$ are computed as

$$G(x, y, z) = G_I + G_z, \qquad (4)$$

where G_I and G_z denote the gradients of image intensity and height difference, respectively. The two gradients can be computed as

$$\begin{aligned} G_I(x, y) &= \|I(x+1, y) - I(x-1, y)\|^2 + \|I(x, y+1) - I(x, y-1)\|^2 \\ G_z(x, y) &= \|DSM(x+1, y) - DSM(x-1, y)\|^2 + \|DSM(x, y+1) - DSM(x, y-1)\|^2, \end{aligned} \qquad (5)$$

where $I(x, y)$ and $DSM(x, y)$ represent the *lab* vector and height corresponding to the pixel at position (x, y), respectively; and $\|.\|$ denotes the L_2 norm. DSM is generated from image-derived 3D point clouds.

All the pixels of the UAV images are associated with the nearest cluster center on the basis of the minimum distance of D_S. The cluster center C_k is then updated by

$$l_k = \frac{\sum_{i=1,i\in\mathbb{R}^{C_k}}^{n_k} l_{i,k}}{n_k}, a_k = \frac{\sum_{i=1,i\in\mathbb{R}^{C_k}}^{n_k} a_{i,k}}{n_k}, b_k = \frac{\sum_{i=1,i\in\mathbb{R}^{C_k}}^{n_k} b_{i,k}}{n_k} \quad (6)$$

$$x_k = \frac{\sum_{i=1,i\in\mathbb{R}^{C_k}}^{n_k} x_{i,k}}{n_k}, y_k = \frac{\sum_{i=1,i\in\mathbb{R}^{C_k}}^{n_k} y_{i,k}}{n_k}, z_k = \frac{\sum_{i=1,i\in\mathbb{R}^{C_k}}^{n_k} z_{i,k}}{n_k},$$

where n_k is the number of pixels that belong to the cluster center C_k. The new cluster center should be moved to the lowest 3D gradient position again on the basis of the values of Equations (4) and (5). The processes of associating all pixels to the nearest cluster center and recomputing the cluster center are iteratively repeated until the convergence of distance D_S.

After all pixels are clustered into the nearest cluster center, a strategy of enforcing connectivity is employed to remove the small disjoint segments and merge the segments in terms of the approximately equal height in each cluster. Therefore, the initial boundaries of ground objects are shaped by connecting the segments in the vicinity. This definition satisfies the constraint in Equation (7), and clusters i and j are regarded to belong to the same ground object.

$$\left| mean\_z(C_i) - mean\_z(C_j) \right| < z\_threshold, \quad (7)$$

where $mean\_z$ represents the average operation of height and $z\_threshold$ is a given height threshold, which is set to 2.5 m in this study.

We use an efficient and effective superpixel segmentation on the basis of the SLIC algorithm, which is regarded as a simple and efficient approach that is suitable for large-image segmentation. 3D space coordinates, rather than a 2D image plane space, are selected as a distance measure to cluster all pixels of an image into superpixels. The algorithm is expressed below, and the comparisons of superpixel segmentation based on the SLIC and 6D-SLIC algorithms are shown in Figure 3. The building areas are identified by vegetation removal and Siamese-typed networks (described in Sections 2.2 and 2.3), except for the regions merging on the basis of height similarity.

Algorithm 1: 6D-SLIC segmentation

Input: 2D image I and DSM.
Parameters: minimum area A_{min}, ground resolution R, compactness m, weight α, maximum number of iterations max_iters, number of iterations n_iters, minimum distance min_dist.
Compute approximately equally sized superpixels $K \leftarrow \frac{N \cdot R^2}{A_{min}}$.
Compute every grid interval $S \leftarrow \sqrt{N/K}$.
Initialize each cluster center $C_k = [l_k, a_k, b_k, x_k, y_k, z_k]^T$.
Perturb each cluster center in a 3×3 neighborhood to the lowest 3D gradient position.
repeat
for each cluster center C_k **do**
Assign the pixels to C_k based on a new distance measure (Equation (2)).
end for
Update all cluster centers based on Equations (5) and (6).
Compute residual error between the previous centers and recomputed centers $e \leftarrow \left| D_S^{prev} - D_S^{cur} \right|$.
Compute n_iters $\leftarrow$ n_iters $+ 1$.
until $e <$ min_dist or n_iters $>$ max_iters
Enforcing connectivity.

Figure 3 depicts that the boundaries of the superpixels at the building edges obtained from the proposed 6D-SLIC algorithm are closer to the true boundaries of buildings than those obtained from the SLIC algorithm are. Additionally, other four state-of-the-art methods (e.g., Entropy Rate Superpixels (ERS) [41], Superpixels Extracted via Energy-Driven Sampling (SEEDS) [42], preemptive SLIC (preSLIC) [43], and Linear Spectral Clustering (LSC) [44]) are used to compare with the 6D-SLIC algorithm, as shown in Figure 4, the four methods do not perform better, and the 6D-SLIC algorithm also shows more similar shapes to the ground-truth maps of the buildings. Moreover, the metrics, e.g., standard boundary recall BR and under-segmentation error USE [45], are used to measure the quality of boundaries between building over-segments and the ground-truth. From the visual assessment and the statistical results of two quantitative metrics in Table 1, it can be inferred that the 6D-SLIC algorithm performs better than the SLIC algorithm and other four state-of-the-art methods do due to the additional height information used for superpixel segmentation in the 3D space instead of a 2D image plane space.

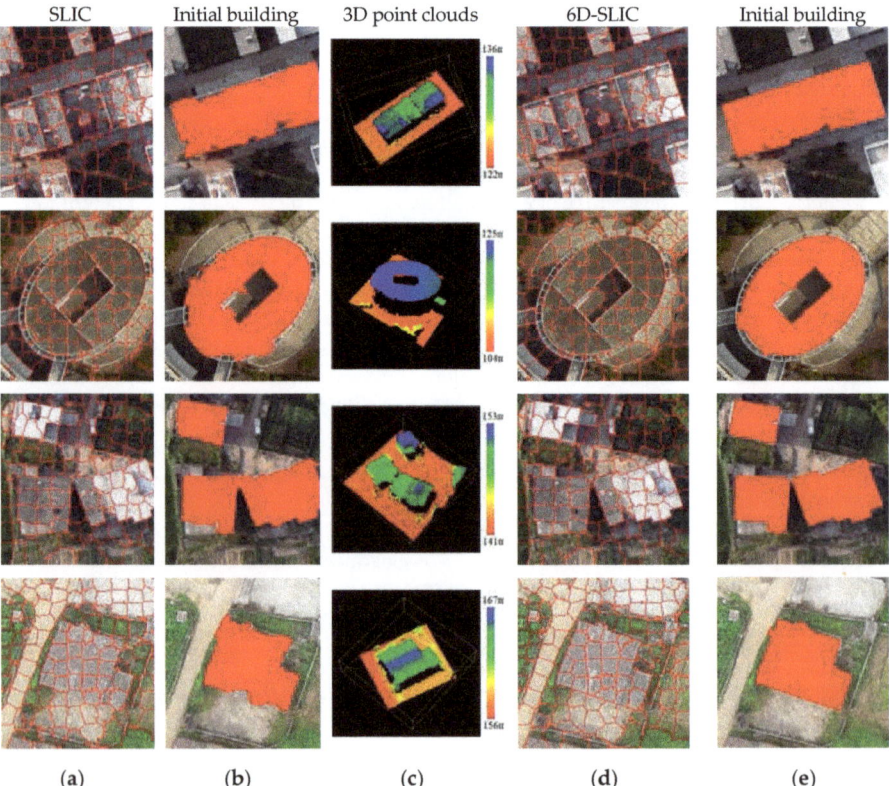

Figure 3. Comparison of building extraction using SLIC and 6D-SLIC algorithms from four building examples corresponding to Figure 2a. Columns (**a**,**d**) are the superpixels obtained from the SLIC and 6D-SLIC algorithms, respectively. Columns (**b**,**e**) are the initial building areas that are shaped by merging superpixels on the basis of approximately equal heights. Column (**c**) shows the 3D point clouds of the four building examples. A high segmentation performance can be achieved when the weight α is set to 0.6.

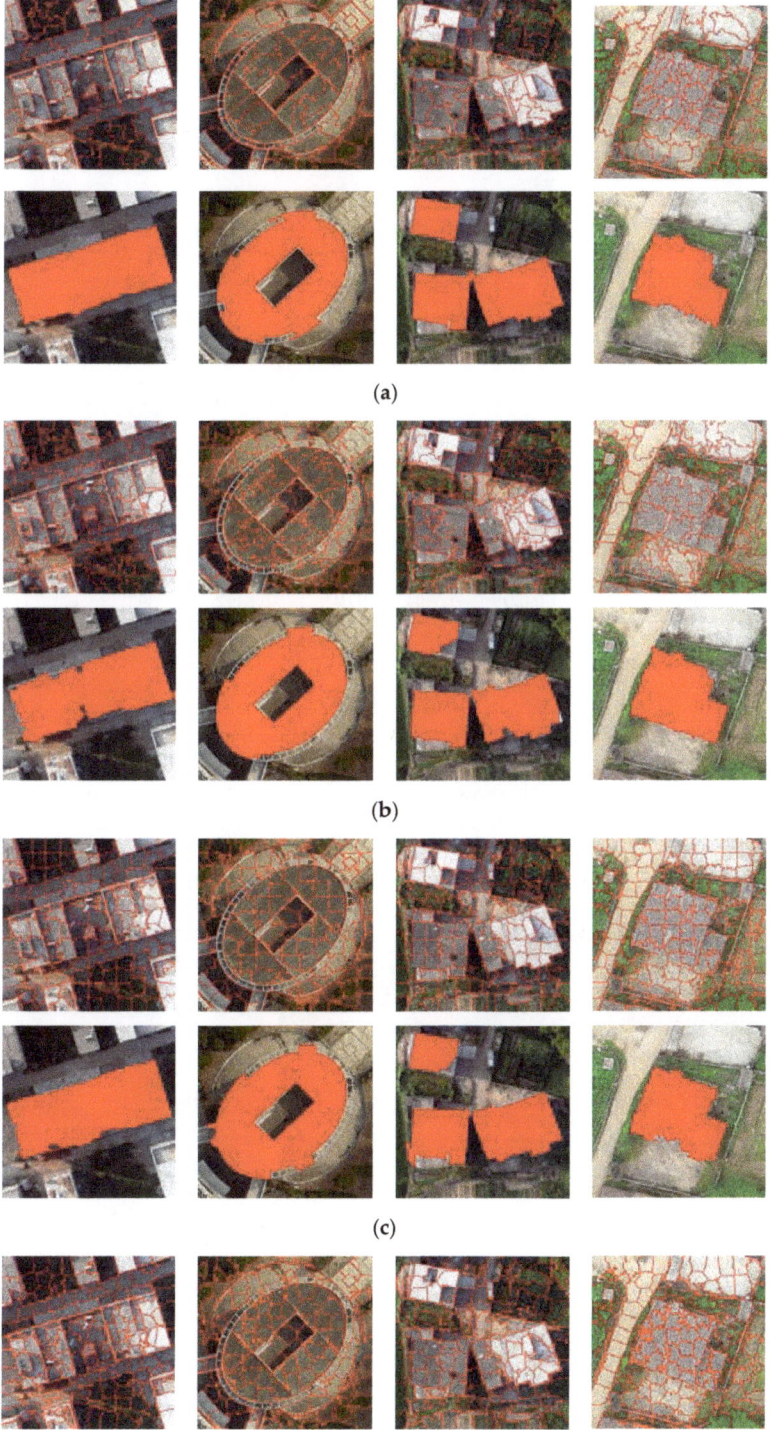

Figure 4. *Cont.*

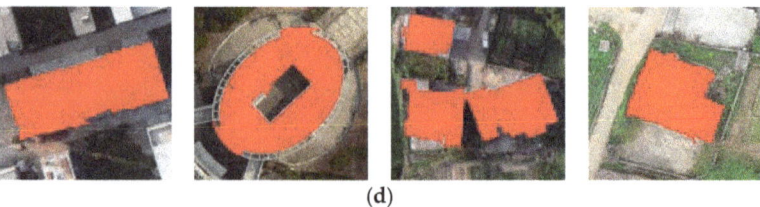

(d)

Figure 4. Building extraction using ERS, SEEDS, preSLIC, and LSC algorithms from four building examples corresponding to Figure 2a. (**a–d**) include the superpixels and the corresponding initial building areas obtained from the ERS, SEEDS, preSLIC, and LSC algorithms, respectively.

Table 1. *BR* and *USE* values of SLIC, ERS, SEEDS, preSLIC, LSC, and 6D-SLIC in the four images in Figure 2a.

Dataset	Metric	SLIC	ERS	SEEDS	preSLIC	LSC	6D-SLIC
(1)	BR	0.7487	0.7976	0.7161	0.8539	0.9039	0.9076
	USE	0.0412	0.0640	0.0378	0.0450	0.0385	0.0231
(2)	BR	0.7152	0.6419	0.7070	0.5769	0.8443	0.9286
	USE	0.1038	0.1213	0.1027	0.1407	0.0654	0.0443
(3)	BR	0.7323	0.8597	0.8608	0.8669	0.8912	0.9629
	USE	0.0681	0.0415	0.0522	0.0539	0.0625	0.0311
(4)	BR	0.7323	0.7918	0.8810	0.8410	0.9313	0.9795
	USE	0.0712	0.0304	0.0413	0.0497	0.0395	0.0325

2.2. Vegetation Removal

In this study, height similarity is not immediately used to merge superpixels for generating initial building boundaries after 6D-SLIC segmentation because the vegetation surrounding buildings with similar heights may be classified as part of these buildings. An example is given in Figure 5. The image-derived 3D point clouds show that the tree canopies have approximately equal heights relative to the nearby buildings; therefore, the surrounding 3D vegetation points are the obstacle and noise for building detection. Vegetation removal is used to truncate vegetation superpixels for further processing to improve the robustness and efficiency of building detection.

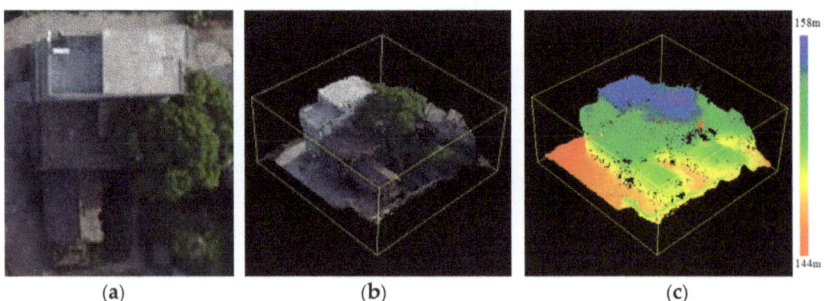

(a) (b) (c)

Figure 5. *Cont.*

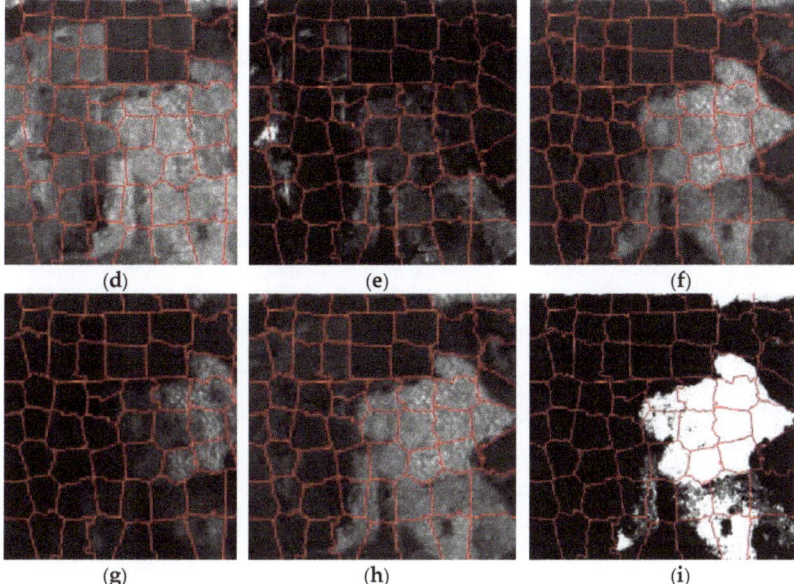

Figure 5. Example to illustrate the vegetation surrounding a building with similar heights. (**a**–**c**) are the orthoimage, 3D point clouds with true color, and 3D point clouds with rendering color, respectively. (**d**–**i**) are the results of NGRDI, VARI, GLI, RI, ExG–ExR, and GGLI. The red lines denote the boundaries of the superpixels.

The NDVI is commonly used to detect vegetation on the basis of near-infrared information, but it is unavailable to 3D image-derived point clouds with true color (RGB) in most UAV remotely-sensed imagery. Thus, many vegetation indices based on the RGB system are proposed, and they include the normalized green–red difference index (NGRDI) [46], visible atmospherically resistant index (VARI) [47], green leaf index (GLI) [48], ratio index (RI) [49], and excess green minus excess red (ExG–ExR) [50]. Figure 4d–h show the extracted vegetation information of Figure 5a using the five vegetation indices. GLI performs better than NGRDI, VARI, GLI, and ExG–ExR do. A suitable intensity threshold is actually difficult to set to separate vegetation from the results of the vegetation index calculation. In [5], a standard SVM classification and a priori training data were employed to extract vegetation from an MFV, which was integrated by the five vegetation indices. However, the method may not achieve a satisfying result when a priori training data are not representative, and the poor vegetation indices also reduce the performance of vegetation extraction. Therefore, in this study, a GGLI is created to extract vegetation by enhancing vegetation intensity and using a self-adaptive threshold. The GGLI is defined as follows:

$$GGLI = 10^{\gamma} \cdot \left(\frac{2G - R - B}{2G + R + B} \right)^{\gamma}, \qquad (8)$$

where γ denotes the gamma value, which is set to 2.5 that is approximately estimated based on the range of 0 to 255 of GGLI value in this study; and R, G, B are the three components of RGB color. Figure 5i shows that the proposed GGLI performs better than the other five vegetation indices do. When the number of pixels belonging to vegetation in the superpixel C_k is more than half of the number of pixels in the superpixel C_k, then the superpixel C_k is considered a vegetation region. The definition satisfies the constraint in Equation (9), and the superpixel C_k is classified into a vegetation region.

$$num\left(I_i; I_i \in v \cap i \in \mathbb{R}^{C_k}\right) > \frac{1}{2} \cdot num\left(I_i; i \in \mathbb{R}^{C_k}\right), \qquad (9)$$

where num denotes the calculation operator of the number of pixels, $I_i \in v$ denotes the pixel I_i belonging to vegetation v, and $\mathbb{R}^{C_k}$ is the region of the superpixel C_k. The $GGLI$ value of a pixel is more than 0.5 times the maximum $GGLI$ value in the entire image, and the pixel is classified into vegetation. Tests using UAV data, including two urban and two rural areas with different vegetation covers, are conducted. Figure 6 shows the receiver operating characteristics (ROCs) of the five popular indices and the proposed GGLI. The true positive rate $TPR = TP/(TP + FN)$ and false positive rate $FPR = FP/(FP + TN)$ of vegetation are computed on the basis of the number of true positives (TP), true negatives (TN), false positives (FP), and false negatives (FN). Over 92.3% of vegetation can be correctly extracted by the proposed GGLI, and the FPs are mainly caused by roads and bare land. Hence, the proposed GGLI achieves the best performance in vegetation detection among all vegetation indices. The vegetation superpixels can be effectively detected and removed with the proposed GGLI, and non-vegetation ground objects are shaped by merging the superpixels on the basis of height similarity.

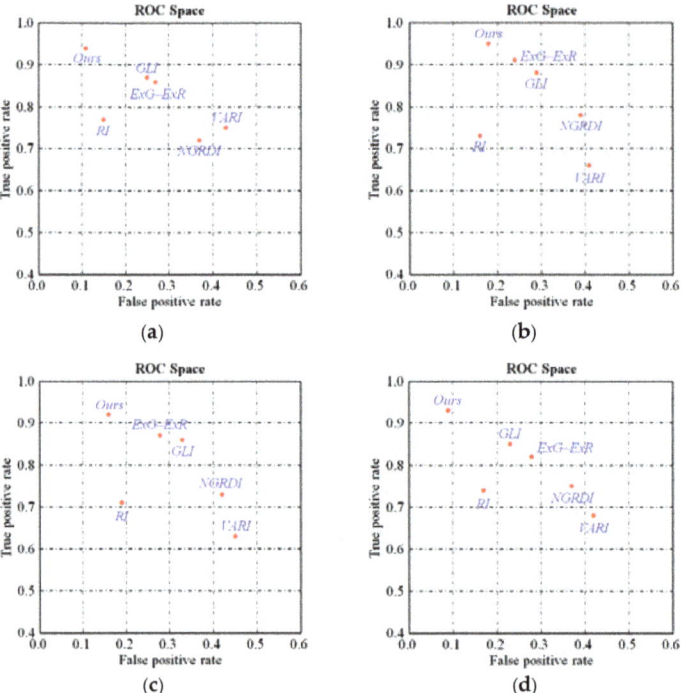

Figure 6. Examples to illustrate the accuracy of vegetation detection by using different datasets. (**a**,**b**) are the results of vegetation detection in two urban areas; (**c**,**d**) are the results of vegetation detection in two rural areas.

2.3. Building Detection Using MSCNs

After the removal of vegetation superpixels, there still exist some non-building superpixels that are meaningless for further delineation of building outlines and should thus be eliminated. Building detection is commonly achieved by classification or recognition of ground objects, in which many types of features, such as color, texture, and geometric structure, are used to directly or indirectly represent building characteristics by feature descriptors. However, most manually designed features remain insufficient to extract buildings from UAV images with high-definition details under various complex backgrounds (e.g., shadow, occlusion, and geometric deformation).

In this paper, we present MSCNs used in building recognition as feature representation using a convolutional network can work efficiently under various complex backgrounds. We aim to learn deep convolutional networks that can discriminate building and non-building ground objects by 2D UAV image and height information. In our case, the discriminative training of buildings does not rely on labels of individual ground objects but on pairs of 2D UAV images and their height information. Multiscale Siamese-typed architecture is suitable for achieving this goal due to three reasons. First, MSCNs are capable of learning generic deep features, which are useful for making predictions on unknown non-building class distributions even when few examples are available in these new distributions. Second, MSCNs are easily trained using a standard optimization technique on the basis of pairs sampled from 2D images and 3D height information. Third, the sizes of buildings in UAV images vary from small neighborhoods to large regions containing hundreds of thousands of pixels. The feature maps displayed in Figure 7 indicate that the small local structures of buildings tend to respond to small convolutional filters, whereas the coarse structures tend to be extracted by large filters. Thus, multiscale convolutional architecture is suitable to extract the detailed and coarse structures of buildings.

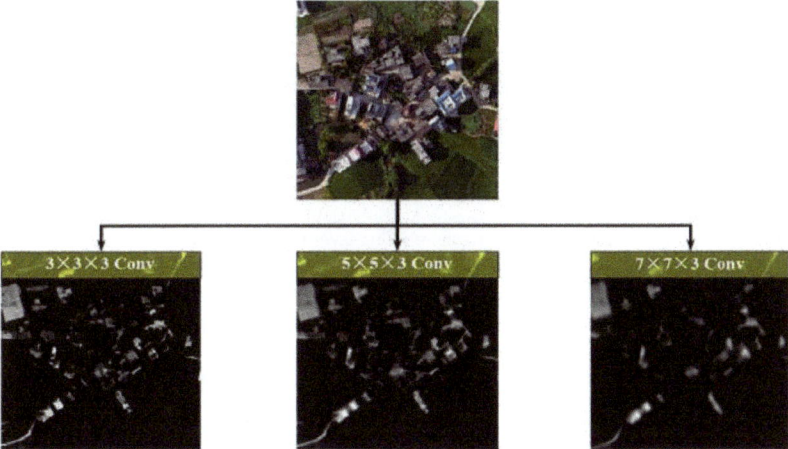

Figure 7. Example to illustrate the feature maps extracted by convolutional filters with three different sizes, which are selected from the first layer of an MSCN model.

The architecture of the proposed MSCNs is shown in Figure 8, and it includes input, feature learning networks, binary decision networks, and output. In this study, input patches are extracted from the merged superpixels. The feature learning network consists of two streams of convolutional and max-pooling layers, three convolutional layers are arranged for feature extraction in each stream, and two max-pooling layers are inserted in between successive convolutional layers to reduce the number of parameters and the computation in MSCNs. Batch normalization [51] is also inserted into each convolutional layer before the activation of neurons. Three subconvolutional layers arranged for the convolutional layers of Conv_x1, Conv_x2, Conv1, and Conv2 are to extract the feature from multiscale space. The convolutional layers Conv1 and Conv2 in two streams share identical weights, whereas Conv_x1 and Conv_x2 do not because of the different inputs of x_1 and x_2. The binary decision network consists of two fully connected layers, and the outputs of MSCNs are predicted as 1 and 0 corresponding to building and non-building regions, respectively.

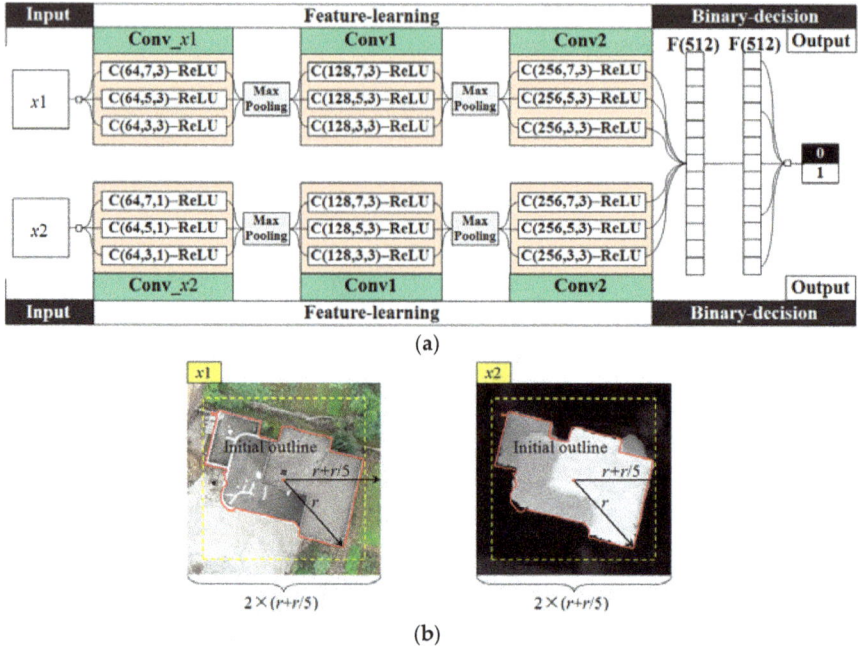

Figure 8. Architecture of MSCNs. In (**a**), C(n, k, m) denotes the convolutional layer with n filters of spatial size k × k of band number m. Each max-pooling layer with a max filter of size 2 × 2 of stride 2 is applied to downsample each feature map. F(n) denotes a fully connected layer with n output units. ReLU represents the activation functions using the rectified linear unit σ(x) = max(0, x). As shown in (**b**), x1 and x2 with same size denote the true-color RGB (m = 3) and height intensity (m = 1) patches, respectively; the extents of x1 and x2 are defined on the basis of the external square and buffer of the initial outline of a ground object, and x1 and x2 are resampled to a fixed size as input, e.g., a fixed size of 127 × 127 pixels used in this study.

In the proposed MSCNs, the output f_j^l of the jth hidden vector in the lth layer via the operators of linear transformation and activation can be expressed as

$$f_j^l = \sigma(z_l) = \sigma\left(\sum_{i \in S_{l-1}} f_i^{l-1} * w_{ij,k}^l + b_j^l\right), \tag{10}$$

where f_i^{l-1} is the ith hidden vector in the (l − 1)th layer; S_{l-1} is the number of hidden vectors in the (l − 1)th layer; w and b represent the weights (or convolution kernels with size k × k in the convolutional layers) and biases, respectively; ∗ is the dot product (or convolution operator in the convolutional layers); and σ(.) denotes the activation function. ReLU is applied to the feature learning and binary decision networks, and sigmoid is used in the output of MSCNs. In this study, discriminative training is prone to achieve the binary output of building and non-building probabilities, which are restricted between 0 and 1. Hence, sigmoid function ($\sigma(x) = \frac{1}{1+e^{-x}}$), instead of ReLU, is used to compute the building and non-building probabilities of a ground object, and the global cost function is an alternative function of the hinge-based loss function with regard to sigmoid output. The proposed MSCNs are trained in a supervised manner by minimizing the global cost function L.

$$L(w,b) = \frac{1}{n}\sum_{i=1}^{n}\left(\frac{1}{2}\left\|h(x^{(i)}) - y^{(i)}\right\|^2\right) + \frac{\lambda}{2}\sum_{l=1}^{n_l-1}\sum_{i=1}^{S_l}\sum_{j=1}^{S_{l+1}}\left(w_{ji,k}^l\right)^2, \quad (11)$$

where $h(x)$ denotes the predicted results of the output layer; y refers to the expected output values (i.e., 0 and 1 in this study) given in a supervised manner; n and n_l are the numbers of trained data and layers, respectively; λ is a weight decay parameter; and S_l and S_{l+1} are the numbers of hidden vectors in layers l and $l+1$, respectively. The optimization of the proposed MSCNs is achieved by using the standard back-propagation algorithm based on stochastic gradient descent. The update rule of weights and biases at epoch T can be written as

$$\begin{aligned} w_{ij,k}^l &= w_{ij,k}^l + \Delta w_{ij,k}^{T,l} \\ \Delta w_{ij,k}^{T,l} &= -\eta \frac{\partial L(w,b)}{\partial w_{ij,k}^l} + \mu \Delta w_{ij,k}^{T-1,l}, \end{aligned} \quad (12)$$

$$\begin{aligned} b_i^l &= b_i^l + \Delta b_i^{T,l} \\ \Delta b_i^{T,l} &= -\eta \frac{\partial L(w,b)}{\partial b_i^l} + \mu \Delta b_i^{T-1,l}, \end{aligned} \quad (13)$$

where η is the learning rate and μ is momentum. We let $\delta_i^{l+1} = \frac{\partial L(w,b)}{\partial z_i^{l+1}}$, and the partial derivatives with respect to the weight and bias between the layer l and the successive layer $l+1$ can be computed by

$$\frac{\partial L(w,b)}{\partial w_{ij,k}^l} = \frac{\partial L(w,b)}{\partial z_i^{l+1}} \frac{\partial z_i^{l+1}}{\partial w_{ij,k}^l} = \delta_i^{l+1} \frac{\partial z_i^{l+1}}{\partial w_{ij,k}^l}, \quad (14)$$

$$\frac{\partial L(w,b)}{\partial b_i^l} = \frac{\partial L(w,b)}{\partial z_i^{l+1}} \frac{\partial z_i^{l+1}}{\partial b_i^l} = \delta_i^{l+1} \frac{\partial z_i^{l+1}}{\partial b_i^l}. \quad (15)$$

The residual errors $\delta_i^{n_l}$ and δ_i^l of the output layer and back propagation in the ith feature map of the lth convolutional layer can be computed as

$$\delta_i^{n_l} = \frac{\partial L(w,b;x,y)}{\partial z_i^{n_l}} = -\left(y_i - f(z_i^{n_l})\right) \cdot f'(z_i^{n_l}), \quad (16)$$

$$\delta_i^l = \left(\sum_{j=1}^{S_{l+1}} w_{ji}^l \delta_i^{l+1}\right) \cdot f'(z_i^l). \quad (17)$$

In this study, the two outputs of MSCNs are considered building probability $p^{(b)}$ and non-building probability $p^{(nb)}$, which are used to define whether a non-vegetation object belongs to a building. The two probabilities satisfy the constraint in Equation (18), and the non-vegetation object is regarded as a building region.

$$\left(p^{(b)} - p^{(nb)}\right) > T_1 \cup p^{(b)} > T_2, \quad (18)$$

where T_1 and T_2 are two given thresholds.

2.4. Building Outline Regularization

Once a building and its initial outline have been determined, the next step is to refine the building outline. An initial outline of a building is shown in Figure 9a. Many points are located in the same line segment, and the building edges are jagged and disturbed by small structures because of pixel-wise segmentation. The initial outline should be optimized by eliminating low-quality vertices and regularizing line segments. For this task, an iterative optimization algorithm, which utilizes the collinear constraint, is applied to regulate the building boundary. This algorithm consists of the following steps:

(1) The Douglas–Peucker algorithm [52,53] is used to optimize building outlines by simplifying the curves that are approximated by a few vertices; the simplified outline is shown in Figure 9b.

(2) The consecutive collinear vertex v_i, which satisfies the condition that the angle $\theta = \angle(\overrightarrow{v_i v_{i-1}}, \overrightarrow{v_i v_{i+1}})$ (as shown in Figure 9c) between two adjacent line segments $\theta \in \left[\frac{11\pi}{12}, \frac{13\pi}{12}\right] \cup \left[0, \frac{\pi}{12}\right]$, is determined. Vertex v_i is added to a candidate point set S_{co} to be eliminated.

(3) Step (2) is repeated by tracking the line segments sequentially from the first vertex to the last vertex until all vertex set V_b of the outline is traversed. The vertices of initial outline belonging to the point sets S_{co} are eliminated from the vertex set V_b, the vertex set V_b is updated, and the candidate point set S_{co} is set as null.

(4) Steps (2) and (3) are repeated until no more consecutive collinear vertex v_i is added to the candidate point set S_{co}.

(5) The vertex set V_b is tracked sequentially from the first vertex to the last vertex; two adjacent vertices v_i and v_{i+1} are considered too close if they satisfy the condition that the distance d (as shown in Figure 9c) between v_i and v_{i+1} is less than a given threshold $d < T_{vv}$ (0.5 m). One of v_i and v_{i+1} is eliminated, and the vertex set V_b is updated.

(6) Step (5) is repeated until no more vertex needs to be eliminated, and the outline is reconstructed by the vertex set V_b.

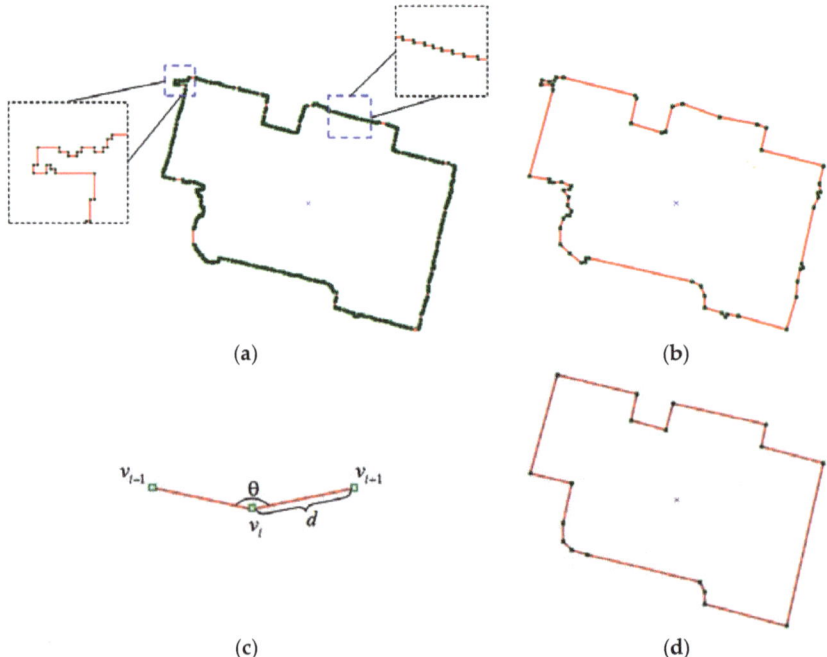

Figure 9. Example to illustrate building outline regularization. (**a**) is an initial outline of the building, with the red lines denoting the line segments and the green dots denoting the vertices. (**b**) is the simplified outline of the building using the Douglas–Peucker algorithm. (**c**) describes the angle of two line segments and the distance of two adjacent vertices. (**d**) is the regulated outline of building obtained from the proposed iterative optimization algorithm.

Figure 9d shows that the proposed iterative optimization algorithm can effectively reduce the superfluous vertices while reconstructing a relatively regular building shape.

3. Experimental Evaluation and Discussion

3.1. Data Description

Two datasets for building extraction are collected by a UAV aerial photogrammetry system, which comprises a UAV platform, one digital camera, a global positioning system, and an inertial measurement unit, to evaluate the performance of the proposed method. The digital camera selected to capture low-altitude UAV remotely-sensed imagery is a SONY ILCE-7RM2 35 mm camera. The test datasets were captured over Zunqiao of Jiangxi Province of China (28° 21′30″N, 117°57′39″E) in the summer of 2016, during which the UAV flew upward for approximately 400 m. These study areas include urban and rural areas, which are characterized by different scales, different roofs, dense residential, tree surrounding, and irregular shape buildings. Structure from motion [54] and bundle adjustment are used to yield high-precision relative orientation parameters of all UAV remotely-sensed images and recover 3D structures from 2D UAV images, which are referenced by using ground control points collected from high-precision GPS/RTK equipment. Dense and precise 3D point clouds with an approximately average point spacing of 0.1 m are derived from corresponding UAV images using a multiview matching method [55] and can thus provide a detailed 3D structure description for buildings. These image-derived 3D point clouds are also used to generate high-resolution UAV orthoimages and DSMs. Two subregions of Zunqiao are selected for building extraction with two datasets of 3501 × 3511 and 1651 × 3511 pixels. The experimental datasets are shown in Figure 10. The two selected regions include not only urban and rural buildings of different materials, different spacings, different colors and textures, different heights, and complex roof structures, but also, complex backgrounds (e.g., topographic relief, trees surrounding buildings, shadow next to buildings, and roads that resemble building roofs).

To facilitate the comparison, the proposed method was also evaluated on an open benchmark dataset, the International Society for Photogrammetry and Remote Sensing (ISPRS) 2D semantic labeling contest (Potsdam), which can be downloaded from the ISPRS official website (http://www2.isprs.org/commissions/comm3/wg4/2d-sem-label-potsdam.html). The dataset contains 38 patches (of the same size, i.e., 6000 × 6000 pixels), each consisting of a very high-resolution true orthophoto (TOP) tile that is extracted from a larger TOP mosaic, and the corresponding DSMs were also provided. The ground sampling distance of both, the TOP and the DSM, is 5 cm. And the buildings were labeled in the ground truth. In this study, to be as consistent as possible with the UAV images, and to evaluate the performance of distinguishing building roof from ground, two very high-resolution true orthophoto tiles that are partially similar in texture and spectral characteristics (e.g., cement road and bare land), are selected to evaluate the proposed method, as shown in Figure 11.

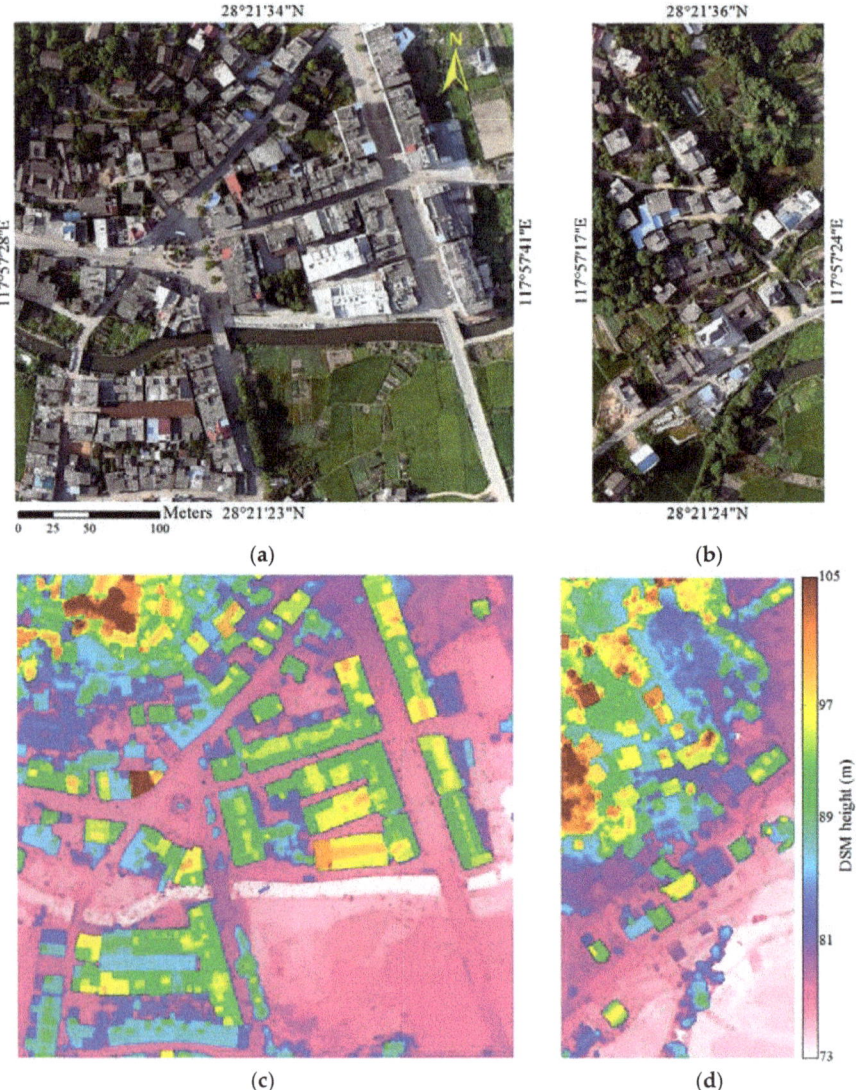

Figure 10. UAV orthoimages for the test regions (**a**,**b**) and the corresponding DSMs (**c**,**d**).

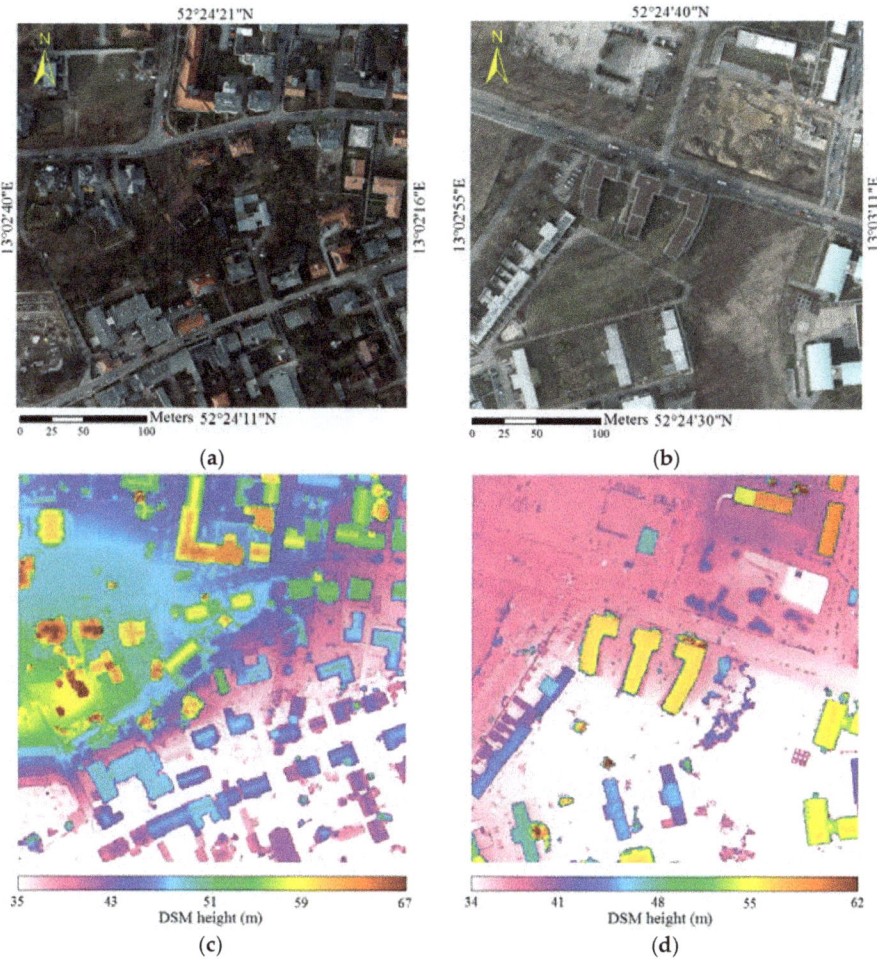

Figure 11. ISPRS true orthophoto tiles for the test regions (**a**,**b**) and the corresponding DSMs (**c**,**d**).

We provide the referenced building outlines, namely, ground-truth building outlines, that are extracted by manually digitizing all recognizable building outlines using ArcGIS software to verify the accuracy of the proposed method and compare it with other state-of-the-art methods. The boundary of each building is difficult to manually interpret by UAV orthoimage alone; therefore, we digitize the boundaries of buildings by the combination of UAV orthoimage and DSM. The two datasets contain 99 and 34 buildings separately. Figure 10a shows many buildings with boundaries that are not rectilinear and not mutually perpendicular or parallel. The ground-truth buildings of the four experimental datasets are given in Figure 12, some buildings with boundaries that are not rectilinear and not mutually perpendicular or parallel are shown in Figure 12a,c,d.

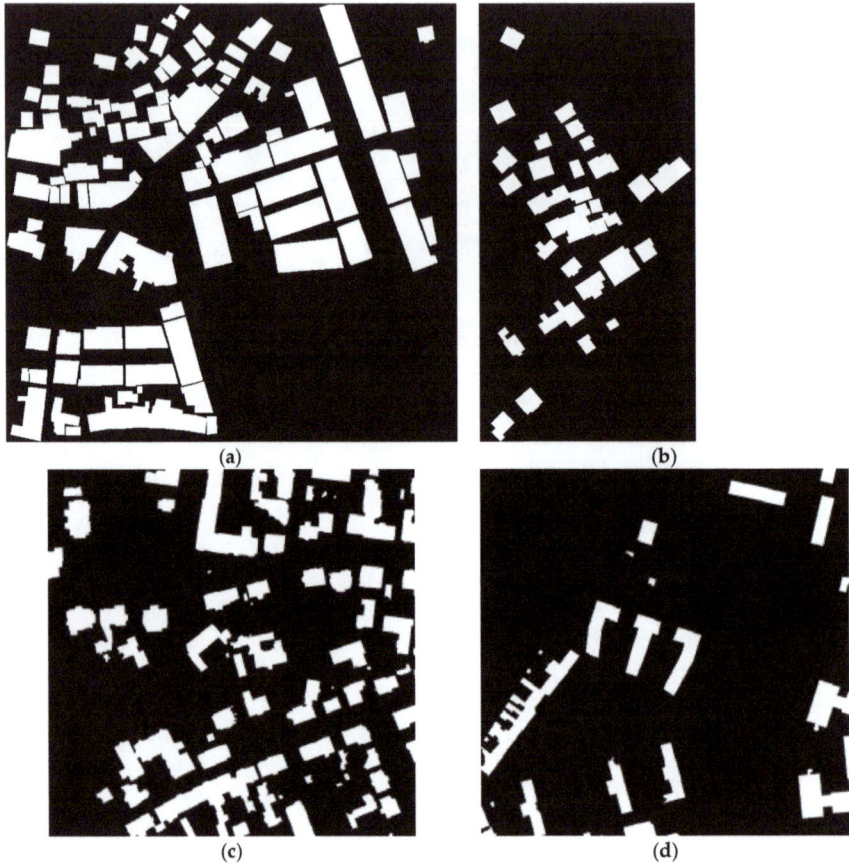

Figure 12. Ground-truth buildings of the four datasets. (**a,b**) are the ground-truth buildings of two UAV datasets. (**c,d**) are the ground-truth buildings collected from the ISPRS dataset. White and black denote building and non-building regions, respectively.

3.2. Evaluation Criteria of Building Extraction Performance

The results of building extraction using the proposed method and other existing methods are evaluated by overlapping with them with the ground-truth maps on the basis of previous reference maps of buildings. Four indicators are used to evaluate the classification performance of buildings and non-buildings: (1) the number of building regions correctly classified as belonging to buildings (i.e., TP), (2) the number of non-building regions incorrectly classified as belonging to buildings (i.e., FP), (3) the number of non-building regions correctly classified as belonging to non-buildings (i.e., TN), and (4) the number of building regions incorrectly classified as belonging to non-buildings (i.e., FN). Three metrics (i.e., completeness, correctness, and quality) are used to assess the results of building detection, which are computed as [56]

$$Comp = \frac{TP}{TP+FN},$$
$$Corr = \frac{TP}{TP+FP}, \quad (19)$$
$$Qual = \frac{TP}{TP+FN+FP},$$

where *Comp* (i.e., completeness) is the proportion of all actual buildings that are correctly identified as buildings, *Corr* (i.e., correctness) is the proportion of the identified buildings that are actual buildings, and *Qual* (i.e., quality) is the proportion of the correctly identified buildings in all actual and identified buildings. The identified building or non-building regions are impossible to completely overlap with the corresponding regions in the reference maps. Therefore, we define two rules to judge whether a region is correctly identified to the corresponding category. First, the identified region that overlaps the reference map belongs to the same category. Second, the percentage of the area of the identified region that overlaps the reference map is more than 60% [9].

Although *Comp*, *Corr*, and *Qual* are the popular metrics to assess the results of building detection, these metrics remain insufficient to measure how good the overlap is between an outline of a building and the corresponding outline in the reference map. Hence, we use three other metrics, i.e., Recall, Precision, and intersection over Union (IoU) [57], to quantitatively evaluate the delineation performance of building outline. As shown in Figure 13, *A* and *B* are respectively the ground truth and the extracted building area, then *Recall*, *Precision*, and *IoU* can be computed as

$$Recall = \frac{Area(A \cap B)}{A}, \qquad (20)$$

$$Precision = \frac{Area(A \cap B)}{B}, \qquad (21)$$

$$IoU = \frac{Area(A \cap B)}{Area(A \cup B)}, \qquad (22)$$

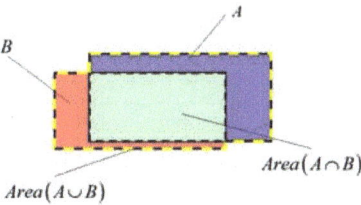

Figure 13. Overlap of a correctly identified building and the corresponding ground truth. The blue area is the ground truth. Green area is the intersection part of *A* and *B*, and the area within the yellow line is the union of *A* and *B*.

3.3. MSCNs Training

The training datasets of MSCNs are generated from UAV orthoimages and DSMs, which are obtained by photogrammetric techniques. The datasets include buildings of multiscale, different colors and heights, and complex roof structures in urban and rural areas. The datasets also contain patches with complex backgrounds, such as shadows, topographic relief, and trees surrounding buildings. A total of 50,000 pairs of patches (half building and half non-building patches) with a fixed size of 127 × 127 pixels are extracted in a supervised manner from the UAV orthoimages and DSMs that do not include the experimental images. The non-building patch examples are generated by two ways. First, we randomly select patches from non-building areas, which are determined by manually masking building areas. Second, some examples that are easily confused with buildings are specially selected from the regions of roads, viaducts, and railways to supplement non-building patches. Furthermore, 150,000 pairs of patches are extended to avoid overfitting by image rotation (e.g., 90°, 180°, and 270°), Gaussian blur, and affine transformation. Therefore, the total number of patch pairs is 200,000, in which 195,000 and 5,000 pairs of patches are randomly selected as training and test datasets, respectively.

At the training stage of MSCNs, a batch size of 100 is used as the input; hence, 1950 iterations exist in each epoch. The MSCNs are trained in parallel on NVIDIA GPUs, and training is forced to terminate when the average value of the loss function is less than 0.001 or the epochs are more than 100. The weights of convolutional and fully connected layers are initialized by random Gaussian distributions [58]. The momentum and weight decay are fixed at 0.9 and 0.0005, respectively. The initial learning rate is set to 0.01 and then gradually reduced by using a piecewise function [25] to accelerate the training of MSCNs. Another metric, namely, overall accuracy (*OA*), is used to evaluate the performance of building and non-building classification for quantitatively assessing the training performance of the proposed MSCNs. *OA* is computed as

$$OA = \frac{TP + TN}{TP + FN + TN + FP},\qquad(23)$$

in which *TP*, *FN*, *TN*, and *FP* are defined in Section 3.2.

We train three Siamese networks, namely, SCNs3, SCNs5, and SCNs7, to evaluate the effects of Siamese networks with and without multiscale. Here, a convolution operator is achieved by using one of the filters with sizes of 3 × 3, 5 × 5, and 7 × 7 in our model. We also evaluate the effect of layer number in our model by adding one convolutional layer to train and test the datasets, namely, MSCNs(layer+). The trained model achieves state-of-the-art results in training and test datasets (Table 2), and Figure 14 shows the changes in *OA* and the losses with increasing epochs during the training of MSCNs. Our network and the deeper network (layer+) achieve higher accuracies than SCNs3, SCNs5, and SCNs7 do. Although the deeper network (layer+) performs slightly better than MSCNs do, the convergence of MSCNs(layer+) is slower than that of MSCNs. MSCNs(layer+) converge at nearly 24 epochs (4.68×10^4 iterations), whereas MSCNs converge at nearly 30 epochs (5.85×10^4 iterations). In addition, MSCNs perform better than SCNs3, SCNs5, and SCNs7 do in terms of *Completeness*, *Correctness*, and *Quality*. The experimental results demonstrate the effective performance of MSCNs given the tradeoff between accuracy and network complexity.

Table 2. Metrics of MSCNs, including *Comp*, *Corr*, *Qual*, and *OA*.

Model	Dataset	*Comp*	*Corr*	*Qual*	*OA*
SCNs3	Training	0.9232	0.9349	0.8674	0.9295
	Test	0.8824	0.9230	0.8219	0.9044
SCNs5	Training	0.9440	0.9584	0.9069	0.9515
	Test	0.9088	0.9385	0.8577	0.9246
SCNs7	Training	0.9530	0.9686	0.9244	0.9610
	Test	0.9226	0.9553	0.8844	0.9397
MSCNs	Training	0.9670	0.9796	0.9479	0.9735
	Test	0.9584	0.9689	0.9298	0.9638
MSCNs(layer+)	Training	0.9672	0.9798	0.9483	0.9736
	Test	0.9594	0.9693	0.9311	0.9645

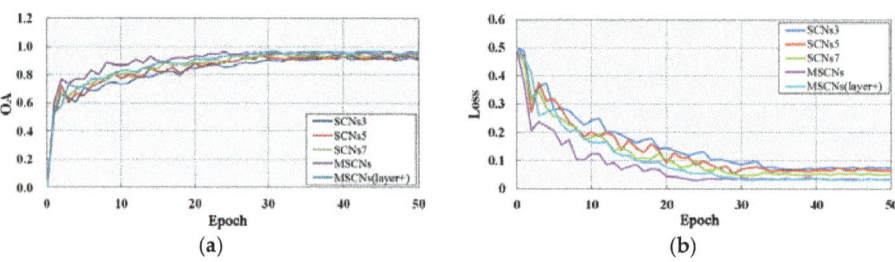

Figure 14. Plots showing the OA (**a**) and loss (**b**) of SCNs3, SCNs5, SCNs7, MSCNs, and MSCNs(layer+) in the training epochs.

3.4. Comparisons of MSCNs and Random Forest Classifier

After vegetation removal and superpixel merging, many non-building regions remain. Postprocessing is needed to further classify building and non-building regions. The identified vegetation and the remaining regions after vegetation removal are shown in Figure 15. A classifier of

MSCNs is designed for building detection in this study due to its capability of non-linear estimation and the robustness of object classification under complex backgrounds. Another classifier, named Random Forest, has been proven to perform efficiently in the classification of building and non-building regions in the literature [59], in which an experiment comparing Random Forest with MSCNs was conducted to test the effectiveness of the MSCN classifier. Multiple features were extracted to classify using Random Forest and compared to deep features. Table 3 provides the details of multiple features and the parameters of the Random Forest classifier. The experimental results of the ISPRS dataset are given in Figure 16, Figure 17 shows the confusion matrices of building and non-building classification obtained from the Random Forest classifier and MSCNs in the four experimental datasets.

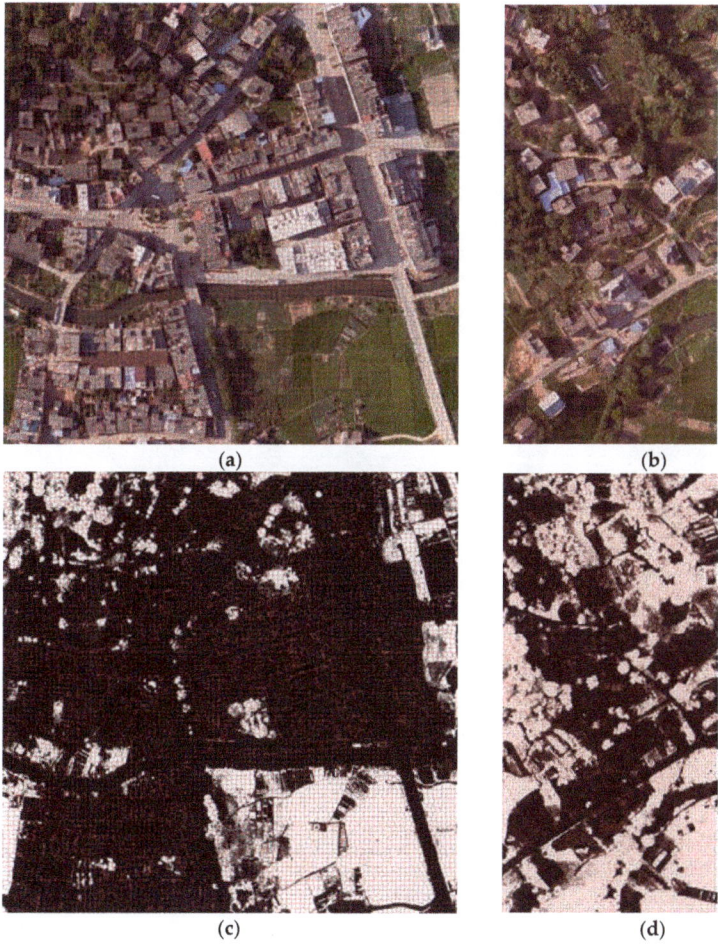

Figure 15. *Cont.*

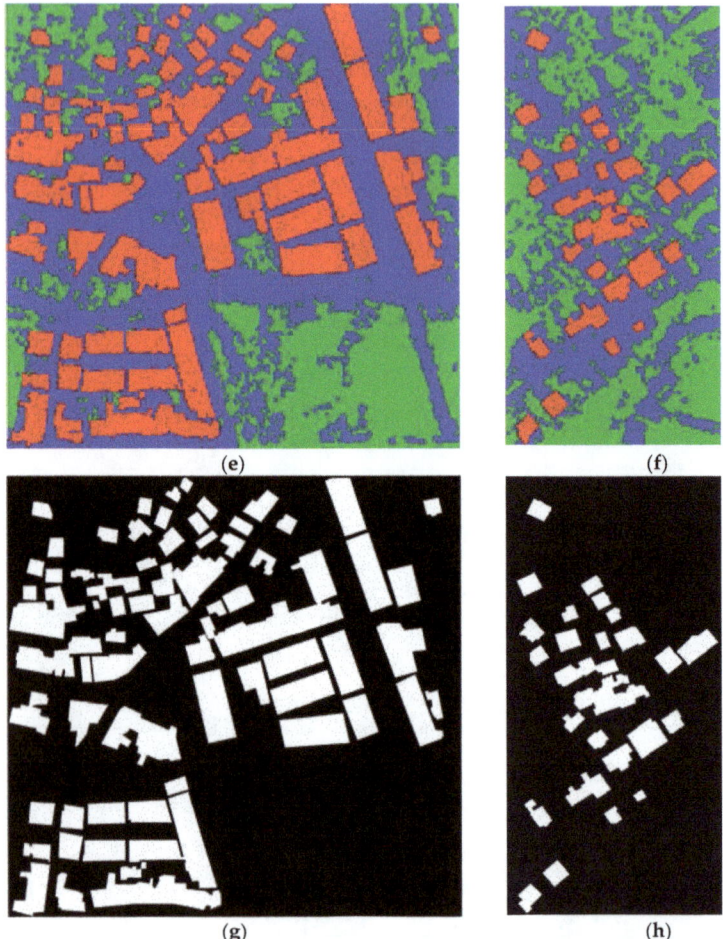

Figure 15. Results of 6D-SLIC, vegetation removal, and superpixel merging in two UAV datasets. Red lines in (**a**,**b**) show the boundaries of superpixels obtained from the 6D-SLIC algorithm. White regions in (**c**,**d**) are the vegetation obtained from the proposed GGLI algorithm. Green, red, and blue in (**e**,**f**) denote the vegetation, building, and non-building regions, respectively. (**g**,**h**) are the buildings extracted by using the proposed framework, and white color denotes the building areas.

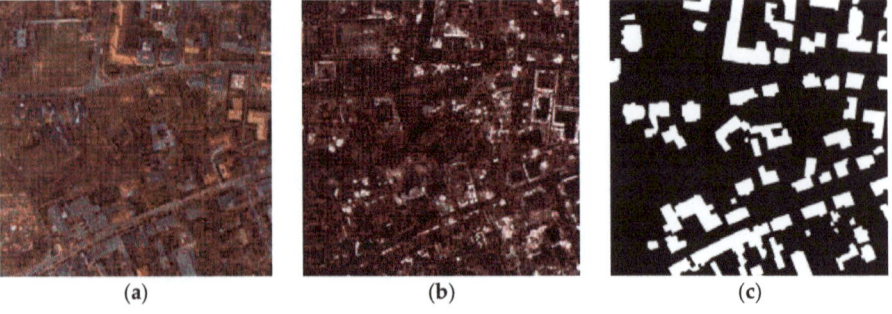

Figure 16. *Cont.*

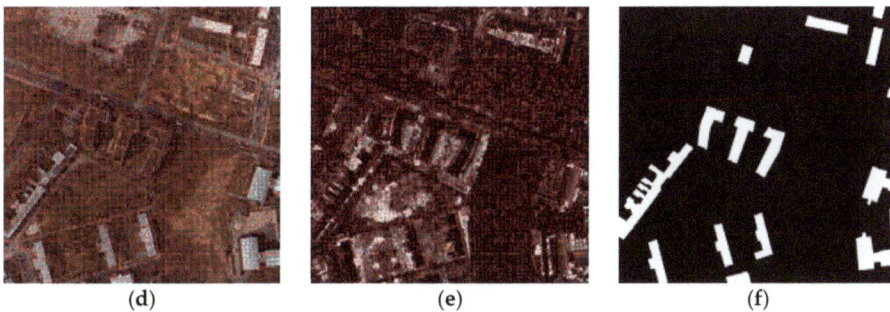

(d)　　　　　　　　　　　(e)　　　　　　　　　　　(f)

Figure 16. The experimental results of the ISPRS dataset using the proposed method. Red lines in (**a,b,d,e**) show the boundaries of superpixels obtained from the proposed 6D-SLIC algorithm. White regions in (**b,e**) are the vegetation obtained from the proposed GGLI algorithm. (**c,f**) are the buildings extracted by using the proposed framework, and white color denotes the building areas.

Table 3. Parameters of Random Forest classifier. SIFT denotes the feature detected by using scale-invariant feature transform. Hog denotes the feature represented by using a feature descriptor, namely, histogram of oriented gradients.

Feature	Parameters	Description
Color histogram	$quantization\_level = 8$ $color_space = $ "lab"	Level of quantization is applied to each image. Image is converted into lab color space.
Bag of SIFT	$vocab\_size = 50$ $dimension = 128$ $smooth\_sigma = 1$ $color_space = $ "grayscale"	Vocabulary size is set as 50. Dimension of descriptor is set as 128. Sigma for Gaussian filtering is set as 1. Image is converted into grayscale.
Hog	$vocab\_size = 50$ $cell\_size = 8$ $smooth\_sigma = 1$ $color_space = $ "rgb"	Vocabulary size is set as 50. Cell size is set as 8. Sigma for Gaussian filtering is set as 1. RGB color space is used.

(a)　　　　　　　(b)　　　　　　　(c)　　　　　　　(d)

Figure 17. Comparison of confusion matrices of building and non-building classification in Random Forest and MSCNs. (**a**–**c**) are the confusion matrices of the Random Forest classifier that uses color histogram, bag of SIFT, and Hog, respectively. (**d**) is the confusion matrix of the proposed MSCNs.

Figure 17 shows that the performance of the proposed MSCNs is better than that of the Random Forest classifier that uses the color histogram, bag of SIFT, and Hog in terms of confusion matrices. Almost all buildings in the two experimental datasets are correctly identified by using the proposed MSCNs, whereas the building identification accuracy of the Random Forest classifier based on color histogram and the bag of SIFT is less than 85%, and that based on Hog is less than 90%. This finding is attributed to two reasons. First, height is combined with spectral information for jointly distinguishing building and non-building ground objects. This approach helps determine a clear gap between building and other ground objects that are similar in texture and spectral characteristics (e.g., cement road and bare land). Second, deep learning-based networks can extract non-linear and high-level semantic features that are not easily affected by image grayscale variations, and they show higher robustness than the other three low-level manually designed features (color histogram, bag of SIFT, and Hog) do. Figure 18 shows the feature representation of the color histogram, bag of SIFT, Hog, and MSCNs. The influence of grayscale variations is given by simulation. Hog is more robust to gray variations than color histogram and SIFT are, the feature vectors extracted by color histogram are easily affected by image grayscale variations, and the feature vectors extracted by SIFT in the dimension of 0 to 20 are different. Compared with the visualization of the three low-level manually designed features, as shown in Figure 18i, that of high-level deep features obtained by the proposed MSCNs shows high similarity to Figure 18a,e under grayscale variations. This result proves that the proposed MSCNs perform with high stability for feature extraction.

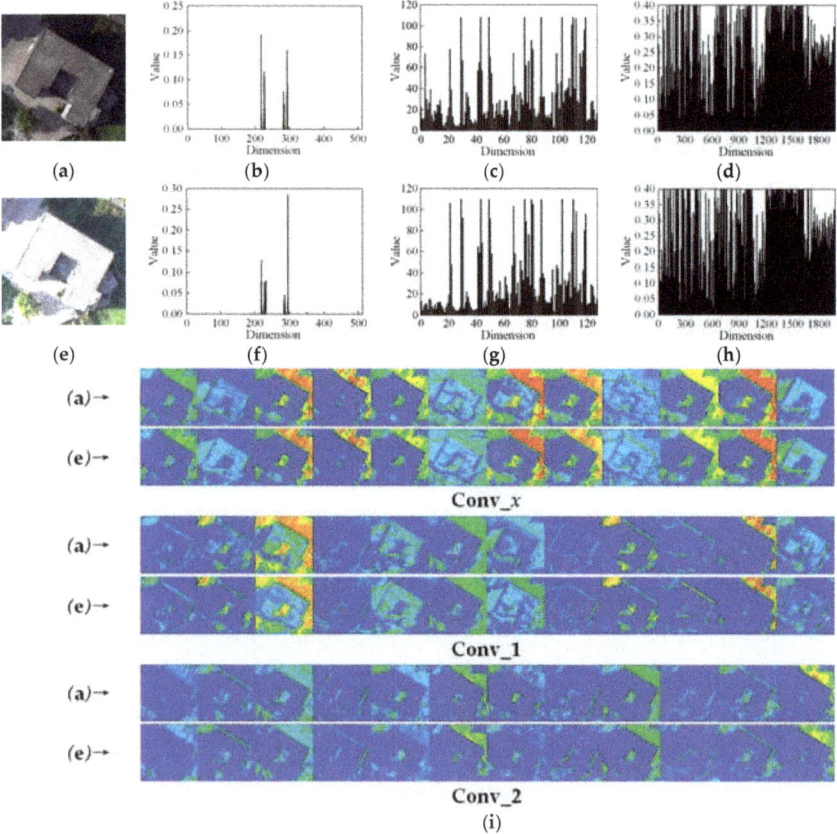

Figure 18. Comparison of features of the color histogram, SIFT, Hog, and MSCNs. (**a**) is an example of a building, and (**e**) denotes the gray variations (e.g., brightness + 50% and contrast + 50%) corresponding to (**a**). (**b**,**f**) are the feature vectors of the color histogram of (**a**,**e**), respectively. (**c**,**g**) are the feature vectors of SIFT of (**a**,**e**), respectively. (**d**,**h**) are the feature vectors of Hog of (**a**,**e**), respectively. (**i**) is the visualization of deep features extracted by MSCNs in the three convolutional layers, i.e., Conv_x, Conv_1, and Conv_2. Only 12 feature maps are provided in each convolutional layer, and (**a**)→ and (**e**)→ denote the corresponding rows to the images (**a**,**e**) that are the feature maps of Conv_x, Conv_1, and Conv_2.

3.5. Comparisons of Building Extraction Using Different Parameters

In the 6D-SLIC-based algorithm, the initial size and compactness of superpixels and the weight of height are the three key parameters that affect the extraction of building boundaries. The metric (i.e., *IoU*) are used to evaluate the effects of building extraction. Figure 19 shows the results of segmentation with different initial sizes of superpixels (e.g., 3, 5, 10, and 15 m^2; i.e., 17 × 17, 22 × 22, 31 × 31, and 38 × 38 pixels inferred by Equation (1)), different compactness values (e.g., 10, 20, 30, and 40), and different weights (e.g., 0.2, 0.4, 0.6, and 0.8).

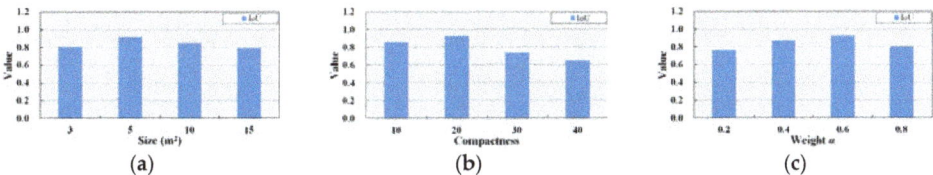

Figure 19. Comparison of the *IoU* values with different initial sizes, compactness of superpixels, and weight of height.

Figure 19a depicts that 6D-SLIC at 5 m^2 initial size of superpixels performs better than it does at other sizes in terms of *IoU*. The superpixel merging of the small size (e.g., 3 m^2) is susceptible to UAV image-derived poor-quality 3D point clouds at the edge of buildings (as shown in Figures 3 and 4) that result in the shrinkage of building boundaries. By contrast, the superpixel merging of the larger size (e.g., 10 and 15 m^2) may be insensitive to building boundary identification because building details are ignored. Therefore, the results of 6D-SLIC at 3, 10, and 15 m^2 initial sizes are worse than those at 5 m^2 initial size. Figure 19b shows a trade-off between spatial proximity and pixel similarity of color and height information when the compactness value is set to 20. A good segmentation performance can be achieved when the weight α is set as 0.6 in Figure 19c, which is also a trade-off of the contribution between *lab* distance d_{lab} and height difference d_h.

3.6. Comparisons of the Proposed Method and State-of-the-Art Methods

Our work uses the proposed 6D-SLIC algorithm as the building outline extractor in the image segmentation part as it allows the full use of the spectral and terrain information of UAV remotely-sensed imagery. The proposed MSCNs with nine layers are then used to classify building and non-building areas. The state-of-the-art results have fewer parameters and involve less computation than the results of two of the most popular networks for image segmentation, i.e., FCN [27] and U-Net [29], do.

To testify the superpixel segmentation performance of the proposed 6D-SLIC algorithm for building extraction, ERS, SEEDS, preSLIC, and LSC are used to extract building from the four experimental datasets. For a fair comparison, the segmented subregions are merged on the basis of the height similarity in the neighborhoods, and the optimal segmentations of ERS, SEEDS, preSLIC, and LSC are achieved through many repeated trials. Also, we select three other state-of-the-art methods, namely, UAV data- (i.e., see Dai [5]), FCN-, and U-Net-based methods, for comparison and analysis to evaluate the proposed building extraction method. The open-source code and pretrained weights of FCN and U-Net are respectively collected from the corresponding GitHub to ensure the repeatability of the experiments. The training samples generated from the UAV images are used for the parameter fine tuning of FCN and U-Net.

Tables 4 and 5 present the comparative results of *Recall*, *Precision*, and *IoU* values using the six superpixel segmentation algorithms (i.e., SLIC, ERS, SEEDS, preSLIC, LSC, and 6D-SLIC) before and after the regularization. 6D-SLIC achieves a better performance than the other five algorithms do in terms of the *Recall*, *Precision*, and *IoU* values. The building outlines obtained from 6D-SLIC are closest to the ground-truth maps, whereas the regions at the building edges with similar colors are easily confused in the other five algorithms and result in poor building extraction. From the comparison of before and after the regularization, it can be inferred that the proposed regularization can also improve the performance of building extraction that uses the six superpixel segmentation algorithms.

Table 4. Superpixel-based building extraction results on the four datasets before regularization stage.

Dataset	Metric	SLIC	ERS	SEEDS	preSLIC	LSC	Ours
Dataset1	Recall	0.8833	0.8933	0.8803	0.9113	0.9153	0.9421
	Precision	0.8927	0.9027	0.9127	0.8977	0.9143	0.9650
	IoU	0.7986	0.8148	0.8119	0.8256	0.8430	0.9109
Dataset2	Recall	0.8994	0.9094	0.8964	0.9304	0.9315	0.9583
	Precision	0.8907	0.9107	0.8804	0.8960	0.9220	0.9675
	IoU	0.8001	0.8349	0.7991	0.8397	0.8635	0.9285
Dataset3	Recall	0.8104	0.8204	0.8077	0.8414	0.8425	0.8890
	Precision	0.8213	0.8413	0.8111	0.8266	0.8526	0.9286
	IoU	0.6889	0.7105	0.6798	0.7152	0.7354	0.8321
Dataset4	Recall	0.8317	0.8417	0.8280	0.8667	0.8628	0.9016
	Precision	0.8446	0.8476	0.8448	0.8493	0.8754	0.9101
	IoU	0.7213	0.7311	0.7187	0.7512	0.7684	0.8279

Table 5. Superpixel-based building extraction results on the four datasets after regularization stage.

Dataset	Metric	SLIC	ERS	SEEDS	preSLIC	LSC	Ours
Dataset1	Recall	0.9233	0.9243	0.8943	0.9223	0.9233	0.9611
	Precision	0.8969	0.9167	0.9237	0.9119	0.9273	0.9656
	IoU	0.8347	0.8527	0.8328	0.8468	0.8609	0.9293
Dataset2	Recall	0.9194	0.9364	0.9165	0.9514	0.9495	0.9683
	Precision	0.8929	0.9227	0.8944	0.9102	0.9330	0.9679
	IoU	0.8281	0.8683	0.8270	0.8698	0.8889	0.9382
Dataset3	Recall	0.8334	0.8474	0.8278	0.8624	0.8605	0.9190
	Precision	0.8393	0.8533	0.8251	0.8408	0.8636	0.9406
	IoU	0.7187	0.7396	0.7042	0.7413	0.7575	0.8740
Dataset4	Recall	0.8577	0.8687	0.8421	0.8897	0.8838	0.9416
	Precision	0.8756	0.8696	0.8638	0.8675	0.9045	0.9321
	IoU	0.7645	0.7685	0.7434	0.7833	0.8084	0.8876

The experimental results of Dai's method, FCN, U-Net are also given in Table 6. A certain height threshold (e.g., 2.5 m is used in the method of Dai [5]) is difficult to give for separating building points; thus, some low-height buildings shown in the results of Dai in Figure 20a,e are incorrectly identified, resulting in a smaller *Recall* value than that achieved by the other three methods. The FCN- and U-Net-based methods, which allow deep neural network-based semantic segmentation that is robust and steady for pixel-wise image classification, work efficiently for building detection, and almost all buildings can be identified, as shown in Table 6 and Figure 20. However, the FCN-based method is sensitive to noise, and it cannot accurately extract the building outlines in some regions, such as the shadows shown in the result of FCN in Figure 20b. In comparison with the FCN-based method, the U-Net-based method performs better in single-house-level building outline extraction, as shown in Figure 20. Overall, our method demonstrates superior performance in terms of major metrics and building outline delineation.

In addition, it can be inferred that the computational cost of our method is much less than FCN- and U-Net-based methods because their architectures include more complex convolutional operations with a high computational cost. During our computational efficiency analysis, our method also shows a significant improvement in computational cost in terms of testing (less than one-fifth of the time consumed by the FCN- and U-Net-based methods operated in parallel on NVIDIA GPUs).

Table 6. *Recall*, *Precision*, and *IoU* values of Dai's method, FCN, U-Net, and our method in the four datasets.

Dataset	Metric	Dai	FCN	U-Net	Ours
Dataset1	Recall	0.7931	0.9306	0.9523	0.9611
	Precision	0.9301	0.8593	0.9547	0.9656
	IoU	0.7485	0.8075	0.9112	0.9293
Dataset2	Recall	0.7971	0.9484	0.9566	0.9683
	Precision	0.9505	0.9533	0.9587	0.9679
	IoU	0.7653	0.9063	0.9187	0.9382
Dataset3	Recall	0.7471	0.8684	0.8836	0.9190
	Precision	0.8805	0.8833	0.9005	0.9406
	IoU	0.6783	0.7790	0.8050	0.8740
Dataset4	Recall	0.7431	0.8506	0.8793	0.9416
	Precision	0.8601	0.7893	0.8965	0.9321
	IoU	0.6630	0.6932	0.7983	0.8876

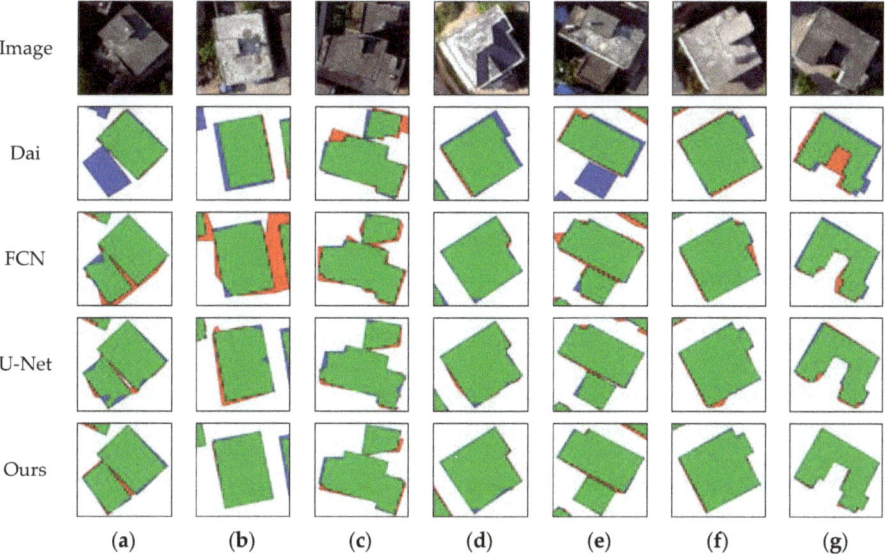

Figure 20. Representative results of single-building-level building extraction from Dai's method, FCN, U-Net, and our method. (**a**–**g**) are the seven examples that are selected to exhibit the experimental results. The green, red, blue, and white channels in the results respectively represent the TP, FP, FN, and TN of building areas.

The experimental results indicate that the proposed framework presents more significant improvements than the other methods do in terms of the effectiveness and efficiency of building extraction, which can be explained by a number of reasons. First, the point clouds provide valuable information for building extraction, the 6D-SLIC algorithm can rapidly cluster pixels into superpixels by utilizing UAV image spectral information and image-derived point clouds; the latter helps accurately delineate the outline of ground objects despite the existence of similar intensity and texture at building edges in Figure 3. Second, the proposed GGLI can significantly remove vegetation and improve the efficiency of building detection. Third, the deep and salient features learned by a Siamese-type network are more useful and stable in classifying building and non-building areas, even in this case of image intensity dramatic variations, in comparison with the manually designed features in Figure 18. Finally, the proposed building outline regularization algorithm integrates the Douglas–Peucker and iterative optimization algorithms that can remove superfluous vertices and small structures, i.e., the pruned processing is useful to improve the precision of building delineation.

In the method of Dai, the height of the off-terrain points is calculated by a certain threshold that is unstable; thus, some buildings that are not in this threshold are incorrectly identified. The assumption that only the geometry of two mutually perpendicular directions exists in buildings, i.e., the building boundary regularization has limitations for accurately delineating non-regular buildings, is referred to. In the FCN-based method, the subsampling and upsampling operations may cause the information loss of input images, and thus, the prediction results of buildings often have blurred and inaccurate boundaries of buildings, as shown in the results of FCN in Figure 20. In the U-Net-based method, despite the skip connections added to achieve superior performance in comparison with the FCN-based method, pixel-wise classification solely relies on the features within a localized receptive field; therefore, it is still insufficient to capture the global shape information of building polygons, and it is sensitive to noisy data. That is, the architectures of FCN and U-Net are not perfect enough, and there are restrictions on performance improvement. As a result, small structures may exist in building boundaries. The experimental results imply that the low-level manually designed features are unsuitable for building detection because of the influences of grayscale variations. FCN- and U-Net-based methods are difficult to use in extracting the regulated boundaries of buildings when noisy data are present. Our method performs better not only because the point clouds provide valuable information but also is much less computational cost in comparison with FCN- and U-Net-based methods.

4. Conclusions

In this paper, we present a framework to effectively extract building outlines by utilizing a UAV image and its image-derived point clouds. First, a 6D-SLIC algorithm is introduced to improve superpixel generation performance by considering the height information of pixels. Initial ground object outlines are delineated by merging superpixels with approximately equal height. Second, GGLI is used to eliminate vegetation for accelerating building candidate detection. Third, MSCNs are designed to directly learn deep features and building confirmation. Finally, the building boundaries are regulated by jointly using the Douglas–Peucker and iterative optimization algorithms. The statistical and visualization results indicate that our framework can work efficiently for building detection and boundary extraction. The framework also shows higher accuracy for all experimental datasets according to qualitative comparisons performed with some state-of-the-art methods for building segmentation, such as UAV data-based method and two semantic segmentation methods (e.g., FCN- and U-Net-based methods). The results prove the high capability of the proposed framework in building extraction from UAV data.

The proposed building extraction framework highly relies on the quality of photogrammetric processing. UAV image-derived poor-quality point clouds at building edges can decrease the accuracy of building boundary extraction. In addition, there are many parameters used in the proposed method, these parameters are referred from literature or determined based on the best trials.

In future studies, we will optimize our framework to achieve the best performance through a collinear constraint and by reducing the dependence on the quality of image-derived point clouds. We will also try to improve the proposed method by reducing the related parameters, and improve the architecture of U-Net to suit for building extraction from RGB bands and the point clouds for further comparing with the proposed method.

Author Contributions: H.H. proposed the framework of extracting buildings and wrote the source code and the paper. J.Z. and M.C. designed the experiments and revised the paper. T.C. and P.C. generated the datasets and performed the experiments. D.L. analyzed the data and improved the manuscript.

Funding: This study was financially supported by the National Natural Science Foundation of China (41861062, 41401526, and 41861052) and Natural Science Foundation of Jiangxi Province of China (20171BAB213025 and 20181BAB203022).

Acknowledgments: The authors thank Puyang An providing datasets. The authors also thank the editor-in-chief, the anonymous associate editor, and the reviewers for their systematic review and valuable comments.

Conflicts of Interest: The authors declare no conflict of interest.

References

1. Gilani, S.A.N.; Awrangjeb, M.; Lu, G. An automatic building extraction and regularisation technique using LiDAR point cloud data and orthoimage. *Remote Sens.* **2016**, *8*, 258. [CrossRef]
2. Wu, G.; Guo, Z.; Shi, X.; Chen, Q.; Xu, Y.; Shibasaki, R.; Shao, X. A boundary regulated network for accurate roof segmentation and outline extraction. *Remote Sens.* **2018**, *10*, 1195. [CrossRef]
3. Castagno, J.; Atkins, E. Roof shape classification from LiDAR and satellite image data fusion using supervised learning. *Sensors* **2018**, *18*, 3960. [CrossRef] [PubMed]
4. Du, S.; Zhang, Y.; Qin, R.; Yang, Z.; Zou, Z.; Tang, Y.; Fan, C. Building change detection using old aerial images and new LiDAR data. *Remote Sens.* **2016**, *8*, 1030. [CrossRef]
5. Dai, Y.; Gong, J.; Li, Y.; Feng, Q. Building segmentation and outline extraction from UAV image-derived point clouds by a line growing algorithm. *Int. J. Digit. Earth* **2017**, *10*, 1077–1097. [CrossRef]
6. Ahmadi, S.; Zoej, M.J.V.; Ebadi, H.; Moghaddam, H.A.; Mohammadzadeh, A. Automatic urban building boundary extraction from high resolution aerial images using an innovative model of active contours. *Int. J. Appl. Earth Obs. Geoinf.* **2010**, *12*, 150–157. [CrossRef]
7. Huang, X.; Zhang, L. A multidirectional and multiscale morphological index for automatic building extraction from multispectral geoeye-1 imagery. *Photogramm. Eng. Remote Sens.* **2011**, *77*, 721–732. [CrossRef]
8. Ghanea, M.; Moallem, P.; Momeni, M. Automatic building extraction in dense urban areas through GeoEye multispectral imagery. *Int. J. Remote Sens.* **2014**, *35*, 5094–5119. [CrossRef]
9. Chen, R.; Li, X.; Li, J. Object-based features for house detection from RGB high-resolution images. *Remote Sens.* **2018**, *10*, 451. [CrossRef]
10. Yang, H.; Wu, P.; Yao, X.; Wu, Y.; Wang, B.; Xu, Y. Building extraction in very high resolution imagery by dense-attention networks. *Remote Sens.* **2018**, *10*, 1768. [CrossRef]
11. Sampath, A.; Shan, J. Segmentation and reconstruction of polyhedral building roofs from aerial Lidar point clouds. *IEEE Trans. Geosci. Remote Sens.* **2010**, *48*, 1554–1567. [CrossRef]
12. Chen, D.; Zhang, L.; Li, J.; Liu, R. Urban building roof segmentation from airborne lidar point clouds. *Int. J. Remote Sens.* **2012**, *33*, 6497–6515. [CrossRef]
13. Xu, B.; Jiang, W.; Shan, J.; Zhang, J.; Li, L. Investigation on the weighted RANSAC approaches for building roof plane segmentation from LiDAR point clouds. *Remote Sens.* **2016**, *8*, 5. [CrossRef]
14. Yan, J.; Shan, J.; Jiang, W. A global optimization approach to roof segmentation from airborne lidar point clouds. *ISPRS J. Photogramm. Remote Sens.* **2014**, *94*, 183–193. [CrossRef]
15. Du, S.; Zhang, Y.; Zou, Z.; Xu, S.; He, X.; Chen, S. Automatic building extraction from LiDAR data fusion of point and grid-based features. *ISPRS J. Photogramm. Remote Sens.* **2017**, *130*, 294–307. [CrossRef]
16. Chen, L.; Zhao, S.; Han, W.; Li, Y. Building detection in an urban area using lidar data and QuickBird imagery. *Int. J. Remote Sens.* **2012**, *33*, 5135–5148. [CrossRef]
17. Awrangjeb, M.; Zhang, C.; Fraser, C.S. Automatic extraction of building roofs using LiDAR data and multispectral imagery. *ISPRS J. Photogramm. Remote Sens.* **2013**, *83*, 1–18. [CrossRef]
18. Tian, J.; Cui, S.; Reinartz, P. Building change detection based on Satellite stereo imagery and digital surface models. *IEEE Trans. Geosci. Remote Sens.* **2014**, *52*, 406–417. [CrossRef]
19. Crommelinck, S.; Bennett, R.; Gerke, M.; Nex, F.; Yang, M.Y.; Vosselman, G. Review of automatic feature extraction from high-resolution optical sensor data for UAV-based cadastral mapping. *Remote Sens.* **2016**, *8*, 689. [CrossRef]
20. Achanta, R.; Shaji, A.; Smith, K.; Lucchi, A.; Fua, P.; Süsstrunk, S. *SLIC Superpixels*; EPFL Technical Report No. 149300; School of Computer and Communication Sciences, Ecole Polytechnique Fedrale de Lausanne: Lausanne, Switzerland, 2010; pp. 1–15.
21. Krizhevsky, A.; Sutskever, I.; Hinton, G.E. ImageNet classification with deep convolutional neural networks. In Proceedings of the Conference on Neural Information Processing Systems (NIPS12), Lake Tahoe, NV, USA, 3–6 December 2012; pp. 1097–1105.

22. Karpathy, A.; Toderici, G.; Shetty, S.; Leung, T.; Sukthankar, R.; Li, F.-F. Large-scale video classification with convolutional neural networks. In Proceedings of the IEEE Conference on Computer Vision and Pattern Recognition (CVPR), Columbus, OH, USA, 23–28 June 2014; pp. 1725–1732.
23. Chen, X.; Xiang, S.; Liu, C.-L.; Pan, C.-H. Vehicle detection in satellite images by hybrid deep convolutional neural networks. *IEEE Trans. Geosci. Remote Sens.* **2014**, *11*, 1797–1801. [CrossRef]
24. Chen, S.; Wang, H.; Xu, F.; Jin, Y.-Q. Target classification using the deep convolutional networks for SAR images. *IEEE Trans. Geosci. Remote Sens.* **2016**, *54*, 4806–4817. [CrossRef]
25. He, H.; Chen, M.; Chen, T.; Li, D. Matching of remote sensing images with complex background variations via Siamese convolutional neural network. *Remote Sens.* **2018**, *10*, 355. [CrossRef]
26. He, H.; Chen, M.; Chen, T.; Li, D.; Cheng, P. Learning to match multitemporal optical satellite images using multi-support-patches Siamese networks. *Remote Sens. Lett.* **2019**, *10*, 516–525. [CrossRef]
27. Long, J.; Shelhamer, E.; Darrel, T. Fully convolutional networks for semantic segmentation. In Proceedings of the IEEE Conference on Computer Vision and Pattern Recognition (CVPR), Boston, MA, USA, 7–12 June 2015; pp. 3431–3440.
28. Bittner, K.; Cui, S.; Reinartz, P. Building extraction from remote sensing data using fully convolutional networks. In Proceedings of the International Archives of the Photogrammetry, Remote Sensing and Spatial Information Sciences, Hannover, Germany, 6–9 June 2017; pp. 481–486.
29. Ronneberger, O.; Fischer, P.; Brox, T. U-Net: Convolutional networks for biomedical image segmentation. In Proceedings of the International Conference on Medical Image Computing and Computer-Assisted Intervention, Munich, Germany, 5–9 October 2015; pp. 234–241.
30. Xu, Y.; Wu, L.; Xie, Z.; Chen, Z. Building extraction in very high resolution remote sensing imagery using deep learning and guided filters. *Remote Sens.* **2018**, *10*, 144. [CrossRef]
31. Spann, M.; Wilson, R. A quad-tree approach to image segmentation which combines statistical and spatial information. *Pattern Recogn.* **1985**, *18*, 257–269. [CrossRef]
32. Roerdink, J.B.; Meijster, A. The watershed transform: Definitions, algorithms and parallelization and strategies. *Fundam. Inform.* **2000**, *41*, 187–228.
33. Baatz, M.; Schäpe, A. Multiresolution segmentation: An optimization approach for high quality multi-scale image segmentation. In *Angewandte Geographische Informationsverarbeitung XII*; Strobl, J., Blaschke, T., Griesebner, G., Eds.; Wichmann-Verlag: Heidelberg, Germany, 2000; pp. 12–23.
34. Kim, M.; Warner, T.A.; Madden, M.; Atkinson, D.S. Multi-scale GEOBIA with very high spatial resolution digital aerial imagery: Scale, texture and image objects. *Int. J. Remote Sens.* **2011**, *32*, 2825–2850. [CrossRef]
35. Liu, J.; Du, M.; Mao, Z. Scale computation on high spatial resolution remotely sensed imagery multi-scale segmentation. *Int. J. Remote Sens.* **2017**, *38*, 5186–5214. [CrossRef]
36. Belgiu, M.; Draguţ, L. Comparing supervised and unsupervised multiresolution segmentation approaches for extracting buildings from very high resolution imagery. *ISPRS J. Photogramm. Remote Sens.* **2014**, *96*, 67–75. [CrossRef] [PubMed]
37. Csillik, O. Fast segmentation and classification of very high resolution remote sensing data using SLIC superpixels. *Remote Sens.* **2017**, *9*, 243. [CrossRef]
38. Farabet, C.; Couprie, C.; Najman, L.; LeCun, Y. Learning hierarchical features for scene labeling. *IEEE Trans. Pattern Anal.* **2013**, *35*, 1915–1929. [CrossRef]
39. Badrinarayanan, V.; Handa, A.; Cipolla, R. SegNet: A deep convolutional encoder-decoder architecture for robust semantic pixel-wise labeling. *arXiv* **2015**, arXiv:1505.07293.
40. Shelhamer, E.; Long, J.; Darrell, T. Fully convolutional networks for semantic segmentation. *IEEE Trans. Pattern Anal.* **2017**, *39*, 640–651. [CrossRef]
41. Lui, M.Y.; Tuzel, O.; Ramalingam, S.; Chellappa, R. Entropy rate superpixel segmentation. In Proceedings of the IEEE Conference on Computer Vision and Pattern Recognition, Colorado Springs, CO, USA, 20–25 June 2011; pp. 2097–2104.
42. Van den Bergh, M.; Boix, X.; Roig, G.; de Capitani, B.; Van Gool, L. SEEDS: Superpixels extracted via energy-driven sampling. In Proceedings of the European Conference on Computer Vision, Florence, Italy, 7–13 October 2012; pp. 13–26.
43. Neubert, P.; Protzel, P. Compact watershed and preemptive SLIC: On improving trade-offs of superpixel segmentation algorithms. In Proceedings of the International Conference on Pattern Recognition, Stockholm, Sweden, 24–28 August 2014; pp. 996–1001.

44. Li, Z.; Chen, J. Superpixel segmentation using linear spectral clustering. In Proceedings of the IEEE Conference on Computer Vision and Pattern Recognition, Boston, MA, USA, 7–12 June 2015; pp. 1356–1363.
45. Neubert, P.; Protzel, P. Superpixel benchmark and comparison. In Proceedings of the Forum Bildverarbeitung 2012; Karlsruher Instituts für Technologie (KIT) Scientific Publishing: Karlsruhe, Germany, 2012; pp. 1–12.
46. Tucker, C.J. Red and photographic infrared linear combinations for monitoring vegetation. *Remote Sens. Environ.* **1979**, *8*, 127–150. [CrossRef]
47. Gitelson, A.A.; Kaufman, Y.J.; Stark, R.; Rundquist, D. Novel algorithms for remote estimation of vegetation fraction. *Remote Sens. Environ.* **2002**, *80*, 76–87. [CrossRef]
48. Booth, D.T.; Cox, S.E.; Meikle, T.W.; Fitzgerald, C. The accuracy of ground-cover measurements. *Rangel. Ecol. Manag.* **2006**, *59*, 179–188. [CrossRef]
49. Ok, A.Ö. Robust detection of buildings from a single color aerial image. In Proceedings of the GEOBIA 2008, Calgary, AB, Canada, 5–8 August 2008; Volume XXXVII, Part 4/C1. p. 6.
50. Meyer, G.E.; Neto, J.C. Verification of color vegetation indices for automated crop imaging applications. *Comput. Electron. Agric.* **2008**, *63*, 282–293. [CrossRef]
51. Ioffe, S.; Szegedy, C. Batch normalization: Accelerating deep network training by reducing internal covariate shift. In Proceedings of the 32nd International Conference on Machine Learning (ICML-15), Lille, France, 6–11 July 2015.
52. Douglas, D.; Peucker, T. Algorithms for the reduction of the number of points required to represent a digitized line or its caricature. *Can. Cartogr.* **1973**, *10*, 112–122. [CrossRef]
53. Saalfeld, A. Topologically consistent line simplification with the Douglas-Peucker algorithm. *Cartogr. Geogr. Inf. Sci.* **1999**, *26*, 7–18. [CrossRef]
54. Snavely, N.; Seitz, S.M.; Szeliski, R. Modeling the world from Internet photo collections. *Int. J. Comput. Vis.* **2008**, *80*, 189–210. [CrossRef]
55. Rothermel, M.; Wenzel, K.; Fritsch, D.; Haala, N. SURE: Photogrammetric surface reconstruction from imagery. In Proceedings of the LC3D Workshop, Berlin, Germany, 4–5 December 2012.
56. Awrangjeb, M.; Fraser, C.S. An automatic and threshold-free performance evaluation system for building extraction techniques from airborne Lidar data. *IEEE J. Sel. Top. Appl. Earth Obs. Remote Sens.* **2014**, *7*, 4184–4198. [CrossRef]
57. Demir, I.; Koperski, K.; Lindenbaum, D.; Pang, G.; Huang, J.; Basu, S.; Hughes, F.; Tuia, D.; Raska, R. DeepGlobe 2018: A challenge to parse the earth through satellite images. In Proceedings of the IEEE/CVF Conference on Computer Vision and Pattern Recognition Workshops (CVPRW), Salt Lake City, UT, USA, 18–22 June 2018; pp. 172–17209.
58. Brown, M.; Hua, G.; Winder, S. Discriminative learning of local image descriptors. *IEEE Trans. Pattern Anal.* **2011**, *33*, 43–57. [CrossRef] [PubMed]
59. Cohen, J.P.; Ding, W.; Kuhlman, C.; Chen, A.; Di, L. Rapid building detection using machine learning. *Appl. Intell.* **2016**, *45*, 443–457. [CrossRef]

© 2019 by the authors. Licensee MDPI, Basel, Switzerland. This article is an open access article distributed under the terms and conditions of the Creative Commons Attribution (CC BY) license (http://creativecommons.org/licenses/by/4.0/).

Article

Building Extraction from High-Resolution Aerial Imagery Using a Generative Adversarial Network with Spatial and Channel Attention Mechanisms

Xuran Pan [1,2], Fan Yang [1,*], Lianru Gao [2], Zhengchao Chen [2], Bing Zhang [2,3], Hairui Fan [1] and Jinchang Ren [4]

1. School of Electronics and Information Engineering, Hebei University of Technology, Tianjin 300401, China; 201611901006@stu.hebut.edu.cn (X.P.); 201731904001@stu.hebut.edu.cn (H.F.)
2. Key Laboratory of Digital Earth Science, Institute of Remote Sensing and Digital Earth, Chinese Academy of Sciences, Beijing 100094, China; gaolr@radi.ac.cn (L.G.); chenzc@radi.ac.cn (Z.C.); zb@radi.ac.cn (B.Z.)
3. College of Resources and Environment, University of Chinese Academy of Sciences, Beijing 100049, China
4. Department of Electronic and Electrical Engineering, University of Strathclyde, Glasgow, G1 1XW, UK; jinchang.ren@strath.ac.uk
* Correspondence: 201621901026@stu.hebut.edu.cn

Received: 18 March 2019; Accepted: 12 April 2019; Published: 15 April 2019

Abstract: Segmentation of high-resolution remote sensing images is an important challenge with wide practical applications. The increasing spatial resolution provides fine details for image segmentation but also incurs segmentation ambiguities. In this paper, we propose a generative adversarial network with spatial and channel attention mechanisms (GAN-SCA) for the robust segmentation of buildings in remote sensing images. The segmentation network (generator) of the proposed framework is composed of the well-known semantic segmentation architecture (U-Net) and the spatial and channel attention mechanisms (SCA). The adoption of SCA enables the segmentation network to selectively enhance more useful features in specific positions and channels and enables improved results closer to the ground truth. The discriminator is an adversarial network with channel attention mechanisms that can properly discriminate the outputs of the generator and the ground truth maps. The segmentation network and adversarial network are trained in an alternating fashion on the Inria aerial image labeling dataset and Massachusetts buildings dataset. Experimental results show that the proposed GAN-SCA achieves a higher score (the overall accuracy and intersection over the union of Inria aerial image labeling dataset are 96.61% and 77.75%, respectively, and the F_1-measure of the Massachusetts buildings dataset is 96.36%) and outperforms several state-of-the-art approaches.

Keywords: high-resolution aerial images; deep learning; generative adversarial network; semantic segmentation; Inria aerial image labeling dataset; Massachusetts buildings dataset

1. Introduction

With the rapid advancement of aerospace remote sensing, the amount and spatial resolution of high-resolution remote sensing images are increasing rapidly. As a result, accurate and automatic semantic labeling of high-resolution remote sensing images is of great significance and receives wide attention [1]. Large intra-class variance and small inter-class differences of higher spatial resolution remote sensing images may cause classification ambiguities, which makes semantic segmentation of high-resolution remote sensing images a challenge. Specific to the buildings in high-resolution aerial images, buildings in different regions have different characteristics. For instance, some regions have small and very dense buildings, whilst some other regions have low-density buildings. This variability brings great challenges to the building segmentation task, and requires strong generalization capabilities of classification techniques [2,3].

Over the last few years, deep learning architectures have made breakthroughs in the image analysis field. Convolutional neural networks (CNNs) have been proposed not only to deal with object detection and whole image classification but also progress fine inference, such as semantic segmentation. Semantic segmentation can accomplish pixel-wise prediction, which is a problem to give each pixel a class label. Long et al. [4] proposed fully convolutional networks (FCNs) to accomplish pixel-wise classification. They replaced the fully connected layers of whole image classification CNNs with convolutional layers and utilized deconvolutional layers to upsample feature maps to score map each class. FCNs created a precedent for pixel-based encoder–decoder architectures. Following this paradigm, many CNN architectures have been proposed and further improved the segmentation performance. In [5], U-Net was proposed to modify FCN by concatenating feature maps of encoder and decoder. Concatenation architecture can take full advantage of both low-level and high-level features. Hence, more precise segmentation results can be obtained. After that, DeepLab V1 [6] and V2 [7] were proposed to mitigate the information loss caused by pooling operations. The authors introduced atrous convolutions to increase receptive field size while maintaining higher resolution of feature maps, and the fully connected conditional random fields (CRFs) were utilized to further improve the segmentation performance as post processor. In [8], Noh et al. proposed DeconvNet which consists of convolution and deconvolution networks. In the deconvolution network, unpooling layers were applied to upscale feature maps and decovoconvolutional layers were followed to densify the initially upscaled sparse feature maps. Badrinarayanan et al. presented SegNet [9] which also included unpooling layers in the decoder stage and with smaller parameterization when compared with DeconvNet.

Although the CNN-based segmentation methods have achieved promising results, they still have drawbacks and can be further improved. The main problem is that the pixel-wise prediction of CNN can guarantee high pixel-wise accuracy, but the relationship between pixels is prone to be ignored. This may lead to discontinuous segmentation results, and the boundaries of objects are usually not accurate enough. Therefore, post-processing methods, e.g., fully connected CRFs or Markov random fields (MRFs), were needed to further improve the raw segmentation results [10–12]. These graphical regularization models coupled both the input images and the predicted score maps of CNN to refine the predictions with the color information and pixel position of the original image. In addition, recurrent neural networks (RNN) can also refine the segmentation results by employing a feedback connection to form a directed cycle [13]. Bergado et al. [14] proposed to incorporate the recurrent approach in the semantic segmentation task (ReuseNet) to learn contextual dependencies in the label space, and further refine the segmentation results. The ReuseNet applied the semantic segmentation operations in R cycles. Each cycle takes the score map of the previous cycle concatenated with the original image as input. Moreover, Generative adversarial networks (GANs) [15] based methods can enforce spatial label contiguity to refine the segmentation results without any time consumption during the testing phase. In [16], Luc et al. first applied adversarial training strategy to semantic segmentation task. A segmentation network and an adversarial network were trained in an alternating fashion to make the generated segmentation results hard to be distinguished from the ground truth. By doing so, the joint distribution of all label variables at each pixel location can be assessed as a whole, and thus, can enforce forms of high-order consistency that cannot be enforced by pixelwise classification or pairwise terms. Xue et al. [17] presented SegAN for medical image segmentation, which is composed of a segmentor and a critic network. The multi-scale L_1 loss function was minimized and maximized alternatively to train these two networks, and the SegAN received better image segmentation performance than the original GAN.

In semantic labeling of high-resolution remote sensing images, deep learning architectures also show excellent performance [18–20]. Saito et al. [21] used patch-based CNN to learn classification maps from high-resolution images and achieved good results on Massachusetts roads and buildings datasets [22]. However the patch-based methods suffer from limited receptive field and large computational overhead, so it was soon surpassed by pixel-based methods.

Maggiori et al. [23] proposed the Inria aerial image labeling dataset that covers different forms of buildings and provided a baseline segmentation result by using an FCN-based architecture combined with multi-layer perceptron. In [24], Bischke et al. introduced a new cascaded multi-task loss to mitigate the poor boundaries of the existing prediction results. Learning with the proposed loss, the performance can achieve certain improvement without any changes in the network architecture. A multi-stage multi-task CNN for building extraction was introduced in [18]. The first stage of the proposed network provided the segmentation results, while the second stage was aimed to give the precise location by two branches. In [25], Khalel et al. proposed a stack of U-Nets to automatically label the buildings from high aerial images, of which each U-Net can be regarded as the post-processor of the previous U-Net. However, the existing results usually suffered from poor boundaries, and the accuracy can be further improved.

In this paper, we propose a generative adversarial network with spatial and channel attention mechanisms (GAN-SCA) for high accurate semantic labeling of buildings in high-resolution aerial images with precise boundaries. The GAN-SCA is composed of a segmentation network and an adversarial network, in which the segmentation network is a semantic segmentation network to predict the pixel-wise labeling results, and the adversarial network is to distinguish whether the inputs are predicted results of the segmentation network or ground truth. Moreover, we embed channel and spatial attention mechanisms into the network to selectively enhance useful information, and further improve the segmentation accuracy.

The main contributions of our work can be summarized as follows:

- A GAN-based network called GAN-SCA is proposed for building extraction from high-resolution aerial imagery. The architecture is composed of a segmentation network and an adversarial network. The segmentation network aims to predict pixel-wise labeling maps that are similar to ground truths, while the adversarial network is set to discriminate different characteristics of different label maps to further enhance the high-frequency continuity of the prediction maps.
- Spatial and channel attention mechanisms are embedded in the proposed GAN-SCA architecture to enable selectively attaching important features from both the spatial dimension and channel relationship.
- The adversarial network and segmentation network are trained to optimize a multi-scale L_1 loss and multiple cross entropy losses combined with a multi-scale L_1 loss alternatively. With no requirements for any post-processing, our proposed network improved the state-of-the-art performance on both the Inria aerial image labeling dataset and Massachusetts buildings dataset.

The rest of this paper is organized as follows. In Section 2, we introduce the architecture and training strategy of the proposed network in detail. The dataset description and experimental setting are presented in Section 3. Section 4 details the experimental results and analyses. Section 5 discusses the effectiveness of the spatial and channel attention mechanisms and the training strategy. The results are drawn in Section 6.

2. Methods

2.1. Proposed Network GAN-SCA

As shown in Figure 1, the proposed GAN-SCA is composed of two parts, i.e., the segmentation network and the adversarial network.

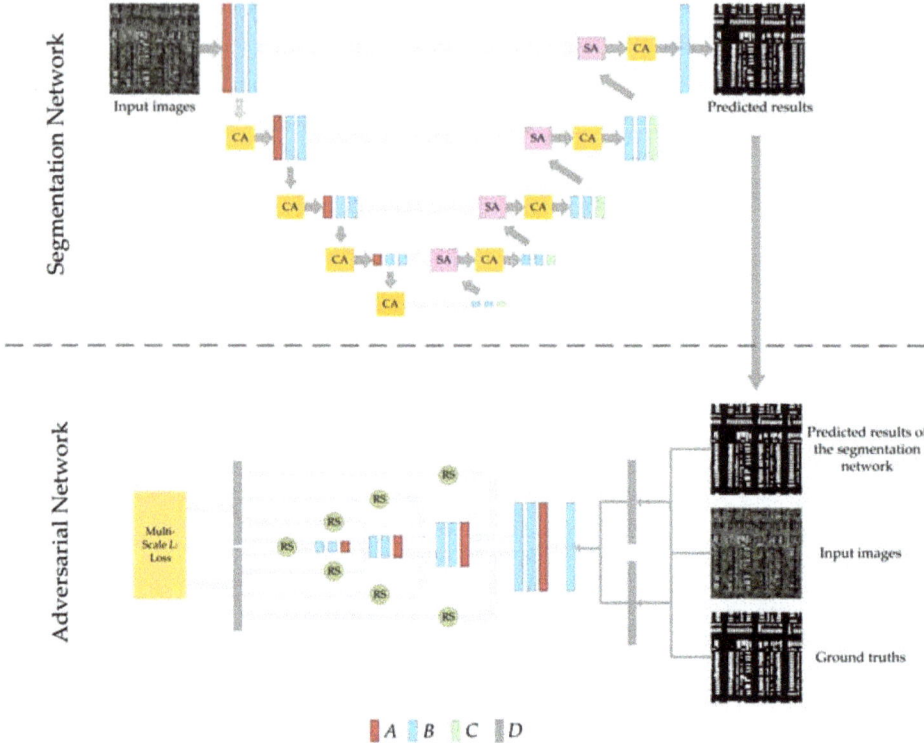

Figure 1. Architecture of the proposed generative adversarial network with spatial and channel attention mechanisms (GAN-SCA). *A* is max pooling layer; *B* are convolutional + batch normalization + rectified linear unit (ReLU) layers; *C* is upsampling layer; *D* is the concatenation operation; SA is the spatial attention mechanism; CA is the channel attention mechanism; RS is the reshape operation.

The segmentation network is a U-Net-based architecture, where spatial and channel attention mechanisms are embedded. U-Net is a powerful CNN architecture for semantic segmentation and has been widely applied in remote sensing image classification field [5]. U-Net was initially designed for binary segmentation of biomedical images with a relatively small number of training samples. As it achieves better performance than other classic semantic segmentation architecture, U-Net is a good choice for the building extraction task in this study. However, these classic deep convolutional neural network (DCNN) architectures for semantic segmentation usually produce a large number of multi-level feature maps but do not perform any feature selection operation throughout the whole process. On the one hand, fusion of the high-level and low-level features without feature selection may result in over-segmentation when the model tends to receive more information from lower layers. On the other hand, the channel-wise information combined by convolutional filters without considering channel-interdependencies might affect the segmentation performance of the network. Therefore, we propose to introduce the attention mechanisms to employ feature selection from the aspect of spatial information and channel relationship.

To mitigate the neglect of inter-pixel relationships caused by the pixel-wise loss function used in the training phase, we propose to refine the segmentation result using the adversarial training. The adversarial network can learn latent higher-order structural features which can be fed into the segmentation network in the training phase, and the segmentation results can be refined without an adversarial network in the testing phase. In contrast with graphical models and recurrent approaches, adversarial training can achieve segmentation refinement without extra time consumption. The architecture of adversarial network we adopt in the proposed GAN-SCA shares a similar structure as the encoder of the segmentation network and is fed with the predicted maps combined with the original images and ground truth maps combined with the original images. In particular, the multi-scale features from a different stage of the adversarial network are reshaped into one-dimensional vectors and concatenated together to compute the multi-scale L_1 loss.

2.1.1. Segmentation Network

The segmentation network of GAN-SCA is based on U-Net architecture. To accomplish feature selection from the aspect of the spatial information and channel-wise relationship, we introduce two kinds of attention mechanisms into the network architecture. The attention mechanism is an effective operation to enable the network to selectively enhance more useful features and has been widely applied in the image analysis field [26]. In this work, we consider both spatial and channel-wise attention mechanisms to improve the segmentation performance. The spatial attention mechanisms are embedded between the contracting path and expanding path of the U-Net, as shown in Figure 1. The U-Net fuses low-level feature maps of the contracting path with the high-level features of the expanding path by concatenation to re-utilize fine details in the low-level features. However, the rough concatenation may result in the over-use of low-level features. Therefore, we can utilize flexible semantic information of the high-level features to assist the selection of low-level information. Usually, the low-level features contain rich details, and we prefer to enhance the hard classified information and suppress the interference information. Figure 2 shows the error map of U-Net prediction result, from which we can observe that building boundaries are prone to be mislabeled in the building extraction task. Inspired by [27], the entropy score map of high-level features has similar characteristics with the mislabeled map, as shown in Figure 2. Therefore, when we compute the entropy score map of high-level features in each decoder stage, and weight the low-level features according to the results of corresponding entropy score map before high-level and low-level feature fusion, we can selectively enhance the hard classified information while suppressing the less useful information of the low-level features. The entropy score map can be computed with Equation (1):

$$E(x) = -\sum_{i=1}^{K} p_i(x) log(p_i(x)) \qquad (1)$$

where $p_i(x)$ denotes the score map of class i, K means the total number of the classes. Figure 2 displays the entropy score maps of four-scale spatial attention mechanisms, from which we can see that the entropy maps have a strong relationship with the error map. Usually, building boundary pixels are prone to being mislabeled, so the entropy maps also share similar characteristics with the boundaries of buildings. Thus, with the spatial attention mechanisms, building boundaries information from lower level features will be highly weighted into the final output fusion feature.

The detailed structure of the spatial attention mechanism is shown in Figure 3a. As can be seen, high-level features are first convoluted by 1×1 convolutions for dimensionality reduction and normalized to [0,1] by using the sigmoid function to generate the score maps. Afterward, the entropy score map is computed to element-wise conducts with low-level features. After that, the high-level features are concatenated with the weighted low-level features to further process. It is worth noting that, the entropy score map has a strong relationship with the building boundaries in the building extraction tasks so the spatial attention mechanisms can bring benefits to the building boundaries segmentation. In particular, we compute four cross entropy losses of each spatial attention mechanism to combine with the overall cross entropy loss to train the segmentation network. The detail of model optimization will be introduced in Section 2.2.

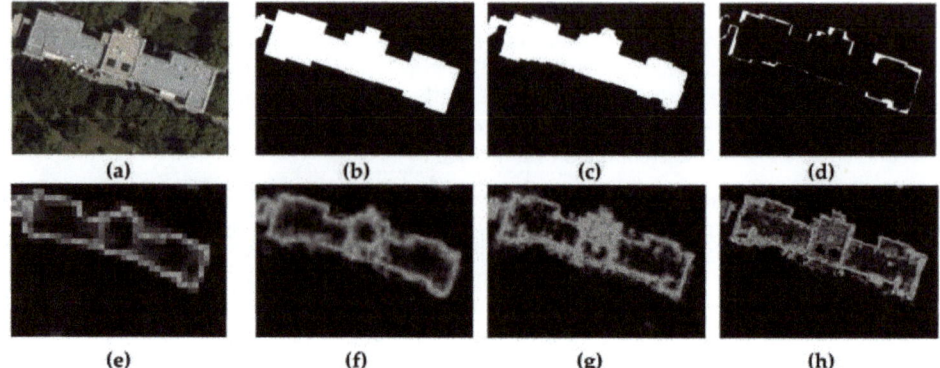

Figure 2. Entropy score maps of four-scale spatial attention mechanisms. (**a**) is the original image; (**b**) is the ground truth; (**c**) is the prediction result; (**d**) is the error map; (**e–h**) are the entropy score maps of the low-to-high scale spatial attention mechanisms.

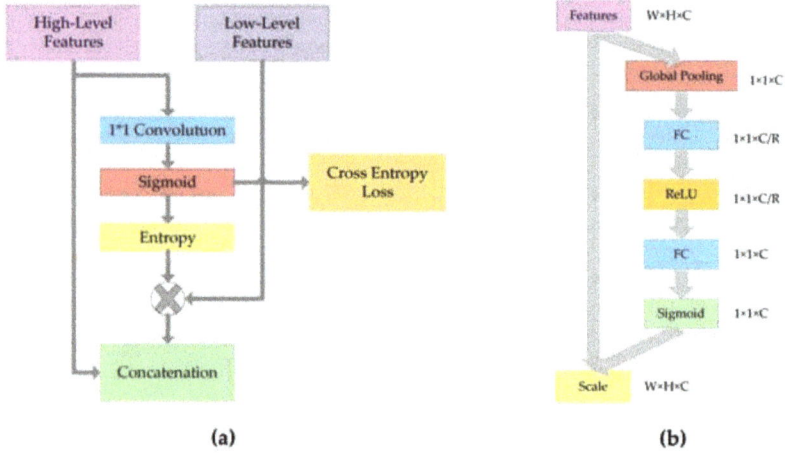

Figure 3. Composition modules in the GAN-SCA. (**a**) Spatial attention mechanism; (**b**) Channel attention mechanism. FC is fully connected layer.

Apart from spatial attention, the proposed architecture also takes advantage of the channel relationship enhancement. Squeeze-and-excitation (SE) block is a computational unit that can re-scale each channel according to its importance adaptively. SE blocks can be stacked together with many existing state-of-the-art CNNs, and bring significant improvements in performance across different datasets with minimal additional computational cost [28]. So we adopt SE blocks as channel attention mechanisms at each stage in both contracting path and expanding path, as shown in Figure 1. The structure of the SE block is depicted in Figure 3b, which can model channel inter-dependencies in two steps, namely, squeeze and excitation. The input features x are first squeezed into channel-wise statistics s by performing global average pooling, and the c-th channel of s can be computed by:

$$s_c(x_c) = \frac{1}{H \times W} \sum_{i=1}^{H} \sum_{j=1}^{W} x_c(i,j) \tag{2}$$

where x_c is the c-th channel of the input feature x, and $H \times W$ denote spatial dimensions of x_c.

To properly capture the information of s to model the channel inter-dependencies, the excitation operation is followed. A fully connected layer is adopted to reduce the dimension of $s_{1\times 1\times C}$ to $s'_{1\times 1\times \frac{C}{R}}$ and a rectified linear unit (ReLU) layer is followed to activate. After that, another fully connected (FC) layer is performed to ascend s' back to the original dimension $1 \times 1 \times C$. By doing so, it can better fit the complex relationship between channels with less computational overhead. The weight of each channel is normalized to [0,1] with a sigmoid activation. The excitation operation can be written as:

$$e = \sigma(W_2 \delta(W_1 s)) \qquad (3)$$

where σ stands for the sigmoid activation, and δ stands for the ReLU function [29]. W_1 and W_2 are two real matrices of size $\frac{C}{R} \times C$ and $C \times \frac{C}{R}$ to limit the complexity and generalization of the channel attention mechanism. This operation is implemented by two FC layers.

The final output of the channel attention mechanism is the re-scaled input features y_c. The re-scaled operation can be expressed by Equation (4) below:

$$y_c = e_c \cdot x_c \qquad (4)$$

2.1.2. Adversarial Network

The adversarial network of GAN-SCA has a similar structure with the encoder in the segmentation network. Two inputs are fed into the adversarial network, namely original images concatenated with predicted label maps and original images concatenated with ground truths. The network starts with a 1×1 convolutional layer to learn to fuse the input images with the predicted label maps/ground truths. Figure 4 shows two visual results of such fusion. Then the fused images are fed into the encoder-like network to extract features, respectively. To capture long- and short- range spatial relations between pixels, we extract multi-scale feature maps from multiple layers and concatenate them together to compute the multi-scale L_1 loss [17], the detailed introduction of loss function will be presented in the next section.

Figure 4. Fusion features of input images (one channel) and the predicted label maps/ground truths. (a) Input images; (b) Fusion results of input images and ground truths; (c) Fusion results of input images and the predicted label maps (5000 iterations).

2.2. Training Strategy

The proposed GAN-SCA is trained in an adversarial fashion. The segmentation network aims to generate the predicted labeling map to deceive the adversarial network, and the adversarial network aims to distinguish the ground truths from the predicted labeling maps generated by the segmentation network. Therefore, the segmentation network and adversarial network are trained alternatively in the training phase [30]. We first fix the parameter of the segmentation network (S) and train adversarial network (A) to minimize the multi-scale L_1 loss (Equation (5)). Then the parameter of A is fixed, and the S is trained by minimizing the cross-entropy losses combined with the negative multi-scale L_1 loss (Equation (7)).

$$L_A = -\frac{1}{N} \sum_{n=1}^{N} l_{mae}(f_A(x_n, S(x_n)), f_A(x_n, y_n)) \quad (5)$$

where $(x_n, S(x_n))$ is the concatenation of input images and the predicted results of (x_n, y_n) is the concatenation of input images and ground truths, $f_A(x)$ denotes hierarchical features extracted from x, l_{mae} is the L_1 distance or mean absolute error (mae), which is defined as:

$$l_{mae}(f_A(x), f_A(x')) = \frac{1}{L} \sum_{i=1}^{L} \|f_A^i(x) - f_A^i(x')\|_1 \quad (6)$$

where L is the total number of the feature scales in the adversarial network, $f_A^i(x)$ is the features in scale i.

$$L_S = -\frac{1}{N} \sum_{n=1}^{N} (y(x_n) \log(S(x_n)) + (1 - y(x_n)) \log(1 - S(x_n))) - L_A + L_{fa} \quad (7)$$

where the L_{fa} is the auxiliary cross entropy loss computed in each spatial attention mechanism, $y(x_n)$ denotes the ground truth of the n-th image in the current batch.

The parameters of the segmentation network and adversarial network are initialized by normally distributed random variables. The initial learning rate is set to 10^{-3} and divided by 2 every 15 epochs. The batch size is set to 5. We crop the training images into size 384 × 384 with 25% overlap, and data augmentation including flip and rotation are also implemented. In the testing phase, to meet the memory constraints, we employ a sliding window with size 1024 × 1024 to accomplish the full tile prediction. We set 75% overlapping size in the testing stage to mitigate inconsistent border phenomenon since the size is proven to give the best results in previous works [11,31].

3. Datasets and Evaluation Metrics

3.1. Datasets

The datasets we used in this work are two open buildings datasets, namely Inria aerial image labeling dataset for buildings and Massachusetts buildings dataset. These two datasets cover various building characteristics, such as shape, size, distribution, and spatial resolution, which can evaluate the generalization ability of networks.

The first dataset we used is the Inria aerial image labeling dataset for buildings [23]. The dataset consists of 360 high-resolution aerial images which over different cities including Austin, Chicago, Kitsap, Western/Eastern Tyrol, Vienna, Bellingham, Bloomington, and San Francisco. These regions cover dissimilar urban buildings, for instance, most buildings in Chicago and San Francisco are densely distributed and usually small in shape, while buildings in Kitsap are scattered. The spatial resolution of images is 30 cm with an image size of 5000 × 5000 pixels, and each image covers a surface of 1500 × 1500 m². Only 180 tiles are provided with ground truths, and the other 180 tiles are preserved for testing. Following a common practice [23], we choose the first five images of each region from the training set for validation.

The second dataset is the Massachusetts buildings dataset [22]. The dataset consists of 151 high-resolution aerial images of urban and suburban areas at Boston. The size of images in this dataset is 1500 × 1500 pixels, and each image covers a surface of 2250 × 2250 m². The dataset is randomly divided into three subsets, namely training set (137 tiles), validation set (4 tiles), and testing set (10 tiles).

3.2. Evaluation Metrics

To make a fair comparison, we compute the same metrics as in other literatures. For the Inria Aerial Image Labeling Dataset, the overall accuracy (Acc.) and intersection over union (IoU) are utilized for quantitative performance evaluation. Acc. is the proportion of the correctly labeled pixels (see Equation (8)). IoU is the intersection of pixels labeled as building in the predicted results and ground truths, divided by the union of pixels labeled as building in the predicted results and ground truths (see Equation (9)).

$$Acc. = \frac{tp + tn}{tp + tn + fp + fn} \tag{8}$$

$$IoU = \frac{tp}{fp + tp + fn} \tag{9}$$

where tp denotes the number of true positive pixels, fp denotes the number of false positive pixels, tn denotes the number of true negative pixels, and fn denotes the number of false negative pixels.

For the Massachusetts buildings dataset, relaxed F_1-measure is used to evaluate the segmentation performance of each network. A relaxed factor ρ is introduced when computing the confusion metrics because the tools producer of this dataset used to generate labels is only accurate up to a few pixels. Following the previous works [23–25], we compute the F_1-measure with a relaxation factor of three, and the F_1-measure without relaxation version ($\rho = 0$) is also reported. The F_1-measure can be written as:

$$F_1 = 2 \times \frac{precision \times recall}{precision + recall} \tag{10}$$

$$precision = \frac{tp}{tp + fp} \tag{11}$$

$$recall = \frac{tp}{tp + fn} \tag{12}$$

4. Experiments

4.1. Ablation Study

In this section, we first evaluate whether the two attention mechanisms can bring benefit to the segmentation performance, so we compare the base architecture (i.e., the standard U-Net) with the U-Net embedded with the attention mechanisms (U-Net-SCA). It should be noted that the U-Net-SCA is the segmentation network of the proposed GAN-SCA. In addition, we employ dense CRFs as the post-processor of U-Net-SCA to further improve the segmentation results (U-Net-SCA+CRFs). We also explore the recurrent approach to achieve label refinement followed the ReuseNet in [14] (U-Net-SCA+Reuse), that applies U-Net-SCA in R cycles. We choose $R = 3$ in this experiment because the U-Net-SCA architecture in three cycles achieves the best performance on the Inria aerial image labeling dataset. Finally, we train the U-Net-SCA combined with an adversarial network (GAN-SCA) in an alternating fashion to see how the adversarial training affects the segmentation results.

The models described above are trained over five independent runs with random initialization, and the average accuracy and IoU with the standard deviation of the experimental results on the validation set of the Inria aerial image labeling dataset are reported in Table 1. As can be observed from Table 1, the proposed U-Net-SCA achieves improvement of 0.19% and 0.72% in terms of the overall accuracy and IoU compared to the standard U-Net. For accuracy and IoU of each region, the U-Net-SCA also outperforms the standard U-Net. Especially for the regions in Chicago and Vienna, where buildings are high-densely distributed, and the proportion of building pixels in the training set is higher, the accuracy increase is more evident. This indicates that the spatial and channel attention mechanisms enable the network to selectively enhance useful features to further improve segmentation accuracy. The U-Net-SCA+CRFs has few improvements over the U-Net-SCA, with the overall accuracy and IoU improved by 0.01% and 0.21%, respectively. By adopting the recurrent approach and adversarial network, the U-Net-SCA+Reuse and GAN-SCA have a similar small improvement of overall accuracy and IoU when compared to the U-Net-SCA. Let us recall that the adversarial training strategy adopted by the proposed GAN-SCA can learn high-order consistency without extra time consumption in the testing phase, whereas the recurrent approach of U-Net-SCA+Reuse is accompanied by the multi-fold increase of trainable weights which increases the computational complexity.

Table 1. Experimental results on Inria aerial image labeling dataset.

Methods	Metrics	Austin	Chicago	Kitsap	Tyrol-w	Vienna	Overall
U-Net	IoU	79.95 ± 0.81	70.18 ± 0.22	68.56 ± 1.49	76.29 ± 2.06	79.92 ± 0.39	76.16 ± 0.21
	Acc.	97.10 ± 0.11	92.67 ± 0.15	99.31 ± 0.03	98.15 ± 0.15	94.25 ± 0.18	96.31 ± 0.07
U-Net-SCA	IoU	80.40 ± 0.43	71.04 ± 0.70	68.25 ± 0.46	76.77 ± 1.86	80.55 ± 0.40	76.88 ± 0.42
	Acc.	97.22 ± 0.04	93.21 ± 0.10	99.30 ± 0.01	98.19 ± 0.15	94.57 ± 0.10	96.50 ± 0.05
U-Net-SCA +CRFs	IoU	80.36 ± 0.76	71.53 ± 0.15	68.40 ± 0.26	77.04 ± 1.99	80.83 ± 0.16	77.09 ± 0.22
	Acc.	97.17 ± 0.10	93.24 ± 0.06	99.31 ± 0.01	98.20 ± 0.15	94.62 ± 0.06	96.51 ± 0.05
U-Net-SCA +Reuse	IoU	80.76 ± 0.23	71.52 ± 0.30	68.08 ± 0.65	78.26 ± 0.51	81.36 ± 0.38	77.48 ± 0.30
	Acc.	97.22 ± 0.04	93.25 ± 0.10	99.30 ± 0.01	98.30 ± 0.03	94.78 ± 0.08	96.57 ± 0.07
GAN-SCA	IoU	**80.82 ± 0.21**	**71.37 ± 0.44**	**68.67 ± 0.18**	**78.68 ± 0.09**	**81.62 ± 0.26**	**77.52 ± 0.19**
	Acc.	**97.24 ± 0.03**	**93.32 ± 0.12**	**99.31 ± 0.01**	**98.33 ± 0.01**	**94.80 ± 0.06**	**96.60 ± 0.02**

Figure 5 shows the segmentation results of methods described above on the Inria aerial image labeling dataset. Figure 5a shows the results of an image patch over Austin, from which we can observe that the standard U-Net is affected by shadows and fail to segment the boundaries of complex structural buildings (upper left part of the figure) correctly. With the help of spatial and channel attention mechanisms, U-Net-SCA achieves better performance when dealing with the same situation, but still mislabels some non-building pixels in shadows as building. The extraction results from U-Net-SCA+CRFs, U-Net-Reuse, and GAN-SCA all seem to have clearer boundaries, especially for complex structural buildings, and this is due to the adopted different label refinement method. In contrast, the extraction result of the proposed GAN-SCA achieves clearer and more accurate outlines of this kind of buildings. In addition, some large buildings are difficult to labeled correctly, due to their edges on the rooftop which have a similar color to roads. Figure 5b shows the results of a large building in the Chicago city, where the results of U-Net suffer from over-segmentation of the inner edges of the detected building, while U-Net-SCA improves the results by using the channel and spatial attention mechanisms to selectively enhance useful features. U-Net-SCA+CRFs smooths the results, yet the improvements seem insignificant. The result of U-Net-SCA+Reuse also shows a slight improvement, but the over-segmentation has not been effectively solved. In contrast, the proposed GAN-SCA labeled the large building more completely. Moreover, buildings with complex shape and multiple colors are prone to be confused by the networks, as shown in the middle of Figure 5c, and most methods mislabel this kind of building as non-buildings, while the proposed GAN-SCA can provide a relatively proper segmentation results.

Original Images

Figure 5. Cont.

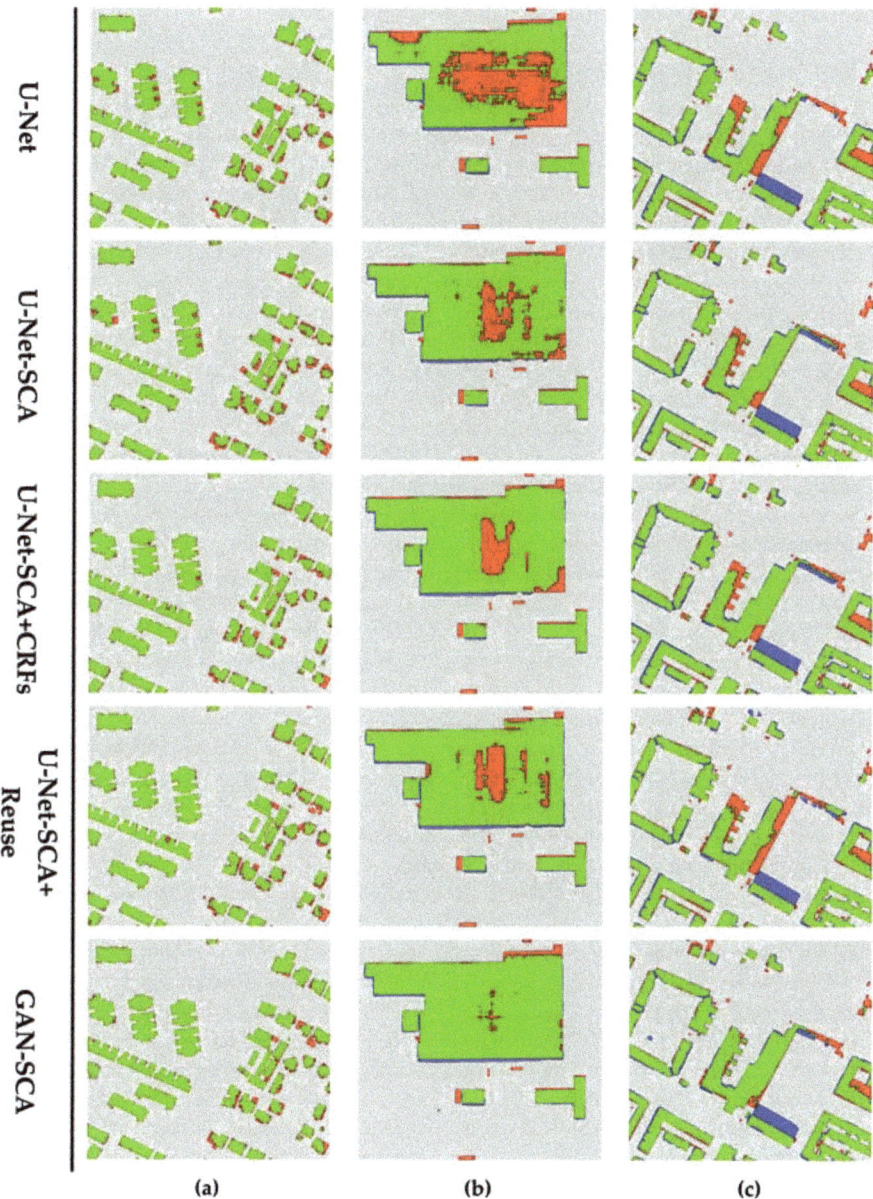

Figure 5. Building extraction results for three image patches of Inria aerial image labeling dataset. (**a**) Image patch over Austin; (**b**) Image patch over Chicago; (**c**) Image patch over Vienna. Green: true positive (tp) pixels; Gray: true negative (tn) pixels; Blue: false positive (fp) pixels; Red: false negative (fn) pixels.

4.2. Comparison to State-of-the-Art Methods

4.2.1. Inria Aerial Image Labeling Dataset

To evaluate the performance on the Inria aerial image labeling dataset, we compare the proposed GAN-SCA (best results we achieved) with some state-of-the-art methods, including the baseline method FCN [23], multi-layer perceptron (MLP) [23], Mask R-CNN [32] performed by Ohleyer et al. [33], SegNet+Multi-Task Loss [24], 2-levels U-Nets [25], and the multi-stage multi-task (MSMT) [34]. FCN and MLP are frameworks proposed by the producers of the Inria aerial image labeling dataset. MLP derived from the base FCN and introduced a multi-layer perceptron to learn how to combine features at different resolutions. Mask R-CNN consisted of a region proposed network (RPN) and an FCN, the RPN took the whole image as input and output the image with bounding box proposals. According to the proposal of RPN, the FCN then performed efficient segmentation. The SegNet+Multi-Task Loss was based on SegNet architecture and trained with an uncertainty based multi-task loss. In particular, one convolutional layer L was followed after the last layer of the decoder to generate the distance classes, and then the output of decoder's last layer was concatenated with the output of L to predict the final segmentation results. 2-Levels U-Nets was proposed in [25], where two U-Net architectures were arranged end-to-end, and the last U-Net was served as the post-processor to the first one. Moreover, the test time augmentation was applied to further improve the segmentation performance. The MSMT architecture was proposed in [34]. Authors proposed an MSMT neural network which had two stages, namely semantic segmentation and localization. The first stage was dedicated to semantic segmentation, while the second stage was designed for localization.

Table 2 presents the accuracy and IoU of different methods on the Inria aerial image labeling dataset. It is worth noting that IoU can take into account both the false alarms and the missing detections that is a more suitable metric than global accuracy on Inria dataset, because this dataset contains large areas of background pixels. It can be seen from Table 2 that MLP outperforms the base FCN [23] by introducing multi-layer perceptron to fuse multi-resolution features. Mask-RCNN is a promising architecture, but it requires very good hyperparameters tuning [33]. Therefore, it achieves better performance in Austin and Tyrol-w but lower in most regions when compared to MLP. SegNet+Multi-Task Loss improves the performance of SegNet by introducing a cascaded multi-task loss, but the improvement is still limited. Although it achieves the best accuracy in regions of Chicago and Vienna, the corresponding IoU is not ideal. 2-Levels U-Nets and MSMT achieve similar accuracy, of which the former approach outperforms the latter one in terms of IoU in all regions. This is mainly because the 2-Levels U-Nets is based on U-Net which is a deeper architecture than that of MSMT. The proposed GAN-SCA is also on top of U-Net. With the help of the attention mechanisms and adversarial training strategy, GAN-SCA outperforms 2-Levels U-Nets in most evaluation metrics and produces the highest IoU in most regions, especially the densely populated cities, such as Austin, Chicago, and Vienna. In terms of the overall accuracy and IoU, the proposed method surpasses all other methods by a considerable margin, which shows that the proposed method can accomplish accurate building segmentation. The qualitative results of the GAN-SCA are shown in Figure 6. It can be seen that the GAN-SCA achieves accurate building segmentation results in each region with smooth outlines.

Table 2. Experimental results on Inria aerial image labeling dataset.

Methods	Metrics	Austin	Chicago	Kitsap	Tyrol-w	Vienna	Overall
FCN [23]	IoU	47.66	53.62	33.70	46.86	60.60	53.82
	Acc.	92.22	88.59	98.58	95.83	88.72	92.79
MLP [23]	IoU	61.20	61.30	51.50	57.95	72.13	64.67
	Acc.	94.20	90.43	98.92	96.66	91.87	94.42
Mask R-CNN [32]	IoU	65.63	48.07	54.38	70.84	64.40	59.53
	Acc.	94.09	85.56	97.32	98.14	87.40	92.49

Table 2. *Cont.*

Methods	Metrics	Austin	Chicago	Kitsap	Tyrol-w	Vienna	Overall
SegNet+Multi-Task Loss [24]	IoU	76.76	67.06	**73.30**	66.91	76.68	73.00
	Acc.	93.21	**99.25**	97.84	91.71	**96.61**	95.73
2-Levels U-Nets [25]	IoU	77.29	68.52	72.84	75.38	78.72	74.55
	Acc.	96.69	92.40	99.25	98.11	93.79	96.05
MSMT [34]	IoU	75.39	67.93	66.35	74.07	77.12	73.31
	Acc.	95.99	92.02	99.24	97.78	92.49	96.06
GAN-SCA	IoU	**81.01**	**71.73**	68.54	**78.62**	**81.62**	**77.75**
	Acc.	**97.26**	93.32	**99.30**	**98.32**	94.84	**96.61**

Figure 6. Building extraction results of Inria aerial image labeling dataset. (**a**) Image patch over Austin; (**b**) Image patch over Chicago; (**c**) Image patch over Vienna. Green: true positive (tp) pixels; Gray: true negative (tn) pixels; Blue: false positive (fp) pixels; Red: false negative (fn) pixels.

4.2.2. Massachusetts Buildings Dataset

We tested the performance of the proposed GAN-SCA on the Massachusetts buildings dataset by using the same metrics as the compared methods. We compared the performance of GAN-SCA with several state-of-the-art methods including Mnih-CNN+CRFs [22], Satio-multi-MA&CIS [21], LG-Seg-ResNet-IL [35], and MTMS [34]. The Mnih-CNN+CRF was proposed by the producers of the Massachusetts building dataset, which belonged to the patch-based category, and CRFs was included as a post-processor. Satio-multi-MA&CIS was based on Mnih-CNN architecture, in which channel-wise inhibited softmax (CIS) loss function and modeled averaging (MA) techniques were used to further enhance the extraction performance. LG-Seg-ResNet-IL is a dual local-global semantic segmentation architecture with residual connections and an intermediate contextual loss (IL), which learned to combine local appearance and global contextual information simultaneously in a complementary way. MTMS is the same method described in Section 4.2.1.

Table 3 compares the F_1-measure of each method, in which ρ denotes the relaxed factor when computing the corresponding recall and precision measures. As shown in Table 3, our GAN-SCA obtains a superior performance than all other methods. With the help of the (CIS) loss function and (MA) with spatial displacement, the Satio-multi-MA&CIS achieves a slight improvement compared to the baseline method Mnih-CNN+CRFs. LG-Seg-ResNet-IL effectively combines the local and global information, which mitigates the problem of the limited receptive field of the patch-based method. So LG-Seg-ResNet-IL achieves a remarkable improvement compared to the first two methods. MTMS is an FCN-based method that introduces a multi-stage multi-task training strategy to enhance segmentation performance. MTMS and GAN-SCA achieve better performance compared to the patch-based methods, which indicates the superiority of the pixel-based method. Thanks to the deeper architecture and the feature selection by adopting attention mechanisms, the GAN-SCA exhibits better performance when compared to MTMS, which further indicates the rationality of the proposed method. Figure 7 exhibits the prediction results of the proposed model for three image patches. It can be seen that our proposed model presents a satisfying performance in challenging areas.

Table 3. Experimental results on Massachusetts buildings dataset.

Method	F_1-Measure	
	$\rho = 0$	$\rho = 3$
Mnih-CNN+CRFs [22]	-	92.11%
Satio-multi-MA&CIS [21]	-	92.30%
LG-Seg-ResNet-IL [35]	-	94.30%
MTMS [34]	83.39%	96.04%
GAN-SCA	84.79%	96.36%

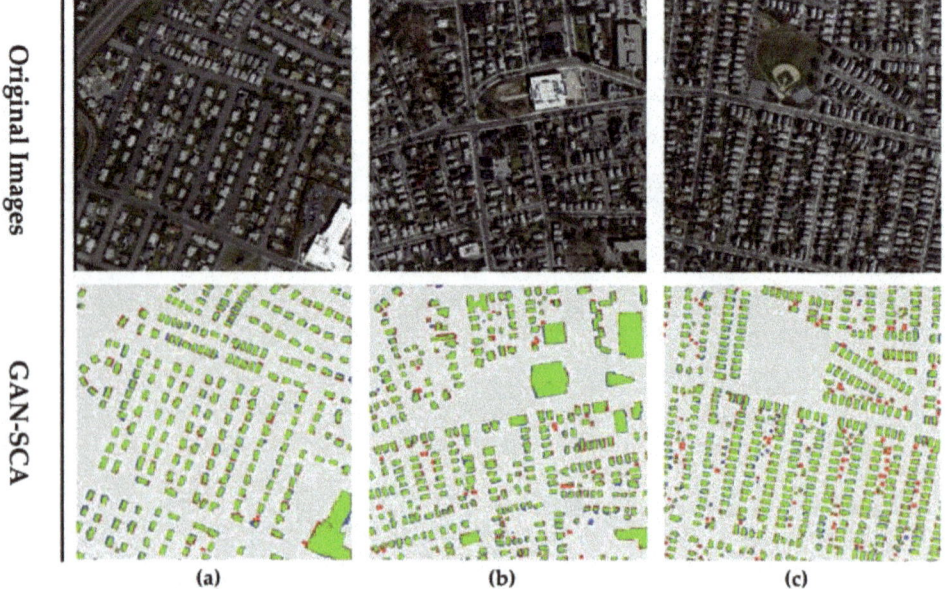

Figure 7. Building extraction results on the Massachusetts buildings dataset. (a–c) prediction results of three image patches in Massachusetts buildings dataset. Green: true positive (tp) pixels; Gray: true negative (tn) pixels; Blue: false positive (fp) pixels; Red: false negative (fn) pixels.

4.3. Experiments on FCN based GAN-SCA

The experiments above adopted U-Net as the baseline of the segmentation network for the proposed GAN-SCA, and achieve a certain improvement when compared with the standard U-Net. In fact, our proposed GAN-SCA can be realized on the top of many other semantic segmentation architectures. In this section, we will explore the GAN-SCA on top of FCN-8s version to further demonstrate the effectiveness of the attention mechanisms and adversarial training in building extraction from high-resolution remote sensing images. Figure 8 shows the architecture of the segmentation network (FCN-8s-SCA), where the channel and spatial attention mechanisms are embedded into the FCN-8s architecture with the VGG-16 [36] architecture as an encoder. Same as the U-Net based GAN-SCA described above, the adversarial network of this version is followed by the encoder of its segmentation network.

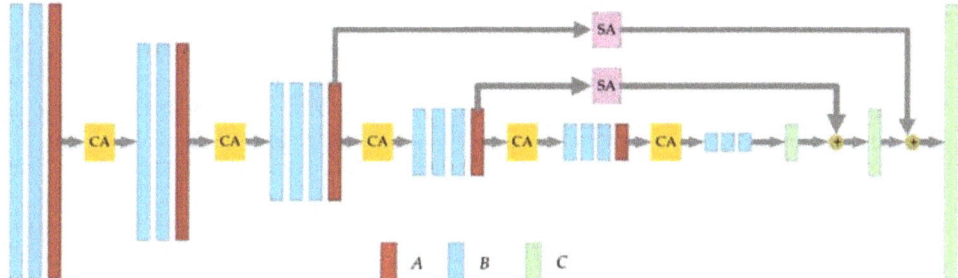

Figure 8. Architecture of FCN-8s-SCA. A is max pooling layer; B are convolutional + Rectified Linear Unit (ReLU) layers; C is the transpose convolutional layer; SA is the spatial attention mechanism; CA is the channel attention mechanism; RS is the reshape operation.

We train the FCN-8s, FCN-8s-SCA, and GAN-SCA on the Inria aerial image labeling dataset using the same training strategy as introduced in Section 2.2. The experimental results are reported in Table 4. Compared with the FCN-8s, the FCN-8s-SCA improved the overall accuracy and IoU by 0.49% and 3.71%, respectively. For the adversarial training strategy, the FCN-8s based GAN-SCA further improved the extraction performance by 0.48% and 3.16% for the overall accuracy and IoU, respectively. We can conclude that the attention mechanisms can improve the segmentation performance by feature selection, and adversarial training can further refine the segmentation result by learning high-order consistency. In addition, the improvement of FCN-8s based GAN-SCA is more significant than the aforementioned U-Net based GAN-SCA. This is because the standard U-Net architecture has already achieved remarkable segmentation performance, as it fused high-level and low-level feature by first concatenating features together and then performing convolutions for dimensionality reduction. The convolutional layers in this process enable the network to learn how to fuse multi-scale features which can be regarded as feature selection to some extent. While FCN-8s has lower segmentation accuracy on this dataset when compared to the standard U-Net, it fused features by adopting element-wise addition, which seems unsuitable without any feature selection. Therefore, FCN-8s can take more advantage of the attention mechanisms.

Table 4. Experimental results on Inria aerial image labeling dataset.

Methods	Metrics	Austin	Chicago	Kitsap	Tyrol-w	Vienna	Overall
FCN-8s	IoU	67.98	63.43	53.17	68.13	72.03	68.05
	Acc.	95.47	91.41	99.01	97.52	92.38	95.16
FCN-8s-SCA	IoU	72.85	69.61	64.97	73.20	73.26	71.76
	Acc.	96.19	92.36	99.25	97.94	92.49	95.65
GAN-SCA	IoU	**78.51**	**70.10**	**66.42**	**76.84**	**77.24**	**74.92**
	Acc.	**96.90**	**92.86**	**99.27**	**98.14**	**93.46**	**96.13**

5. Discussion

The experimental results reported in Section 4 prove that the proposed approach achieved state-of-the-art performance on both Inria and Massachusetts buildings datasets. Furthermore, the GAN-SCA can also be employed on top of other semantic segmentation architectures with better performance. The effectiveness of our proposed method comes from the feature selection in spatial and channel dimensions, and the label refinement by learning high-order structural features. First, the adoption of spatial and channel attention mechanisms helps with enhancing the useful features while suppressing the interference information, improving the segmentation performance around building borders, and mitigating over-segmentation. Second, the adversarial training strategy learns the latent high-order structural information in the training phase and achieves label refinement in the testing phase without extra time consumption. Especially, the segmentation network and adversarial network of our architecture were optimized by multi-scale feature loss to better capture multi-range spatial relationships between pixels. These factors make the proposed GAN-SCA have a better feature extraction capability and better segmentation performance.

Although the proposed approach performs well as a fully supervised method, it relies on a large number of manual labeling samples. Further researches are needed to alleviate the task of manual annotation. Possible directions that can be explored include data augmentation techniques and adversarial learning for semi-supervised semantic segmentation. Data augmentation techniques can increase the number of training samples and improve the generalization ability of models. We have explored some standard data augmentation techniques including flip and rotation in this work and previous works to mitigate overfitting. More data augmentation strategies will be explored in our future work. In addition, adversarial learning for semi-supervised semantic segmentation is also an interesting research direction, which can take advantage of unlabeled data to generate self-taught signal to refine the segmentation network. These approaches will be highly relevant in fields, such as remote sensing images analysis, in which large datasets are expensive to obtain.

6. Conclusions

This paper presented an effective GAN-based approach for building extraction from high-resolution remote sensing images. The adopted architecture consists of two parts: the segmentation network and the adversarial network, which are, in turn, used to generate segmentation maps of buildings and to discriminate the ground truths and the predicted results of the segmentation network, respectively. To enable the segmentation network to focus on more useful information, spatial and channel attention mechanisms are embedded into the standard U-Net. The adversarial network architecture is similar to the encoder of the segmentation network, where the extracted multi-layer features are considered when computing the multi-scale L_1 loss in the adversarial training phase.

The experiments were conducted on the Inria aerial image labeling dataset for buildings as well as the Massachusetts buildings dataset. The experimental results show that the spatial and channel attention mechanisms can selectively enhance useful features to improve the segmentation performance, while adversarial training can further refine the segmentation results with little time consumption during the testing stage. Compared with the state-of-the-art methods on both the datasets, the proposed GAN-SCA achieved higher overall accuracy (96.61%), IoU (77.75%) for the Inria aerial image labeling dataset and F_1-Measure (96.36%) for the Massachusetts buildings dataset. Especially for samples with dense-distributed buildings, the improvement was more evident.

In future studies, we will explore adversarial network architecture optimization and loss function improvement to take full advantage of the adversarial training. We will also research the data augmentation techniques and semi-supervised semantic segmentation.

Author Contributions: Conceptualization, F.Y.; Methodology, X.P. and L.G.; Software, H.F.; Supervision, B.Z.; Validation, Z.C. and H.F.; Writing-Original Draft Preparation, X.P.; Writing-Review & Editing, L.G., J.R. and Z.C.

Funding: This research was supported by the Strategic Priority Research Program of the Chinese Academy of Sciences under Grant No. XDA19080302, and by the National Natural Science Foundation of China under Grant No. 91638201.

Conflicts of Interest: The authors declare no conflict of interest.

References

1. Rees, W.G. *Physical Principles of Remote Sensing*; Cambridge University Press: Cambridge, UK, 2013.
2. Alshehhi, R.; Marpu, P.R.; Wei, L.W.; Mura, M.D. Simultaneous extraction of roads and buildings in remote sensing imagery with convolutional neural networks. *ISPRS J. Photogramm. Remote Sens.* **2017**, *130*, 139–149. [CrossRef]
3. Yang, H.L.; Yuan, J.; Lunga, D.; Laverdiere, M.; Rose, A.; Bhaduri, B. Building extraction at scale using convolutional neural network: Mapping of the United States. *IEEE J. Sel. Top. Appl. Earth Obs. Remote Sens.* **2018**, *11*, 2600–2614. [CrossRef]
4. Long, J.; Shelhamer, E.; Darrell, T. Fully convolutional networks for semantic segmentation. In Proceedings of the 2015 IEEE Conference on Computer Vision and Pattern Recognition (CVPR), Boston, MA, USA, 7–12 June 2015; pp. 3431–3440.
5. Ronneberger, O.; Fischer, P.; Brox, T. U-net: Convolutional networks for biomedical image segmentation. In Proceedings of the International Conference on Medical Image Computing and Computer-Assisted Intervention, Munich, Germany, 5–9 October 2015; Springer: Cham, Switzerland, 2015; pp. 234–241.
6. Chen, L.C.; Papandreou, G.; Kokkinos, I.; Murphy, K.; Yuille, A.L. Semantic Image Segmentation with Deep Convolutional Nets and Fully Connected CRFs. *Computer Sci.* **2014**, *4*, 357–361.
7. Chen, L.C.; Papandreou, G.; Kokkinos, I.; Murphy, K.; Yuille, A.L. DeepLab: Semantic Image Segmentation with Deep Convolutional Nets, Atrous Convolution, and Fully Connected CRFs. *IEEE Trans. Pattern Anal. Mach. Intell.* **2016**, *40*, 834–848. [CrossRef] [PubMed]
8. Noh, H.; Hong, S.; Han, B. Learning deconvolution network for semantic segmentation. In Proceedings of the IEEE International Conference on Computer Vision (ICCV), Santiago, Chile, 13–16 December 2015; pp. 1520–1528.
9. Badrinarayanan, V.; Kendall, A.; Cipolla, R. Segnet: A deep convolutional encoder-decoder architecture for image segmentation. *IEEE Trans. Pattern Anal. Mach. Intell.* **2017**, *39*, 2481–2495. [CrossRef] [PubMed]
10. Liu, Y.; Piramanayagam, S.; Monteiro, S.; Saber, E. Dense semantic labeling of very-high-resolution aerial imagery and LiDAR with fully-convolutional neural networks and higher-order crfs. In Proceedings of the IEEE Conference on Computer Vision and Pattern Recognition Workshops (CVPRW), Honolulu, HI, USA, 21–26 July 2017; pp. 1561–1570.
11. Liu, Y.; Minh Nguyen, D.; Deligiannis, N.; Ding, W.; Munteanu, A. Hourglass-shape network based semantic segmentation for high resolution aerial imagery. *Remote Sens.* **2017**, *9*, 522. [CrossRef]
12. Pan, X.; Gao, L.; Marinoni, A.; Zhang, B.; Yang, F.; Gamba, P. Semantic labeling of high resolution aerial imagery and LiDAR data with fine segmentation network. *Remote Sens.* **2018**, *10*, 743. [CrossRef]
13. Maggiori, E.; Charpiat, G.; Tarabalka, Y.; Alliez, P. Recurrent Neural Networks to Correct Satellite Image Classification Maps. *IEEE Trans. Geosci. Remote Sens.* **2017**, *55*, 4962–4971. [CrossRef]
14. Bergado, J.R.; Persello, C.; Stein, A. "Recurrent Multiresolution Convolutional Networks for VHR Image Classification,". *IEEE Trans. Geosci. Remote Sens.* **2018**, *56*, 6361–6374. [CrossRef]
15. Goodfellow, I.; Pouget-Abadie, J.; Mirza, M.; Xu, B.; Warde-Farley, D.; Ozair, S.; Courville, A.; Bengio, Y. Generative adversarial nets. In Proceedings of the Advances in Neural Information Processing Systems, Montreal, Qc, Canada, 8–13 December 2014; pp. 2672–2680.

16. Luc, P.; Couprie, C.; Chintala, S.; Verbeek, J. Semantic segmentation using adversarial networks. *arXiv*, 2016; arXiv:161108408. Available online: https://arxiv.org/abs/1611.08408 (accessed on 1 April 2018).
17. Xue, Y.; Xu, T.; Zhang, H.; Long, L.R.; Huang, X. SegAN: Adversarial network with multi-scale L_1 loss for medical image segmentation. *Neuroinformatics* **2017**, *6*, 1–10. [CrossRef] [PubMed]
18. Pan, X.; Gao, L.; Zhang, B.; Yang, F.; Liao, W. High-resolution aerial imagery semantic labeling with dense pyramid network. *Sensors* **2018**, *18*, 3774. [CrossRef] [PubMed]
19. Sherrah, J. Fully convolutional networks for dense semantic labelling of high-resolution aerial imagery. *arXiv*, 2016; arXiv:1606.02585v1. Available online: https://arxiv.org/abs/1606.02585 (accessed on 1 April 2018).
20. Volpi, M.; Tuia, D. Dense semantic labeling of subdecimeter resolution images with convolutional neural networks. *IEEE Trans. on Geosci. Remote Sens.* **2017**, *55*, 881–893. [CrossRef]
21. Saito, S.; Yamashita, T.; Aoki, Y. Multiple Object Extraction from Aerial Imagery with Convolutional Neural Networks. *Electronic Imag.* **2016**, *60*, 10401–10402.
22. Mnih, V. Machine Learning for Aerial Image Labeling. Ph.D. Thesis, University of Toronto, Toronto, ON, Canada, 2013.
23. Maggiori, E.; Tarabalka, Y.; Charpiat, G.; Alliez, P. Can semantic labeling methods generalize to any city? The Inria aerial image labeling benchmark. In Proceedings of the IEEE International Geoscience and Remote Sensing Symposium (IGARSS), Fort Worth, TX, USA, 23–28 July 2017; pp. 3226–3229.
24. Bischke, B.; Helber, P.; Folz, J.; Borth, D.; Dengel, A. Multi-task learning for segmentation of building footprints with deep neural networks. *arXiv*, 2017; arXiv:1709.05932. Available online: https://arxiv.org/abs/1709.05932 (accessed on 27 April 2018).
25. Khalel, A.; El-Saban, M. Automatic pixelwise object labeling for aerial imagery using stacked u-nets, arXiv 2018, arXiv:1803.04953. Available online: https://arxiv.org/abs/1803.04953 (accessed on 27 April 2018).
26. Itti, L.; Koch, C. Computational modelling of visual attention. *Nat. Rev. Neurosci.* **2001**, *2*, 194–203. [CrossRef] [PubMed]
27. Wang, H.; Wang, Y.; Zhang, Q.; Xiang, S.; Pan, C. Gated convolutional neural networkk for semantic segmentation in high-resolution images. *Remote Sens.* **2017**, *9*, 446. [CrossRef]
28. Hu, J.; Shen, L.; Sun, G. Squeeze-and-excitation networks. In Proceedings of the IEEE/CVF Conference on Computer Vision and Pattern Recognition, Salt Lake City, UT, USA, 18–23 June 2018; pp. 7132–7141.
29. Glorot, X.; Bordes, A.; Bengio, Y. Deep Sparse Rectifier Neural Networks. In Proceedings of the International Conference on Artificial Intelligence and Statistics (AISTATS), Fort Lauderdale, FL, USA, 11–13 April 2011; pp. 315–323.
30. Kingma, D.; Ba, J. Adam: A method for stochastic optimization. In Proceedings of the International Conference on Learning Representations (ICLR), Banff, AB, Canada, 14–16 April 2014.
31. Audebert, N.; Le Saux, B.; Lefèvre, S. Semantic segmentation of earth observation data using multimodal and multi-scale deep networks. In Proceedings of the Asian Conference on Computer Vision (ACCV), Taipei, Taiwan, 21–23 November 2016; pp. 180–196.
32. He, K.; Gkioxari, G.; Dollar, P.; Girshick, R. Mask R-CNN. In Proceedings of the IEEE International Conference on Computer Vision (ICCV), Venice, Italy, 22–29 October 2017; pp. 2980–3288.
33. Building segmentation on satellite images. Available online: https://project.inria.fr/aerialimagelabeling/files/2018/01/fp_ohleyer_compressed.pdf (accessed on 1 April 2018).
34. Marcu, A.; Costea, D.; Slusanschi, E.; Leordeanu, M. A Multi-stage Multi-task neural network for aerial scene interpretation and geolocalization. *arXiv*, 2018; arXiv:1804.01322v1. Available online: https://arxiv.org/abs/1804.01322 (accessed on 27 April 2018).
35. Marcu, A.; Leordeanu, M. Object contra context: Dual local-global semantic segmentation in aerial images. In Proceedings of the Thirty-First AAAI Conference on Artificial Intelligence (AAAI-17), San Francisco, CA, USA, 4–9 February 2017; pp. 146–152.
36. Simonyan, K.; Zisserman, A. Very deep convolutional networks for large-scale image recognition. In Proceedings of the International Conference on Machine Learning (ICML), San Diego, CA, USA, 7–9 May 2015.

© 2019 by the authors. Licensee MDPI, Basel, Switzerland. This article is an open access article distributed under the terms and conditions of the Creative Commons Attribution (CC BY) license (http://creativecommons.org/licenses/by/4.0/).

Article

Semantic Segmentation-Based Building Footprint Extraction Using Very High-Resolution Satellite Images and Multi-Source GIS Data

Weijia Li [1,2,†], Conghui He [3,4,†], Jiarui Fang [3], Juepeng Zheng [1,2,5], Haohuan Fu [1,2,*] and Le Yu [1,2]

1. Ministry of Education Key Laboratory for Earth System Modeling, Department of Earth System Science, Tsinghua University, Beijing 100084, China; liwj14@mails.tsinghua.edu.cn (W.L.); 1351177@tongji.edu.cn (J.Z.); leyu@tsinghua.edu.cn (L.Y.)
2. Joint Center for Global Change Studies (JCGCS), Beijing 100084, China
3. Department of Computer Science, Tsinghua University, Beijing 100084, China; heconghui@gmail.com (C.H.); fjr14@mails.tsinghua.edu.cn (J.F.)
4. Tencent, Shenzhen 518000, China
5. College of Surveying and Geo-Informatics, Tongji University, Shanghai 200092, China
* Correspondence: haohuan@tsinghua.edu.cn; Tel.: +86-010-62798365
† These authors contributed equally to this work.

Received: 8 January 2019; Accepted: 13 February 2019; Published: 16 February 2019

Abstract: Automatic extraction of building footprints from high-resolution satellite imagery has become an important and challenging research issue receiving greater attention. Many recent studies have explored different deep learning-based semantic segmentation methods for improving the accuracy of building extraction. Although they record substantial land cover and land use information (e.g., buildings, roads, water, etc.), public geographic information system (GIS) map datasets have rarely been utilized to improve building extraction results in existing studies. In this research, we propose a U-Net-based semantic segmentation method for the extraction of building footprints from high-resolution multispectral satellite images using the SpaceNet building dataset provided in the DeepGlobe Satellite Challenge of IEEE Conference on Computer Vision and Pattern Recognition 2018 (CVPR 2018). We explore the potential of multiple public GIS map datasets (OpenStreetMap, Google Maps, and MapWorld) through integration with the WorldView-3 satellite datasets in four cities (Las Vegas, Paris, Shanghai, and Khartoum). Several strategies are designed and combined with the U-Net–based semantic segmentation model, including data augmentation, post-processing, and integration of the GIS map data and satellite images. The proposed method achieves a total F1-score of 0.704, which is an improvement of 1.1% to 12.5% compared with the top three solutions in the SpaceNet Building Detection Competition and 3.0% to 9.2% compared with the standard U-Net–based method. Moreover, the effect of each proposed strategy and the possible reasons for the building footprint extraction results are analyzed substantially considering the actual situation of the four cities.

Keywords: building extraction; deep learning; semantic segmentation; data fusion; high-resolution satellite images; GIS data

1. Introduction

High-resolution remote sensing images have been increasingly popular and widely used in many geoscience applications, including automatic mapping of land use or land cover types, and automatic detection or extraction of small objects such as vehicles, ships, trees, roads, buildings, etc. [1–6]. As one of these geoscience applications, the automatic extraction of building footprints from high-resolution imagery is beneficial for urban planning, disaster management, and environmental management [7–10]. The spatial distributions of buildings are also essential for monitoring urban settlements, modeling urban demographics, updating the geographical database, and many other aspects [11,12]. Due to the diversity of buildings (e.g., in color, shape, size, materials, etc.) in different regions and the similarity of buildings to the background or other objects [9], developing reliable and accurate building extraction methods has become an important and challenging research issue receiving greater attention.

Over the past few decades, many building extraction studies were based on traditional image processing methods, such as shadow-based methods, edge-based methods, object-based methods, and more [13–15]. For instance, Belgiu and Drăguţ [16] proposed and compared supervised and unsupervised multi-resolution segmentation methods combined with the random forest (RF) classifier for building extraction using high-resolution satellite images. Chen et al. [17] proposed edge regularity indices and shadow line indices as new features of building candidates obtained from segmentation methods, and employed three machine learning classifiers (AdaBoost, RF, and support vector machine (SVM)) to identify buildings. Huang and Zhang [18] proposed the morphological shadow index (MSI) to detect shadows (used as a spatial constraint of buildings) and proposed dual-threshold filtering to integrate the information from the morphological building index with the one from MSI. Ok et al. [19] proposed a novel fuzzy landscape generation method that models the directional spatial relationship of the building and its shadow for automatic building detection. These studies were based on traditional methods and focused on extracting buildings in a relatively small study region. However, the methods have not been evaluated in complex regions with a high diversity of buildings.

In recent years, deep learning methods have been broadly utilized in various remote sensing image–based applications, including object detection [2,3,20], scene classification [21,22], land cover, and land use mapping [23,24]. Since it was proposed in 2014, deep convolutional neural network (CNN)-based semantic segmentation algorithms [25] have been applied to many pixel-wise remote sensing image analysis tasks, such as road extraction, building extraction, urban land use classification, maritime semantic labeling, vehicle extraction, damage mapping, weed mapping, and other land cover mapping tasks [5,6,26–31]. Several recent studies used semantic segmentation methods for building extraction from remote sensing images [9–12,32–38]. For example, Shrestha et al. [10] proposed a fully connected network-based building extraction approach combined with the exponential linear unit (ELU) and conditional random fields (CRFs) using the Massachusetts building dataset. Lu et al. [32] employed the richer convolutional features network–based approach to detect building edges using the Massachusetts building dataset. Xu et al. [12] proposed a building extraction method based on the Res-U-Net model combined with guided filters using the ISPRS (International Society for Photogrammetry and Remote Sensing) 2D semantic labeling dataset. Sun et al. [7] proposed a building extraction method that combines the SegNet model with the active contour model using the ISPRS Potsdam dataset and the proposed Marion dataset. These existing studies demonstrated the excellent performance of the semantic segmentation algorithms for building extraction tasks.

As an essential part of the semantic segmentation algorithms, the public semantic labeling datasets used in previous state-of-the-art building extraction studies can be summarized as follows: (1) The Massachusetts building dataset [39] (used in References [10,32,35]) contains 151 aerial images (at 100 cm spatial resolution, with red/green/blue (RGB) bands, each with a size of 1500 × 1500 pixels) of the Boston area. (2) The ISPRS Vaihingen and Potsdam datasets [40] (used in References [7,12]) contain 38 image patches (at 5 cm resolution, each at a size of around 6000 × 6000 pixels) and 33 image patches (at 9 cm resolution, each with a size of around 2500 × 2500 pixels) with the near infrared, red, and green bands and the corresponding digital surface model (DEM) data. (3) The Inria dataset [41] (used in References [36,37]) contains aerial images covering 10 regions in the USA and Austria (at 30 cm resolution, with RGB bands). (4) The WHU (Wuhan University) building dataset [42] (used in Reference [38]) includes an aerial dataset containing 8189 image patches (at 30 cm resolution, with RGB bands, each with a size of 512 × 512 pixels) and a satellite dataset containing 17,388 image patches (at 270 cm resolution, with the same bands and size as the aerial dataset). (5) The AIRS (Aerial Imagery for Roof Segmentation) dataset [43] contains aerial images covering the area of Christchurch city in New Zealand (at 7.5 cm resolution, with RGB bands).

In this study, our proposed building extraction method is trained and evaluated based on the SpaceNet building dataset [44] proposed in 2017 and further explored in the 2018 DeepGlobe Satellite Image Understanding Challenge [11]. The SpaceNet building dataset provided in the DeepGlobe Challenge contains WorldView-3 multispectral imagery and the corresponding building footprints of four cities (Las Vegas, Paris, Shanghai, and Khartoum) located on four continents. The buildings in the SpaceNet dataset are much more diverse compared with the five datasets mentioned above. Details of the SpaceNet dataset are described in Section 2.

In addition, many studies employed data-fusion strategies that integrate different data to improve the building extraction results. Airborne light detection and ranging (LiDAR) data are among the most broadly utilized data in numerous building extraction studies [7,45–53]. For instance, Awrangjeb et al. [52] proposed a rule-based building roof extraction method from a combination of LiDAR data and multispectral imagery. Pan et al. [53] proposed a semantic segmentation network–based method for semantic labeling of the ISPRS dataset using high-resolution aerial images and LiDAR data. However, public and free LiDAR datasets are still very limited. On the other hand, GIS data (e.g., OpenStreetMap) has been utilized in several building extraction and semantic labeling studies [54–57] as either the reference map of the labeled datasets [54,55] or auxiliary data combined with satellite images [56,57]. For instance, Audebert [56] investigated different ways of integrating OpenStreetMap data and semantic segmentation networks for semantic labeling of aerial and satellite images. Du et al. [57] proposed an improved random forest method for semantic classification of urban buildings, which combines high-resolution images with GIS data. Nevertheless, OpenStreetMap data still cannot provide enough building information for many places in the world, including the selected regions in Las Vegas, Shanghai, and Khartoum of the SpaceNet building dataset used in our study.

In this research, we propose a semantic segmentation–based building footprint extraction method using the SpaceNet building dataset provided in the CVPR 2018 DeepGlobe Satellite Challenge. Several public GIS map datasets (OpenStreetMap [58], Google Maps [59], and MapWorld [60]) are integrated with the provided WorldView-3 satellite datasets to improve the building extraction results. The proposed method obtains an overall F1-score of 0.704 for the validation dataset, which achieved fifth place in the DeepGlobe Building Extraction Challenge. Our main contributions can be summarized as follows:

(1) To the best of our knowledge, this is the first attempt conducted to explore the combination of multisource GIS map datasets and multispectral satellite images for building footprint extraction in four cities that demonstrates great potential for reducing extraction confusion caused by overlapping objects and improving the extraction of building outlines.

(2) We propose a U-Net–based semantic segmentation model for building footprint extraction. Several strategies (data augmentation, post-processing, and integration of GIS map data and satellite images) are designed and combined with the semantic segmentation model, which increases the F1-score of the standard U-Net–based method by 3.0% to 9.2%.

(3) The effect of each proposed strategy, the final building footprint extraction results, and the potential causes are analyzed comprehensively based on the actual situation of four cities. Even compared with the top three solutions in the SpaceNet Building Detection Competition, our proposed method improves the total F1-score by 1.1%, 6.1%, and 12.5%.

The rest of the paper is organized as follows. Section 2 introduces the study area and the datasets of this research, including the SpaceNet building dataset provided in the DeepGlobe Challenge and the auxiliary GIS map data. Section 3 introduces our proposed method, including data preparation and augmentation, the semantic segmentation model for building footprint extraction, and the integration and post-processing of results. Section 4 describes the building footprint extraction results of the proposed method. Section 5 discusses and analyzes the building footprint extraction results obtained from different methods and proposed strategies, and the potential causes for each city. Section 6 summarizes the conclusions of this research.

2. Study Area and Datasets

2.1. SpaceNet Building Dataset Provided in the DeepGlobe Challenge

In this research, we used the SpaceNet building dataset provided in the CVPR 2018 DeepGlobe Satellite Challenge. The study area of this dataset includes four cities (Las Vegas, Paris, Shanghai, and Khartoum), which covers both urban and suburban regions. The whole labeled dataset contains 24,586 image scenes in which each has a size of 200 m × 200 m. A total of 302,701 building footprint polygons were fully annotated in the whole study area by a GIS team at the DigitalGlobe. In the DeepGlobe challenge, a total of 10,593 image scenes were publicly provided with labeled files (in geojson format). For the other image scenes, the labeled files were not published in the challenge and the prediction results could only be evaluated during the challenge. Thus, we selected the 10,593 image scenes with labeled files as the dataset for this study. Table 1 shows the number of image scenes and annotated building footprint polygons of each city. The image scenes of each city were further divided randomly into 70% training samples and 30% validation samples for training and evaluation of the proposed method.

Table 1. Number of image scenes and annotated building footprint polygons of each city.

City	Las Vegas	Paris	Shanghai	Khartoum	Total
Number of images	3851	1148	4582	1012	10,593
Number of buildings	108,328	16,207	67,906	25,046	217,487

The source dataset of this study is WorldView-3 satellite imagery, including the original single-band panchromatic imagery (0.3 m resolution, 650 pixels × 650 pixels), the 8-band multi-spectral imagery (1.24 m resolution, 163 pixels × 163 pixels), and the Pan-sharpened 3-band RGB and 8-band multispectral imagery (0.3 m resolution, 650 pixels × 650 pixels). We selected the Pan-sharpened 8-band multispectral imagery as the satellite dataset for our proposed method. The annotation dataset contains a summary file of the spatial coordinates of all annotated building footprint polygons and geojson files corresponding to each image scene. These files were converted into single-band binary images as the labeled dataset for our proposed method, in which values of 0 and 1 indicate that pixels belong to nonbuilding and building areas, respectively. In the SpaceNet building dataset provided in the DeepGlobe Challenge, small building polygons with an area equal to or smaller than 20 pixels were discarded because these were actually artifacts generated from the image tiling process (e.g., one building divided into multiple parts by a tile boundary). Examples of the satellite images and annotated building footprints can be found in Figure 1.

2.2. Auxiliary Data Used in Our Proposed Method

Besides the multispectral satellite imagery, we also used several public GIS map datasets as the auxiliary data for our proposed method because of the extra useful information they provide for building footprint extractions. Contrary to previous studies that used single-source auxiliary GIS data, we selected the map dataset with the most abundant information from several public GIS map datasets for each city. For Las Vegas, we selected the Google Maps dataset [59], which contains more information than the OpenStreetMap [58]. For Paris, we selected the popular OpenStreetMap dataset because of its abundant information. For Shanghai, we selected the MapWorld dataset [60] because it contains abundant information on buildings and there is no coordinate shifting between that dataset and the satellite imagery. For Khartoum, we selected the OpenStreetMap dataset, which is slightly more informative than the Google Maps dataset but still lacks building information for most areas. All of the map datasets were collected in a raster image format, according to the geospatial information of their corresponding satellite images (i.e., longitude, latitude, and spatial resolution) and resized into 650 × 650 pixels for further integration with the satellite imagery. Examples of the multi-source GIS map images and corresponding satellite images can be found in Figure 1.

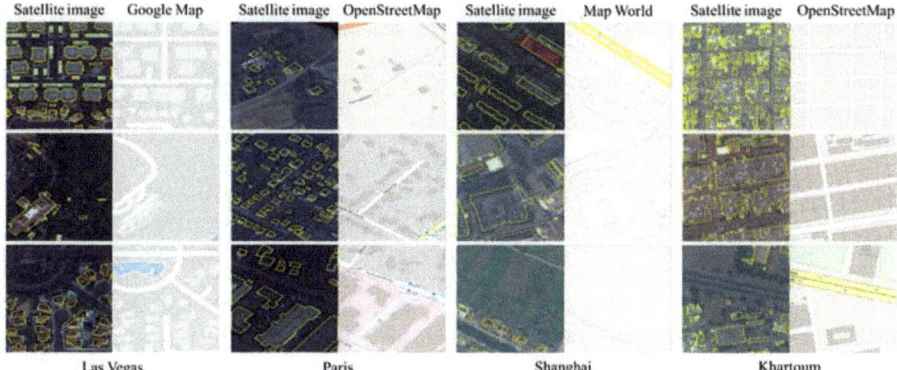

Figure 1. Examples of WorldView-3 satellite images, annotated building footprints (denoted by yellow polygons), and multi-source geographic information system (GIS) map images of four cities.

3. Materials and Methods

In this study, we designed a semantic segmentation–based approach for building footprint extraction. Figure 2 shows the overall flowchart of the proposed approach. It consists of 3 main stages including data preparation and augmentation, semantic segmentation for building footprint extraction, and integration and post-processing of results. In the first stage, we designed a data fusion method to make full use of both the satellite images and the extra information of GIS map data. We applied data augmentation (rescaling, slicing, and rotation) to our dataset in order to avoid potential problems (e.g., overfitting), which resulted from insufficient training samples, and to improve the generalization ability of the model. In the second stage, we trained and evaluated the U-Net–based semantic segmentation model, which is widely used in many remote sensing image segmentation studies. In the third stage, we applied the integration and post-processing strategies for further refinement of the building extraction results. Details of each stage are described in the following sections.

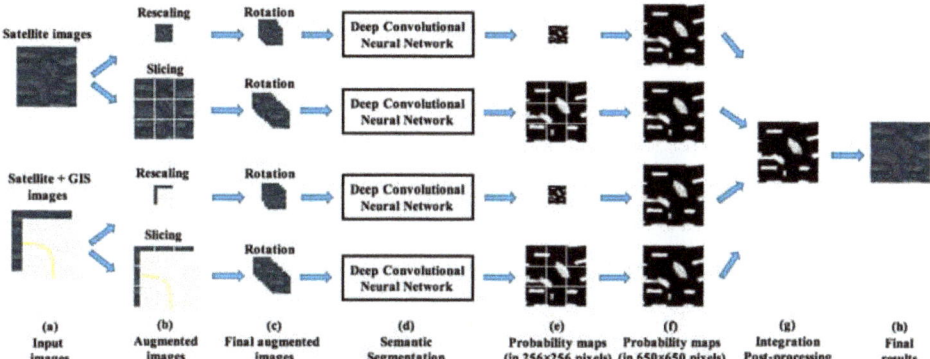

Figure 2. Overall flowchart of the proposed approach for building extraction, including (**a–c**) data preparation and augmentation, (**d**) semantic segmentation for building footprint extraction, and (**e–h**) integration and post-processing of results.

3.1. Data Preparation and Augmentation

3.1.1. Integration of Satellite Data and GIS Map Data

As mentioned in Section 2, besides the WorldView-3 multispectral satellite imagery provided in the SpaceNet dataset, we also used multiple public GIS map datasets as the auxiliary data for our proposed method. Although these public GIS map datasets provide extra information for building footprint extraction, it is unreasonable to train a separate deep neural network using the 3-band map datasets. The main reason is that many buildings are not displayed on the map image (especially tiny buildings and those in Khartoum city). In many regions, the building areas or outlines displayed in map images are not consistent with the ground truth buildings annotated based on the satellite images.

In this research, the training and validation datasets were preprocessed into two collections for each city. The first collection contained the eight-band multi-spectral satellite images while the second collection integrated the multi-spectral satellite images and the GIS map dataset. In order to unify the structure of the semantic segmentation network for the 2 dataset collections and enable the model trained by one dataset collection to be used as the pre-trained model for the other, we stacked the first 5 bands (red, red edge, coastal, blue, and green) of each WorldView-3 satellite image with the 3 bands (red, green, and blue) of its corresponding map image to generate an 8-band integrated image.

3.1.2. Data Augmentation

Data augmentation was proven to be an effective strategy to avoid potential problems (e.g., overfitting) resulting from insufficient training samples and to improve the generalization ability of deep learning models in many previous studies [9,10,32]. Considering the large number of hyper-parameters in the semantic segmentation model and the relatively small number of training samples in the SpaceNet building dataset (fewer than 5000 samples for each city), we applied the following data augmentation strategy (rescaling, slicing, and rotation) in order to increase the quantity and diversity of training samples and semantic segmentation models. Each dataset collection described in Section 3.1.1 was further preprocessed into 2 formats of input images for the training of each semantic segmentation model. First, each image with a size of 650 × 650 pixels was rescaled into an image of 256 × 256 pixels. Second, each image with a size of 650 × 650 pixels was sliced into 3 × 3 sub-images of 256 × 256 pixels. Moreover, we further augmented the training dataset through four 90° rotations. Consequently, we obtained 4 collections of preprocessed and augmented input datasets for each city, which we used for training and evaluating each deep convolutional neural network.

3.2. Semantic Segmentation Model for Building Footprint Extraction

3.2.1. Architecture of Semantic Segmentation Model for the Building Extraction

In this study, the semantic segmentation model for the building extraction is based on the U-Net architecture [61]. U-Net is a popular deep convolutional neural network architecture for semantic segmentation and has been used in several satellite image segmentation studies [5,12,30,62]. Since U-Net was initially designed for the binary segmentation of biomedical images with a relatively small number of training samples, it is a good choice for the building extraction task in this study as well. We modified the size of layers in the U-Net architecture to fit our building extraction task. We also added a batch normalization layer behind each convolutional layer.

Figure 3 shows the architecture of the semantic segmentation model for our building extraction task, including the name and size of each layer. It consists of the following 6 parts: (1) the convolutional layers for feature extraction through multiple 3 × 3 convolution kernels (denoted by Convolution); (2) the batch normalization layer for accelerating convergence during the training phase (denoted by Batch Normalization); (3) the activation function layer for nonlinear transformation of the feature maps, in which we used the widely used rectified linear unit (ReLU) in this study (denoted by Activation); (4) the max-pooling layer for downsampling of the feature maps (denoted by Max-pooling);

(5) the upsampling layer for recovering the size of the feature maps that are downsampled by the max-pooling layer (denoted by Upsampling); and (6) the concatenation layer for combining the upsampled feature map in deep layers with the corresponding feature map from shallow layers (denoted by Concatenation).

For the last batch-normalized layer of the semantic segmentation model (in the same size as the input image), we applied the sigmoid function as the activation function layer and obtained the pixel-wise probability map (indicating the probability that a pixel belonged to the building type). Lastly, we binarized the probability map using a given threshold (0.5 in common cases) to obtain the predicted building footprint extraction result (the output of the semantic segmentation network), and vectorized the output image to obtain a list of predicted building polygons.

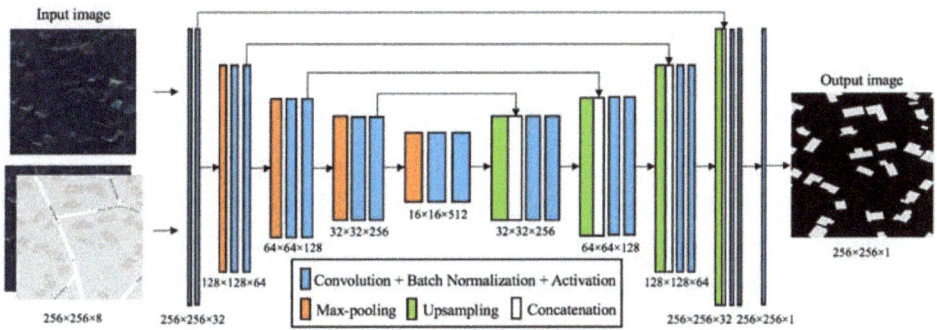

Figure 3. Architecture of semantic segmentation model for building extraction.

3.2.2. Training and Evaluation of Semantic Segmentation Model

To train the semantic segmentation model, we selected Adam as the optimization method and the binary cross entropy as the loss function. Due to the limited size of GPU memory, the batch size in the training phase was set to 8 in this study. The learning rate was set to 0.001 and the maximum number of epochs was set to 100. Moreover, we monitored the average Jaccard coefficient as an indicator for early stopping in order to avoid the potential problem of overfitting. Formula (1) shows the calculation process of the average Jaccard coefficient (denoted by J), in which $y_{gt}^{(i)}$ denotes the ground truth label of the ith pixel, $y_{pred}^{(i)}$ denotes the predicted label of the ith pixel, and n denotes the total number of pixels. The training phase was terminated before reaching the maximum number of epochs if the average Jaccard coefficient had no improvement for more than 10 epochs.

$$J = \frac{1}{n}\sum_{i=1}^{n}(y_{gt}^{(i)} \times y_{pred}^{(i)} / (y_{gt}^{(i)} + y_{pred}^{(i)} - y_{gt}^{(i)} \times y_{pred}^{(i)})) \qquad (1)$$

During the training phase, the semantic segmentation model was evaluated by the validation dataset at the end of each epoch. Besides the pixel-based accuracy that is commonly used in semantic segmentation tasks, we also recorded the object-based accuracy of the validation dataset in each epoch since it was the evaluation metric of the DeepGlobe challenge. For pixel-based accuracy, we compared the binarized building extraction image results predicted from the semantic segmentation model with the rasterized ground truth image. For object-based accuracy, we compared the vectorized building extraction image results (a list of predicted building polygons) with the ground truth building polygons (details are described in Section 3.4). As described in Section 3.1, for each city, 4 preprocessed and augmented dataset collections were used for the training and evaluation of the semantic segmentation model. For each dataset collection, the predicted building extraction results with the highest object-based accuracy were used for further integration and post-processing, which is described in the following section.

3.3. Integration and Post-Processing of Results

After training and evaluating the semantic segmentation model based on each of the 4 dataset collections, we obtained 4 groups of probability maps (each with a size of 256 × 256 pixels) for each validation sample. The value of each pixel in the probability map indicates the predicted probability that the pixel belongs to the building area. For each validation sample, the 4 groups of probability maps were obtained from (1) the satellite image with a rescaling strategy, (2) the satellite image with a slicing strategy, (3) the satellite + map image with a rescaling strategy, and (4) the satellite + map image with a slicing strategy, respectively. For the first and third groups, we rescaled the single probability map into the one at the original sample size. For the second and fourth groups, we combined 9 probability maps into a single map corresponding to the complete image. As a result, we obtained 4 probability maps (each with a size of 650 × 650 pixels) for each validation sample.

We proposed a 2-level integration strategy for integrating the results obtained from each model into the final building footprint extraction results. At the first level, for both the satellite and satellite + map image–based dataset collections, we averaged the pixel values of 2 probability maps (obtained from 2 preprocessing methods) into an integrated probability map. At the second level, the 2 integrated probability maps (obtained from the 2 dataset collections) were further averaged into the final building probability map.

After obtaining the integrated building probability map, we applied 2 post-processing strategies to optimize the final predicted results. In the first strategy, we adjusted the threshold of the probability (indicating whether a pixel belongs to a building area or a nonbuilding area) from 0.45 to 0.55 for each city. The optimized probability threshold was then used for vectorizing the probability map into the binary building extraction image result. In the second strategy, in order to filter out potential noise in the building extraction image results, we adjusted the threshold of the polygon size (indicating the minimal possible size of a building polygon) from 90 to 240 pixels for each city. The optimized thresholds of probability and polygon size of the validation dataset were also applied to the test dataset for each city.

3.4. Evaluation Metric

The building extraction results can be evaluated by several methods including the pixel-based and object-based methods that are the most broadly used in existing building extraction studies [7,63]. In the pixel-based evaluation method (used in References [9,10,12]), the binary building extraction image result (predicted from the semantic segmentation network) is directly compared with the binary ground truth image. In the object-based evaluation method (often used in building edge or footprint detection studies, such as in Reference [32]), the building extraction image result needs to be converted into the predicted building polygons for comparison with the ground truth building polygons. The DeepGlobe challenge selected the object-based method to evaluate the building footprint extraction results. Compared with the pixel-based method, the object-based method emphasizes not only the importance of accurate detection of building areas, but also the complete identification of building outlines.

In the DeepGlobe challenge, the ground truth dataset for evaluating building extraction results contained the spatial coordinates of the vertices corresponding to each annotated building footprint polygon. Thus, we needed to convert the single-band building extraction image results (the output of the semantic segmentation network) into a list of building polygons (in the same format as the ground truth dataset). Formula (2) shows the definition of the IoU (intersection over union) for evaluating whether a detected building polygon is accurate, which is equal to the intersection area of a detected building polygon (denoted by A) and a ground truth building polygon (denoted by B) divided by the union area of A and B. If a detected building polygon intersects with more than one ground truth building polygon, then the ground truth building with the highest IoU value will be selected.

$$\mathrm{IoU} = \frac{\mathrm{Area}(\mathbf{A} \cap \mathbf{B})}{\mathrm{Area}(\mathbf{A} \cup \mathbf{B})} \qquad (2)$$

The precision, recall, and F1-score were calculated according to Formulas (3)–(5), where true positive (TP) indicates the number of building polygons that are detected correctly, false positive (FP) indicates the number of other objects that are detected as building polygons by mistake, and false negative (FN) indicates the number of building polygons not detected. A building polygon will be scored as correctly detected if the IoU between the detected building polygon and a ground truth building polygon is larger than 0.5. The results of each city were evaluated independently and the final F1-score is the average value of F1-scores for each city.

$$\text{Precision} = \frac{\text{TP}}{\text{TP} + \text{FP}} \tag{3}$$

$$\text{Recall} = \frac{\text{TP}}{\text{TP} + \text{FN}} \tag{4}$$

$$\text{F1-score} = \frac{2 \times \text{Precision} \times \text{Recall}}{(\text{Precision} + \text{Recall})} = \frac{2 \times \text{TP}}{(2 \times \text{TP} + \text{FP} + \text{FN})} \tag{5}$$

4. Experimental Results Analysis

4.1. Experiment Setting and Semantic Segmentation Results

In this study, training and evaluation of the semantic segmentation network was based on the Keras deep learning framework [64] and the NVIDIA Titan V GPU hardware platform. The image scenes of each city were randomly divided into 70% training samples and 30% validation samples for the semantic segmentation networks. The number of training and validation samples for each city can be found in Table 2. Considering the significant differences between the four cities, the semantic segmentation network of each city was trained and evaluated independently based on its own training and validation samples.

Table 2. Number of training and validation samples in four cities.

Number	Las Vegas	Paris	Shanghai	Khartoum
Training samples	2695	803	3207	708
Validation samples	1156	345	1375	304

As shown in Figure 2, the semantic segmentation networks were trained and evaluated based on four dataset collections for each city: the original satellite dataset (Satellite-org), the augmented satellite dataset (Satellite-aug), the original satellite dataset combined with the GIS map dataset (Satellite-Map-org), and the augmented satellite dataset combined with the GIS map dataset (Satellite-Map-aug). Table 3 shows the validation accuracies of the semantic segmentation network in four cities when using different types of datasets. We find that the validation accuracies of the four cities are all over 93% and vary slightly among the cities and the types of datasets, which indicates accurate detection of building areas of the semantic segmentation network. Moreover, the average validation accuracy of the four cities is the highest when using the augmented satellite dataset combined with the GIS map dataset (Satellite-Map-aug). The evaluation of the building footprint extraction results is described in Section 4.2.

Table 3. Validation accuracies of semantic segmentation networks in four cities.

Type of Dataset	Las Vegas	Paris	Shanghai	Khartoum	Average
Satellite-org	0.9684	0.9752	0.9610	0.9386	0.9608
Satellite-aug	0.9646	**0.9776**	0.9613	0.9399	0.9609
Satellite-Map-org	0.9681	0.9772	0.9677	0.9371	0.9625
Satellite-Map-aug	**0.9692**	0.9772	**0.9681**	**0.9420**	**0.9641**

4.2. Building Footprint Extraction Results of the Proposed Method

Table 4 shows the building footprint extraction results of the proposed method evaluated by the validation dataset in the four cities in terms of TP, FP, FN, precision, recall, and the F1-score. There are significant differences between the results in different cities. Our method obtains the highest F1-score of 0.8911 for Las Vegas and the lowest F1-score of 0.5415 for Khartoum. Table 5 shows the results of our proposed method in the final phase of the CVPR 2018 DeepGlobe Satellite Challenge, which are evaluated by an unlabeled dataset selected from other regions in the four cities. The evaluation results in the final phase can only be seen through the online submission, and each team has only five submission chances. The experimental results demonstrate that our proposed method achieves similar F1-scores for the validation dataset and the dataset provided in the final phase. Figure 4 shows some examples of the building footprint extraction results of our proposed method in which the green, red, and yellow polygons denote correctly extracted buildings (TP), other objects extracted as buildings by mistake (FP), and ground truth buildings that are not extracted correctly by the proposed method (FN), respectively. The building footprint extraction results of the four cities are analyzed in detail, according to the actual situation of each city in Section 5.3.

Figure 4. Examples of building footprint extraction results of our proposed method in (**a,b**) Las Vegas, (**c,d**) Paris, (**e,f**) Shanghai, and (**g,h**) Khartoum. Green, red, and yellow polygons denote correctly extracted buildings (TP), other objects extracted as buildings by mistake (FP), and ground truth buildings that were not extracted correctly by the proposed method (FN), respectively.

Table 4. Results of the proposed method evaluated by the validation dataset. TP, true positive. FP, false positive. FN, false negative.

Index	Las Vegas	Paris	Shanghai	Khartoum
TP	27,526	3097	11,323	3495
FP	1629	564	3835	1968
FN	5098	1441	9661	3951
Precision	0.9441	0.8459	0.7470	0.6398
Recall	0.8437	0.6825	0.5396	0.4694
F1-score	0.8911	0.7555	0.6266	0.5415

Table 5. Results of proposed method evaluated by the dataset provided in the final phase.

Index	Las Vegas	Paris	Shanghai	Khartoum
TP	30,068	4056	11,674	4031
FP	1912	844	4132	2106
FN	5187	2006	8974	4443
Precision	0.9402	0.8278	0.7386	0.6568
Recall	0.8529	0.6601	0.5654	0.4757
F1-score	0.8944	0.7400	0.6250	0.5518

5. Discussion

5.1. Comparison of Building Footprint Extraction Results Obtained from Different Methods

In this section, we compare the building footprint extraction results obtained from our proposed method with those achieved from the top three solutions in the SpaceNet Building Detection Competition (round 2) [11]. Table 6 shows the final F1-scores of the four cities obtained from our proposed method and from the top three solutions (XD_XD, wleite, and nofto, the competitors' usernames). The numbers in bold type indicate the highest F1-scores. The solution proposed by the XD_XD is based on an ensemble of U-Net models, which combines multi-spectral satellite images with OpenStreetMap data. Different from our proposed method, XD_XD's solution uses the OpenStreetMap as the only auxiliary data for all cities, and the OpenStreetMap vector layers (each layer represents a single land use type) are rasterized into four or five bands to integrate with the multi-spectral satellite image. Wleite and nofto use a similar approach, including traditional feature extraction (e.g., Sobel filter-based edge detection, average, variance, and skewness for small neighborhood squares around each evaluated pixel) and two random forest classifiers (one for predicting whether a pixel belongs to the border and the other one for predicting whether a pixel is inside a building).

Compared with the winning solution (XD_XD), the F1-score of our proposed method increased significantly (by 3%) for Shanghai and by 1.1% and 0.6% for Paris and Las Vegas. The F1-score decreased slightly (by 0.2%) for Khartoum. This method improved the total F1-score by 1.1%, 6.1%, and 12.5% compared with the top three solutions in the competition. All four methods performed best in Las Vegas, second best in Paris, third best in Shanghai, and worst in Khartoum. Possible reasons for this phenomenon are analyzed in Section 5.3.

Table 6. F1-scores obtained from different methods.

Method	Las Vegas	Paris	Shanghai	Khartoum	Total
Ours	**0.891**	**0.756**	**0.627**	0.542	**0.704**
XD_XD	0.885	0.745	0.597	**0.544**	0.693
wleite	0.829	0.679	0.581	0.483	0.643
nofto	0.787	0.584	0.520	0.424	0.579

5.2. Building Extraction Results Obtained from Different Strategies of Our Proposed Method

In this section, we compare and analyze the effects of each strategy in our proposed method on the building footprint extraction results in different cities. Table 7 shows the precision, recall, and F1-score of the four cities after applying the different strategies. The numbers in bold type indicate the highest values. Baseline refers to training the semantic segmentation model using the rescaled satellite images. Data-aug (data augmentation) refers to training the semantic segmentation model using the augmented satellite images. Post-proc (post-processing) refers to applying the post-processing strategy to the integrated results of the baseline and data-aug. Add-map (adding GIS map data) refers to integrating the results obtained from the satellite image–based dataset collection with those from the combined satellite and GIS map image–based dataset collection. The F1-scores obtained after applying the different strategies are summarized in Figure 5.

Table 7. Results obtained after applying different strategies of our proposed method.

Strategy	Index	Las Vegas	Paris	Shanghai	Khartoum
Baseline	Precision	0.8849	0.7370	0.5973	0.4885
	Recall	0.8384	0.6342	0.4831	0.4248
	F1-score	0.8611	0.6817	0.5342	0.4544
Data-aug	Precision	0.8896	0.7474	0.5649	0.5338
	Recall	**0.8570**	**0.6911**	0.5304	0.4589
	F1-score	0.8730	0.7181	0.5471	0.4935
Post-proc	Precision	0.9308	0.8272	0.6875	0.6141
	Recall	0.8464	0.6666	0.5163	0.4525
	F1-score	0.8866	0.7383	0.5897	0.5210
Add-map	Precision	**0.9441**	**0.8459**	**0.7470**	**0.6398**
	Recall	0.8437	0.6825	**0.5396**	**0.4694**
	F1-score	**0.8911**	**0.7555**	**0.6266**	**0.5415**

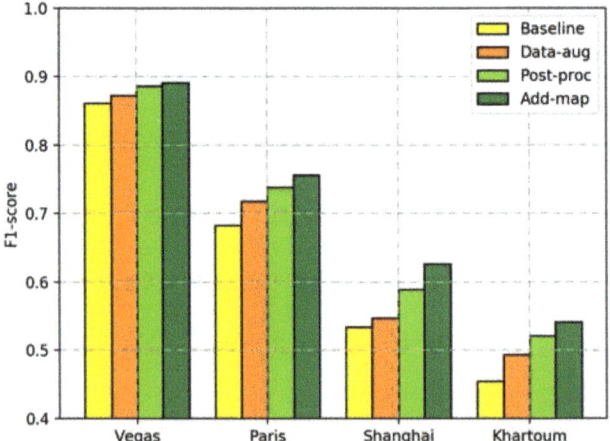

Figure 5. F1-scores obtained after applying different strategies of our proposed method.

Compared with the baseline, our proposed method improved the F1-score by 3.01%, 7.38%, 9.24%, and 8.71% for Las Vegas, Paris, Shanghai, and Khartoum, respectively. The improvement is much more significant for Paris, Shanghai, and Khartoum than for Las Vegas, which had an F1-score of 0.8849 using the baseline model. For the data augmentation strategy, the F1-score improvements for Paris and Khartoum (3.64% and 3.91%) are more remarkable than for Las Vegas and Shanghai (1.19% and 1.29%). We can conclude that, for cities with fewer initial training samples, the data augmentation strategy significantly improves the F1-score. The post-processing strategy was more beneficial for Shanghai and Khartoum, with relatively low F1-scores compared to Las Vegas and Paris, with relatively high F1-scores. The strategy of integrating satellite data with GIS map data improved the F1-score more for Shanghai than for the other three cities, which might be due to the relatively poor building extraction results of the baseline model and the substantial building information of the MapWorld datasets. It is worth noting that the F1-score of Khartoum increased by 2.05% after the add-map strategy even though the OpenStreetMap dataset lacked building information for most areas in Khartoum. We can conclude that other information in the map data (e.g., many roads and other land use types) might also contribute to the improved building extraction results.

Figures 6–9 show some examples of the building footprint extraction results after applying the different strategies in which green, red, and yellow polygons denote correctly extracted buildings (TP), other objects extracted as buildings by mistake (FP), and ground truth buildings that were not extracted correctly (FN), respectively. The experimental results demonstrate that the proposed strategies led to remarkable improvements in the building footprint results in many aspects. For instance, we could obtain more complete building outlines (e.g., the top images in Figures 6–8), and the neighboring buildings were more likely to be successfully extracted separately (e.g., the bottom images in Figures 8 and 9). Moreover, there was less confusion between tiny buildings and noise in the results (e.g., top images in Figure 6 and bottom images in Figure 8). Analysis about the results regarding the actual situation in different cities is demonstrated in the following section.

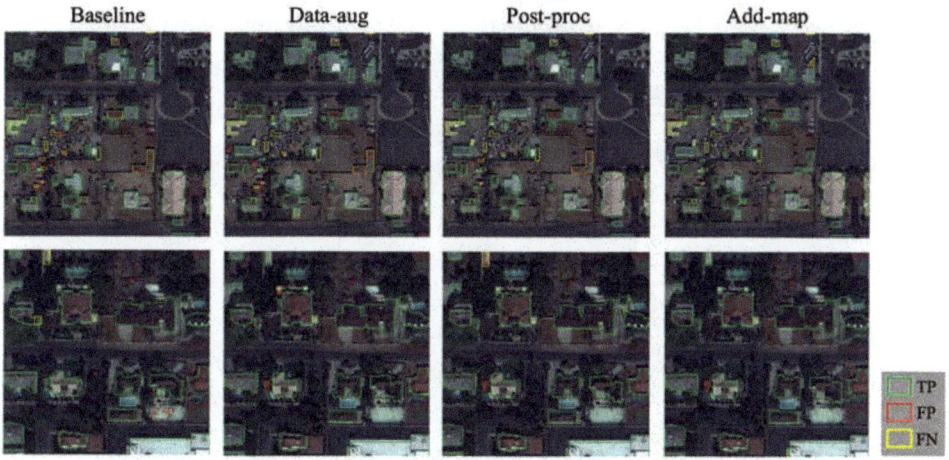

Figure 6. Examples of building extraction results for Las Vegas using different strategies.

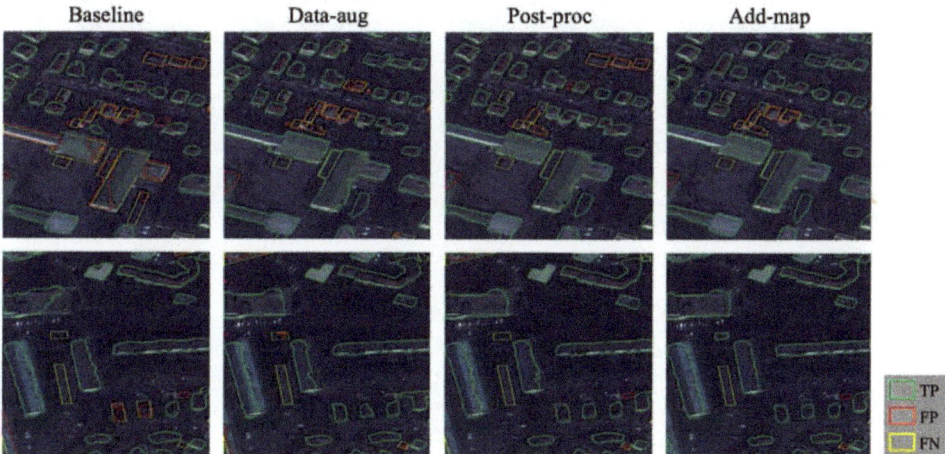

Figure 7. Examples of building extraction results for Paris using different strategies.

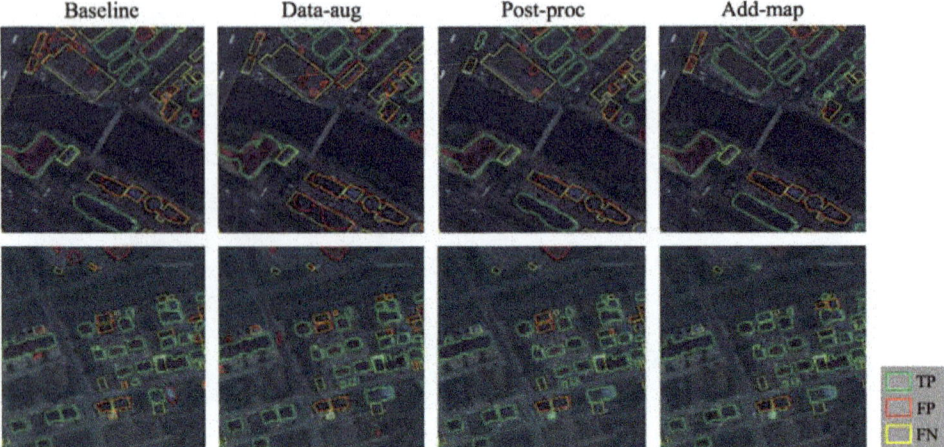

Figure 8. Examples of building extraction results for Shanghai using different strategies.

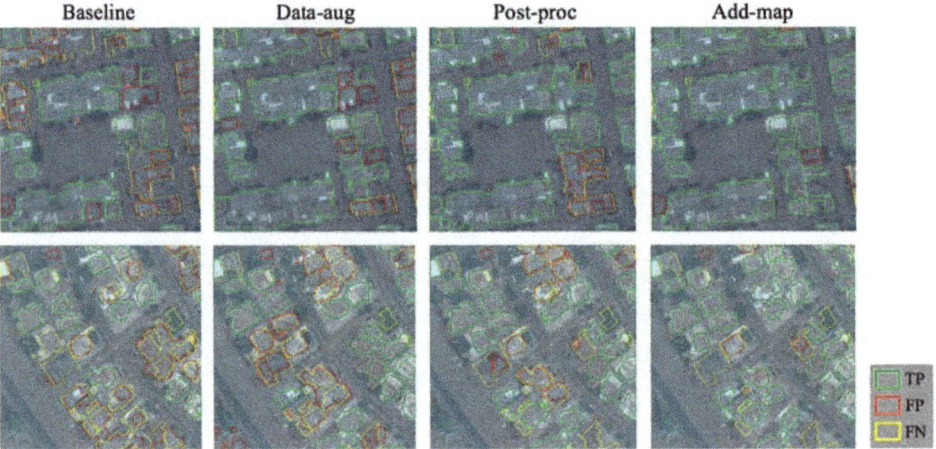

Figure 9. Examples of building extraction results for Khartoum using different strategies.

5.3. Analysis of Building Footprint Extraction Results for Different Cities

Figure 10 shows typical examples of the footprint extraction results obtained from our proposed method in the four cities. The two left columns of the images are selected examples with good results. The two right columns are selected examples with inferior results. The results of our proposed method are analyzed based on the specific situation of each city as follows.

Our method achieved the best results for Las Vegas. Most of the satellite images in the Las Vegas dataset are collected from residential regions. Compared with the other three cities, the buildings in Las Vegas have a more unified architectural style. Buildings partly covered by trees can also be successfully extracted by our proposed method for most regions (e.g., buildings on the left of Figure 10a,b). Tiny buildings and buildings of a similar color as the background region are relatively harder to extract correctly using the proposed method (e.g., FN buildings denoted by yellow polygons in Figure 10c,d).

Our method obtained the second highest F1-score for Paris. The satellite images are collected from the western part of Paris. Similar to Las Vegas, the buildings in Paris have a relatively unified architectural style. However, more buildings in Paris are a similar color as the background (e.g., trees and roads), which are difficult to correctly detect compared with those in Las Vegas. The proposed method also had difficulty identifying the outlines of two neighboring buildings separately and completely extracting large buildings that consist of several parts (e.g., buildings in the bottom of Figure 10g,h).

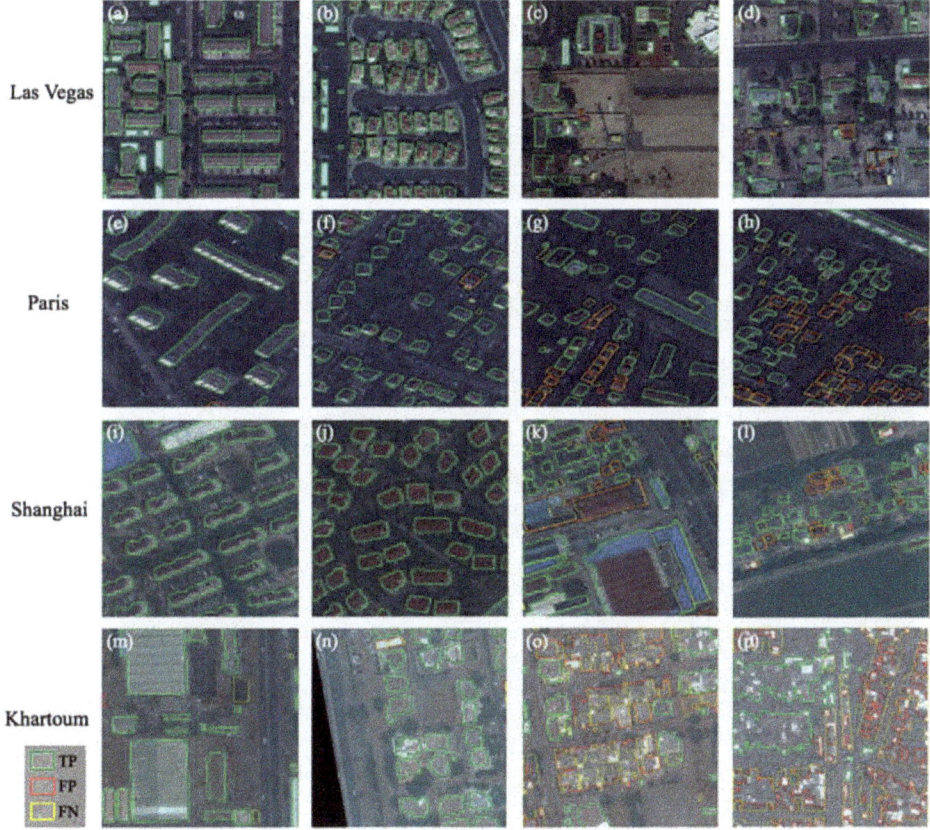

Figure 10. Examples of footprint extraction results obtained from our proposed method in four cities. The two left columns show selected examples with good building extraction results. The two right columns show selected examples with inferior building extraction results.

Our method obtained the second lowest F1-score for Shanghai. Most of the satellite images are collected from suburban regions of Shanghai. Compared with the other three cities, buildings in the Shanghai dataset are more diverse in many aspects, including the construction area, the building height, the architectural style, etc. There are more high-rise buildings in Shanghai with a larger distance between the roof and the footprint polygons on the satellite images (e.g., Figure 4e). Buildings located in residential areas (e.g., Figure 10i,j) are relatively easier to extract correctly by the proposed method than those located in agricultural areas, industrial areas, gardens, etc. (e.g., Figure 10k,l). Moreover, our proposed method had difficulty correctly extracting buildings with green roofs, of a similar color as the background, partly covered by trees, or of extremely small size, etc. (e.g., FN buildings denoted by yellow polygons in Figure 10k,l), even though the integration of satellite and map data solved the above problems to a great extent when compared with using only the provided satellite datasets (see Section 5.2).

Our method obtained the lowest F1-score for Khartoum. Most of the satellite images in the Khartoum dataset are collected from residential regions, where the buildings have great variance in structural organization and construction area. There are many building groups in Khartoum, and it is hard to judge, even by the human eye, whether a group of neighboring buildings should be extracted entirely or separately in many regions (e.g., Figure 10o,p). To the best of our knowledge, all of the existing public GIS map datasets show very limited building information in Khartoum. All of these aspects might result in inferior performance of building footprint extraction in Khartoum.

6. Conclusions

In this study, we proposed a U-Net–based semantic segmentation method for building footprint extraction from high-resolution satellite images using the SpaceNet building dataset provided in the DeepGlobe Challenge. Multisource GIS map datasets (OpenStreetMap, Google Maps, and MapWorld) are explored to improve the building extraction results in four cities (Las Vegas, Paris, Shanghai, and Khartoum). In our proposed method, we designed a data fusion and augmentation method for integrating multispectral WorldView-3 satellite images with selected GIS map datasets. We trained and evaluated four U-Net–based semantic segmentation models based on augmented and integrated dataset collections. Lastly, we integrated the results obtained from the semantic segmentation models and employed a post-processing method to further improve the building extraction results.

The experimental results show that our proposed method improves the total F1-score by 1.1%, 6.1%, and 12.5% when compared with the top three solutions in the SpaceNet Building Detection Competition. The F1-scores of Las Vegas, Paris, Shanghai, and Khartoum are 0.8911, 0.7555, 0.6266, and 0.5415, respectively. The significant difference in the results is due to many possible aspects, including the consistency or the diversity of buildings in a city (e.g., construction area, building height, and architectural style), the similarity between buildings and background, and the number of training samples. We also analyze the effects of proposed strategies on the building extraction results. Our proposed strategies improved the F1-score by 3.01% to 9.24% for the four cities compared with those obtained from the baseline method, which achieved precise building outlines and less confusion between tiny buildings and noise. The data augmentation strategy improves the F1-scores greatly for Paris and Khartoum, with fewer training samples, and slightly for Las Vegas and Shanghai, with more training samples. The post-processing strategy brings more improvement for Shanghai and Khartoum, with lower initial F1-scores, than for Las Vegas and Paris, with higher initial F1-scores. The strategy of integrating satellite and GIS data brings the most improvement for Shanghai, with a low initial F1-score and substantial building information in GIS map data. In our future research, we will try to combine the semantic segmentation model with other image processing algorithms (e.g., traditional image segmentation and edge detection algorithms) to further improve the extraction of building outlines. We will also explore different data fusion strategies for combining satellite images and GIS data, and other state-of-the-art semantic segmentation models for building footprint extraction using the SpaceNet building dataset.

Author Contributions: Conceptualization, W.L., C.H. and J.F.; Data curation, W.L.; Formal analysis, W.L.; Funding acquisition, H.F.; Investigation, W.L., C.H. and J.F.; Methodology, W.L., C.H. and J.F.; Project administration, H.F.; Resources, H.F.; Software, W.L., C.H. and J.F.; Supervision, H.F.; Validation, W.L. and C.H.; Visualization, W.L., C.H., and J.Z.; Writing—original draft, W.L.; Writing—review & editing, H.F. and L.Y.

Funding: This research was supported in part by the National Key R&D Program of China (Grant No. 2017YFA0604500 and 2017YFA0604401), by the National Natural Science Foundation of China (Grant No. 51761135015), and by the Center for High Performance Computing and System Simulation, Pilot National Laboratory for Marine Science and Technology (Qingdao).

Acknowledgments: The authors would like to thank the editors and reviewers for their valuable comments.

Conflicts of Interest: The authors declare no conflict of interest.

References

1. Zhang, B.; Wang, C.; Shen, Y.; Liu, Y. Fully Connected Conditional Random Fields for High-Resolution Remote Sensing Land Use/Land Cover Classification with Convolutional Neural Networks. *Remote Sens.* **2018**, *10*, 1889. [CrossRef]
2. Li, W.; Fu, H.; Yu, L.; Cracknell, A. Deep learning based oil palm tree detection and counting for high-resolution remote sensing images. *Remote Sens.* **2016**, *9*, 22. [CrossRef]
3. Li, W.; Dong, R.; Fu, H.; Le, Y. Large-Scale Oil Palm Tree Detection from High-Resolution Satellite Images Using Two-Stage Convolutional Neural Networks. *Remote Sens.* **2019**, *11*, 11. [CrossRef]
4. Tang, T.; Zhou, S.; Deng, Z.; Lei, L.; Zou, H. Arbitrary-oriented vehicle detection in aerial imagery with single convolutional neural networks. *Remote Sens.* **2017**, *9*, 1170. [CrossRef]
5. Xu, Y.; Xie, Z.; Feng, Y.; Chen, Z. Road Extraction from High-Resolution Remote Sensing Imagery Using Deep Learning. *Remote Sens.* **2018**, *10*, 1461. [CrossRef]
6. Audebert, N.; Le Saux, B.; Lefèvre, S. Segment-before-detect: Vehicle detection and classification through semantic segmentation of aerial images. *Remote Sens.* **2017**, *9*, 368. [CrossRef]
7. Sun, Y.; Zhang, X.; Zhao, X.; Xin, Q. Extracting building boundaries from high resolution optical images and LiDAR data by integrating the convolutional neural network and the active contour model. *Remote Sens.* **2018**, *10*, 1459. [CrossRef]
8. Tian, J.; Cui, S.; Reinartz, P. Building change detection based on satellite stereo imagery and digital surface models. *IEEE Trans. Geosci. Remote Sens.* **2014**, *52*, 406–417. [CrossRef]
9. Li, L.; Liang, J.; Weng, M.; Zhu, H. A Multiple-Feature Reuse Network to Extract Buildings from Remote Sensing Imagery. *Remote Sens.* **2018**, *10*, 1350. [CrossRef]
10. Shrestha, S.; Vanneschi, L. Improved Fully Convolutional Network with Conditional Random Fields for Building Extraction. *Remote Sens.* **2018**, *10*, 1135. [CrossRef]
11. Demir, I.; Koperski, K.; Lindenbaum, D.; Pang, G.; Huang, J.; Basu, S.; Hughes, F.; Tuia, D.; Raska, R. Deepglobe 2018: A challenge to parse the earth through satellite images. In Proceedings of the IEEE Conference on Computer Vision and Pattern Recognition Workshops, Salt Lake City, UT, USA, 18–22 June 2018; pp. 238–241.
12. Xu, Y.; Wu, L.; Xie, Z.; Chen, Z. Building Extraction in Very High Resolution Remote Sensing Imagery Using Deep Learning and Guided Filters. *Remote Sens.* **2018**, *10*, 144. [CrossRef]
13. Cheng, G.; Han, J. A survey on object detection in optical remote sensing images. *ISPRS J. Photogramm.* **2016**, *117*, 11–28. [CrossRef]
14. Ziaei, Z.; Pradhan, B.; Mansor, S.B. A rule-based parameter aided with object-based classification approach for extraction of building and roads from WorldView-2 images. *Geocarto Int.* **2014**, *29*, 554–569. [CrossRef]
15. Ok, A.O. Automated detection of buildings from single VHR multispectral images using shadow information and graph cuts. *ISPRS J. Photogramm.* **2013**, *86*, 21–40. [CrossRef]
16. Belgiu, M.; Drăguţ, L. Comparing supervised and unsupervised multiresolution segmentation approaches for extracting buildings from very high resolution imagery. *ISPRS J. Photogramm.* **2014**, *96*, 67–75. [CrossRef] [PubMed]
17. Chen, R.; Li, X.; Li, J. Object-based features for house detection from RGB high-resolution images. *Remote Sens.* **2018**, *10*, 451. [CrossRef]
18. Huang, X.; Zhang, L. Morphological building/shadow index for building extraction from high-resolution imagery over urban areas. *IEEE J. Sel. Top. Appl. Earth Obs. Remote Sens.* **2012**, *5*, 161–172. [CrossRef]
19. Ok, A.O.; Senaras, C.; Yuksel, B. Automated detection of arbitrarily shaped buildings in complex environments from monocular VHR optical satellite imagery. *IEEE Trans. Geosci. Remote Sens.* **2013**, *51*, 1701–1717. [CrossRef]
20. Ding, P.; Zhang, Y.; Deng, W.J.; Jia, P.; Kuijper, A. A light and faster regional convolutional neural network for object detection in optical remote sensing images. *ISPRS J. Photogramm.* **2018**, *141*, 208–218. [CrossRef]
21. Hu, F.; Xia, G.S.; Hu, J.; Zhang, L. Transferring deep convolutional neural networks for the scene classification of high-resolution remote sensing imagery. *Remote Sens.* **2015**, *7*, 14680–14707. [CrossRef]
22. Liu, Y.; Zhong, Y.; Fei, F.; Zhu, Q.; Qin, Q. Scene Classification Based on a Deep Random-Scale Stretched Convolutional Neural Network. *Remote Sens.* **2018**, *10*, 444. [CrossRef]
23. Li, W.; Fu, H.; Yu, L.; Gong, P.; Feng, D.; Li, C.; Clinton, N. Stacked autoencoder-based deep learning for remote-sensing image classification: A case study of African land-cover mapping. *Int. J. Remote Sens.* **2016**, *37*, 5632–5646. [CrossRef]

24. Huang, B.; Zhao, B.; Song, Y. Urban land-use mapping using a deep convolutional neural network with high spatial resolution multispectral remote sensing imagery. *Remote Sens. Environ.* **2018**, *214*, 73–86. [CrossRef]
25. Long, J.; Shelhamer, E.; Darrell, T. Fully convolutional networks for semantic segmentation. In Proceedings of the IEEE Conference on Computer Vision and Pattern Recognition, Boston, MA, USA, 7–12 June 2015; pp. 3431–3440.
26. Li, W.; He, C.; Fang, J.; Fu, H. Semantic Segmentation based Building Extraction Method using Multi-source GIS Map Datasets and Satellite Imagery. In Proceedings of the IEEE Conference on Computer Vision and Pattern Recognition Workshops, Salt Lake City, UT, USA, 18–22 June 2018; pp. 238–241.
27. Cao, R.; Zhu, J.; Tu, W.; Li, Q.; Cao, J.; Liu, B.; Zhang, Q.; Qiu, G. Integrating Aerial and Street View Images for Urban Land Use Classification. *Remote Sens.* **2018**, *10*, 1553. [CrossRef]
28. Lin, H.; Shi, Z.; Zou, Z. Maritime semantic labeling of optical remote sensing images with multi-scale fully convolutional network. *Remote Sens.* **2017**, *9*, 480. [CrossRef]
29. Piramanayagam, S.; Saber, E.; Schwartzkopf, W.; Koehler, F. Supervised Classification of Multisensor Remotely Sensed Images Using a Deep Learning Framework. *Remote Sens.* **2018**, *10*, 1429. [CrossRef]
30. Bai, Y.; Mas, E.; Koshimura, S. Towards Operational Satellite-Based Damage-Mapping Using U-Net Convolutional Network: A Case Study of 2011 Tohoku Earthquake-Tsunami. *Remote Sens.* **2018**, *10*, 1626. [CrossRef]
31. Sa, I.; Popović, M.; Khanna, R.; Chen, Z.; Lottes, P.; Liebisch, F.; Nieto, J.; Stachniss, C.; Walter, A.; Siegwart, R. WeedMap: A large-scale semantic weed mapping framework using aerial multispectral imaging and deep neural network for precision farming. *Remote Sens.* **2018**, *10*, 1423. [CrossRef]
32. Lu, T.; Ming, D.; Lin, X.; Hong, Z.; Bai, X.; Fang, J. Detecting building edges from high spatial resolution remote sensing imagery using richer convolution features network. *Remote Sens.* **2018**, *10*, 1496. [CrossRef]
33. Yang, H.; Wu, P.; Yao, X.; Wu, Y.; Wang, B.; Xu, Y. Building Extraction in Very High Resolution Imagery by Dense-Attention Networks. *Remote Sens.* **2018**, *10*, 1768. [CrossRef]
34. Wu, G.; Guo, Z.; Shi, X.; Chen, Q.; Xu, Y.; Shibasaki, R.; Shao, X. A boundary regulated network for accurate roof segmentation and outline extraction. *Remote Sens.* **2018**, *10*, 1195. [CrossRef]
35. Alshehhi, R.; Marpu, P.R.; Woon, W.L.; Dalla Mura, M. Simultaneous extraction of roads and buildings in remote sensing imagery with convolutional neural networks. *ISPRS J. Photogramm.* **2017**, *130*, 139–149. [CrossRef]
36. Huang, B.; Lu, K.; Audebert, N.; Khalel, A.; Tarabalka, Y.; Malof, J.; Boulch, A.; Le Saux, B.; Collins, L.; Bradbury, K.; et al. Large-scale semantic classification: Outcome of the first year of Inria aerial image labeling benchmark. In Proceedings of the IEEE International Geoscience and Remote Sensing Symposium (IGARSS), Valencia, Spain, 22–27 July 2018.
37. Li, X.; Yao, X.; Fang, Y. Building-A-Nets: Robust Building Extraction from High-Resolution Remote Sensing Images with Adversarial Networks. *IEEE J. Sel. Top. Appl. Earth Obs. Remote Sens.* **2018**, *99*, 3680–3687. [CrossRef]
38. Ji, S.; Wei, S.; Lu, M. A scale robust convolutional neural network for automatic building extraction from aerial and satellite imagery. *Int. J. Remote Sens.* **2018**, 1–15. [CrossRef]
39. Mnih, V. Machine Learning for Aerial Image Labeling. Ph.D. Thesis, University of Toronto, Toronto, ON, Canada, 2013.
40. Rottensteiner, F.; Sohn, G.; Jung, J.; Gerke, M.; Baillard, C.; Benitez, S.; Breitkopf, U. The ISPRS benchmark on urban object classification and 3D building reconstruction. *ISPRS Ann. Photogramm. Remote Sens. Spat. Inf. Sci.* **2012**, *1*, 293–298. [CrossRef]
41. Maggiori, E.; Tarabalka, Y.; Charpiat, G.; Alliez, P. Can semantic labeling methods generalize to any city? The Inria aerial image labeling benchmark. In Proceedings of the IEEE International Symposium on Geoscience and Remote Sensing (IGARSS), Fort Worth, TX, USA, 23–28 July 2017.
42. Ji, S.; Wei, S.; Lu, M. Fully Convolutional Networks for Multisource Building Extraction from an Open Aerial and Satellite Imagery Data Set. *IEEE Trans. Geosci. Remote Sens.* **2018**, *99*, 1–13. [CrossRef]
43. Chen, Q.; Wang, L.; Wu, Y.; Wu, G.; Guo, Z.; Waslander, S.L. Aerial imagery for roof segmentation: A large-scale dataset towards automatic mapping of buildings. *ISPRS J. Photogramm.* **2019**, *147*, 42–55. [CrossRef]
44. Van Etten, A.; Lindenbaum, D.; Bacastow, T.M. Spacenet: A remote sensing dataset and challenge series. *arXiv* **2018**, arXiv:1807.01232.

45. Qin, R.; Tian, J.; Reinartz, P. Spatiotemporal inferences for use in building detection using series of very-high-resolution space-borne stereo images. *Int. J. Remote Sens.* **2016**, *37*, 3455–3476. [CrossRef]
46. Du, S.; Zhang, Y.; Zou, Z.; Xu, S.; He, X.; Chen, S. Automatic building extraction from LiDAR data fusion of point and grid-based features. *ISPRS J. Photogramm.* **2017**, *130*, 294–307. [CrossRef]
47. Gilani, S.A.N.; Awrangjeb, M.; Lu, G. An automatic building extraction and regularisation technique using lidar point cloud data and orthoimage. *Remote Sens.* **2016**, *8*, 258. [CrossRef]
48. Sohn, G.; Dowman, I. Data fusion of high-resolution satellite imagery and LiDAR data for automatic building extraction. *ISPRS J. Photogramm.* **2007**, *62*, 43–63. [CrossRef]
49. Tournaire, O.; Brédif, M.; Boldo, D.; Durupt, M. An efficient stochastic approach for building footprint extraction from digital elevation models. *ISPRS J. Photogramm.* **2010**, *65*, 317–327. [CrossRef]
50. Wang, Y.; Cheng, L.; Chen, Y.; Wu, Y.; Li, M. Building point detection from vehicle-borne LiDAR data based on voxel group and horizontal hollow analysis. *Remote Sens.* **2016**, *8*, 419. [CrossRef]
51. Lee, D.H.; Lee, K.M.; Lee, S.U. Fusion of lidar and imagery for reliable building extraction. *Photogramm. Eng. Remote Sens.* **2008**, *74*, 215–225. [CrossRef]
52. Awrangjeb, M.; Ravanbakhsh, M.; Fraser, C.S. Automatic detection of residential buildings using LIDAR data and multispectral imagery. *ISPRS J. Photogramm.* **2010**, *65*, 457–467. [CrossRef]
53. Pan, X.; Gao, L.; Marinoni, A.; Zhang, B.; Yang, F.; Gamba, P. Semantic Labeling of High Resolution Aerial Imagery and LiDAR Data with Fine Segmentation Network. *Remote Sens.* **2018**, *10*, 743. [CrossRef]
54. Huang, Z.; Cheng, G.; Wang, H.; Li, H.; Shi, L.; Pan, C. Building extraction from multi-source remote sensing images via deep deconvolution neural networks. In Proceedings of the IEEE International Geoscience and Remote Sensing Symposium (IGARSS), Beijing, China, 10–15 July 2016; pp. 1835–1838.
55. Yuan, J.; Cheriyadat, A.M. Learning to count buildings in diverse aerial scenes. In Proceedings of the 22nd ACM SIGSPATIAL International Conference on Advances in Geographic Information Systems, Dallas, TX, USA, 4–7 November 2014; pp. 271–280.
56. Audebert, N.; Le Saux, B.; Lefèvre, S. Joint learning from earth observation and openstreetmap data to get faster better semantic maps. In Proceedings of the EARTHVISION 2017 IEEE/ISPRS CVPR Workshop on Large Scale Computer Vision for Remote Sensing Imagery, Honolulu, HI, USA, 21–26 July 2017.
57. Du, S.; Zhang, F.; Zhang, X. Semantic classification of urban buildings combining VHR image and GIS data: An improved random forest approach. *ISPRS J. Photogramm.* **2015**, *105*, 107–119. [CrossRef]
58. OpenStreetMap Static Map. Available online: http://staticmap.openstreetmap.de/ (accessed on 15 April 2018).
59. Google Map Static API. Available online: https://developers.google.com/maps/documentation/static-maps/ (accessed on 15 April 2018).
60. MapWorld Static API. Available online: http://lbs.tianditu.gov.cn/staticapi/static.html (accessed on 15 April 2018).
61. Ronneberger, O.; Fischer, P.; Brox, T. U-net: Convolutional networks for biomedical image segmentation. In Proceedings of the International Conference on Medical Image Computing and Computer-Assisted Intervention, Munich, Germany, 5–9 October 2015; pp. 234–241.
62. Iglovikov, V.; Mushinskiy, S.; Osin, V. Satellite imagery feature detection using deep convolutional neural network: A kaggle competition. *arXiv* **2017**, arXiv:1706.06169.
63. Wang, X.; Liu, S.; Du, P.; Liang, H.; Xia, J.; Li, Y. Object-Based Change Detection in Urban Areas from High Spatial Resolution Images Based on Multiple Features and Ensemble Learning. *Remote Sens.* **2018**, *10*, 276. [CrossRef]
64. Chollet, F. *Deep Learning with Python*; Manning Publications Co.: Shelter Island, NY, USA, 2017.

© 2019 by the authors. Licensee MDPI, Basel, Switzerland. This article is an open access article distributed under the terms and conditions of the Creative Commons Attribution (CC BY) license (http://creativecommons.org/licenses/by/4.0/).

Article

An Automatic Morphological Attribute Building Extraction Approach for Satellite High Spatial Resolution Imagery

Weixuan Ma [1], Youchuan Wan [1], Jiayi Li [1,*], Sa Zhu [1] and Mingwei Wang [2]

1   School of Remote Sensing and Information Engineering, Wuhan University, Wuhan 430072, China; weixuanma@whu.edu.cn (W.M.); ychwan@whu.edu.cn (Y.W.); sazhu_rs@163.com (S.Z.)
2   Institute of Geological Survey, China University of Geosciences, Wuhan 430074, China; wangmingwei@cug.edu.cn
*   Correspondence: zjjerica@163.com; Tel.: +86-135-5407-5012

Received: 21 December 2018; Accepted: 6 February 2019; Published: 8 February 2019

Abstract: A new morphological attribute building index (MABI) and shadow index (MASI) are proposed here for automatically extracting building features from very high-resolution (VHR) remote sensing satellite images. By investigating the associated attributes in morphological attribute filters (AFs), the proposed method establishes a relationship between AFs and the characteristics of buildings/shadows in VHR images (e.g., high local contrast, internal homogeneity, shape, and size). In the pre-processing step of the proposed work, attribute filtering was conducted on the original VHR spectral reflectance data to obtain the input, which has a high homogeneity, and to suppress elongated objects (potential non-buildings). Then, the MABI and MASI were calculated by taking the obtained input as a base image. The dark buildings were considered separately in the MABI to reduce the omission of the dark roofs. To better detect buildings from the MABI feature image, an object-oriented analysis and building-shadow concurrence relationships were utilized to further filter out non-building land covers, such as roads and bare ground, that are confused for buildings. Three VHR datasets from two satellite sensors, i.e., Worldview-2 and QuickBird, were tested to determine the detection performance. In view of both the visual inspection and quantitative assessment, the results of the proposed work are superior to recent automatic building index and supervised binary classification approach results.

Keywords: building detection; building index; feature extraction; mathematical morphology; morphological attribute filter; morphological profile

1. Introduction

Buildings are one of the most important types of artificial targets in the urban environment. Due to the high frequency of changes in buildings, understanding their current distribution is important for urban planning, change detection, urban environmental investigations, and urban monitoring applications [1]. The use of a new generation of very high spatial resolution sensors, such as Ikonos, QuickBird, and Worldview, has broadened the application of remote sensing technology [2]. A great amount of spatial and thematic information on land cover at local and national scales is contained in VHR data [3], and this information clearly gives buildings identifiable shape and texture features. In view of this, VHR images are suitable for building feature extraction tasks. However, the high intra-class variance and the low inter-class variances in the spectral statistics of VHR images greatly reduce the distinguishing ability of small land-cover areas in these images [4]. To address this problem, numerous studies have focused on the extraction of spatial and structural information in images and the use of this information as a supplement to improve the recognition ability [5]. Researchers have indicated that importing spatial features significantly improves the accuracy of VHR image classification [6–8]. For building feature extraction applications, current works mainly use supervised machine-learning approaches [9–13]. However, such methods require a large number of training samples and a high time cost in the sample selection stage. In recent years, some automatic building detection methods for high-resolution satellite imagery have been proposed. Different strategies, such as automatic building boundary extraction [14], automatic building feature extraction combined with an existing geodatabase [15], and the use of LiDAR data [16], have been employed in these studies. In addition, a number of building feature indexes have been proposed to characterize potential buildings [17] or exclude confused non-building features, such as vegetation [18], water [19], and shadows [20].

In recent years, a combination of the morphological building index (MBI) [21] with the morphological shadow index (MSI) [22] has been proposed to automatically detect buildings in VHR images. By modeling the local contrast, building-directivity, and granulometry with a series of multiscale morphological profiles (MPs) [23], the MBI and its variants [24–26] have proven to be effective tools for building detection tasks. However, MPs do not fully exploit spectral information, which restricts the extraction performance to some extent.

Concerning the above restriction, morphology attribute profiles (APs) [27] are proposed as an extension of MPs. As a more flexible way than MPs to model information from high-resolution images, the transformations in APs can extract features based on either the geometrical or spectral characteristics of objects. According to the different attributes considered in the morphological attribute transformation, different features can be obtained from a VHR image. Classification [28], building feature extraction [29], and change detection task [30] results have suggested that the use of APs is an effective way to model spatial information from VHR images. However, instead of acting as an automatic image-processing index, APs often work as ancillary features of the spectral characteristics in supervised learning. That is, the intrinsic land-cover recognition ability of APs may be underestimated, prompting researchers to continue to study it.

In this paper, a novel morphological attribute building index (MABI), as well as the morphological attribute shadow index (MASI), are proposed, and the study contributions can be summarized as follows:

(1) In the pre-processing step, APs were used to maintain the homogeneity of the original image. In addition, a new strategy to eliminate bright narrow and long non-building artificial objects, such as bright paths, road and narrow open ground, is proposed.

(2) A new building feature index based on APs, the MABI, is proposed for automatic building feature extraction. By the sequential application of attribute filters (AFs), multilevel characterization of the VHR image was obtained to model the structural information of buildings. Considering the different reflectance characteristics of buildings in the VHR image, features of bright buildings and dark buildings were extracted separately in the MABI to reduce the omission rate caused by the absence of dark roofs.

(3) Furthermore, in the post-processing step, the MASI, which is derived from the MABI, is proposed for the automatic shadow detection task. With the aid of the spatial co-occurrence between buildings and shadows, some confused flat features, such as regular bare land and open ground, could be filtered out.

The rest of this article is organized as follows. Section 2 introduces the morphological attribute building and shadow index. The experimental analysis and comparison results, are presented in Section 3. The parameter analysis is in Section 4. Section 5 concludes the paper.

2. Morphological Attribute Building Index

The flowchart of the proposed framework is shown in Figure 1. There are three main parts contained in the proposed framework: pre-processing, building feature extraction, and post-processing. Before jumping into the steps in detail, the APs, the basic mathematical foundation, are presented at first (Section 2.1). The pre-processing step is then presented (Section 2.2). The proposed building and shadow indexes, MABI and MASI, are calculated to obtain the building and shadow features (Sections 2.3 and 2.4, respectively). To better detect buildings from the obtained feature images, a post-processing framework is designed (SubSection 2.5). The variable notations used in this article are defined in Table 1.

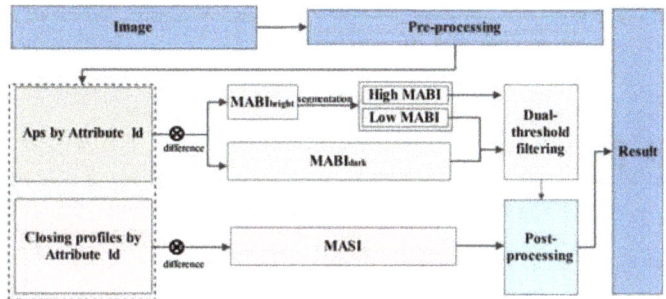

Figure 1. Flowchart of the proposed framework.

Table 1. Notations used in this paper.

Notation	Description
$f = \{b_1, b_2 \ldots, b_n\}$	The n bands of image f
$\gamma / \widetilde{\gamma} / \varphi$	Opening/thinning/closing operator
γ_DAP / φ_DAP	The differential attribute profile (DAP) obtained by the opening/closing profile in the attribute profile (APs)
$\widetilde{\gamma}\_EAP$	The stack of thinning profiles in EAP (the extension of the APs)
$T = \{t_1, t_2, \ldots, t_m\}$	Ordered set of m criteria/attributes
$\gamma^t_{Attr} / \varphi^t_{Attr} / DAP^t_{Attr}$	The opening profile/closing profile/DAP obtained by $Attr$ with t
λ_{Attr}	The filter parameter of attribute $Attr$

2.1. Attribute Profiles

APs are multi-scale features obtained by conducting the sequential application of AFs. AFs [31] are morphologically connected filters that act on the image by merging the connected components that compose the image according to the filter criterion. The connected components represent the regions that are composed of the spatially connected isointensity pixels in the image. According to the filter criterion of AFs, the transformation evaluates the value measured for each connected component in the image of an arbitrary attribute against a given filter parameter. For example, the filter criterion: means that, given the attribute *Attr*, the attribute value calculated on the connected component *C* is compared against the given reference value. The merging rule of AF is as follows: The regions that fulfill the given criterion remain unaltered, while the regions that do not fulfill the criterion are merged with darker or brighter (according to the grayscale value) adjacent regions corresponding to the extensive (i.e., thickening) or anti-extensive (i.e., thinning) transformation, respectively. These two transformations can be further subdivided into increasing (for the increasing criteria, one connected component satisfies the criterion and the subset components also meet this condition) and non-increasing categories according to the attribute selected in the filtering criterion. The non-increasing operation is not uniquely defined when dealing with grayscale images because it obtains different results according to the selected filter criterion [32]. For the non-increasing criterion, the basic operators are thinning and thickening, while the operators for the increasing criterion correspond to opening and closing. As two basic AF operators, multiscale thinning (or opening) and thickening (or closing) transformations can detect dark and bright objects, respectively.

For the grayscale image b, the APs obtained according to a sequence of ordered criteria with m attributes are defined as

$$APs(b) = \{\varphi_m(b), \varphi_{m-1}(b), \ldots, \varphi_1(b), b, \gamma_1(b), \ldots, \gamma_{m-1}(b), \gamma_m(b)\} \quad (1)$$

where φ_m and γ_m are the m attribute closing and attribute opening operators according to criterion T, respectively. The EAPs are the extension of the APs in multi-band images. The EAPs obtained from the multi-band image f can be defined as

$$EAPs(f) = \{AP(b_1), AP(b_2) \ldots, AP(b_n)\} \quad (2)$$

where b_n is the nth band of image f.

Progressive filtering residuals at multiple scales can be used for describing the structural composition of image contents [33]. Each obtained profile is associated with a specific scale. By computing the derivative of the profiles, a differential attribute profile (DAP) generated by an ordered set of criteria $T = \{t_1, t_2, \ldots, t_m\}$ is

$$DAP(f) = \left\{ \Delta_i : \left\langle \begin{array}{l} \Delta_i = \Delta_{\varphi_{m-i+1}}, \forall i \in [1, m] \\ \Delta_i = \Delta_{\gamma_{i-m}}, \forall i \in [m+1, 2m] \end{array} \right. \right\} \quad (3)$$

where Δ_φ and Δ_γ are the differential closing and opening profiles, respectively. To better understanding the multiscale DAP, we took the attribute named the diagonal of the minimum enclosing rectangle (*ld*), a measure of the object size, as an instance to describe the multiscale approach, where five scales with size T = {10,30,50,70,90}. Given a grayscale image, as shown in Figure 2j, opening profiles on *ld* at each elements are presented in Figure 2a–e in sequence. Furthermore, the different operation of APs between adjacent scales was computed to capture the components in the range of specific scales. Differences between each profile are shown in Figure 2f–i.

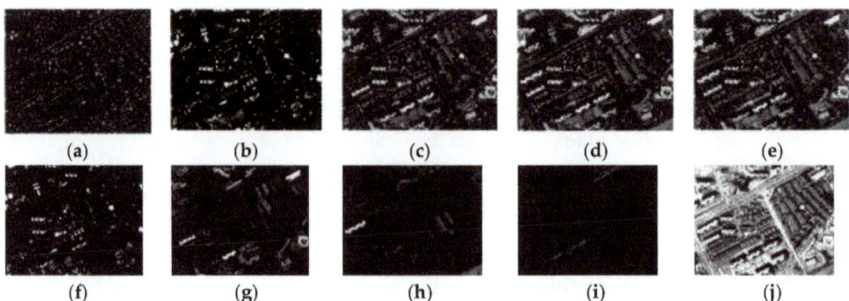

Figure 2. The attribute profiles (APs) and differential attribute profiles (DAPs) obtained by attribute *ld* on threshold 10,30,50,70, and 90. The example grayscale image is in Figure 4j. (**a**–**e**) are the opening profiles obtained on threshold 10,30,50,70, and 90, respectively. (**f**–**i**) are the DAPs obtained between adjacent scales.

To enhance the efficiency of attribute filtering, an effective data structure named max-tree [34] is used in building APs. The image filtering processing comprises three procedures: First, the image is represented by a hierarchical tree. For the grayscale image, the depth of the tree represents the number of gray levels of the image after threshold decomposition. The number of nodes is associated with the number of connected components of the binary image on the current graylevel. The tree is then pruned by evaluating the reference value λ at each node. The filtering process is performed by removing the nodes that do not satisfy the filtering criterion. Finally, the pruned tree is converted back into an image. The max-tree is particularly applicable for the computation of multiple filtering, e.g., profiles and granulometries, because the structure completes filtering with different criteria by creating the tree only once. The attribute values are calculated for all regions in the image before the image filtering step, and the filters then prune the tree according to the defined criterion.

In this paper, every EAP feature is calculated using Profattran software, which was kindly provided by the authors of the article [35].

2.2. Pre-Processing

Pre-processing consists of two steps: image denoising and elongated non-building object detection. The entire pre-processing flow chart is shown in Figure 3.

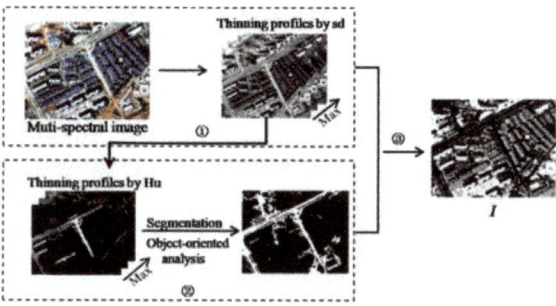

Figure 3. Pre-processing flowchart.

2.2.1. Image Denoising

The diverse materials of building roofs in a VHR image show different reflectivities, while the interior of building rooftops usually presents as a region with high spectral homogeneity. In view of this, the high contrast between the interior homogeneous section and its surroundings is often utilized as a basic principle of morphological operator-based building feature extraction strategies. However, variations in the bright image of VHR, which is calculated as the maximum value of each spectral band and acts as the basic unit for MBI-like processing, may lead to the incomplete extraction of building features. To maintain homogeneity and remove the small amount of dark noise inside bright homogeneous regions, an image denoising process based on AFs is applied to the original spectral reflectance image. This step corresponds to Box ① in Figure 3.

The standard deviation of the pixels belonging to each region (denoted by *sd*) is chosen as the filtering attribute in the image denoising task. This attribute is used to measure the spectral homogeneity of the intensity values of the pixels in the region. Equation (1) shows that APs are generated by a sequence of closing and opening profiles. For the APs built on a region (a set of pixels treated as a basic unit of the filters), all pixels in the region are located in either the closing or opening profiles. In fact, dark regions are obtained in the closing profiles and bright regions are obtained in the opening profiles. To keep the bright homogeneous regions and remove the small amount of dark noise, the opening operator is employed. Since *sd* is a non-increasing attribute, the opening operator corresponding to *sd* is attribute thinning. The stack of thinning profiles built on *sd* by the criterion $T(C) = sd(C) > \lambda_{sd}$ from the multispectral image f is obtained according to Equation (2). The maximum value of corresponding pixels in each obtained thinning profiles is then calculated, denoted by $\widetilde{\gamma}\_EAP_{sd}$. After image denoising, bright regions with high homogeneity in the original image remain in the maximum result, and the small dark structures are filtered out. It should be noted that AFs only process the image by suppressing the regions that do not meet the criterion without edge blur. With the virtue of maintaining edges for following building geometrical characteristic descriptions, AF is an effective tool, as a pre-processing step following building detection. By calculating the maximum values of each profile, the obtained regions with high reflectivity and homogeneity correspond to potential buildings.

2.2.2. Elongated Non-Building Object Detection

Buildings in dense urban areas are often easily confused with adjacent non-building landcovers, such as open parking lots, bare soil, roads, and small paths. This confusion is mostly attributed to the similar spectral characteristics of these land covers to buildings in the VHR image. Since these non-building land-covers may result in false alarms in the building feature extraction results, it is necessary to identify them independently. By analyzing the shape characteristics of roads and open areas surrounding buildings, it was found that these features generally present as elongated and curve-shaped regions. In this study, these objects are named elongated non-building objects. The elongated non-building object detection strategy is shown in Box ② of Figure 3 and is divided into two steps: a) elongated feature extraction and b) elongated feature segmentation.

(a) Elongated Feature Extraction

Despite the varying shape of buildings, the compactness of buildings is generally higher than that of roads and paths. Therefore, the attribute that measures the compactness of objects is considered able to separate building and non-building objects. In this part of the paper, a geometric attribute, i.e., the first moment invariant of Hu [36], denoted by Hu, is considered the filter attribute in the attribute filters. This attribute describes the ductility of a region relative to its centroid, which indicates the degree of non-compactness of an object, and the indexes in Hu are invariant to translation, rotation, and scaling [37]. The value of Hu is small for the compact region and gradually increases for the elongated regions. Since Hu is a non-increasing attribute, the thinning profiles filtered by Hu are used to detect bright and elongated non-building objects.

The elongated feature is calculated by the following steps: First, the stack of the thinning profiles is obtained by conducting a thinning operation on each profile in $\widetilde{\gamma}\_EAP_{sd}$, which is obtained in the previous image denoising step, with attribute Hu according to criterion λ_{Hu}. To detect structures with a high reflectance, the maximum of the profiles obtained in the first step is then calculated and acts as the input in the next segmentation step.

(b) Elongated Feature Segmentation

Since buildings also show elongated shape characteristics to some extent, object-oriented analysis is carried out to prevent potential buildings from being missed. Meanshift [38] segmentation is employed to obtain the image objects. To better identify buildings from the other landcovers, an over-segmentation strategy is preferred here. Because the main difference between building and non-building objects in the elongated feature image lies in the different degree of the object that approximates to the rectangle, the rectangular fit (*RcFit*), which is calculated by the ratio of the area of the object to the area of the smallest circumscribed rectangle of the object, is employed to filter out potential building objects. Objects with a high *RcFit* value are more likely than objects with a low *RcFit* value to be buildings. Giving the threshold λ_{RcFit}, the objects satisfying $RcFit(obj) < \lambda_{RF}$ are reserved to compose the resulting map.

Finally, by removing the obtained objects in Box ② from the result in Box ① (shown as Step ③ in Figure 3), a new basic image, denoted as *I*, is obtained. *I* acts as the input image in the following building feature extraction steps.

2.3. Morphological Attribute Building Index

Since buildings in high-resolution images are variable in size and orientation, a multiscale strategy is performed in the building detection task. Considering the regular shape of buildings, the length of the diagonal of the minimum enclosing rectangle, referred to as *ld*, is used to measure the scale characteristic of the objects. Both the attribute *area* and *ld* in attribute filtering can be used to measure the scale of objects. The *ld* rather than the *area* is chosen because attribute opening using *ld* retains more grain boundary segments than that using *area* [31]. In addition, the rectangular shape of buildings makes *ld* more suitable than *area* to measure the scale characteristics in the building detection task. The DAP can be built with an increasing criterion of attribute *ld* to obtain scale information.

In a VHR image, building roofs can be divided into two parts according to the difference in their spectral contrast with surrounding regions: local bright buildings and local dark buildings. To reduce the omission rate caused by dark roofs, these two types of buildings are detected separately in the MABI. The bright and dark building features in the MABI are recorded as MABI$_{bright}$ and MABI$_{dark}$, respectively.

The procedures for calculating MABI$_{bright}$ from *I* are as follows. Since *ld* is an increasing attribute, the opening profiles obtained from *I* by attribute *ld* according to criterion *t* is denoted by γ_{ld}^t. Considering the complex spatial patterns of the building, granulometry is conducted by building the DAP of the opening profiles obtained by attribute *ld* with an ordered set of criteria $T = \{t^{min}, \ldots, t, \ldots, t^{max}\}$, and the MABI$_{bright}$ is calculated as

$$\text{MABI}_{bright} = \max(\gamma\_DAP_{ld}), \text{ where } \begin{cases} \gamma\_DAP_{ld} = \left\{\gamma\_DAP_{ld}^{t^{min}}, \ldots, \gamma\_DAP_{ld}^{t}, \ldots, \gamma\_DAP_{ld}^{t^{max}}\right\} \\ \gamma\_DAP_{ld}^{t} = \left|\gamma_{ld}^{t+\Delta t} - \gamma_{ld}^{t}\right| \\ t^{min} < t < t^{max} \end{cases} \quad (4)$$

where Δt is the interval of threshold *T*, and max represents the max value of the corresponding pixels in all profiles. Through the above steps, the spectral characteristics (homogeneity and contrast) and spatial characteristics (size and shape) are addressed.

The procedures presented above are straightforwardly extended to MABI$_{dark}$ by replacing the opening profiles with closing φ in Equation (4), and the MABI$_{dark}$ is calculated as

$$\text{MABI}_{dark} = \max(\varphi\_DAP_{ld}), \text{ where } \begin{cases} \varphi\_DAP_{ld} = \left\{\varphi\_DAP_{ld}^{t^{min}}, \ldots, \varphi\_DAP_{ld}^{t}, \ldots, \varphi\_DAP_{ld}^{t^{max}}\right\} \\ \varphi\_DAP_{ld}^{t} = \left|\varphi_{ld}^{t+\Delta t} - \varphi_{ld}^{t}\right| \\ t^{min} < t < t^{max} \end{cases} \quad (5)$$

Since shadows also present as relatively dark regions in VHR images, some shadows may be contained in MABI$_{dark}$. To remove potential shadows, the spectral value of the pixels in the original image is considered. Because of the low reflectivity of the shadow in each visible band of the original image, the bright image is calculated by the max value of the pixels in all visible bands. The pixels in MABI$_{dark}$ that satisfy *bright* > λ_{bright} are saved as MABI$_{dark}$. With regard to the characteristics of buildings as homogeneous and continuous areas, pixels with high MABI values are more likely than those with low MABI values to be buildings.

2.4. Morphological Attribute Shadow Index

The spectral and geometrical characteristics of shadows are opposite and similar, respectively, to the corresponding characteristics of adjacent buildings. A shadow presents as a homogeneous dark area with geometrical characteristics similar to those of the adjacent building. Considering the high homogeneity, low spectral reflectance, and shape characteristics of shadows, the procedures for building the MASI are similar to those for building the MABI$_{dark}$ to obtain the dark structures in *I*. Furthermore, considering the different scale characteristics between buildings and shadows in the satellite image, the threshold value of *ld* in shadow detection is smaller than that in dark building feature extraction.

Due to the low spectral reflectance of shadows, the MASI is calculated by transforming the max operator in Equation (5) into the average value of the DAP feature:

$$MASI = \text{mean}(\varphi\_DAP_{ld}) \quad (6)$$

The pixels with large values are more likely than those with small values to be shadows in the MASI. Finally, the pixels that satisfy the conditions *bright* < λ_{bright}, $NDVI < \lambda_{NDVI}$, and $MASI \geq$

T_{MASI} are treated as shadows, where λ_{NDVI}, T_{MASI} indicate the threshold of the vegetation index (NDVI) and the MASI, respectively. The threshold of brightness is used to remove structures that have a high reflectance but are darker than the surrounding structures.

2.5. Building Extraction Framework of the Proposed Method

Extracting buildings by the dual threshold segmentation of the MABI may cause high commission errors (CEs) and omission errors (OEs). The CEs mainly come from the land covers that have similar characteristics with buildings, such as bare soil and roads, while the OEs are often related to dark roofs. To address these problems, a building feature extraction framework is conducted via the following steps.

First, the $MABI_{bright}$ image is divided into two parts: Given a threshold T_{MABI}, the high-MABI and low-MABI regions are separated. Pixels that satisfy the T_{MABI} in each part are assigned a value of one, and other pixels are assigned a value of zero. Object-oriented analysis can be performed on the obtained binary image. The objects belonging to the high-MABI region are analyzed with a relatively low shape threshold to prevent the bright irregular buildings from being missed, while objects in the low-MABI and $MABI_{dark}$ regions are analyzed by more strict geometric constraints. The *RcFit* and shape index (*SI*) values are utilized to measure the shape characteristics of objects. The *SI* is calculated by the boundary length of an object divided by four times the square root of its area. *SI* measures the smoothness of the object boundary, and more fragmented objects tend to have a high *SI* value.

According to [22], the distance between shadows and buildings is considered to suppress non-building objects. Different distance thresholds are set to objects in the high-MABI and low-MABIcategories, respectively. The thresholds on $MABI_{dark}$ are the same as the low-MABI thresholds. To present the entire processing flow more intuitively, a small region acting as an instance is shown in Figure 4.

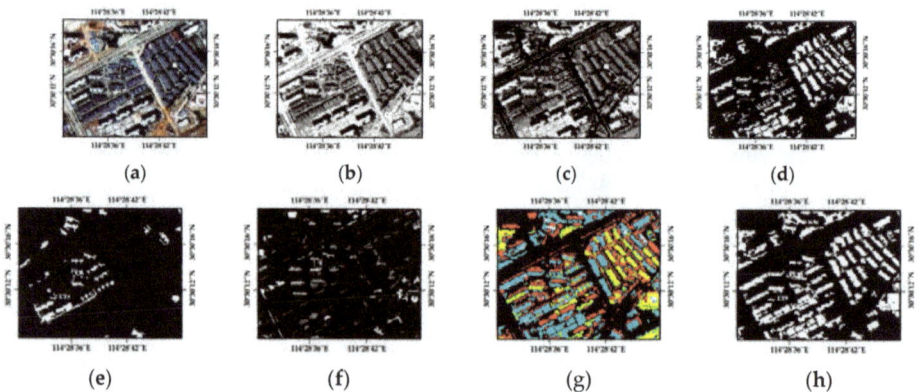

Figure 4. Example showing the steps of the proposed strategy: (**a**) example image; (**b**) the image obtained after image denoising; (**c**) the input image *I*; (**d,e**) the building maps obtained from $MABI_{bright}$ and $MABI_{dark}$, respectively; (**f**) MASI feature image; (**g**) overlay image of the obtained buildings and shadows, with high-MABI in yellow, low-MABI and $MABI_{dark}$ in blue, and shadows in red; (**h**) the final results of the proposed method.

Bright bare soil, roads, and small paths are easily confused with buildings. Figure 4b,c are images resulting from the two steps in the pre-processing step, respectively. (b) is the image obtained after image denoising, and (c) is the input image *I*. The two images show that, although the bright roads at the top of the image have spectral properties similar to those of the surrounding buildings, these roads and buildings are separated by their different shape characteristics in the elongated object detection step. After removing non-building objects, the false alarms in the input image *I* are reduced; for example, the bright open ground and small paths in the top left corner of (b) are removed in (c). The building maps obtained from $MABI_{bright}$ and $MABI_{dark}$ are presented in (d) and (e), and the MASI feature image is displayed in (f). The parameter setting in this dataset is the same as the datasets in the experiment section. A detailed analysis is provided in the following parameter analysis section. (g) is the overlapping image of the buildings and shadows obtained by the proposed method. Buildings in the high-MABI part are colored in yellow, and the low-MABI and $MABI_{dark}$ parts are colored in blue; shadows are colored in red. The building feature extraction result obtained by measuring the distance between the shadows and buildings is shown in (h). (h) shows that the buildings are retained and backgrounds are removed in comparison with (g).

3. Building Feature Extraction Experiments

3.1. Datasets and Experimental Strategy

3.1.1. Dataset Description

The proposed building feature extraction framework was applied to three high-resolution remote sensing images, which are radiometrically and geometrically calibrated in this section. These VHR images and the corresponding reference images are displayed in Figure 5. The ground truth images were manually delineated by field investigation and visual interpretation. Some representative subgraphs, which are marked with red (Images I1, I3, and I5) and blue (Images I2, I4, and I6) rectangular boxes in Figure 5, were chosen for detailed comparison and analysis. The basic information of the three datasets is listed in Table 2.

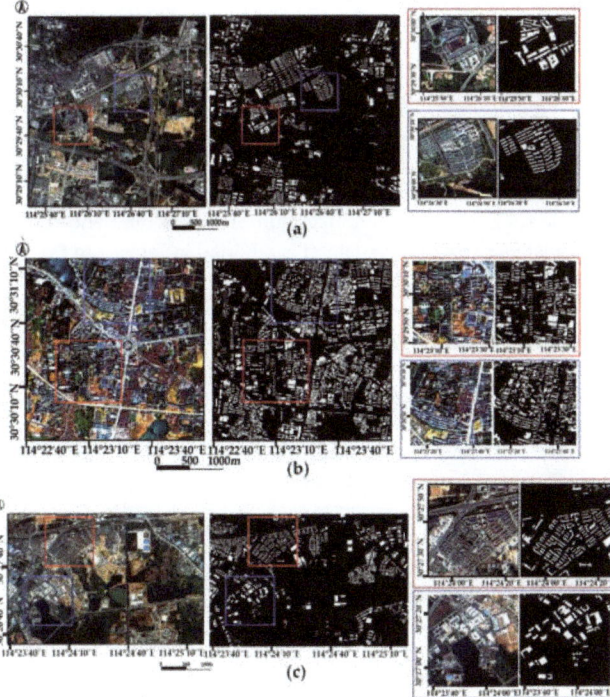

Figure 5. Three test datasets and the corresponding ground truth maps: (**a**) Dataset 1 and Subgraphs I1 (in the red box) and I2 (in blue box); (**b**) Dataset 2 and Subgraphs I3(in the red box) and I4 (in the blue box); (**c**) Dataset 3 and Subgraphs I5 (in the red box) and I6 (in the blue box).

Table 2. Details of the test datasets.

Dataset	Sensor	Resolution	Size	Major Land Cover Types
Dataset 1	WorldView-2	2.0	2000 × 2000	Building: 428,674 pixels. Background (vegetation, road, baresoil, path): 3,571,326 pixels.
Dataset 2	QuickBird	2.4	1100 × 1100	Building: 290,403pixels. Background (vegetation, road, baresoil, path, water): 919,597pixels.
Dataset 3	QuickBird	2.4	1060 × 1600	Building: 184,034 pixels. Background (vegetation, asphalt road, bare soil, open area): 1,511,966 pixels.

3.1.2. Experimental Set-Up

A comparative study between the MABI and MBI was performed to investigate the effectiveness of the proposed method. The recommended values in [22] were selected as the thresholds for the MBI. To obtain a fair comparison result, the same NDVI threshold and object-oriented analysis processes were conducted on both the MABI and MBI. The effectiveness of the pre-processing and shadow verification step in the proposed framework was explored by comparing the results obtained by the MABI and MBI under different conditions.

To further verify the effectiveness of the proposed algorithm, two widely used classifiers including support vector machines (SVM) [39] and random forest (RF) [40] were also used for comparison. In addition to the original spectral information of the image, there are two spatial characteristics used for classification in the above two supervised classifiers. The first comprises the multiscale and multidirectional DMPs that are used to compute the MBI. By feeding spectral bands and the DMPs into the SVM and RF, the binary classifiers DMP-SVM [41] and DMP-RF divide the test image into buildings and non-buildings. The second is the object-oriented SVM and the object-oriented RF. Employing object-based methods on VHR images can generate spectral and shape information to improve the accuracy of building feature extraction. In this study, the meanshift algorithm was used for segmentation. The spectral features of the object employed in the object-oriented SVM were the brightness and the spectral standard deviation of the object, and the spatial features were the length–width ratio, area, border length, *RcFit*, and SI. The parameters for the SVM and RF were set according to specific suggestions [39,40]. The number of training and test samples used in the supervised classification algorithms of each dataset is reported in Table 3. In this study, an SVM, which was implemented with the help of the LibSVM package, was used as a supervised binary classification to label each pixel in a high spatial resolution image as building/non-building (i.e., background). The nonlinear SVM with radial basis kernel was used and is abbreviated as SVM in the revised manuscript. All parameters in this SVM were tuned by five-fold cross validation. Except for the SVM-related work, which was implemented with the help of the LibSVM package using C++, processes were performed using MATLAB R2014a on a computer with a single i5-24003.10 GHz processer and 8.0 Gb of RAM.

Table 3. Training and test samples for the three datasets.

	Dataset 1		Dataset 2		Dataset 3	
Methods	No. of Training Samples	No. of Test Samples	No. of Training Samples	No. of Test Samples	No. of Training Samples	No. of test Samples
Building	858	427,816	1,275	289,128	1,147	182,887
Background	1,184	3,570,142	1,835	917,762	1,562	1,510,404

The parameters used in the proposed method and their suggested range are summarized in Table 4. The parameter sensitivity is further analyzed in the discussion section, and several issues should be noted. First, appropriate ranges of parameters for the proposed framework were analyzed in this study. Second, most of the parameters could be kept the same for different datasets, and the parameters were fixed for all three datasets in this paper. The accuracy statistics were calculated according to the correctly classified pixels in the building feature extraction map of each method. The building detection accuracy was evaluated by the following four statistical measures: overall accuracy (OA), Kappa coefficient (Kc), omission errors (OEs), and omission errors (CEs) [42]. The first two indexes were computed based on the confusion matrix [43], and the remaining two indexes measure the accuracies of classification.

Table 4. Parameters and the suggested range of the proposed method.

Feature Extraction Parameters			Parameters in Post-Processing		
Variables	Fixed Value in This Study	Suggested Range	Variables	Fixed Value in This Study	Suggested Range
λ_{bright}	0.35	[0.1,0.5]	t_{sd}	7	[5,8]
NDVI	0.58	[0.1,0.6]	t_{Hu}	0.7	[0.7,0.9]
RcFit	0.7	[0.5,0.7]	TMABI	0.25	[0.1,0.4]
SI	1.1	[1,1.5]	TMASI	0.4	[0.1,0.4]
Dist	0 in high-MABI, 10 in low-MABI	0 in high-MABI, 10 in low-MABI	ld in MABI	From 10 to 100, interval is 5	[10,200]
			ld in MASI	From 4 to 28, interval is 4	[2,50]

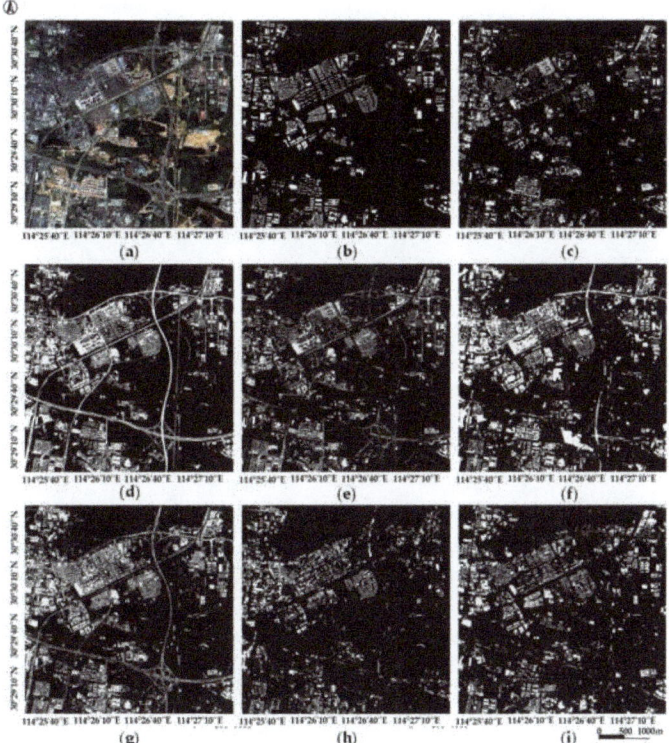

Figure 6. Building feature extraction results for Dataset 1: (a,b) the RGB image and the ground truth map; (c) the building detection resultof the MBI; (d–f) the building maps with the results of the pixel-based SVM, DMP-SVM, and object-oriented SVM, respectively; (g,h) the building detection results of DMP-RF and object-oriented RF; (i) the results of the proposed framework.

3.2. Experimental Results

3.2.1. General Results and Analysis of the Datasets

The building detection results of the three datasets are given in Figures 6–8, respectively, in which the detected buildings are in white pixels, and the background is in black pixels. Three datasets of urban areas have their own characteristics. There is a dense road network in Dataset 1. The difficulty of this dataset lies in the similarity between the spectral characteristics of roads and buildings. Compared with the buildings in Dataset 1, Dataset 2 has a high-density urban area. The varying spectral characteristics of building roofs and the existence of certain building groups increase the difficulty of analyzing Dataset 2. To carry out a comprehensive experiment, an image containing a large number of non-buildings was chosen as Dataset 3. This image has a large area of bare ground and vegetation, which poses a challenge to the building feature extraction task.

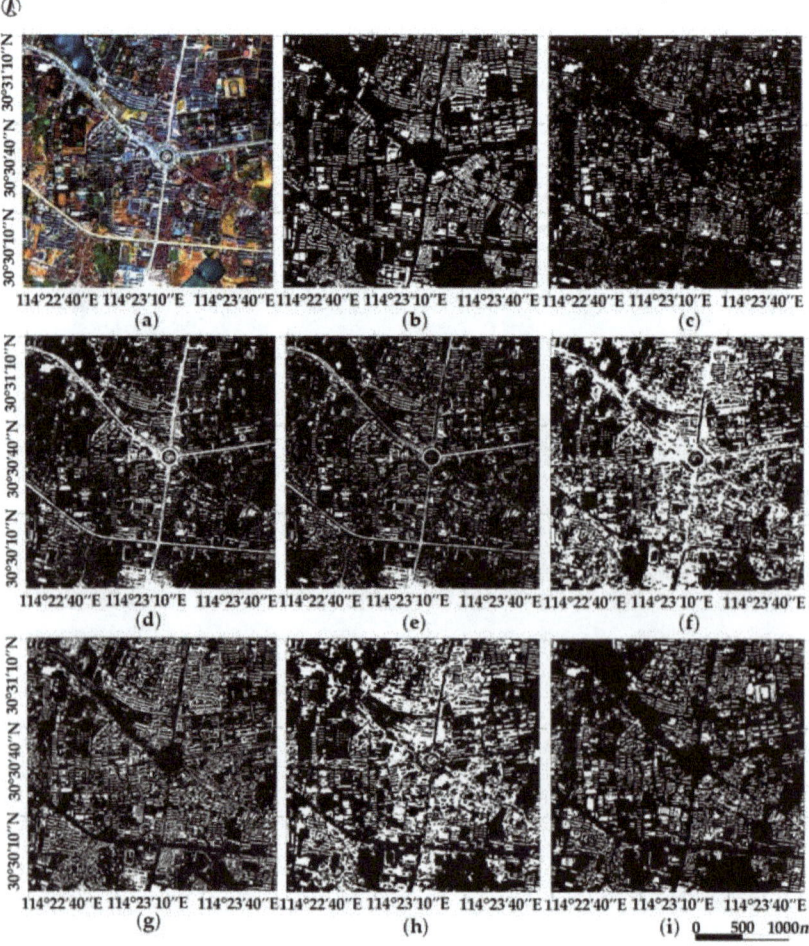

Figure 7. Building feature extraction results for Dataset 2: (**a,b**) the RGB image and the ground truth map; (**c**) the building detection result of the MBI; (**d–f**) the building maps with the results of the pixel-based SVM, DMP-SVM, and object-oriented SVM, respectively; (**g,h**) the building detection results of DMP-RF and object-oriented RF; (**i**) the results of the proposed framework.

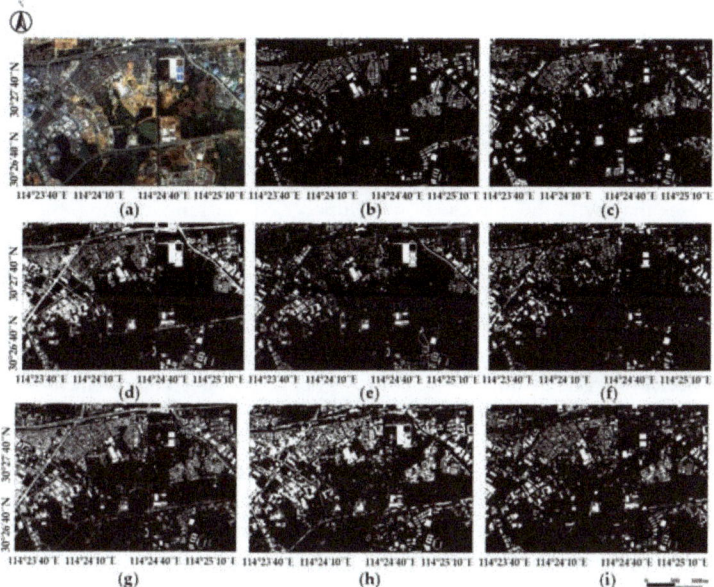

Figure 8. Building feature extraction results for Dataset 3: (**a,b**) the RGB image and the ground truth map; (**c**) the building detection result of the MBI; (**d–f**) the building maps with the results of the pixel-based SVM, DMP-SVM, and object-oriented SVM, respectively; (**g,h**) the building detection results of DMP-RF and object-oriented RF; (**i**) the results of the proposed framework.

The quantitative results of the different algorithms are reported in Table 5. The statistical accuracy and the visual inspection ((d) in Figures 6–8) show that the pixel-based SVM leads to unreliable results in the three datasets. This inferior performance is mainly due to the poor discriminatory ability of using only the spectral value of the original image. By joining the spatial information, the remaining algorithms obtain more acceptable results according to the statistical values in Table 5. Furthermore, in most cases, the proposed framework obtains competitive results. Detailed analysis of the results of the MBI, DMP-SVM, DMP-RF, object-oriented SVM, object-oriented RF, and the proposed method are as follows.

Table 5. Building detection accuracies of the test datasets.

Method	Dataset 1				Dataset 2				Dataset 3			
	OA	OE	CE	Kc	OA	OE	CE	Kc	OA	OE	CE	Kc
MBI	88.81	49.56	52.09	0.62	81.56	57.54	31.24	0.61	89.60	35.83	48.34	0.66
Pixel-Based SVM	71.07	19.3	75.64	0.51	62.38	11.11	62.1	0.51	76.43	16.1	70.56	0.56
DMP-SVM	85.81	45.9	61.53	0.59	77.14	47.17	47.17	0.56	87.08	53.55	58.53	0.59
Object-Oriented SVM	88.32	21.89	52.73	0.66	72.58	19.13	54.04	0.58	89.45	51.85	48.54	0.63
DMP-RF	85.03	15.25	59.48	0.64	78.34	30.16	46.25	0.62	84.11	20.81	61.34	0.62
Object-OrientedRF	89.91	49.92	48.22	0.63	80.51	13.99	43.87	0.66	85.17	7.57	58.27	0.65
Proposed	90.27	26.52	46.65	0.69	84.53	36.32	30.67	0.68	91.13	27.09	42.90	0.70

The MBI performed well for all three datasets. The OA of the MBI was second only to that of the proposed method in most cases, according to Table 5. Compared to the MBI OA, the OA of the proposed method increased by 1.46%, 2.97%, and 1.53% for the three datasets. The Kappa coefficient, increased from 0.62, 0.61, and 0.66 to 0.69, 0.68, and 0.7, respectively. The MBI was subject to a high CE rate in Datasets 1 and 3 due to the misclassification of non-buildings in the scenes. Regarding both the OE and CE, the proposed method obtainedbetter results than the MBI. For example, the OE and

CE decreased by 23.04% and 5.44%, respectively, in Dataset 1 and by 8.74% and 5.44%, respectively, in Dataset 3. The improvement of the CE in the proposed framework can be ascribed to the removal of non-buildings in the input image. In Datasets 1 and 3, there were many building blocks that were darker than the surrounding backgrounds. These buildings were excluded from the MBI results, causing the increase in the OE. The proposed MABI compensated for the missing buildings by a separate consideration of dark buildings.

The analysis of the outcomes of the DMP-SVM, DMP-RF, object-oriented SVM, and object-oriented RF demonstrates that, with the introduction of supervised machine learning, the two algorithms obtained competitive results. In particular, the OA of the two object-oriented methods for Datasets 1 and 3 is comparable to that of the proposed framework. Table 5 shows that the object-oriented SVM and RF obtained the lowest OE but were subject to severe omission problems. This problem was particularly noticeable in the dense building area in Dataset 2. A large area of asphalt roads that have similar spectral characteristics as the buildings in Datasets 1 and 3 caused an increase in false alarms in the results of the object-oriented classifiers. Although the object-oriented methods increased the efficiency and identification ability of the supervised classifier, the accuracy was dependent on the choice of representative training samples. The same problem also existed in the DMP-SVM and DMP-RF. The OA of these two methods in Dataset 2 was obvious lower than that in Datasets 1 and 3. However, from the result in Datasets 1 and 3, it was found that the discrimination power of the SVM was obviously increased by feeding the multi-scales and the multidirectional DMP feature. Compared with the pixel-based SVM that used only the spectral features of the image, the OA significantly increased in the three datasets. Nevertheless, supervised classification algorithms are time-consuming. An analysis of the above experiment results shows that the proposed MABI is more suitable than the other methods for the feature extraction of buildings in large and complex urban areas.

The running times of the different algorithms are reported in Table 6. The pixel-based SVM and MBI were the most efficient, followed by the proposed method. The other supervised methods still had a much higher cost than these two unsupervised ones, except for the cost of the training sample collection. Regarding the two unsupervised methods, in view of the detection superiority of the proposed work over MBI, it was considered that the proposed one is generally preferable.

Table 6. Running time (second) of all building detection methods used in this study.

Method	Dataset 1	Dataset 2	Dataset 3
MBI	146.35	55.34	72.91
Pixel-Based SVM	130.57	45.46	66.97
DMP-SVM	624.85	145.53	193.65
Object-Oriented SVM	1434.39	184.59	241.93
DMP-RF	1648.25	413.67	579.21
Object-Oriented RF	1581.41	185.42	252.43
Proposed	217.72	101.58	132.09

3.2.2. Visual Comparisons of the Representative Patches

The results of the representative patches in each test image are reported in Figure 9 (show Images I1 and I2), Figure 10 (show Images I3 and I4), and Figure 11 (show Images I5 and I6), respectively. The results obtained by the proposed framework are the most complete and precise in most scenes. The object-oriented SVM was subject to false alarms in the dense urban area, and the DMP-SVM was affected by the omission phenomenon, especially for heterogeneous buildings. The results of each representative patch are discussed as follows.

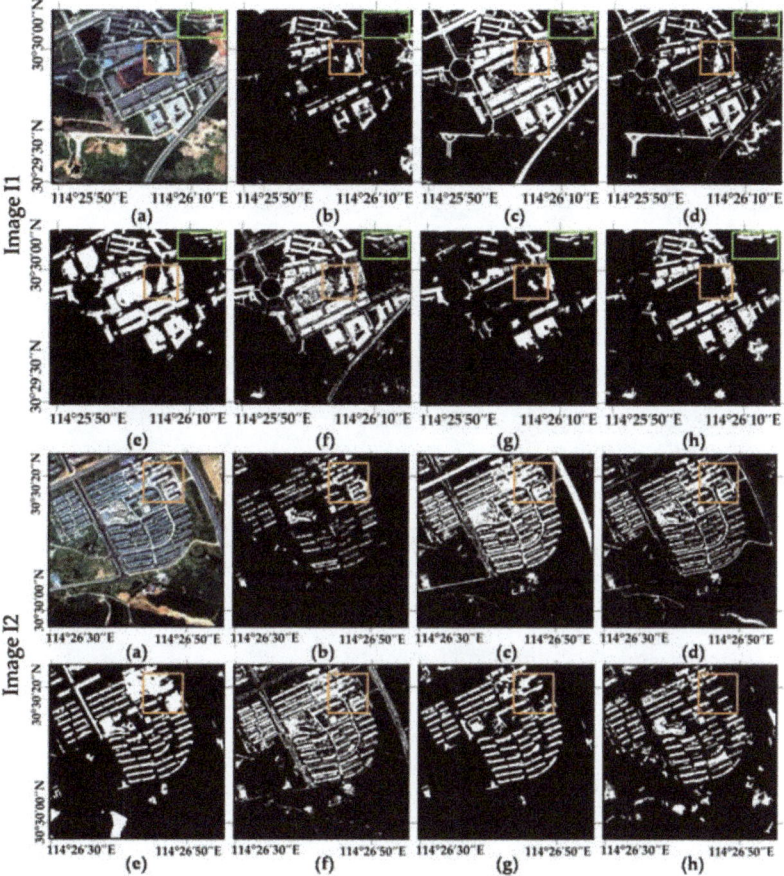

Figure 9. Building detection results of Test Patches I1 and I2. (**a**) RGB image; (**b**) MBI results; (**c**–**e**) the building maps with the results of the pixel-based SVM, DMP-SVM, and object-oriented SVM, respectively; (**f**,**g**) DMP-RF and object-oriented RF results; (**h**) the proposed method results.

The buildings in I1 and I2 in Figure 9 are surrounded by vegetation and bare soil. All detectors filtered out most of the vegetation, but, except for the proposed method, some bare soil and open ground information (yellow rectangles in I1 and I2) was incorrectly extracted. Some buildings with poor internal homogeneity (green rectangles in I1) were excluded by the MBI and DMP-SVM. The two object-oriented classifiers and the proposed MABI correctly extracted these building features by increasing the internal homogeneity of image objects before the building feature extraction step via segmentation and the proposed image denoising step, respectively. Patches I3 and I4 in Figure 10 show dense building areas, and the paths adjacent to buildings (green rectangle in I3 and yellow rectangle in I6 in Figure 11) were detected as buildings in the MBI and all supervised methods. As for the proposed framework, the paths were detected and removed in the pre-processing step. The bare ground (yellow rectangle in I3 and I4 in Figure 10), which was well removed with the constraint of shadows in both the MBI and the proposed method, was wrongly identified by all supervised methods. In the green rectangle in I4, the similarity between the spectral characteristics of buildings and the surrounding backgrounds made it difficult to identify buildings while excluding the backgrounds. A large number of buildings in this region were missed in most result maps, but the proposed method still identified the highest number of correct buildings. Patch I5 in Figure 11 shows a building block with low reflectivity and internal homogeneity. The heterogeneity of building roofs led to some omission phenomena in the results of the MBI and DMP-SVM. The DMP-SVM and object-oriented RF extracted the building features completely, but was still subject to under- and overestimation, respectively. The false alarms, such as the roads with spectral characteristics similar to those of the surrounding buildings were extracted in the object-oriented RF. Because the attribute filtering in the proposed method smooths the image while keeping the original boundaries, the buildings in the results of the proposed method had a more precise outline than those in the object-oriented RF. In summary, the results of these representative patches show that the proposed framework obtains better results than the comparison algorithms in different types of scenes.

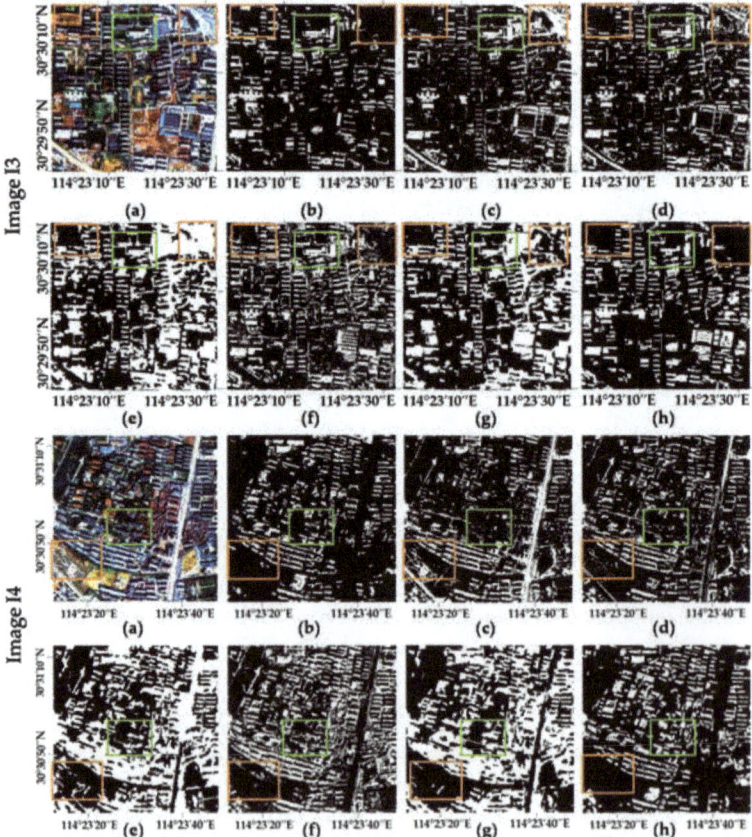

Figure 10. Building detection results of Test Patches I3 and I4. (**a**) RGB image; (**b**) MBI results; (**c–e**) the building maps with the results of the pixel-based SVM, DMP-SVM, and object-oriented SVM, respectively; (**f,g**) DMP-RF and object-oriented RF results; (**h**) the proposed method results.

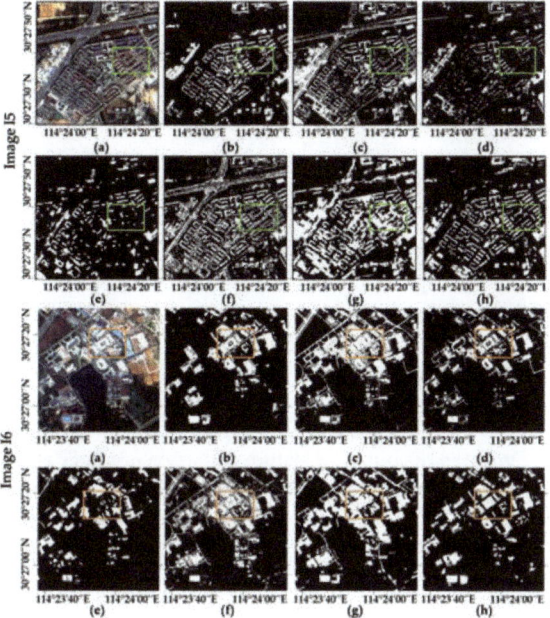

Figure 11. Building detection results of Test Patches I5 and I6. (**a**) RGB image; (**b**) MBI results; (**c**–**e**) the building maps with the results of the pixel-based SVM, DMP-SVM, and object-oriented SVM, respectively; (**f,g**) DMP-RF and object-oriented RF results; (**h**) the proposed method results.

4. Discussion

In this section, we first discuss the role of each step of the proposed method and then conduct parameter sensitivity analysis to verify the relative robustness of the proposed method.

4.1. Step Analysis of the Proposed Work

4.1.1. Effects of Denoising in Preprocessing: Analysis on MBI and MABI

To show the efficiency of image denoising in the pre-processing step, denoting the image obtained after image denoising step as I', the MBI and MABI features were calculated based on the bright image (marked as MBI and MABI (bright) in Figure 12) and I' (marked as MBI(I') and MABI in Figure 12). Each statistical result table in Figure 12a–c is composed of 320,000 randomly selected pixels from all datasets. The diagram displays the classification accuracy of the building and background areas in MBI, MABI (bright), MBI(I'), and MABI. To ensure a fair comparison, the MABI feature considered here is the high-MABI part calculated by the application of binary segmentation on the MABI according to the TMABI given in Table 4. The thresholds in the MBI are set according to values suggested in [21]. The classification accuracy is a statistic from the results without the shadow constraint.

As shown in Figure 12, both the MBI and MABI can extract most of the building features from the bright image and I', respectively, but the proposed method extracts the most accurate building information while filtering out false alarms. The OA of the buildings in the three tables is slightly improved from left to right. Specifically, after replacing the input image from the bright image to I', the increase in the OA of the MBI is more obvious than that of the MABI in tables (a) and (c). Due to the improvement in both the MBI and MABI, I' is more suitable than the bright image as the input image for building feature extraction. Furthermore, the observable increase in the correct backgrounds in the results based on I' also shows the good effect of I' on suppressing background noise in the building detection task.

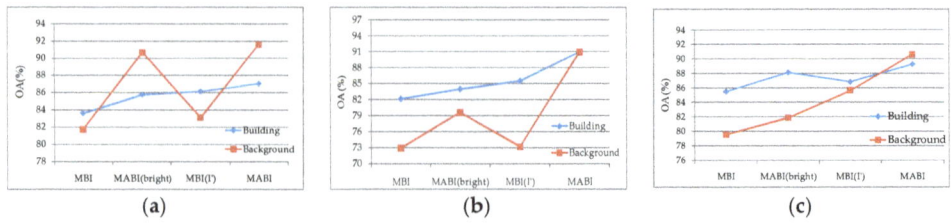

Figure 12. The OA of the building feature detection results of the MBI and MABI based on different input images: the bright image and I'. (**a**–**c**) are the statistical results of Dataset 1, Dataset 2, and Dataset 3, respectively.

A representative patch I5 is chosen for further comparisons. Again, the results displayed in Figure 13 confirm that using I' as the input image can effectively suppress false alarms in the building feature extraction results. For example, the highlighted vegetation and inhomogeneous bare land in the green box and the roads in the yellow box were removed by changing the input image from a bright image to I'. The improvement in the building feature extraction accuracy is attributed to the increase in the homogeneity of image I'; in addition, both statistical tables and images show that the MABI obtained a more accurate result than the MBI under identical conditions. For both the bright image and I', the proposed MABI achieves more accurate results than the MBI, and the most appropriate combination is the proposed one.

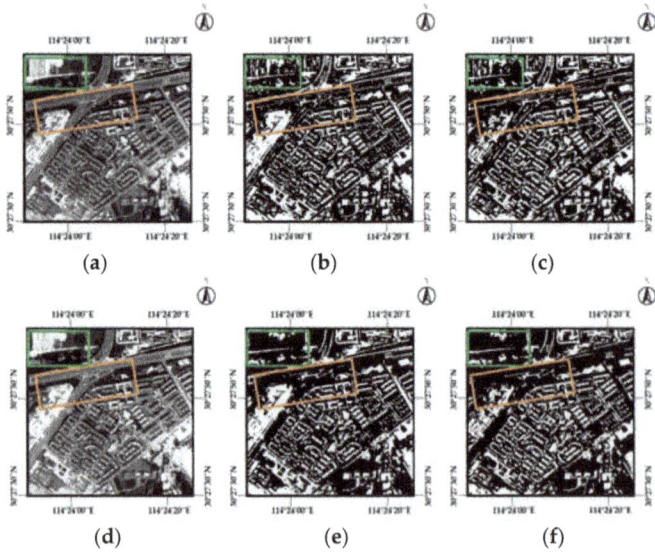

Figure 13. The MBI and MABI feature results based on the bright image and I' for Patches I1 and I5: (**a**) bright image; (**b**) results of the MBI based on the bright image; (**c**) results of the MABI based on bright image; (**d**) image I'. (**e**,**f**) are the results of MBI and MABI, respectively, based on I'.

4.1.2. Functions of Elongated Non-Building Object Detection and Dark Building Feature Extraction

The first step was utilized to reduce the non-building objects in the input image I before building feature extraction. The dark building feature extraction step was conducted to account for missing dark roofs. To illustrate the role of these two processes, the quantitative results for each step of the three datasets in Table 7 and three patches of a dense urban area in Figure 14 were utilized for statistical and visual comparisons, respectively.

Table 7. Accuracies of the building feature extraction results for each step of the proposed framework.

Step	Dataset 1				Dataset 2				Dataset 3			
	OA	OE	CE	Kc	OA	OE	CE	Kc	OA	OE	CE	Kc
$MABI_{bright}(I')$	81.07	32.95	68.15	0.57	71.82	39.51	55.96	0.54	86.33	34.22	58.26	0.62
$MABI_{bright}(I)$	89.71	31.64	48.37	0.66	82.11	37.48	38.24	0.65	90.94	33.79	42.47	0.68
MABI	90.6	26.18	45.92	0.68	83.72	35.55	33.22	0.67	90.22	26.51	42.78	0.68

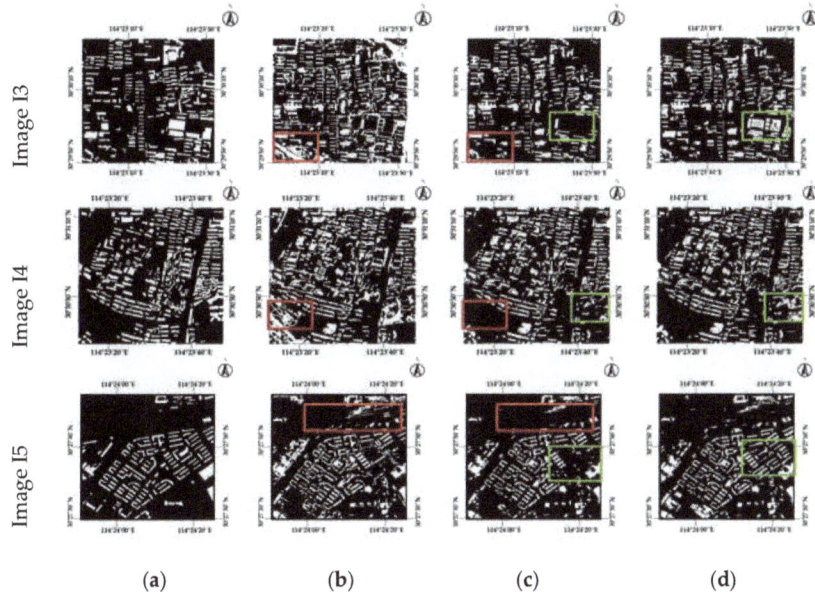

(a)   (b)   (c)   (d)

Figure 14. Building feature extraction results of Patches I3, I4, and I5 for step analysis of the proposed method: (**a**) ground truth image; (**b**) result of $MABI_{bright}$ without non-building object detection; (**c**) result of $MABI_{bright}(I)$; (**d**) result of the MABI without shadow constraint. The red and green regions emphasize the performance for elongated objects and dark building, respectively.

The values in Line 2 of the $MABI_{bright}$ of the proposed methodhavean obviously lower CE compared with the results of the $MABI_{bright}$ feature without eliminating the elongated objects for the three datasets. This improvement reflects that removing easily confused non-building objects in the input image can effectively reduce the false alarms in the final result. The red regions in Figure 14b show that the regular road in I3, the open ground in I4, and the small paths in I5 are filtered out in (c). This improvement demonstrates that detecting these objects is necessary to reduce false alarms that cannot be recognized in post-processing. Line 3 of Table 7 represents the accuracy of the MABI that combines the results of $MABI_{bright}$ and $MABI_{dark}$ before shadow constraint. For Datasets 1 and 2, the four MABI statistics are better than the results in Line 2. As for Dataset 3, due to the large area of dark backgrounds, the CE in Line 3 is slightly increased compared to that in Line 2 after the feature extraction of dark buildings, which also led to a slight decrease in the OA. Nevertheless, the decrease in the OE of Dataset 3 was the largest of the three data sets. This result can be viewed visually in Patch I5 of Figure 14. The missing buildings in the green region in (c) were supplemented in (d). Moreover, a slight increase in the CE is acceptable when compared with a substantial decrease in the OE, and false alarms can be further removed with the shadow constraint.

4.1.3. The Usage of Proposed Shadow Detection: Analysis on MSI and MASI

Shadow constraint was used to filter out the non-buildings from the obtained building map in the post-processing step. Since the omission of shadow should lead to an increase in the OE value, and the false-positive shadows may cause an increase in the CE value, the accuracy of four results from a pairwise combination of two shadow detection and two building feature extraction results (MBI and MABI) are given in Table 8 to compare the shadow detection results of the MSI and the proposed MASI. Lines 1 and 2 in Table 8 are the building detection results of the MBI with the shadow constraints of the MSI and MASI, respectively. Line 3 lists the building detection results of the MABI with the shadow results of the MSI. The combination of building maps with the proposed MASI (in Lines 2 and 4) obtained a higher OA than that with MSI (in Lines 1 and 3) for the three datasets. The reduction in CE and OE values also proves the effectiveness of the MASI. The comparison of these results shows that the most accurate combination is the proposed work.

Table 8. Accuracy of the building detection results with different shadow constraints.

Method	Dataset 1				Dataset 2				Dataset 3			
	OA	OE	CE	Kc	OA	OE	CE	Kc	OA	OE	CE	Kc
MBI	88.81	48.56	52.09	0.62	81.56	57.54	31.24	0.61	89.6	35.83	48.34	0.66
MBI+MASI	89.1	48.18	50.88	0.63	81.6	57.36	31.17	0.61	89.65	35.12	48.07	0.66
MABI+MSI	90.17	27.31	45.06	0.68	84.23	36.31	31.66	0.68	91.11	27.89	41.7	0.7
Proposed	91.02	26.44	44.71	0.7	84.54	36.2	30.67	0.68	91.13	27.09	41.7	0.7

4.2. Parameter Analysis

In this section, the values of some important parameters of the proposed method are discussed.

4.2.1. Pre-Processing Parameters

The thresholds for the attributes sd and Hu used in the pre-processing step are analyzed here. Attribute sd was employed to increase the homogeneity of the original image. A high value of sd corresponds to a high object homogeneity. Analyzing the gray histogram of the filtering results with different thresholds shows that, when the threshold value is greater than 20, most objects in the complex urban image are removed after filtering, and the effect of the AF is not obvious when the threshold is below 5. Therefore, the threshold values in [5,20] are discussed here. Figure 15a,b show the relationship between the value of sd and the building feature extraction precision of Dataset 2. The OE and CE are more balanced when the threshold is between 5 and 8, and a satisfactory and stable OA and Kappa coefficient rate are also obtained in this interval. When the proposed framework was applied to images with a high, medium, and low building density, the threshold value of sd in [5,8] possessed good generality and stability for the different scenes. Furthermore, a relatively small threshold is recommended for dense building areas, and a relatively large threshold can be selected for images containing a high amount of background. The suggested threshold for attribute sd in shadow detection is the same as that of the parameters in building feature extraction since shadows and the surrounding buildings have similar characteristics.

The Hu attribute was used to detect the elongated non-building objects in the pre-processing step. Hu indicates the non-compactness degree of the objects and ranges from 0 to 1. The value is gradually increased from compact to elongated objects. Since buildings are compact objects in the image, a small value of Hu can filter out some buildings, so Hu values below 0.5 are not considered here. Figure 15c,d show the relationship between the accuracies of building detection and the threshold value of Hu at [0.5,0.9] of Dataset 2. The four statistical values show an improvement as the value of Hu increases from 0.7 to 0.9. In general, when the threshold is in the interval of 0.7–0.9, the proposed framework achieves a more accurate result. Since Hu is only related to the geometrical characteristics of objects, the thresholds can be safely applied to different images.

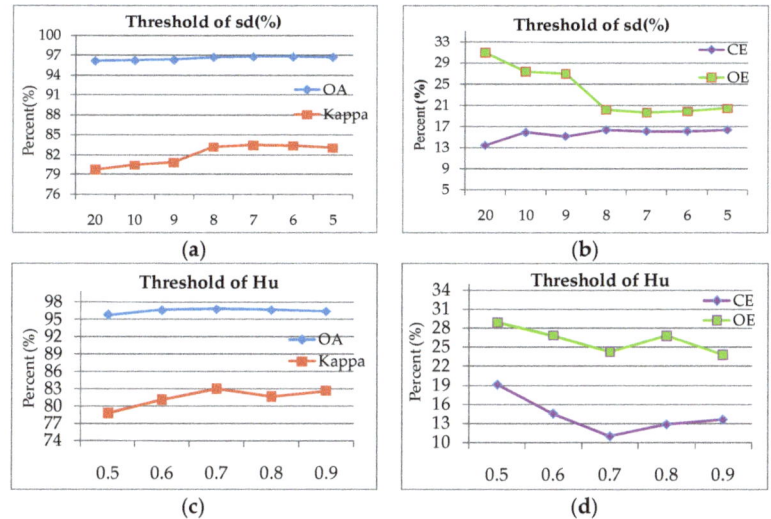

Figure 15. Relationship between building detection accuracies and the thresholds of attributes *sd* and Hu for Dataset 2.

4.2.2. Parameters in the Building Feature Extraction Steps

Threshold values of attribute *ld* in the MABI were arbitrarily selected in terms of the scale of the buildings. The OA of the building detection results (calculated from the MABI$_{bright}$) of Dataset 2 obtained by different intervals of *ld* is visualized in Figure 16. The vertical axis represents the OA values, and the horizontal axis represents the *ld* intervals. *ld* intervals less than 10 are 2,6,10, with a step of 5 after 10. The OA is obviously decreased after the upper limit of *ld* exceeds 200 and the minimum lower limit is 20. The accuracies decrease slowly when the upper limit of *ld* is in the interval [100,200]. According to Equation (4), the value of *ld* is selected based on the building scale; therefore, an *ld* value in the interval of [2,100] is suggested for the VHR image of the urban area.

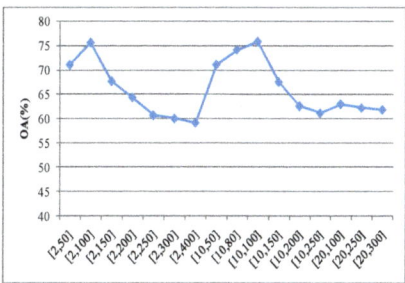

Figure 16. Relationship between overall accuracies of building detection and the thresholds of attribute *ld* in Dataset 2.

In the proposed framework, many non-building objects are removed in the pre-processing step, so a small threshold value of the high and low MABI is recommended to prevent the removal of some dark roofs. As the MABI ranges from 0 to 1, the suggested threshold is within the range of 0.1–0.4, where the quality scores are stable.

For the threshold value of the distance between buildings and shadows, the NDVI, building area, and SI have been discussed in detail in [22]. The value of the *RcFit* ranges from 0 to 1, and the larger the value, the more the object approximates the rectangle. For objects in the high MABI region, the *RcFit* value is between 0.5 and 0.6, while the *RcFit* value for objects in the low MABI region is between 0.6 and 0.7.

5. Conclusions

In this paper, a new building index, i.e., the MABI, and a new shadow index, i.e., the MASI, are proposed based on morphological attribute operators. An analysis of the existing MBI showed that the building feature extraction algorithm based on morphological operators is subject to some OEs and CEs. The OEs occur when the extraction misses some dark roofs and due to noise in building objects, and the CEs are caused by certain types of land cover, such as roads, bare ground, and open ground, which have spectral and shape characteristics similar to those of buildings. Our work aimed at improving these issues, and the contributions of this study are as follows: First, a thinning operator based on the attribute standard deviation was conducted to increase the homogeneity of the original image. Then, elongated non-building objects were detected to decrease the effect of interference objects in the input image before the building detection process. In the building feature extraction step, dark buildings were considered independently with the MABI to further reduce the OE. By jointly using the MABI and MASI in an object-oriented framework, false alarms were further reduced.

The proposed method was conducted on three VHR images. A comparison of the building detection results of the proposed framework with those of the MBI, DMP-SVM, pixel- and object-based SVM, DMP-RF, and object-oriented RF shows that the proposed method is the most effective at increasing the OA and reducing the OE and CE, especially for images with few buildings and large path and bare ground areas. The parameters of the proposed framework were analyzed, and the threshold selection conclusions can be summarized as follows: *sd* is used to remove small dark structures and to increase the homogeneity of an image. To maintain the details in the image, the choice of a small threshold is recommended, especially for dense urban areas. The attribute Hu is employed to measure the elongated degree of objects; therefore, a large value of Hu is recommended to better indicate non-building objects. The MABI threshold was used to distinguish buildings from other land cover types. Since a large number of easily confused objects were removed in the pre-processing step in the proposed framework, a small threshold value is recommended to avoid the erroneous removal of buildings.

In future studies, more attributes will be considered to better model the spectral and structural information of scenes for building feature extraction tasks, and automatic threshold selection research is also planned.

Author Contributions: W.M., Y.W. and J.L. conceived and conducted the experiments, and performed the data analysis; S.Z. and M.W. provided advice and helped with the revision of the manuscript. W.M. wrote the article.

Funding: The research was supported by the National Key R & D Program under Grant 2018YFD1100405, the National Natural Science Foundation of China under Grant 41701382, and the Hubei Provincial Natural Science Foundation Project under Grant 220100039.

Acknowledgments: The authors are very grateful for the Profattran software, which was kindly provided by Marpu, et al. (the authors of article [35]).

Conflicts of Interest: The authors declare no conflict of interest.

References

1. Pesaresi, M.; Guo, H.; Blaes, X.; Ehrlich, D.; Ferri, S.; Gueguen, L.; Halkia, M.; Kauffmann, M.; Kemper, T.; Lu, L. A Global Human Settlement Layer From Optical HR/VHR RS Data: Concept and First Results. *IEEE J. Sel. Top. Appl. Earth Obs. Remote Sens.* **2013**, *6*, 2102–2131. [CrossRef]
2. Gamba, P.; Dell'Acqua, F.; Stasolla, M.; Trianni, G.; Lisini, G. Limits and Challenges of Optical Very-High-Spatial-Resolution Satellite Remote Sensing for Urban Applications. In *Urban Remote Sensing: Monitoring, Synthesis and Modeling in the Urban Environment*; John Wiley & Sons, Ltd.: Hoboken, NJ, USA, 2011; pp. 35–48.

3. Aplin, P.; Atkinson, P.M.; Curran, P.J. Fine spatial resolution satellite sensors for the next decade. *Int. Remote Sens.* **1997**, *18*, 3873–3881. [CrossRef]
4. Bruzzone, L.; Carlin, L. A Multilevel Context-Based System for Classification of Very High Spatial Resolution Images. *IEEE Trans. Geosci. Remote Sens.* **2006**, *44*, 2587–2600. [CrossRef]
5. Huang, X.; Zhang, L. An SVM Ensemble Approach Combining Spectral, Structural, and Semantic Features for the Classification of High-Resolution Remotely Sensed Imagery. *IEEE Trans. Geosci. Remote Sens.* **2013**, *51*, 257–272. [CrossRef]
6. Johnson, B.; Xie, Z. Classifying a high resolution image of an urban area using super-object information. *ISPRS J. Photogramm. Remote Sens.* **2013**, *83*, 40–49. [CrossRef]
7. Yan, W.Y.; Shaker, A.; Zou, W. Panchromatic IKONOS Image Classification using Wavelet Based Features. In Proceedings of the 2009 IEEE Toronto International Conference Science and Technology for Humanity (TIC-STH), Toronto, ON, Canada, 26–27 September 2009; pp. 456–461.
8. Kuffer, M.; Barrosb, J. Urban Morphology of Unplanned Settlements: The Use of Spatial Metrics in VHR Remotely Sensed Images. *Procedia Environ. Sci.* **2011**, *7*, 152–157. [CrossRef]
9. Pesaresi, M.; Benediktsson, J.A. A new approach for the morphological segmentation of high-resolution satellite imagery. *IEEE Trans. Geosci. Remote Sens.* **2001**, *39*, 309–320. [CrossRef]
10. Bian, L. Retrieving Urban Objects Using a Wavelet Transform Approach. *Photogramm. Eng. Remote Sens.* **2003**, *69*, 133–141. [CrossRef]
11. Huang, X.; Zhang, L.; Li, P. Classification and Extraction of Spatial Features in Urban Areas Using High-Resolution Multispectral Imagery. *IEEE Geosci. Remote Sens. Lett.* **2007**, *4*, 260–264. [CrossRef]
12. Vakalopoulou, M.; Karantzalos, K.; Komodakis, N.; Paragios, N. Building Detection in Very High Resolution Multispectral Data with Deep Learning Features. In Proceedings of the 2015 IEEE International Geoscience and Remote Sensing Symposium (IGARSS), Milan, Italy, 26–31 July 2015; pp. 1873–1876.
13. Zhang, Y. Optimisation of building detection in satellite images by combining multispectral classification and texture filtering. *ISPRS J. Photogramm. Remote Sens.* **1999**, *54*, 50–60. [CrossRef]
14. Ahmadi, S.; Zoej, M.J.V.; Ebadi, H.; Moghaddam, H.A.; Mohammadzadeh, A. Automatic urban building boundary extraction from high resolution aerial images using an innovative model of active contours. *Int. J. Appl. Earth Obs. Geoinf.* **2010**, *12*, 150–157. [CrossRef]
15. Bouziani, M.; Goïta, K.; He, D.C. Automatic change detection of buildings in urban environment from very high spatial resolution images using existing geodatabase and prior knowledge. *ISPRS J. Photogramm. Remote Sens.* **2010**, *65*, 143–153. [CrossRef]
16. Sohn, G.; Dowman, I. Data fusion of high-resolution satellite imagery and LiDAR data for automatic building extraction. *ISPRS J. Photogramm. Remote Sens.* **2007**, *62*, 43–63. [CrossRef]
17. Pesaresi, M.; Gerhardinger, A.; Kayitakire, F. A Robust Built-Up Area Presence Index by Anisotropic Rotation-Invariant Textural Measure. *IEEE J. Sel. Top. Appl. Earth Obs. Remote Sens.* **2009**, *1*, 180–192. [CrossRef]
18. Aytekin, O.; Ulusoy, I.; Erener, A.; Duzgun, H.S.B. Automatic and unsupervised building extraction in complex urban environments from multi spectral satellite imagery. In Proceedings of the International Conference on Recent Advances in Space Technologies, Istanbul, Turkey, 11–13 June 2009; pp. 287–291.
19. Feyisa, G.L.; Meilby, H.; Fensholt, R.; Proud, S.R. Automated Water Extraction Index: A new technique for surface water mapping using Landsat imagery. *Remote Sens. Environ.* **2014**, *140*, 23–35. [CrossRef]
20. Ok, A.O. Automated detection of buildings from single VHR multispectral images using shadow information and graph cuts. *ISPRS J. Photogramm. Remote Sens.* **2013**, *86*, 21–40. [CrossRef]
21. Huang, X. A Multidirectional and Multiscale Morphological Index for Automatic Building Extraction from Multispectral GeoEye-1 Imagery. *Photogramm. Eng. Remote Sens.* **2011**, *77*, 721–732. [CrossRef]
22. Huang, X.; Zhang, L. Morphological Building/Shadow Index for Building Extraction From High-Resolution Imagery Over Urban Areas. *IEEE J. Sel. Top. Appl. Earth Obs. Remote Sens.* **2012**, *5*, 161–172. [CrossRef]
23. Chuvpilo, S.; Jankevics, E.; Tyrsin, D.; Akimzhanov, A.; Moroz, D.; Jha, M.K.; Schulze-Luehrmann, J.; Santner-Nanan, B.; Feoktistova, E.; König, T. A new approach for the morphological segmentation of high-resolution satellite imagery. *IEEE Trans. Geosci. Remote Sens.* **2002**, *39*, 309–320.
24. You, Y.; Wang, S.; Ma, Y.; Chen, G.; Wang, B.; Shen, M.; Liu, W. Building Detection from VHR Remote Sensing Imagery Based on the Morphological Building Index. *Remote Sens.* **2018**, *10*, 1287. [CrossRef]

25. Huang, X.; Yuan, W.; Li, J.; Zhang, L. A New Building Extraction Postprocessing Framework for High-Spatial-Resolution Remote-Sensing Imagery. *IEEE J. Sel. Top. Appl. Earth Obs. Remote Sens.* **2017**, *10*, 654–668. [CrossRef]
26. Zhang, Q.; Huang, X.; Zhang, G. A Morphological Building Detection Framework for High-Resolution Optical Imagery Over Urban Areas. *IEEE Geosci. Remote Sens. Lett.* **2016**, *13*, 1388–1392. [CrossRef]
27. Mura, M.D.; Benediktsson, J.A.; Waske, B.; Bruzzone, L. Morphological Attribute Profiles for the Analysis of Very High Resolution Images. *IEEE Trans. Geosci. Remote Sens.* **2010**, *48*, 3747–3762. [CrossRef]
28. Ghamisi, P.; Mura, M.D.; Benediktsson, J.A. A Survey on Spectral—Spatial Classification Techniques Based on Attribute Profiles. *IEEE Trans. Geosci. Remote Sens.* **2015**, *53*, 2335–2353. [CrossRef]
29. Mura, M.D.; Benediktsson, J.A.; Bruzzone, L. Modeling structural information for building extraction with morphological attribute filters. In Proceedings of the SPIE—The International Society for Optical Engineering, Berlin, Germany, 31 August–3 September 2009.
30. Falco, N.; Mura, M.D.; Bovolo, F.; Benediktsson, J.A.; Bruzzone, L. Change Detection in VHR Images Based on Morphological Attribute Profiles. *IEEE Geosci. Remote Sens. Lett.* **2013**, *10*, 636–640. [CrossRef]
31. Breen, E.J.; Jones, R. Attribute Openings, Thinnings, and Granulometries. *Comput. Vis. Image Underst.* **1996**, *64*, 377–389. [CrossRef]
32. Salembier, P.; Oliveras, A.; Garrido, L. Antiextensive connected operators for image and sequence processing. *IEEE Trans. Image Process.* **1998**, *7*, 555–570. [CrossRef]
33. Ouzounis, G.K.; Soille, P. Differential Area Profiles. In Proceedings of the 20th International Conference on Pattern Recognition, Istanbul, Turkey, 23–26 August 2010; pp. 4085–4088.
34. Westenberg, M.A.; Roerdink, J.B.T.M.; Wilkinson, M.H.F. Volumetric attribute filtering and interactive visualization using the Max-Tree representation. *IEEE Trans. Image Process.* **2007**, *16*, 2943–2952. [CrossRef]
35. Marpu, P.R.; Pedergnana, M.; Mura, M.D.; Benediktsson, J.A.; Bruzzone, L. Automatic Generation of Standard Deviation Attribute Profiles for Spectral-Spatial Classification of Remote Sensing Data. *IEEE Geosci. Remote Sens. Lett.* **2013**, *10*, 293–297. [CrossRef]
36. Hu, M. Visual pattern recognition by moment invariants. *IEEE Trans. Inf. Theory* **1962**, *8*, 179–187.
37. Gonzalez, R.C.; Woods, R.E. *Digital Image Processing*, 2nd ed.; Addison-Wesley Longman Publishing Co.: Boston, MA, USA, 1987; pp. 186–191.
38. Comaniciu, D.; Meer, P. Mean shift: A robust approach toward feature space analysis. *IEEE Trans. Pattern Anal. Mach. Intell.* **2002**, *24*, 603–619. [CrossRef]
39. Camps-Valls, G.; Bruzzone, L. Kernel-based methods for hyperspectral image classification. *IEEE Trans. Geosci. Remote Sens.* **2005**, *43*, 1351–1362. [CrossRef]
40. Pal, M. Random forest classifier for remote sensing classification. *Int. J. Remote Sens.* **2005**, *26*, 217–222. [CrossRef]
41. Fauvel, M.; Benediktsson, J.A.; Chanussot, J.; Sveinsson, J.R. Spectral and Spatial Classification of Hyperspectral Data Using SVMs and Morphological Profiles. *IEEE Trans. Geosci. Remote Sens.* **2008**, *46*, 3804–3814. [CrossRef]
42. Foody, G. Assessing the Accuracy of Remotely Sensed Data: Principles and Practices. *Photogramm. Rec.* **2010**, *25*, 204–205. [CrossRef]
43. Congalton, R.G. A review of assessing the accuracy of classifications of remotely sensed data. *Remote Sens. Environ.* **1998**, *37*, 270–279. [CrossRef]

© 2019 by the authors. Licensee MDPI, Basel, Switzerland. This article is an open access article distributed under the terms and conditions of the Creative Commons Attribution (CC BY) license (http://creativecommons.org/licenses/by/4.0/).

Article

Comparison of Digital Building Height Models Extracted from AW3D, TanDEM-X, ASTER, and SRTM Digital Surface Models over Yangon City

Prakhar Misra [1,*], Ram Avtar [2] and Wataru Takeuchi [1]

1   Institute of Industrial Science, The University of Tokyo, Tokyo 153-8505, Japan; wataru@iis.u-tokyo.ac.jp
2   Graduate School of Environmental Earth Science, Hokkaido University, Sapporo 060-0808, Japan; ram@ees.hokudai.ac.jp
*   Correspondence: mprakhar@iis.u-tokyo.ac.jp or prakharmisra90@gmail.com; Tel.: +81-070-481-32297

Received: 25 October 2018; Accepted: 8 December 2018; Published: 11 December 2018

Abstract: Vertical urban growth in the form of urban volume or building height is increasingly being seen as a significant indicator and constituent of the urban environment. Although high-resolution digital surface models can provide valuable information, various places lack access to such resources. The objective of this study is to explore the feasibility of using open digital surface models (DSMs), such as the AW3D30, ASTER, and SRTM datasets, for extracting digital building height models (DBHs) and comparing their accuracy. A multidirectional processing and slope-dependent filtering approach for DBH extraction was used. Yangon was chosen as the study location since it represents a rapidly developing Asian city where urban changes can be observed during the acquisition period of the aforementioned open DSM datasets (2001–2011). The effect of resolution degradation on the accuracy of the coarse AW3D30 DBH with respect to the high-resolution AW3D5 DBH was also examined. It is concluded that AW3D30 is the most suitable open DSM for DBH generation and for observing buildings taller than 9 m. Furthermore, the AW3D30 DBH, ASTER DBH, and SRTM DBH are suitable for observing vertical changes in urban structures.

Keywords: digital building height; 3D urban expansion; land-use; DTM extraction; open data; developing city; accuracy analysis

1. Introduction

Urban areas in the 21st century are facing growing challenges from natural and man-made crises. These include chronic stresses, like environmental pollution and climate change, and acute shocks, like floods and earthquakes. Urban risk assessment maps and appropriate land-use profiles are needed to increase the resilience of our cities to these disasters [1]. Vertical urban growth or urban volume is one such evolving measure of an urban land-use profile [2]. Traditionally, building heights were assessed from maps showing the floor-area ratio derived from land transaction cases and land-use update surveys [3], statistical yearbooks [4], aerial photos, and local agency-supplied maps [5]. Increasingly, digital building height models (DBHs) generated from remote sensing techniques are becoming a popular technique for monitoring the urban environment. Digital building heights have several applications, such as modeling urban expansion [5], extracting and reconstructing buildings [6], simulating air pollution dispersion [7], estimating energy consumption [8] and solar potential [9], observing heat islands [10], flood hazard zoning [11], assessing GPS performance [12], and many others. Furthermore, if building heights from different time periods are available, they can also provide information about policy effects on horizontal and vertical urban growth [3,4,13].

Digital building height (DBH) is extracted from a digital surface model (DSM). A DSM is obtained from airborne laser scanning [14], high-resolution stereo image pairs [15], or interferometric

SAR (synthetic aperture radar) pairs [16]. Of these technologies, airborne laser scanning (ALS) has the highest accuracy in parameterizing building morphology, ranging from simple footprint identification [17] to complicated 3D structure and roof plane modeling [14,18]. State-of-the-art ALS approaches have also achieved very high accuracy in complex urban environments by integrating aerial imagery [19], city administrative data [20], architectural knowledge [21], and the Big Data approach [22].

Despite these promising results, there have been relatively few published studies on such methods being applied to large areas [23]. Furthermore, ALS data sources and aerial images are often under the control of government ministries, and, due to high operational costs, they are not available in many parts of the world [24]. Since several such regions are also undergoing rapid urban growth and will potentially face the associated adverse environmental impacts and safety concerns, it is necessary to monitor their urban volumes or building heights. At the same time, the quality and quantity of satellite images as well as the capabilities of sophisticated algorithms for DSM and DBH computations have increased dramatically in recent years [25]. Although such high-resolution satellite datasets are available for a fraction of the cost compared with ALS data, they are prohibitively expensive to obtain at the global scale. Despite various applications for building height data, there is still no such global dataset available that is comparable to the 'Global Rural-Urban Mapping Project (GRUMP) Urban Extents Grid, v1' [26,27] or the 'Global Urban Heat Island (UHI)' dataset [28]. Being able to derive building heights at a global scale is crucial not only for places that lack access to such data but also for global climate model simulations Zhang et al. [29], population distribution mapping [30], and other useful applications. Fortunately, freely available but coarse-resolution global DSMs, such as SRTM (Shuttle Radar Topography Mission), ASTER (Advanced Spaceborne Thermal Emission and Reflection Radiometer), and AW3D30 (ALOS (Advanced Land Observing Satellite) World 3D 30 m Resolution DSM) also exist, and they present an attractive opportunity to explore their application to urban areas. Presently, open DSMs are used mostly for large-scale regional geomorphological studies, such as earthquake and flood inundation modeling, among others. [31]. Studies have compared and revealed the relative merits of open DSMs of diverse geomorphological terrains [32–35] as well as urban areas [36]. However, the effectiveness of using open DSMs for urban applications is largely unexplored, so it is unclear whether the established accuracies of open DSMs would translate to similar accuracies of their corresponding DBHs.

The possibility of extracting building heights from open DSMs was first alluded to by Nghiem et al. [37] with regard to using SRTM for large-scale area mapping. When the SRTM DSM was examined in Los Angeles by Gamba et al. [38] and in Baltimore City by Quartulli and Datcu [39], they concluded that the SRTM digital elevation model (DEM) could be used for detecting tall buildings. Since then, other global DSMs, including AW3D30, TanDEM-X (TerraSAR-X Add-On for Digital Elevation Measurements), and ASTER Global DEM (GDEM), have also been shared publicly, but it is not clear whether they are suitable for the detection and estimation of DBHs. Each DSM has a different acquisition period and acquisition method. For example, AW3D and TanDEM-X were acquired around the early 2010s, while ASTER and SRTM were acquired in the early 2000s. This affects what can be 'seen' in these DSMs. Given what is available, this decadal period could be significant for studying 3D urban changes in rapidly growing cities [40–42]. Although the DSMs AW3D30 and TanDEM-X are known to be vertically more accurate than ASTER and SRTM, a DBH comparison is needed to establish the extracted height accuracy and its limitations. With sufficient accuracy, the models could be deployed at a global scale. To the best of the authors' knowledge, only Wang et al. [43] has attempted to address these challenges so far.

Objective

The objective of this study was to compare the accuracy of digital building heights (DBHs) extracted from open DSMs (AW3D30, ASTER, and SRTM).

2. Methodology

The flowchart of the datasets and methodology used for DBH estimation and comparison is shown in Figure 1. Briefly, the AW3D5 digital building height model (DBH) was validated with respect to the GeoEye DBH and TanDEM-X DBH, the degradation of the height accuracy from the fine-resolution AW3D5 DBH to the coarse-resolution AW3D30 DBH was assessed, and the terrain model was compared with the AW3D30, ASTER, and SRTM DBHs.

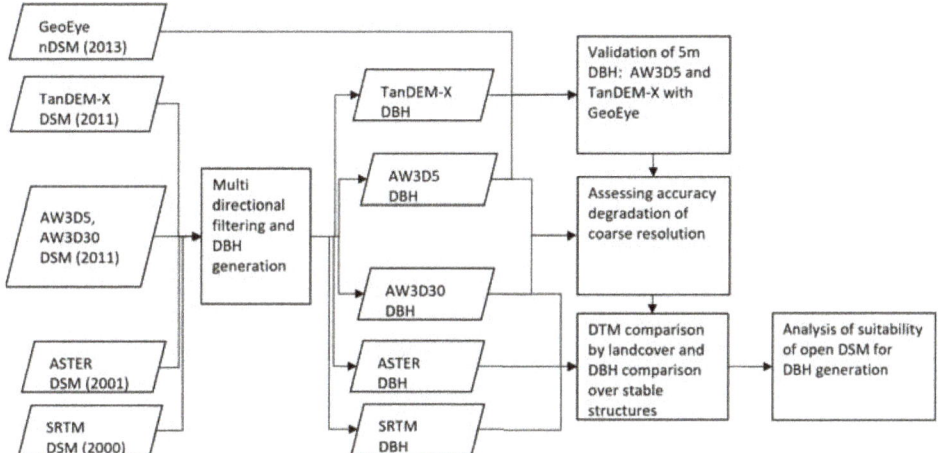

Figure 1. Flowchart outlining DSM data and processing.

2.1. Study Site

Yangon city (Figure 2a), the former capital of Myanmar, was selected as the study area due to its intense urban expansion within the last two decades. As per the 2014 Myanmar population census [44], urban Yangon has 5.16 million inhabitants. This is an increase of 85% over 1983 estimates. In roughly the same period between 1979 to 2009, Yangon's urban area experienced about a 5-fold expansion [45], most of which took place within the last decade. Apart from this, Yangon lies in one of the world's most disaster-prone countries. Yangon is situated on hilly terrain surrounded by a river and is at high risk of earthquakes and floods. The country was affected by Cyclone Nargis in 2008 and the Shan State Earthquake in 2011, which displaced several thousand people. Alarmingly, simulations of future urban expansion have shown that development will continue in flood-prone and earthquake-risk areas [46]. A land cover map of Yangon that shows built-up areas, water-bodies, vegetation, and fallow land for the year 2015 is presented in Figure 2b. Land cover types were classified using cloud-free Landsat-8 surface reflectance imagery available in Google Earth Engine [47]. In this paper, the fallow-land class refers to non-cultivated agricultural land and other bare lands, while the vegetation class refers to both forests and agricultural land with crops. Central Yangon has seen vertical expansion in the form of the construction of several new buildings alongside the older industrial, residential areas and colonial buildings. Rapid horizontal expansion has taken place from the center to periphery, stretching the built-up boundary.

(a) Yangon, Myanmar

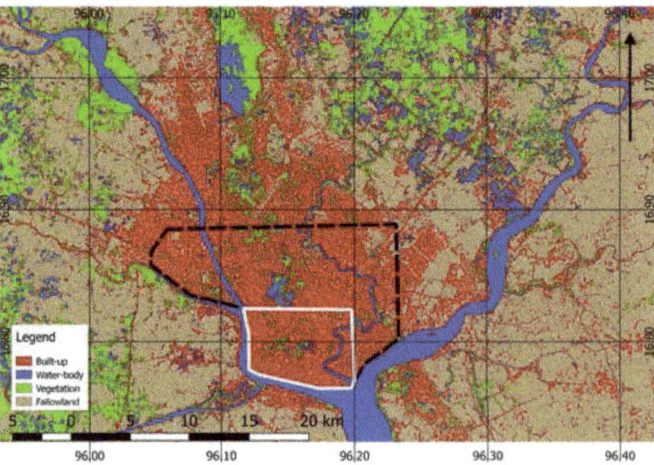

(b) Land cover map in 2015, generated from Landsat 8

Figure 2. Location of study site (**a**) Yangon, Myanmar. In (**b**), the central Yangon region is shown in the solid white polygon. The total region within the black dashed polygon and solid white polygon was used for dataset comparison.

2.2. Data Used

SRTM: The Shuttle Radar Topography Mission (SRTM) DEM was an international effort led by NASA and NGA (US National Geospatial Agency). The DSM was processed from C-band and X-band radar imagery collected from two antennae atop the Space Shuttle in an 11-day mission in February 2000 [48] and had an absolute vertical accuracy of less than 9 m [49]. Until 2014, the global dataset was available at a 3-arcsecond posting for regions outside of the US. In 2015, the LP DAAC (Land Processes Distributed Active Archive Center) released the NASA SRTM Version 3.0 Global 1-arcsecond dataset (SRTMGL1) [50]. At a global scale, the 1-arcsecond version (SRTMGL1) has the same root-mean-square error (RMSE) of 10.3 m as its 3-arcsecond version [51]. Its RMSE ranges from 5.9 m in urban areas to 10.4 m in bushland [32,52]. In this research, the 1-arcsecond (approximately 30 m at the equator)

SRTMGL1 was used and is subsequently referred to as SRTM. It is available from NASA's Earth Explorer website [53].

ASTER: Advanced Spaceborne Thermal Emission and Reflection Radiometer (ASTER) Global Digital Elevation Model Version 2 (GDEM V2) dataset is a DSM from NASA and Japan's Ministry of Economy, Trade and Industry (METI). It is freely available at a 1-arcsecond posting from NASA's Earth Explorer. The DSM was generated from nadir and backward-looking visible and near-infrared imagery from the ASTER sensor aboard NASA's Terra satellite. It was compiled from over 1.5 million scenes acquired between 2000 and 2009 and released in 2011 [54]. GDEM V2 is an improved version of the earlier GDEM V1 in terms of spatial resolution and coverage, water body mask, and horizontal and vertical accuracy [55]. Still, it contains disturbances in the values due to an increased frequency of noise on account of using a smaller correlation kernel to enhance the horizontal resolution. The RMSE accuracy of the ASTER GDEM changes with location [32,56] and is influenced by the land cover type, varying from 15.1 m in forested mountainous areas [54] to 23.3 m in urban areas [57]. In this study, ASTER GDEM V2 was used and is further referred to as ASTER.

TanDEM-X: TanDEM-X (TerraSAR-X Add-On for Digital Elevation Measurements) was launched in 2010 by the German Aerospace Center (DLR) with the aim of generating WorldDEM, a consistent global DSM. Its identical twin, TerraSAR-X, was launched earlier in 2007, and both satellites collect microwave imagery with X-band single-polarized SAR antennae. A uniqueness of this mission is that data collection takes place in a bistatic mode, in which both the satellites orbit with a short baseline and acquire data at the same location and same time. This helps to greatly reduce the effects of atmospheric disturbances. Marconcini et al. [58] demonstrated promising results of building height extraction over the Yellow River Delta, China using preliminary TanDEM-X DEM. Wessel et al. [59] validated the 12 m resolution TanDEM-X DEM with GPS measurements scattered over the United States and established its RMSE accuracy for urban (1.4 m) and vegetation areas (1.8 m). Its vertical RMSE over the mostly urban Tokyo was evaluated as 3.2 m [60], with higher errors occurring over built-up and vegetation classes. The final WorldDEM is publicly available at a 90 m resolution. The 12 m and 30 m resolution versions are freely available for research proposals (through DLR) and are priced for commercial use (through Airbus Defence and Space company). As part of a research project, a pair of TanDEM-X HH polarization images in ascending orbit were acquired in StripMap mode (ground spatial resolution between 2 and 3 m) for 6 September 2011. The incidence angle of the master image was 44.57° with a height of ambiguity of 50.14 m. A 12 m TanDEM-X InSAR DSM was generated in [60] and upsampled to a 5 m resolution for comparison with other DSM products.

AW3D: The ALOS World 3D (AW3D©JAXA) DSM, publicly released by JAXA in 2016, is the most recent DSM considered in this paper. The AW3D DSM was generated using images from PRISM's (Panchromatic Remote-Sensing Instrument for Stereo Mapping) front, nadir, and backward-looking panchromatic bands aboard ALOS (Advanced Land Observing Satellite). PRISM sensors were in operation between 2006 and 2011 and acquired imagery at a 2.5 m resolution which was processed with a 5 m grid spacing to generate a global elevation dataset, AW3D [61]. The AW3D DSM is commercially distributed at a 5 m resolution, while a 30 m downsampled dataset (known as 'AW3D30') is publicly available. The AW3D DSM generally meets the 5 m RMSE target height accuracy as per its producers [61]. However, Takaku et al. [61] found slope-dependent errors, with errors greater than 5 m occurring for slope angles larger than 30 degrees. Using longitudinal profiles of airport runways, Caglar et al. [62] found that AW3D30 has an RMSE of 1.78 m and contains an elevation anomaly due to sensor noise and the processing algorithm. Takaku et al. [61] found a mostly positive bias, while Caglar et al. [62] identified a negative bias in elevation estimation. In the Philippines, AW3D30's RMSE varies from 4.3 m in urban areas to 6.8 m in areas with dense vegetation [32]. Estoque et al. [63] found that heights filtered from the AW3D5 DSM are more accurate for lower buildings (e.g., ground truth building height <100 m) in less dense cities than for high-rise buildings and denser cities. In this research, a commercial 5 m DSM [64] was obtained as part of the research project, while the freely distributed 30 m AW3D DSM was downloaded from [65]. The 5 m resolution and 30 m resolution AW3D DSMs are henceforth referred to as AW3D5 and AW3D30, respectively.

Reference data: Ideally, the heights obtained from ground control points should be used as references. A higher-resolution surface model can also be used as a reference when ground control data are unavailable [66]. A high-resolution DSM was generated from 0.5 m resolution commercial GeoEye-1 stereo image pairs acquired in 2013 over Yangon. The DSM was then resampled to 4 m using PCI Geomatica 2015 software. The digital terrain model (DTM) was extracted by the in-built Wallis filter, which is a local adaptive filter that is useful for areas with significant shadow. The DSM generated with GeoEye-1 image pairs has a vertical RMSE accuracy ranging from 0.57 m in flat areas to 0.87 m in urban areas [67]. The completeness of the DSM in urban areas is 63.23% due to occlusion resulting from a high base/height (B/H) ratio (ratio of the image-pair distance to the height of the sensor) and the convergence angle of the imaging geometry [67]. In the pair used in this research, the stereo images also had different acquisition times that affected the quality of the generated DSM over some locations. For example, inaccurate matching was generated over the pagodas constructed with metallic roof plates, as they appeared differently in the stereo-pair due to the changed sun-view angle. This led to improper registration and erroneous height estimation.

Stable structures: Since open DSMs (AW3D30, ASTER, and SRTM) were acquired in different years, their DBHs cannot be compared directly in a fast-developing city like Yangon. To overcome this limitation, 'stable structures' were identified for comparison. These structures are those buildings that were consistently present between 2003 and 2011 and can be identified visually from historical imagery in Google Earth Pro software. The year 2003 is the earliest year for which high-resolution optical imagery is available. Care was taken to select only those structures that appear without any errors in the GeoEye DBH. In total, 52 'stable structures' were identified, which included large pagodas and temples, colonial buildings, a palace, government offices, a sports complex, large hotels, and residential apartments. Some examples are shown in Figure 3. A polygon was drawn manually around each stable structure's footprint.

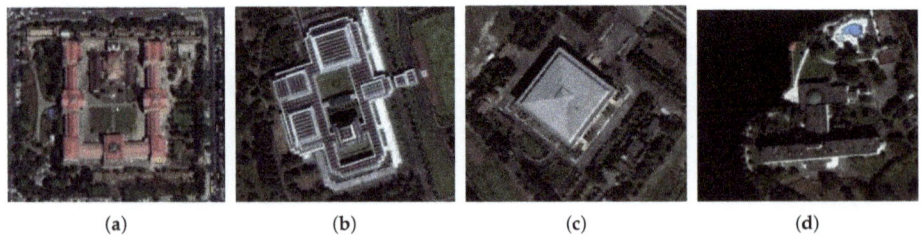

Figure 3. Examples of 'stable structures' in Yangon whose heights were compared among DSMs from different years: (**a**) Secratariat Office, (**b**) Parliament building, (**c**) Thuwunna Stadium, and (**d**) Inya Lake Hotel. The structures of these buildings remained consistent throughout our study duration, 2000–2011.

All DSMs used in this research are summarized in Table 1. All DSMs and DBHs were referenced to the World Geodetic System (WGS84) horizontal datum and Earth Gravitational Model 1996 (EGM96) vertical (geoid) datum. A highly accurate image registration that is precise to each pixel is desirable for comparison. Since the DSMs were originally not georegistered with each other, we co-registered each DSM and DBH with the reference GeoEye DSM. Thirty ground control points for high-resolution DSMs and 15 tie-points spread evenly over the study area were selected for each co-registration. This was performed in the map registration module of the software ENVI4.7 (Exelis Visual Information Solutions, Boulder, CO, USA) using a rotation, scaling, and translation technique, followed by cubic convolution resampling. Separate co-registration of DSMs and DBHs was done to prevent the influence of interpolation on height estimation.

Table 1. Summary of DSMs used in this research. Vertical accuracy refers to the accuracy of the DSM reported by other studies over all classes. For GeoEye DSMs, vertical accuracy was strictly over urban areas. SRTM: Shuttle Radar Topography Mission, ASTER: Advanced Spaceborne Thermal Emission and Reflection Radiometer.

DSM	Resolution	Acquisition Period	Vertical Accuracy (m)	Remarks
SRTM	30 m	2000	5.9–10.3 m	Open dataset acquired with InSAR
ASTER	30 m	2000–2009	15.1–23.2 m	Open dataset acquired with stereo photogrammetry
TanDEM-X	12 m	6 September 2011	1.6–6.2 m	Closed dataset acquired with InSAR, open for research purposes
AW3D	5 m, 30 m	2006–2011	1.7–6.8 m	Commercial and open dataset available, generated with stereo photogrammetry
GeoEye	0.5 m	16 November 2013	0.57–0.87 m	Generated from commercial high-resolution stereo-pairs

2.3. DBH Generation

There are several types of building extraction based on the desired or possible details, ranging from building footprints to building roof contours [23]. As per the study objective and data limitations, the focus was on building height extraction. A DBH is different from a digital building model (DBM), which is a more comprehensive 3D representation of buildings and includes all aspects of the building geometry [6]. DBH is considered a normalized DSM (nDSM) over built-up class pixels. An nDSM is calculated as the difference in elevation values between the DSM and DTM (digital terrain model, also known as a bare earth model). The extraction of an nDSM requires distinguishing ground from non-ground pixels by generating a DTM. Most algorithms first generate the DTM from a photogrammetric DSM by identifying pixels which are part of the local terrain [68]. There are several methods for identifying non-ground pixels, but they often assume that the terrain is smooth and that a large height difference exists between neighboring ground and non-ground points [69]. Deep learning approaches have resulted in high-accuracy building extraction (overall accuracy > 95%), with very high resolution imagery [70,71]. However, these networks are designed for small-sized images (e.g., 256 × 256 pixels, 512 × 512 pixels, etc.) to prevent memory overloading, which can produce discontinuous artifacts [72]. Many such models rely on a fully connected neural network [73], which is a pre-trained model using an RGB image repository (Imagenet [74]) and exploit similar features between the RGB intensities and the depth images, such as edges, corner, and end-points [72]. In the case of a coarse-resolution DSM, such features are not clearly visible, and we were skeptical of their performance with coarse resolution. Recognizing these possible limitations, a morphological approach—a multi-directional processing and slope-dependent filtering technique called 'MSD filtering' [75]—was used for DTM generation in light of its consideration of the terrain slope and overall simplicity in implementation [69]. The MSD filtering technique is an extension of a similar technique developed for an ALS DSM [76]. MSD filtering is effective over hilly terrains with slopes for extracting a DTM with a sub-meter high-resolution DSM [75]. An enhancement of MSD filtering, the 'network of ground points' technique, also exists [77] and does not need to consider the slope angle. However, as admitted by Mousa et al. [77], this probably holds true only for very high resolution DSMs. Therefore, we implemented the MSD method instead of the 'network of ground points' method. MSD filtering has also been used to generate a DTM for the alignment of high-resolution optical and SAR images in urban areas [78].

The MSD filtering technique requires four parameters to generate a DTM: the Gaussian smoothing kernel size, the scanline filter extent, the height threshold, and the slope threshold. Each DSM pixel was checked to determine whether it should be considered ground by comparing it with other pixels within the predefined neighborhood scanline filter extending in eight directions. If the pixel was identified as a ground pixel in more than five directions, it was labeled as a terrain pixel by the majority voting method. To draw the comparison, a local reference terrain slope was first generated by 2D Gaussian smoothing. Then, the pixel's height was compared with the lowest elevated pixel within the scanline filter extent. If this height difference was more than the height threshold parameter, the pixel was classified as a non-ground pixel. Then, if the slope difference between the current and the successive pixel in the scanline direction was greater than the slope threshold, it was labeled as a non-ground pixel. If the slope was positive and less than the slope threshold, then that pixel was given the same label as its previous pixel. Otherwise, that pixel was labeled as ground. This resulted in a raster with only ground points and holes, the latter being locations where non-ground points exist. Thereafter, a linear interpolation technique from the 'SciPy' module of Python [79] was used to fill the holes for generating the DTM. The nDSM was generated by subtracting the DTM from its DSM.

Parameter Selection

The GeoEye nDSM was used as a reference to choose suitable parameter values for the scanline extent, height threshold, and slope threshold. The parameters for the Gaussian smoothing filter were set to a 100 m kernel size and a 25 m standard deviation to generate the initial local terrain. After trying various combinations of height difference thresholds and slope thresholds, 3 m and 30° were chosen, respectively, as they captured the greatest number of structures. A 3 m height difference threshold approximately corresponds to a one-story construction. A lower value of the height difference threshold leads to underestimation, while higher values lead to an overestimation of the ground terrain. One drawback to the MSD scanline approach arises when no ground pixels lie within the eight directional scanlines [77]. This can happen when a structure is contiguous and larger than the scanline extent. The neighborhood scanline filter extent parameter was stretched beyond 100 m for a greater chance of successfully 'finding' a ground pixel. This ensured more chances to observe a ground pixel within the scanline since any contiguous urban structure is unlikely to be larger than 100 m in all scanline directions.

The AW3D5 nDSM was generated with a scanline extent of 300 m, a height threshold of 3 m, and a slope threshold of 30°. Setting a lower value for the scanline filter extent (<300 m) underestimated the structures' footprints and also their heights, e.g., a scanline extent of 150 m resulted in a lesser overall mean height estimation by 0.2 m when compared with the DBH generated with a scanline filter extent of 300 m. This was more pronounced for tall structures. Similarly, the TanDEM-X nDSM was generated with a scanline extent of 100 m, a height threshold of 3 m, and a slope threshold of 30°. The same parameters used for AW3D5 were deemed fit to extract the nDSM from AW3D30 and ASTER GDEM v2. Due to the low differentiation between ground and non-ground points in SRTM, the height threshold parameter was lowered to 2 m. In the AW3D5 and TanDEM-X nDSMs thus generated, about 10% of the pixels had negative heights, out of which 90% of the values were between −1 m and 0 m. In the SRTM, ASTER, and AW3D30 nDSMs, 20% of the pixels had negative values, out of which 90% were between −2 m and 0 m. These negative heights were removed.

2.4. Vertical Accuracy Assessment

There are several accuracy metrics for roof level and roof plane level evaluations [80]. Recent additions include shape similarity and positional accuracy metrics [81] and a threshold-free metric based on the overlap between extracted and reference roof planes [80]. However, the coarse DBH imposes limitations due to which such advanced metrics cannot be applied. For example, in a 30 m gridded DBH, roof planes are not visible except on very large structures that span several hundred meters. Therefore, pixel-based and object-based height accuracies were evaluated with conventional statistical metrics. Object-based heights were derived as mean pixel heights within the footprint polygon of each stable structure. The vertical accuracy of the estimated datasets (DBH and DTM) was analyzed by calculating the descriptive statistics of the difference between the estimated height and the reference height. These statistics were the root-mean-square error (RMSE), mean error (ME), mean absolute error (MAE), and standard deviation (SD). The RMSE describes how much the estimated dataset differs from the reference dataset in terms of deviation from zero. The ME describes the bias toward underestimation (negative ME) or overestimation (positive ME) with respect to the reference dataset. The SD represents the distribution of errors from the mean error (for normally distributed errors, the mean error is zero). So, a low SD value means less variation in error magnitudes. For any DBH or DTM, Z_D was extracted from DSM D with an image containing n pixels or objects, and its error metrics with respect to the reference DBH or DTM Z_{ref} were calculated as shown in Equations (1)–(4).

$$RMSE = \sqrt{\frac{\sum_{i=1}^{n}(Z_D - Z_{ref})^2}{n}} \quad (1)$$

$$ME = \frac{\sum_{i=1}^{n}(Z_{D_i} - Z_{ref_i})}{n} \quad (2)$$

$$MAE = \frac{\sum_{i=1}^{n}|(Z_{D_i} - Z_{ref_i})|}{n} \quad (3)$$

$$SD = \sqrt{\frac{\sum_{i=1}^{n}(Z_{D_i} - Z_{ref_i} - ME)^2}{n-1}} \quad (4)$$

Finally, in accordance with Rutzinger et al. [82], the correspondence of a building footprint within the stable structure polygon was checked pixel-wise. For this method, true positive (*TP*, when the footprint exists in the reference as well as in the DBH), false negative (*FN*, when the footprint is incorrectly identified as ground), true negative (*TN*, when the footprint is correctly identified as ground), and false positive (*FP*, when a ground pixel is identified as a footprint) pixels within each stable structure polygon were identified. The completeness and correctness was computed according to Equations (5) and (6).

$$completeness = \frac{||TP||}{||TP|| + ||FN||} \quad (5)$$

$$correctness = \frac{||TP||}{||TP|| + ||FP||} \quad (6)$$

where $||.||$ denotes the number of pixels.

3. Results and Discussion

3.1. Comparison of AW3D5 and TanDEM-X DBH

The AW3D5 DBH and TanDEM-X DBH along with the GeoEye DBH are shown in Figure 4. Also, the profiles at three example locations (A, B, and C) for the DSMs and DBHs from AW3D5, TanDEM-X, and the reference DBH are shown in Figure 5. Location A contains low-height buildings, while location C contains relatively taller buildings. Location B in Figure 5c shows a hilly terrain consisting of Yangon's most important historic landmark, the Shwedagon Pagoda (distance mark: 580 m). It is a conical structure situated atop the highest elevated location. A visual comparison showed that compared with the GeoEye DBH, the AW3D5 DBH underestimated the height of single-story buildings, which were generally 3–4 m tall (in the Eastern portion of Figure 4c), but it better captured the heights of tall structures (in the Southern portion of Figure 4c). Generally, the TanDEM-X DBH had a similar profile trend to that of the AW3D5 DBH but did not capture the heights of tall structures. In the TanDEM-X DSM, and consequently, in its DBH, some skyscrapers and pagodas appeared as a tall inclined wall immediately followed by a hole. Since such locations were not flagged by TanDEM-X's data consistency mask, it can be assumed that this is due to the local phase unwrapping errors that result in shadows or noise [83]. Many shadow pixels were also observed in tall buildings due to layover that resulted in incomplete footprints, 'ramps', and overall height underestimation. Also, buildings located along the azimuth direction were severely affected by layover issues. On the other hand, short structures were devoid of such artifacts. The reference GeoEye DBH also showed buildings that were missed due to the incorrect registration of high-resolution stereo-pairs over tall structures on account of occlusion, shadows, or high parallax error [84,85], although there were far fewer in number than those missed by the TanDEM-X DBH. The accuracies of the AW3D5 and TanDEM-X DBHs with respect to the GeoEye DBH are shown in Table 2. For the AW3D5 DBH, built-up areas had an RMSE of 3.55 m. This RMSE almost corresponds to the single-story high structures often seen in residential areas. The AW3D5 DBH had a negative ME (-1.55 m), which points to an underestimation bias by AW3D5. This implies that height estimation from AW3D5 over residential areas is unreliable. Further, some locations with tall buildings had large height differences on account of the low GeoEye accuracy, as seen in Figures 4 and 5e (distance mark: 580 m). So, it is possible that the RMSE of the AW3D5 DBH could be even lower than 3.55 m if a more accurate reference DBH were used. Nonetheless, this value is within the desired producer RMSE of 5 m. The TanDEM-X DBH captured more short height structures (height < 5 m) than did the AW3D5 DBH. Its accuracy measures were slightly better than those of the AW3D5 DBH in all respects except the SD (Table 2). Its almost zero ME but positive MAE could also point to a higher occurrence of random errors. Overall, the AW3D DBH provided a slightly higher RMSE (by 0.21 m), lower ME (by 1.51 m), lower MAE (by 0.12 m), and lower SD (0.15 m). The AW3D5 DBH also identified more tall buildings in densely built-up areas than did TanDEM-X DBH. However, this comparison is based on the TanDEM-X DSM generated from a single pair of images. With the use of TanDEM-X's final DEM which combines multiple acquisitions, better results can be expected.

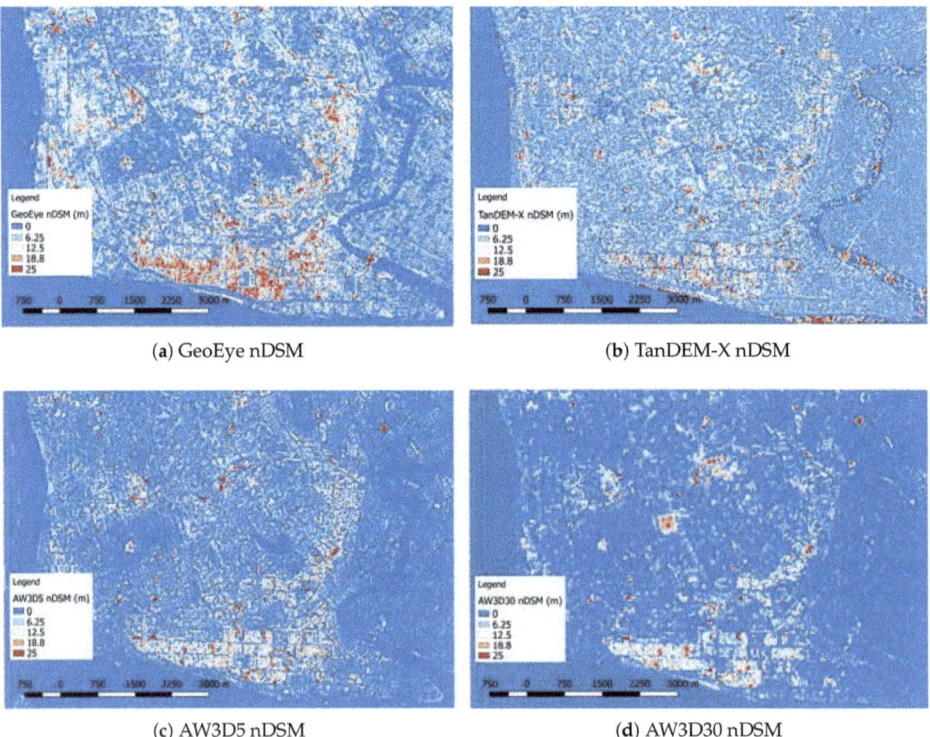

(a) GeoEye nDSM (b) TanDEM-X nDSM

(c) AW3D5 nDSM (d) AW3D30 nDSM

Figure 4. *Cont.*

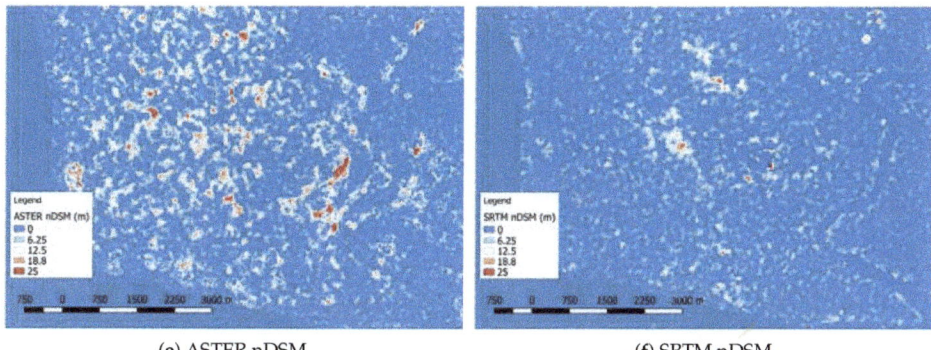

(e) ASTER nDSM (f) SRTM nDSM

Figure 4. The nDSMs extracted from (**a**) GeoEye stereo-pairs, (**b**) TanDEM-X, (**c**) AW3D5, (**d**) AW3D30, (**e**) ASTER, and (**f**) SRTM over central Yangon.

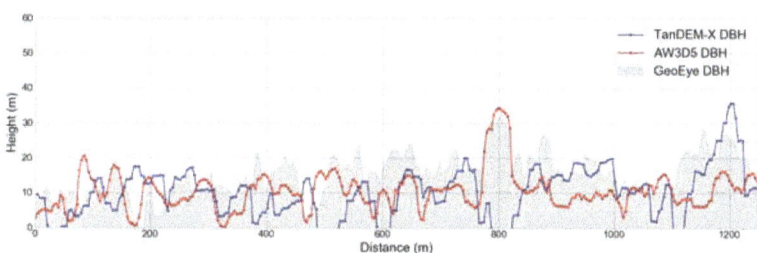

(a) Height profile at location A

(b) Optical image at location A

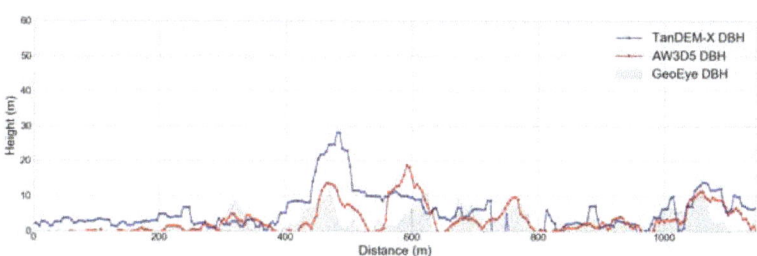

(c) Height profile at location B

(d) Optical image at location B

Figure 5. *Cont.*

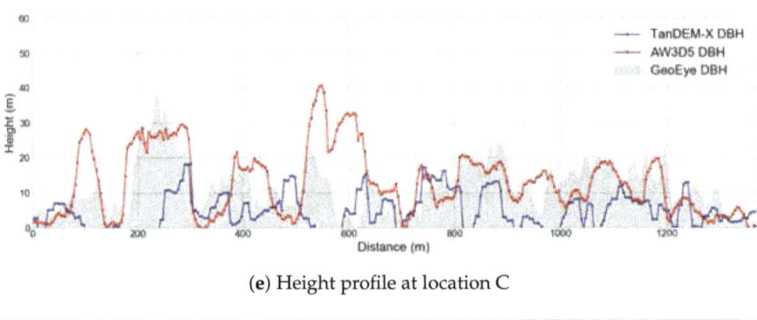

(e) Height profile at location C

(f) Optical image at location C

Figure 5. Digital building height model (DBH) profiles extracted from the AW3D5 digital surface model (DSM; red) and TanDEM-X (blue) were compared with GeoEye (solid gray) at three locations: (a) A, (c) B, and (e) C. (b) Location A contains low height buildings, (d) location B contains hilly terrain with tall structures, and (f) location C contains tall buildings.

Table 2. Descriptive statistics for the built-up class for the AW3D5 DBH and TanDEM-X DBH with respect to the reference GeoEye DBH.

DSM Source	RMSE (m)	ME (m)	MAE (m)	SD (m)
AW3D5	3.55	−1.55	1.99	3.20
TanDEM-X	3.35	−0.04	1.87	3.35

One interesting aspect is the base/height (B/H) ratio, which is defined as the ratio of the stereo-pair separation to the height of the sensor. A high B/H leads to improved vertical accuracy [86]. However, in a dense urban setting, a high B/H leads to increased occlusion and poor matching [87]. The PRISM imagery (from which the AW3D DSM was generated) had a higher B/H ratio of 1.0 than the GeoEye imagery, whose B/H ratio varied from 0.54 to 0.83. GeoEye also has a high off-nadir look angle (10°–35°) to provide fast and varied acquisition, which is likely to result in slanted buildings due to the perspective view. This suggests that in the AW3D5 DBH, a higher vertical accuracy is achieved at the cost of more occlusion, while the reverse is true for the GeoEye DBH.

3.2. Accuracy Loss in AW3D DBH with Resolution Degradation

Profile sections of the AW3D5 and AW3D30 DBHs derived from the original DSM (not co-registered with the GeoEye DBH to preserve original height) are shown in Figure 6. The AW3D30 DBH showed similar height variation to that of AW3D5 but in a much coarser fashion. Due to the coarseness of the AW3D30 DBH, fewer ground points were preserved, especially in street canyons between buildings. This inhibits identification of individual buildings compared with the case of the AW3D5 DBH. This can be seen in Profile A (Figure 6) at the 100 m and 500 m distance marks. Pixels over those locations are more likely to be considered non-ground, leading to a height overestimation when there is a sudden steep change in ground elevation. This is why at location B (Figure 6), the AW3D30 DBH showed a higher height immediately preceding (distance mark: 550 m) or following (distance mark: 700 m) a relatively tall structure (Shwedagon Pagoda, distance mark: 600 m).

Another concern of note is the impact of mixed pixels arising from the pixel grid when AW3D5 is downsampled to AW3D30. Several instances were observed when buildings in the AW3D5 DBH with a ground footprint of approximately 30 m or less was split into adjacent 30 m resolution pixels, each with a lower height than the original. An example is shown (Figure 7), where a 30 × 30 m^2 sized building was split into two pixels in the AW3D30 DBH. This is an unavoidable consequence of downsampling, and the split buildings appear to have less height in the 30 m resolution compared with the 5 m resolution. This can be seen in the original DSMs as well. This suggests that tall adjacent pixels seen in the AW3D30 DBH may have a smaller ground footprint than is estimated by the model.

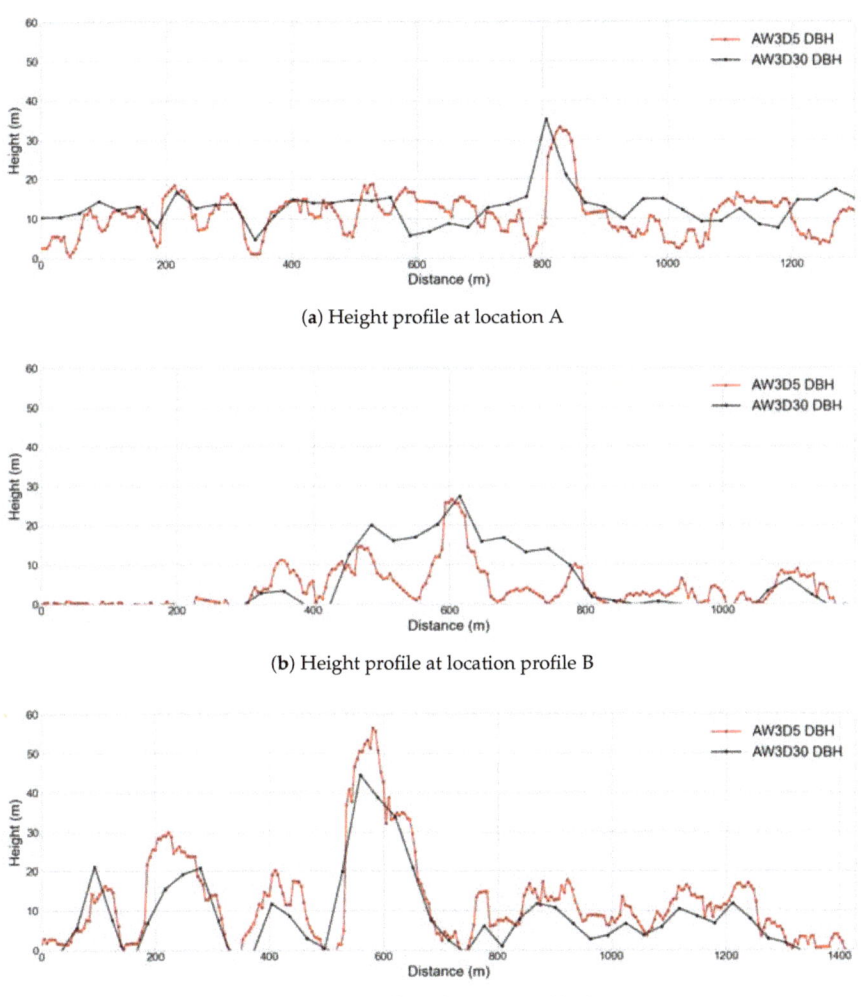

(a) Height profile at location A

(b) Height profile at location profile B

(c) Height profile at location profile C

Figure 6. DBH profiles extracted from AW3D products at a 5 m (AW3D5, in red) and 30 m (AW3D30, in black) resolution. Locations (**a**) A, (**b**) B, and (**c**) C refer to the locations shown in Figure 5.

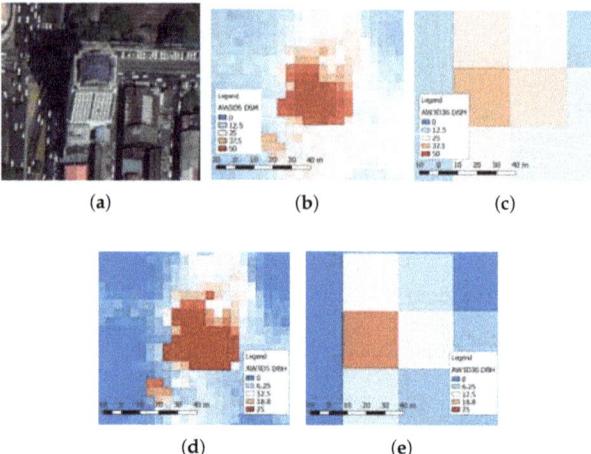

(a)　　　　　　　　(b)　　　　　　　　(c)

(d)　　　　　　　　(e)

Figure 7. Effect of mixed pixels on estimated height in AW3D30 over a tall structure (location: 16°46.74′N, 96°9.56′E) with a ground footprint of less than 30 m (**a**). The figures show that the height observed with AW3D5 fine resolution DSM (**b**) and DBH (**d**) is split into pixels with uneven heights with AW3D30 coarse resolution DSM (**c**) and DBH (**e**).

The descriptive statistics for the AW3D30 DBH with the AW3D5 DBH downsampled with a mean filter as a reference highlighted the effect of coarse resolution (Table 3). The RMSE of the AW3D30 DBH was impacted only slightly (by 0.79 m) over the original AW3D5 DBH. The ME and MAE were −0.03 m and 0.18 m, respectively, which points to a minor underestimation by the AW3D30 DBH. This could be due to the mixed-pixel issue highlighted in the previous paragraph. The non-visibility of several street canyons in the AW3D30 DBH could also have contributed to the SD.

Table 3. Descriptive statistics for the low-resolution AW3D30 DBH with the downsampled AW3D5 DBH as a reference.

	RMSE (m)	ME (m)	MAE (m)	SD (m)
AW3D30	0.79	−0.03	0.18	0.78

3.3. Comparison of DTMs from AW3D30, ASTER, and SRTM

The different acquisition time periods of the AW3D30, ASTER, and SRTM DSMs resulted in dissimilar DBHs due to surface changes over time. To ensure a robust comparison, the DTMs extracted from the DSMs were compared to assess the agreement between the three DTMs. This was performed by land cover types: built-up, vegetation, and fallow land. Only the DTM values for those pixels with the same land cover types in both the Landsat 7 (2001) and Landsat 8 (2015) images were considered. The DTM over the complete area shown in Figure 2b was considered so as to ensure sufficient representation of all land cover types. Comparison plots are shown in Figure 8. In general, low-elevation pixels (≤10 m) mostly belonged to fallow land and vegetation, mid-elevation pixels (10–30 m) belonged to built-up areas, and high-elevation areas (≥30 m) were mixed between vegetation and built-up areas. The SRTM and AW3D30 DTMs were fairly consistent with each other, having a low overall RMSE (1.85 m) with a high correlation of 0.97. Comparatively, the ASTER DTM showed a higher overall RMSE with AW3D30 (3.12 m) and SRTM (4.03 m) and a lower correlation with AW3D30 (0.88) and SRTM (0.87). From Figure 8b, it can be seen that ASTER overestimated low-elevation pixels but underestimated high-elevation pixels. The ASTER DTM over built-up pixels located at a higher elevation was underestimated compared with AW3D30 and could be a cause of the DBH inaccuracy in those regions. This suggests the presence of systematic errors, which could be locally resolved by the calibration of the ASTER DTM. Comparison of the DTMs over vegetation was inconsistent, resulting in the highest RMSE compared with the other classes. In the SRTM DTM, this was due to the C-band SAR sensor, which can penetrate the leaf foliage, resulting in a lower DSM elevation than that in the DSM from optical sensors. In built-up areas, the DTMs of AW3D30 and SRTM were more consistent with each other than they were with the ASTER DTM, as can be seen by their RMSE values in Table 4. The ASTER DTM had a lower RMSE (3.18 m) compared with vegetation (RMSE: 3.72 m) and fallow land (RMSE: 3.24 m) with respect to the AW3D30 DTM.

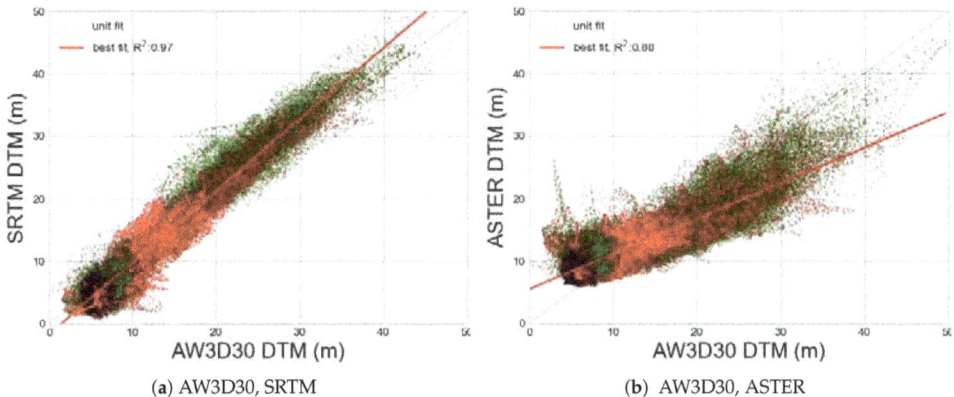

Figure 8. Comparison of AW3D30 DTM with (**a**) ASTER DTM and (**b**) SRTM DTM by class: built-up (red dot), vegetation (green dot), and fallow land (black dot).

Table 4. Descriptive statistics for AW3D30, ASTER, SRTM DTMs over the built-up land class.

DTM	RMSE (m)	ME (m)	MAE (m)	SD (m)	Correlation
AW3D30-SRTM	1.91	0.34	1.47	1.88	0.97
SRTM-ASTER	4.09	−1.08	3.27	3.95	0.87
AW3D30-ASTER	3.18	−0.75	2.46	3.09	0.88

3.4. Comparison of DBHs from AW3D30, ASTER, and SRTM

The DSM and DBH profiles of the SRTM, ASTER, AW3D30, and GeoEye DBHs are shown in Figure 9. To ensure comparison at the 30 m resolution, the GeoEye DBH was downsampled using a mean kernel and is henceforth referred to as GeoEye$_{mean}$. The SRTM DBH had a similar trend to the AW3D30 DBH with large height underestimations. Although, originally, the SRTM DSM was intended to identify natural topography, and man-made features are mostly absent, some buildings can still be identified in the SRTM DBH. The SRTM DBH mostly hovered around 0 m except when a large structure or several tall buildings with heights greater than 10 m in GeoEye$_{mean}$ DBH were present. The ASTER DBH also underestimated structure heights but was closer to the AW3D30 DBH and GeoEye$_{mean}$ DBH than it was to the SRTM DBH. It can be seen in Figure 9b that over the Shwedagon Pagoda stable structure (distance mark: 600 m), the maximum height was estimated by the AW3D30 DBH (31 m), followed by the ASTER DBH (14 m) and SRTM DBH (11 m). There were several locations where the ASTER DBH estimation was higher than that of the AW3D30 DBH and GeoEye$_{mean}$ DBH. This can be seen in Figure 9b at the distances marked 400 m and at 1100 m. In the former case, it is due to the nDSM generating the algorithm; in the latter, it is the presence of noise in the original ASTER DSM itself that contributed to overestimation.

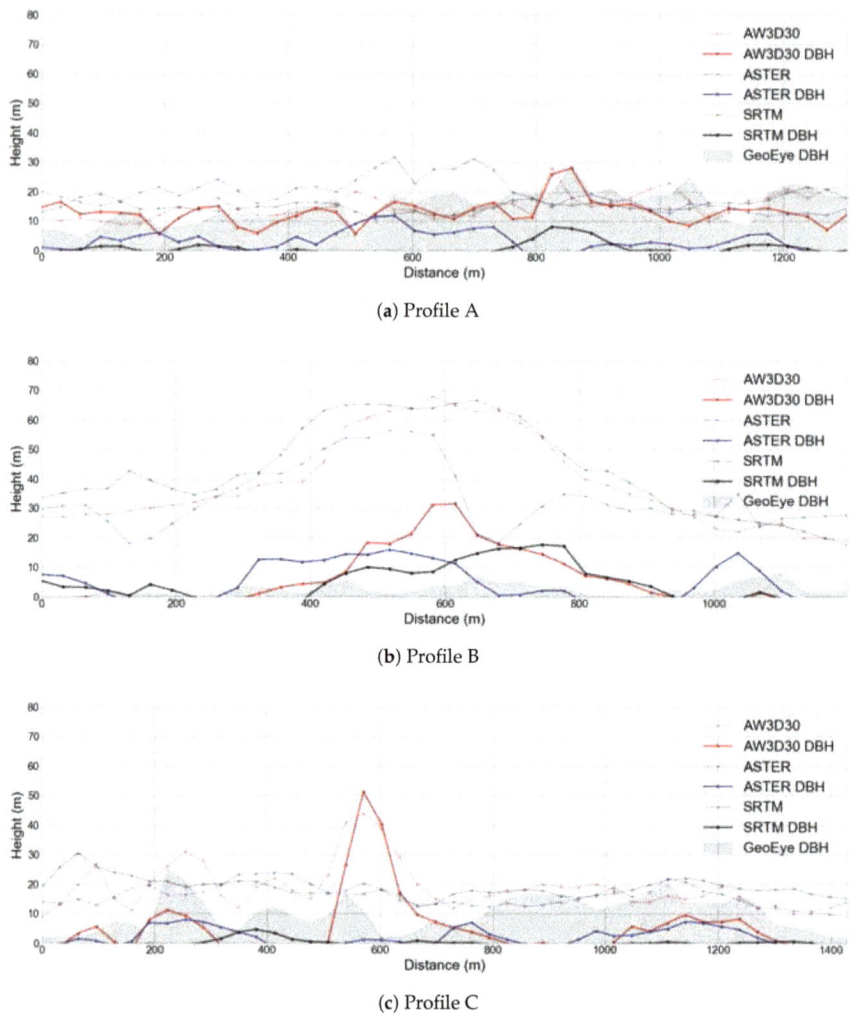

Figure 9. DSM and DBH profiles for AW3D30 (red), ASTER (blue), SRTM (black), and GeoEye (solid gray). Profiles (**a**) A, (**b**) B, and (**c**) C refer to the locations shown in Figure 5.

The statistical comparison of DBHs over stable structure pixels from the open DSMs and GeoEye is shown in Table 5. It is noteworthy that the mean pixel height of the AW3D30 DBH (ME: 9.14 m) was significantly higher than that of SRTM (ME: 3.10 m) and ASTER (ME: 5.49 m). The SDs of the AW3D30 DBH (6.40 m) and ASTER DBH (6.08 m) were closer to that of the GeoEye$_{mean}$ DBH (8.19 m) and much higher than that of SRTM (2.24 m), implying that the AW3D30 DBH and ASTER DBH show good variation in building heights. However, the high SD for the ASTER DBH may also be on account of noise generating anomalous heights. Along with the results of the previous section, we can conclude that a good agreement between a pair of DTMs does not imply a good agreement between their DBHs, e.g., SRTM and AW3D30 had a high correlation (0.97) but vastly different SD values. When pixel-based comparisons were made with GeoEye$_{mean}$ DBH, the ME values of the AW3D30 DBH and ASTER DBH were much lower than their MAE values. This, along with the sign of the ME, indicates that the AW3D30 DBH and ASTER DBH consistently underestimated the reference heights. This can also be seen in the scatterplots in Figure 10a–c. The coefficient of determination was low for the SRTM DBH (R^2: 0.14) and ASTER DBH (R^2: 0.22) but relatively high for the AW3D30 DBH (R^2: 0.60). This suggests that ASTER building heights were indeed noise artifacts, while the AW3D30 DBH represented reference building heights with some underestimation. The locations where the SRTM DBH was low (<0.3 m) despite high values in the GeoEYE DBH (>25 m) were checked. Such pixels belonged to dense neighborhoods with sloped metal-roofed buildings. On the other hand, a high SRTM DBH (>8 m) was estimated over large buildings with red brick or tiled rooftops. It is possible that shadowing and layover originating from sloped rooftops resulted in underestimation. The pixel locations where the AW3D30 DBH overestimated height compared with GeoEye$_{mean}$ were also probed. Such pixels belonged to locations with dense tall buildings and also those buildings where the footprint in GeoEye$_{mean}$ was smaller than the actual footprint. Interestingly, the RMSE of the AW3D30 DBH (8.69 m) with respect to GeoEye$_{mean}$ was not much degraded over its 5 m resolution version, i.e., the AW3D5 DBH (RMSE: 5.04 m). Assuming that a single-story building is about 3 m high, the RMSE of the AW3D30 DBH suggests that the AW3D30 DBH pixels can detect the height of buildings taller than 9 m or three stories.

Some instances of minor spatial misregistration (shift of 1–2 pixels) between the DBH datasets persisted despite several georegistration attempts. To overcome this limitation, object-based comparison of the mean height of each individual stable structure was performed. The summary of the statistics is shown in Table 5, and the scatterplot is shown in Figure 10d. The RMSE and MAE improved by about 2 m for each DBH when object-level statistics were computed. In Figure 10d, four outlying building heights can be seen for AW3D30, where the GeoEye DBH is between 7 and 12 m and the AW3D30 DBH is between 18 and 22 m. On closer inspection, it was found that these buildings were affected by blurring and small footprints in the GeoEye DBH. From the original GeoEye stereo image, we noticed that, in reality, these buildings were 6–10 stories high. This was determined by visually identifying the number of windows in the vertical direction. This suggested that the AW3D30 DBH was possibly correct over those locations, and the removal of these outlying building estimates increased the R^2 to 0.62. The object-based RMSE of the AW3D30 DBH (6.92 m) suggests that if building footprints are already known, the AW3D30 DBH is suitable for the height estimation of buildings taller than two stories.

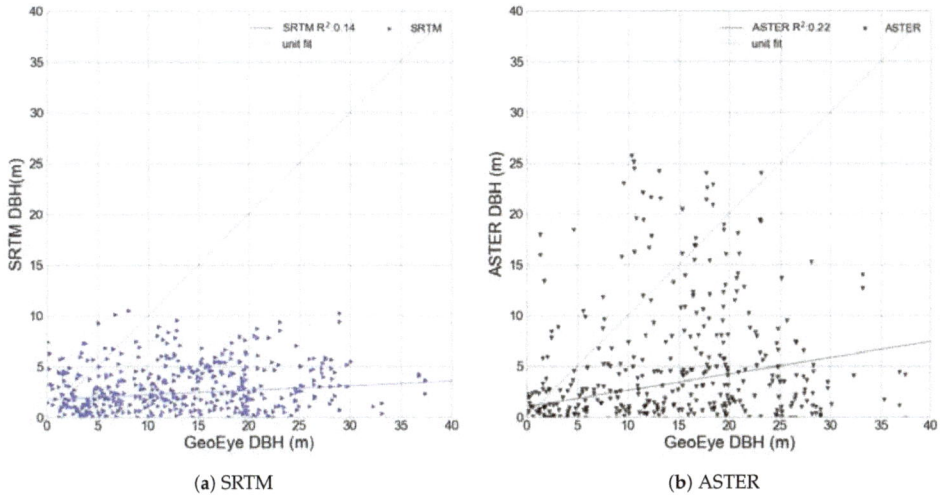

Figure 10. *Cont.*

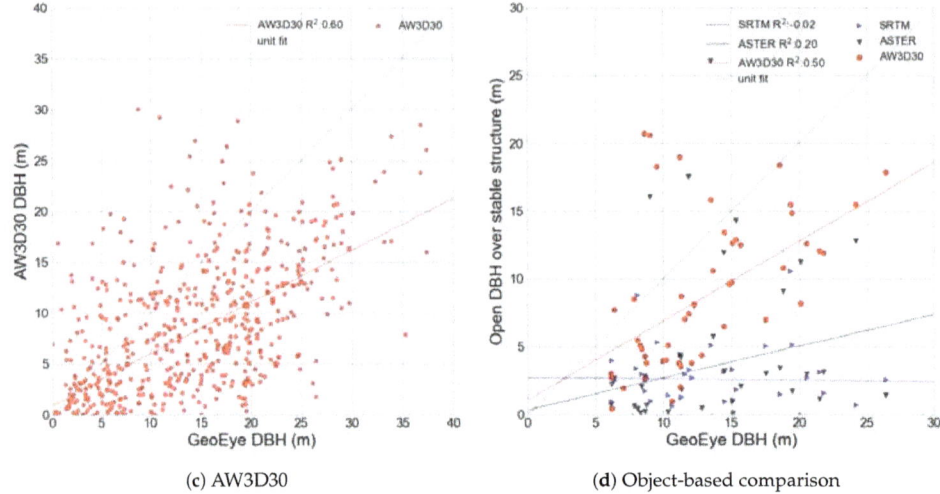

(c) AW3D30

(d) Object-based comparison

Figure 10. Scatterplot of DBHs of stable structures from open DSMs, with GeoEye DBH as a reference. Per-pixel comparisons were made for (**a**) SRTM, (**b**) ASTER, and (**c**) AW3D30. (**d**) Per-object height comparison for all the DSMs.

Table 5. Pixel-based and object-based descriptive statistics over stable structures for SRTM, ASTER, AW3D30, and GeoEye DBHs. All units are in meters. Minimum and maximum heights are denoted by 'min' and 'max', respectively. GeoEye$_{mean}$ refers to the GeoEye DBH downsampled to a 30 m resolution using a 'mean' filter. Statistical measures denoted by '–' are not applicable.

	Min	Max	RMSE	ME	MAE	SD
DBH						
SRTM	0.04	10.52	–	3.10	–	2.24
ASTER	0.01	25.71	–	5.49	–	6.08
AW3D30	0.02	30.06	–	9.14	–	6.40
GeoEye$_{mean}$	0.03	37.41	–	13.06	–	8.19
Pixel-based						
SRTM-GeoEye$_{mean}$	−35.96	7.46	13.50	−10.65	11.06	8.31
ASTER-GeoEye$_{mean}$	−37.55	16.60	13.35	−9.88	10.96	8.99
AW3D30-GeoEye$_{mean}$	−27.40	21.23	8.88	−5.61	7.27	6.89
Object-based						
SRTM-GeoEye$_{mean}$	−23.86	4.30	11.94	−10.34	10.53	6.04
ASTER-GeoEye$_{mean}$	−24.98	12.58	11.68	−9.55	10.54	6.80
AW3D30-GeoEye$_{mean}$	−16.06	12.08	6.92	−4.31	6.06	6.92

Regarding the average completeness of the building footprint, buildings in the AW3D30 DBH were 85.34% complete, followed by SRTM (82.12%), and ASTER (64.82%). This suggests that the AW3D30 DBH detected building footprints with a sufficiently high degree of completeness. Surprisingly, the SRTM DBH had a similar completeness rate to that of AW3D30, both of which were much better than that of ASTER DBH. This suggests the utility of the SRTM DBH in estimating the presence of buildings, although at a much lower height estimation. Upon examining the buildings with low footprint completeness, it was found that these mainly consisted of buildings with complicated rooftop structures and conical pagodas. However, each DBH product was affected to different extents by these complications. This is likely due to feature and intensity matching errors from complicated rooftop reflectance during the production of the original DSM. The use of the three images was effective in reducing such errors during DSM generation compared with using only two images [88]. The ASTER

DBH could be severely affected by this factor, as its DSM was generated from only paired images compared with the triplet images used for AW3D DSM generation. It was not possible to compute the correctness satisfactorily, as buildings in the coarse DBH were clearly separated from each other, resulting in unrealistically high FP values.

Since the DBH generated from the DSMs were acquired during different time periods, this points to the possibility that vertical growth can be observed where it has taken place. A tall building at the 580 m distance mark in Figure 9c is clearly visible in the AW3D30 DBH but is absent in the ASTER and SRTM DBHs. After visual inspection of historical imagery with Google Earth, it was found that this particular building was constructed after 2005. Another tall building can be seen at the 850 m distance mark in Figure 9a in the AW3D30 DBH and SRTM DBH but not in the ASTER DBH. It is one of the stable structures, as identified from Google Earth's historical imagery. Furthermore, there were several buildings that were present in the AW3D30 DBH and ASTER DBH but absent in the SRTM DBH. They can be seen at the distance marked from 500 to 700 m in Figure 9a and at the distance marked from 200 to 300 m in Figure 9c. These buildings were constructed after 2004, which explains their absence from the SRTM DBH. These examples also suggest that the ASTER DBH contained buildings built between the years 2004 and 2005. However, if a building seems to be present in the SRTM DBH but absent from the ASTER DBH, then the building is indeed present on the ground, and its absence from the ASTER DBH can be explained by errors in the ASTER DSM itself.

3.5. Discussion

In this paper, a simple approach to the extraction and comparison of DBHs without the use of any supporting datasets was used. It is possible that other methodologies may result in significantly better estimates of RMSE. So far, only one other study has extracted building heights from the same DSM sources as those used in this study. Wang et al. [43] derived building heights at a country-wide scale in the UK by using ALS building heights as training data in a random forest classifier. Their approach indicated that the highest accuracy was found with SRTM (R^2: 0.67), followed by ASTER (R^2: 0.66) and AW3D30 (R^2: 0.63), each with an RMSE lower than 1.9 m. This is interesting for two reasons: first, their coefficient of determination (R^2) for the AW3D30 DBH was quite close to that obtained in this study (R^2: 0.60) with a simpler technique. Second, Wang et al. [43] also achieved high accuracy for DBHs derived from SRTM and ASTER, which suggests that these datasets should not be discounted for DBH estimation. The results of this study could be biased on two clear accounts. First, the height accuracy was compared only with the DBH filtered from GeoEye, which has its own shortcomings. Furthermore, photogrammetric DSMs have limitations owing to spatial resolution, atmospheric errors, and matching errors, leading to erroneous heights at times. A more robust benchmarking of the results for rooftops could be achieved with a high-quality dense DBH obtained from an airborne ALS sensor, as it has a small footprint and high point density [83]. Second, the comparisons were performed with only 52 stable structures. These structures were mostly either tall buildings or prominent religious or colonial structures located in relatively less dense areas. It is possible that these structures biased the accuracy judgment in favor of the open DSMs. Comparison over a sufficiently large number of diverse stable structures could help to overcome this limitation.

The DSMs used in this study were based on images that are about a decade to two old. Even the latest available DSM used (TanDEM-X) has images from the year 2011. The relative importance of these datasets depends on the location under study, e.g., they are useful for studying urban growth in cities such as Yangon, Manila, and Shanghai that have seen rapid expansion during this period compared with already established cities, such as Tokyo and Seoul [89]. Nonetheless, future open DSM mission proposals, such as Tandem-L [90], are in the pipeline and shall further enhance the value of past open DBHs. Based on the current findings, there is no reason to suggest the use of open DSMs in lieu of DBHs from very high resolution imagery and ALS to obtain highly accurate models. However, there is clear potential for the use of open DSMs in mapping urban heights and studying vertical expansion on a large scale and from past periods if high-resolution data are not available.

This research also highlights the technical improvements needed in open DSMs so that they can better represent building heights, thus enhancing their value. Frequent identification of road features as non-ground points in coarse-resolution nDSMs led to the lower estimation of surrounding structure heights. There is a need for a filtering algorithm that can identify street canyons between buildings as ground pixels. Road networks mostly lie on the ground surface and can be used as additional terrain reinforcement information. Possible approaches could be devised that use road vector masks or Open Street Map datasets to identify them as ground points. Roads narrower than 30 m may still not be identified, but this should help to identify occlusions and matching inaccuracies arising from imaging geometry over dense urban areas. Furthermore, ASTER and SRTM could be fused such that more representative building heights are estimated for the period around the early 2000s. If cadastral maps from that period are available, that information could be used to derive building heights for tall structures. In the current research, this is a limitation due to the lower accuracy of ASTER and SRTM DBHs with respect to the AW3D30 DBH and the non-availability of older reference DBMs from other sources. By resolving these limitations, open DBHs will be of great use in assessing the vertical growth component of land-use change in rapidly growing cities.

A direct next step is the generation of a global DBH from these open DSMs and the benchmarking of their accuracy by comparison studies in cities with high-quality reference datasets. Some countries, like India, have open 30 m DEMs ('CartoDEM') which may be used for such a comparison. Conclusions from such studies will be useful for DBH preparation over regions that lack such maps. The DBHs can also be employed to identify land-use agglomerations by considering human-activity-describing datasets, such as night-time light [42]. Such agglomeration maps could be useful for characterizing the impact of urban vertical expansion on the environment. For example, Zhang et al. [29] recently found that urbanization increased the extreme flood event probability by several magnitudes in Houston and suggested including the effect of urbanization on extreme precipitation in climate models. Open DBHs can also help to identify factories and tall buildings on a large scale, which could be useful for updating emission inventories. The understanding of the impact of urbanization on other environmental issues, like air pollution transport in environmentally deteriorating Asian cities [91], stands to benefit from building height datasets, as such information can support evidence-based policies.

4. Conclusions

Open DSMs, like AW3D30, ASTER, and SRTM, are already valuable for use in morphological studies. However, digital building heights (DBHs) derived from them could be useful for several applications in cities without building height maps. To assess the suitability of extracting DBH from an open DSM, DBHs extracted from several high-resolution and coarse-resolution DSMs were compared. It was found that the RMSE of the AW3D5 DBH was comparable to that of the TanDEM-X DBH and demonstrated accuracy with an RMSE of 3.79 m. On using the coarser AW3D30, the RMSE did not degrade significantly over the finer AW3D5 DBH. A good correlation among digital terrain models does not guarantee a good agreement among the DBHs, as was observed between AW3D30 and SRTM. Furthermore, height comparison over stable structures showed that the AW3D30 DBH has a much higher accuracy than that of the ASTER DBH and SRTM DBH and was able to capture variation in building heights. It is concluded that AW3D30 is the most suitable open DSM for DBH generation and for observing buildings taller than 9 m in height. Further, different acquisition periods of the available open DSMs could be exploited for studying vertical land-use changes at regional and global scales. Such applications will be useful for policy studies addressing environmental impacts and disaster mitigation.

Author Contributions: Conceptualization, P.M. and W.T.; methodology, P.M.; software, P.M.; validation, P.M., R.A. and W.T.; formal analysis, P.M.; investigation, P.M.; resources, R.A., W.T.; data curation, W.T.; writing—original draft preparation, P.M.; writing-review and editing, P.M., R.A. and W.T.; visualization, P.M.; supervision, W.T.; project administration, W.T.

Funding: This research received no external funding.

Acknowledgments: This research was supported by Science and Technology Research Partnership for Sustainable Development (SATREPS), Japan Science and Technology Agency (JST) and Japan International Cooperation Agency (JICA). The TanDEM-X data were provided by the DLR through DLR scientific projects, avtar_XTI_VEGE6813. We are also indebted to the three anonymous reviewers for their fruitful criticisms.

Conflicts of Interest: The authors declare no conflict of interest.

Abbreviations

The following abbreviations are used in this manuscript:

ALS	airborne laser scanning
ASTER	Advanced Spaceborne Thermal Emission and Reflection Radiometer
AW3D5	ALOS world 3D 5 m resolution DSM
AW3D30	ALOS world 3D 30 m resolution DSM
DSM	digital surface model
DTM	ditial terrain model
DBH	digital building height model
MAE	mean absolute error
ME	mean error
MSD	multi-directional slope filtering
nDSM	normalized digital surface model
RMSE	root mean square error
SAR	synthetic aperture radar
SD	standard deviation
SRTM	Shuttle Radar Topography Mission
TanDEM-X	TerraSAR-X add on for Digital Elevation Measurements
VHR	very high-resolution

References

1. UN Habitat. Resilience. Available online: https://unhabitat.org/urban-themes/resilience/ (accessed on 19 November 2018).
2. Handayani, H.H.; Estoque, R.C.; Murayama, Y. Estimation of built-up and green volume using geospatial techniques: A case study of Surabaya, Indonesia. *Sustain. Cities Soc.* **2018**, *37*, 581–593. [CrossRef]
3. Jiang, F.; Liu, S.; Yuan, H.; Zhang, Q. Measuring urban sprawl in Beijing with geo-spatial indices. *J. Geogr. Sci.* **2007**, *17*, 469–478. [CrossRef]
4. Shi, L.; Shao, G.; Cui, S.; Li, X.; Lin, T.; Yin, K.; Zhao, J. Urban three-dimensional expansion and its driving forces—A case study of Shanghai, China. *Chin. Geogr. Sci.* **2009**, *19*, 291–298. [CrossRef]
5. Clarke, K.C.; Gaydos, L.J. Loose-coupling a cellular automaton model and GIS: Long-term urban growth prediction for San Francisco and Washington/Baltimore. *Int. J. Geogr. Inf. Sci.* **1998**, *12*, 699–714. [CrossRef] [PubMed]
6. Haala, N.; Brenner, C. Extraction of buildings and trees in urban environments. *ISPRS J. Photogramm. Remote Sens.* **1999**, *54*, 130–137. [CrossRef]
7. Ratti, C.; Ratti, C.; Di Sabatino, S.; Di Sabatino, S.; Britter, R.; Britter, R.; Brown, M.; Burian, S.; Caton, F.; Caton, F.; et al. Analysis of 3-D urban databases with respect to pollution dispersion for a number of European and American cities. *Water Air Soil Pollut. Focus* **2002**, *2*, 459–469. [CrossRef]
8. Ratti, C.; Baker, N.; Steemers, K. Energy consumption and urban texture. *Energy Build.* **2005**, *37*, 762–776. [CrossRef]
9. Hofierka, J.; Kaňuk, J. Assessment of photovoltaic potential in urban areas using open-source solar radiation tools. *Renew. Energy* **2009**, *34*, 2206–2214. [CrossRef]
10. Buyantuyev, A.; Wu, J. Urban heat islands and landscape heterogeneity: Linking spatiotemporal variations in surface temperatures to land-cover and socioeconomic patterns. *Landsc. Ecol.* **2010**, *25*, 17–33. [CrossRef]
11. Fernández, D.S.; Lutz, M.A. Urban flood hazard zoning in Tucumán Province, Argentina, using GIS and multicriteria decision analysis. *Eng. Geol.* **2010**, *111*, 90–98. [CrossRef]

12. Costa, E. Simulation of the effects of different urban environments on gps performance using digital elevation models and building databases. *IEEE Trans. Intell. Transp. Syst.* **2011**, *12*, 819–829. [CrossRef]
13. Zhang, W.; Li, W.; Zhang, C.; Ouimet, W.B. Detecting horizontal and vertical urban growth from medium resolution imagery and its relationships with major socioeconomic factors. *Int. J. Remote Sens.* **2017**, *38*, 3704–3734. [CrossRef]
14. Kim, C.; Habib, A.; Chang, Y.C. Automatic Generation of Digital Building Models for Complex Structures From Lidar Data. *Int. Arch. Photogramm. Remote Sens. Spat. Inf. Sci.* **2008**, *37*, 463–468.
15. Sirmacek, B.; Taubenböck, H.; Reinartz, P.; Ehlers, M. Performance evaluation for 3-{D} city model generation of six different DSMs from air-and spaceborne sensors. *IEEE J. Sel. Top. Appl. Earth Obs. Remote Sens.* **2012**, *5*, 59–70. [CrossRef]
16. Gamba, P.; Houshmand, B.; Saccani, M. Detection and extraction of buildings from interferometric SAR data. *IEEE Trans. Geosci. Remote Sens.* **2000**, *38*, 611–618. [CrossRef]
17. Rutzinger, M.; Rüf, B.; Höfle, B.;Vetter, M. Change Detection of Building Footprints from Airborne Laser Scanning Acquired in Short Time Intervals. In *ISPRS Technical Commission VII Symposium—100 Years ISPRS*; Wagner, W., Székely, B., Eds.; Institute of Photogrammetry and Remote Sensing, Vienna University of Technology: Vienna, Austria, 2010; Volume XXXVIII, pp. 475–480.
18. Wu, B.; Yu, B.; Wu, Q.; Yao, S.; Zhao, F.; Mao, W.; Wu, J. A graph-based approach for 3d building model reconstruction from airborne lidar point clouds. *Remote Sens.* **2017**, *9*, 92. [CrossRef]
19. Sun, Y.; Zhang, X.; Zhao, X.; Xin, Q. Extracting building boundaries from high resolution optical images and LiDAR data by integrating the convolutional neural network and the active contour model. *Remote Sens.* **2018**, *10*, 1459. [CrossRef]
20. Bonczak, B.; Kontokosta, C.E. Large-scale parameterization of 3D building morphology in complex urban landscapes using aerial LiDAR and city administrative data. *Comput. Environ. Urban Syst.* **2018**, *73*, 126–142. [CrossRef]
21. Chen, K.; Lu, W.; Xue, F.; Tang, P.; Li, L.H. Automatic building information model reconstruction in high-density urban areas: Augmenting multi-source data with architectural knowledge. *Autom. Constr.* **2018**, *93*, 22–34. [CrossRef]
22. Aljumaily, H.; Laefer, D.F.; Cuadra, D. Big-Data Approach for Three-Dimensional Building Extraction from Aerial Laser Scanning. *J. Comput. Civ. Eng.* **2016**, *30*, 04015049. [CrossRef]
23. Tomljenovic, I.; Höfle, B.; Tiede, D.; Blaschke, T. Building extraction from Airborne Laser Scanning data: An analysis of the state of the art. *Remote Sens.* **2015**, *7*, 3826–3862. [CrossRef]
24. Alobeid, A.; Jacobsen, K.; Heipke, C. Comparison of Matching Algorithms for DSM Generation in Urban Areas from Ikonos Imagery. *Photogramm. Eng. Remote Sens.* **2010**, *76*, 1041–1050. [CrossRef]
25. Tian, J.; Cui, S.; Reinartz, P. Building Change Detection Based on Satellite Stereo Imagery and Digital Surface Models. *IEEE Trans. Geosci. Remote Sens.* **2014**, *52*, 406–417. [CrossRef]
26. Balk, D.; Deichmann, U.; Yetman, G.; Pozzi, F.; Hay, S.; Nelson, A. Determining Global Population Distribution: Methods, Applications and Data. *Adv. Parasitol.* **2006**, *62*, 119–156. [PubMed]
27. Center for International Earth Science Information Network—CIESIN—Columbia University; International Food Policy Research Institute (IFPRI); The World Bank; Centro Internacional de Agricultura Tropical. *Global Rural-Urban Mapping Project, Version 1 (GRUMPv1): Urban Extents Grid*; NASA Socioeconomic Data and Applications Center (SEDAC): Palisades, NY, USA, 2011.
28. Center for International Earth Science Information Network—CIESIN—Columbia University. *Global Urban Heat Island (UHI) Data Set, 2013*; NASA Socioeconomic Data and Applications Center (SEDAC): Palisades, NY, USA, 2016.
29. Zhang, W.; Villarini, G.; Vecchi, G.A.; Smith, J.A. Urbanization exacerbated the rainfall and flooding caused by hurricane Harvey in Houston. *Nature* **2018**, *563*, 384–388. [CrossRef] [PubMed]
30. Lwin, K.K.; Murayama, Y. Estimation of Building Population from LIDAR Derived Digital Volume Model. In *Spatial Analysis and Modeling in Geographical Transformation Process*; Murayama, Y., Thapa, R., Eds.; Springer: Dordrecht, The Netherlands, 2011; Volume 100.
31. Yamazaki, D.; Ikeshima, D.; Tawatari, R.; Yamaguchi, T.; O'Loughlin, F.; Neal, J.C.; Sampson, C.C.; Kanae, S.; Bates, P.D. A high-accuracy map of global terrain elevations. *Geophys. Res. Lett.* **2017**, *44*, 5844–5853. [CrossRef]

32. Santillan, J.R.; Makinano-Santillan, M. Vertical accuracy assessment of 30-M resolution ALOS, ASTER, and SRTM global DEMS over Northeastern Mindanao, Philippines. *Int. Arch. Photogramm. Remote Sens. Spat. Inf. Sci. ISPRS Arch.* **2016**, *41*, 149–156. [CrossRef]
33. Jain, A.O.; Thaker, T.; Chaurasia, A.; Patel, P.; Singh, A.K. Vertical accuracy evaluation of SRTM-GL1, GDEM-V2, AW3D30 and CartoDEM-V3.1 of 30-m resolution with dual frequency GNSS for lower Tapi Basin India. *Geocarto Int.* **2018**, *33*, 1237–1256. [CrossRef]
34. Purinton, B.; Bookhagen, B. Validation of digital elevation models (DEMs) and comparison of geomorphic metrics on the southern Central Andean Plateau. *Earth Surf. Dyn.* **2017**, *5*, 211–237. [CrossRef]
35. Acharya, T.D.; Yang, I.T.; Lee, D.H. Comparative Analysis of Digital Elevation Models between AW3D30, SRTM30 and Airborne LiDAR: A case of Chuncheon, South Korea. *J. Korean Soc. Surv. Geod. Photogramm. Cartogr.* **2018**, *36*, 17–24.
36. Alganci, U.; Besol, B.; Sertel, E. Accuracy Assessment of Different Digital Surface Models. *ISPRS Int. J. Geo-Inf.* **2018**, *7*, 114. [CrossRef]
37. Nghiem, S.; Balk, D.; Small, C.; Deichmann, U.; Wannebo, A.; Blom, R.; Sutton, P.; Yetman, G.; Chen, R.; Rodriguez, E.; et al. *Global Infrastructure: The Potential of SRTM Data to Break New Ground*; White Paper; NASA-JPL and CIESIN: Columbia University: New York, NY, USA, 2001; p. 20.
38. Gamba, P.; Acqua, F.D.; Houshmand, B. SRTM data Characterization in urban areas. *Int. Arch. Photogramm. Remote Sens. Spat. Inf. Sci.* **2002**, *34*, 55–58.
39. Quartulli, M.; Datcu, M. Information fusion for scene understanding from interferometric SAR data in urban environments. *IEEE Trans. Geosci. Remote Sens.* **2003**, *41*, 1976–1985. [CrossRef]
40. Schneider, A.; Woodcock, C.E. Compact, dispersed, fragmented, extensive? A comparison of urban growth in twenty-five global cities using remotely sensed data, pattern metrics and census information. *Urban Stud.* **2008**, *45*, 659–692. [CrossRef]
41. Seto, K.C.; Fragkias, M.; Güneralp, B.; Reilly, M.K. A Meta-Analysis of Global Urban Land Expansion. *PLoS ONE* **2011**, *6*, e23777. [CrossRef] [PubMed]
42. Small, C.; Elvidge, C.D. Night on earth: Mapping decadal changes of anthropogenic night light in Asia. *Int. J. Appl. Earth Obs. Geoinf.* **2013**, *22*, 40–52. [CrossRef]
43. Wang, P.; Huang, C.; Tilton, J.C. Mapping Three-dimensional Urban Structure by Fusing Landsat and Global Elevation Data. *arXiv* **2018**, arXiv:1807.04368.
44. Department of Population. *The 2014 Myanmar Population and Housing Census*; The Union Report, Census Report Volume 2; Ministry of Immigration and Population: Nay Pyi Taw, Myanmar, 2015.
45. Sritarapipat, T.; Takeuchi, W. Urban Growth Modeling based on the Multi-centers of the Urban Areas and Land Cover Change in Yangon, Myanmar. *J. Remote Sens. Soc. Jpn.* **2017**, *37*, 248–260.
46. Sritarapipat, T.; Takeuchi, W. Land Cover Change Simulations in Yangon Under Several Scenarios of Flood and Earthquake Vulnerabilities with Master Plan. *J. Disaster Res.* **2018**, *13*, 50–61. [CrossRef]
47. Gorelick, N.; Hancher, M.; Dixon, M.; Ilyushchenko, S.; Thau, D.; Moore, R. Google Earth Engine: Planetary-scale geospatial analysis for everyone. *Remote Sens. Environ.* **2017**, *202*, 18–27. [CrossRef]
48. Farr, T.; Rosen, P.; Caro, E.; Crippen, R.; Duren, R.; Hensley, S.; Kobrick, M.; Paller, M.; Rodriguez, E.; Roth, L.; et al. The shuttle radar topography mission. *Rev. Geophys.* **2007**, *45*, 1–33. [CrossRef]
49. Rodríguez, E.; Morris, C.S.; Belz, J.E. A Global Assessment of the SRTM Performance. *Photogramm. Eng. Remote Sens.* **2006**, *72*, 249–260. [CrossRef]
50. U.S. Geological Survey. Societal Benefits of Higher Resolution SRTM Products. Available online: https://lpdaac.usgs.gov/societal_benefits_higher_resolution_srtm_products (accessed on 19 November 2018).
51. Mukul, M.; Srivastava, V.; Mukul, M. Accuracy analysis of the 2014–2015 global shuttle radar topography mission (SRTM) 1 arc-sec C-Band height model using international global navigation satellite system service (IGS) network. *J. Earth Syst. Sci.* **2016**, *125*, 909–917. [CrossRef]
52. Satge, F.; Denezine, M.; Pillco, R.; Timouk, F.; Pinel, S.; Molina, J.; Garnier, J.; Seyler, F.; Bonnet, M.P. Absolute and relative height-pixel accuracy of SRTM-GL1 over the South American Andean Plateau. *ISPRS J. Photogramm. Remote Sens.* **2016**, *121*, 157–166. [CrossRef]
53. U.S. Geological Survey. USGS Earth Explorer. Available online: https://earthexplorer.usgs.gov/ (accessed on 19 November 2018).

54. Tachikawa, T.; Hat, M.; Kaku, M.; Iwasaki, A. Characteristics of ASTER GDEM version 2. In Proceedings of the 2011 IEEE International Geoscience and Remote Sensing Symposium (IGARSS), Vancouver, BC, Canada, 24–29 July 2011; pp. 3657–3660.
55. NASA JPL. ASTER Global Digital Elevation Map Announcement. Available online: https://asterweb.jpl.nasa.gov/gdem.asp (accessed on 19 November 2018).
56. Jing, C.; Shortridge, A.; Lin, S.; Wu, J. Comparison and validation of SRTM and ASTER GDEM for a subtropical landscape in Southeastern China. *Int. J. Digit. Earth* **2014**, *7*, 969–992. [CrossRef]
57. Colosimo, G.; Crespi, M.; De Vendictis, L.; Jacobsen, K. Accuracy evaluation of SRTM and ASTER DSMs. In Proceedings of the 29th EARSeL Symposium, Chania, Greece, 15–18 June 2009.
58. Marconcini, M.; Marmanis, D.; Esch, T.; Felbier, A. A novel method for building height estmation using TanDEM-X data. In Proceedings of the 2014 IEEE Geoscience and Remote Sensing Symposium (IGARSS), Quebec City, QC, Canada, 13–18 July 2014; pp. 4804–4807.
59. Wessel, B.; Huber, M.; Wohlfart, C.; Marschalk, U.; Kosmann, D.; Roth, A. Accuracy assessment of the global TanDEM-X Digital Elevation Model with GPS data. *ISPRS J. Photogramm. Remote Sens.* **2018**, *139*, 171–182. [CrossRef]
60. Avtar, R.; Yunus, A.P.; Kraines, S.; Yamamuro, M. Evaluation of DEM generation based on Interferometric SAR using TanDEM-X data in Tokyo. *Phys. Chem. Earth Parts A/B/C* **2015**, *83–84*, 166–177. [CrossRef]
61. Takaku, J.; Tadono, T.; Tsutsui, K.; Ichikawa, M. Validation of 'AW3D' Global DSM Generated From ALOS PRISM. *ISPRS Ann. Photogramm. Remote Sens. Spat. Inf. Sci.* **2016**, *III-4*, 25–31. [CrossRef]
62. Caglar, B.; Becek, K.; Mekik, C.; Ozendi, M. On the vertical accuracy of the ALOS world 3D-30m digital elevation model. *Remote Sens. Lett.* **2018**, *9*, 607–615. [CrossRef]
63. Estoque, R.C.; Murayama, Y.; Ranagalage, M.; Hou, H.; Subasinghe, S.; Gong, H.; Simwanda, M.; Handyani, H.H.; Zhang, X. Validating ALOS PRISM DSM-derived surface feature height: Implications for urban volume estimation. *Tsukuba Geoenviron. Sci.* **2017**, *13*, 13–22.
64. NTT DATA; RESTEC. High-Resolution Digital 3D Map Covering the Entire Global Land Area. Available online: https://www.aw3d.jp/en/products/standard/ (accessed on 19 November 2018).
65. Earth Obervation Research Center JAXA. ALOS Global Digital Surface Model "ALOS World 3D—30m (AW3D30)". Available online: https://www.eorc.jaxa.jp/ALOS/en/aw3d30/ (accessed on 19 November 2018).
66. Grohmann, C.H. Evaluation of TanDEM-X DEMs on selected Brazilian sites: comparison with SRTM, ASTER GDEM and ALOS AW3D30. *Remote Sens. Environ.* **2018**, *212*, 121–133. [CrossRef]
67. Aguilar, M.A.; del Mar Saldana, M.; Aguilar, F.J. Generation and Quality Assessment of Stereo-Extracted DSM From GeoEye-1 and WorldView-2 Imagery. *IEEE Trans. Geosci. Remote Sens.* **2014**, *52*, 1259–1271. [CrossRef]
68. Beumier, C.; Idrissa, M. Digital terrain models derived from digital surface model uniform regions in urban areas. *Int. J. Remote Sens.* **2016**, *37*, 3477–3493. [CrossRef]
69. Özcan, A.H.; Ünsalan, C.; Reinartz, P. Ground filtering and DTM generation from DSM data using probabilistic voting and segmentation. *Int. J. Remote Sens.* **2018**, *39*, 2860–2883. [CrossRef]
70. Xu, Y.; Wu, L.; Xie, Z.; Chen, Z. Building extraction in very high resolution remote sensing imagery using deep learning and guided filters. *Remote Sens.* **2018**, *10*, 144. [CrossRef]
71. Gevaert, C.M.; Persello, C.; Nex, F.; Vosselman, G. A deep learning approach to DTM extraction from imagery using rule-based training labels. *ISPRS J. Photogramm. Remote Sens.* **2018**, *142*, 106–123. [CrossRef]
72. Bittner, K.; Cui, S.; Reinartz, P. Building extraction from remote sensing data using fully convolutional networks. *Int. Arch. Photogramm. Remote Sens. Spat. Inf. Sci. ISPRS Arch.* **2017**, *42*, 481–486. [CrossRef]
73. Long, J.; Shelhamer, E.; Darrell, T. Fully convolutional networks for semantic segmentation. In Proceedings of the 2015 IEEE Conference on Computer Vision and Pattern Recognition (CVPR), Boston, MA, USA, 7–12 June 2015; pp. 3431–3440.
74. Deng, J.; Dong, W.; Socher, R.; Li, L.J.; Li, K.; Li, F.-F. ImageNet: A large-scale hierarchical image database. In Proceedings of the 2009 IEEE Conference on Computer Vision and Pattern Recognition, Miami, FL, USA, 20–25 June 2009; Volume 20, pp. 248–255.
75. Perko, R.; Raggam, H.; Gutjahr, K.H.; Schardt, M. Advanced DTM generation from very high resolution satellite stereo images. *ISPRS Ann. Photogramm. Remote Sens. Spat. Inf. Sci.* **2015**, *II-3/W4*, 165–172. [CrossRef]
76. Meng, X.; Wang, L.; Silván-Cárdenas, J.L.; Currit, N. A multi-directional ground filtering algorithm for airborne LIDAR. *ISPRS J. Photogramm. Remote Sens.* **2009**, *64*, 117–124. [CrossRef]

77. Mousa, Y.A.K.; Helmholz, P.; Belton, D. New DTM extraction approach from airborne images derived DSM. *Int. Arch. Photogramm. Remote Sens. Spat. Inf. Sci. ISPRS Arch.* **2017**, *42*, 75–82. [CrossRef]
78. Auer, S.; Schmitt, M.; Reinartz, P. Automatic alignment of high resolution optical and SAR images for urban areas. In Proceedings of the 2017 IEEE International Geoscience and Remote Sensing Symposium (IGARSS), Fort Worth, TX, USA, 23–28 July 2017; pp. 5466–5469.
79. Jones, E.; Olopihant, T.; Pearu, P. SciPy: Open Source Scientific Tools for Python. Available online: http://www.scipy.org/ (accessed on 10 December 2018).
80. Awrangjeb, M.; Fraser, C.S. An automatic and threshold-free performance evaluation system for building extraction techniques from airborne LIDAR data. *IEEE J. Sel. Top. Appl. Earth Obs. Remote Sens.* **2014**, *7*, 4184–4198. [CrossRef]
81. Truong-Hong, L.; Laefer, D.F. Quantitative evaluation strategies for urban 3D model generation from remote sensing data. *Comput. Graph. (Pergamon)* **2015**, *49*, 82–91. [CrossRef]
82. Rutzinger, M.; Rottensteiner, F.; Pfeifer, N. A Comparison of Evaluation Techniques for Building Extraction From Airborne Laser Scanning. *IEEE J. Sel. Top. Appl. Earth Obs. Remote Sens.* **2009**, *2*, 11–20. [CrossRef]
83. Rossi, C.; Gernhardt, S. Urban DEM generation, analysis and enhancements using TanDEM-X. *ISPRS J. Photogramm. Remote Sens.* **2013**, *85*, 120–131. [CrossRef]
84. Capaldo, P.; Crespi, M.; Fratarcangeli, F.; Nascetti, A.; Pieralice, F. DSM generation from high resolution imagery: Applications with WorldView-1 and Geoeye-1. *Eur. J. Remote Sens.* **2012**, *44*, 41–53. [CrossRef]
85. Zeng, C.; Wang, J.; Shi, P. A stereo image matching method to improve the DSM accuracy inside building boundaries. *Can. J. Remote Sens.* **2013**, *39*, 308–317. [CrossRef]
86. Hasegawa, H.; Matsuo, K.; Koarai, M.; Watanabe, N.; Masaharu, H. DEM Accuracy and the Base to Height (B/H) Ratio of Stereo Images. *Int. Arch. Photogramm. Remote Sens.* **2000**, *33*, 356–359.
87. Arai, X.; Ozawa, M.; Terayama, Y. Optimization method for B/H ratio determination taking occlusion effect and MTF degradation due to atmosphere into account. In Proceedings of the IGARSS '94—1994 IEEE International Geoscience and Remote Sensing Symposium, Pasadena, CA, USA, 8–12 August 1994; Volume 3, pp. 1464–1466.
88. Zhang, L.; Gruen, A. Multi-image matching for DSM generation from IKONOS imagery. *ISPRS J. Photogramm. Remote Sens.* **2006**, *60*, 195–211. [CrossRef]
89. Schneider, A.; Mertes, C.M.; Tatem, A.J.; Tan, B.; Graves, S.J.; Patel, N.N.; Horton, J.A.; Gaughan, A.E.; Rollo, J.T.; Schelly, I.H.; et al. A new urban landscape in East–Southeast Asia, 2000–2010. *Environ. Res. Lett.* **2015**, *10*, 034002. [CrossRef]
90. DLR. Tandem-L Mission Description. Available online: https://www.tandem-l.de/mission-description/ (accessed on 19 November 2018).
91. Misra, P.; Fujikawa, A.; Takeuchi, W. Novel decomposition scheme for characterizing urban air quality with MODIS. *Remote Sens.* **2017**, *9*, 812. [CrossRef]

© 2018 by the authors. Licensee MDPI, Basel, Switzerland. This article is an open access article distributed under the terms and conditions of the Creative Commons Attribution (CC BY) license (http://creativecommons.org/licenses/by/4.0/).

Article

Hierarchical Regularization of Building Boundaries in Noisy Aerial Laser Scanning and Photogrammetric Point Clouds

Linfu Xie [1,2,3], Qing Zhu [1,2,4,*], Han Hu [1,2,3,*], Bo Wu [3], Yuan Li [3], Yeting Zhang [1] and Ruofei Zhong [5]

1. State Key Laboratory of Information Engineering in Surveying Mapping and Remote Sensing, Wuhan University, Wuhan 430079, China; linfuxie@whu.edu.cn (L.X.); zhangyeting@263.net (Y.Z.)
2. Key Laboratory of Urban Land Resources Monitoring and Simulation, Ministry of Land and Resources, Shenzhen 518040, China
3. Department of Land Surveying and Geo-Informatics, The Hong Kong Polytechnic University, Hung Hom, Kowloon 999077, Hong Kong; bo.wu@polyu.edu.hk (B.W.); y-uan.li@connect.polyu.hk (Y.L.)
4. Faculty of Geosciences and Environmental Engineering, Southwest Jiaotong University, Chengdu 611756, China
5. Beijing Advanced Innovation Center for Imaging Technology, Capital Normal University, Beijing 100875, China; zrfsss@163.com
* Correspondence: zhuq66@263.net (Q.Z.); huhan8807@gmail.com (H.H.); Tel.: +86-138-0908-0727 (Q.Z.)

Received: 29 October 2018; Accepted: 7 December 2018; Published: 10 December 2018

Abstract: Aerial laser scanning or photogrammetric point clouds are often noisy at building boundaries. In order to produce regularized polygons from such noisy point clouds, this study proposes a hierarchical regularization method for the boundary points. Beginning with detected planar structures from raw point clouds, two stages of regularization are employed. In the first stage, the boundary points of an individual plane are consolidated locally by shifting them along their refined normal vector to resist noise, and then grouped into piecewise smooth segments. In the second stage, global regularities among different segments from different planes are softly enforced through a labeling process, in which the same label represents parallel or orthogonal segments. This is formulated as a Markov random field and solved efficiently via graph cut. The performance of the proposed method is evaluated for extracting 2D footprints and 3D polygons of buildings in metropolitan area. The results reveal that the proposed method is superior to the state-of-art methods both qualitatively and quantitatively in compactness. The simplified polygons could fit the original boundary points with an average residuals of 0.2 m, and in the meantime reduce up to 90% complexities of the edges. The satisfactory performances of the proposed method show a promising potential for 3D reconstruction of polygonal models from noisy point clouds.

Keywords: point clouds; boundary extraction; regularization; building reconstruction

1. Introduction

With the rapid developments in aerial laser scanning (ALS) and aerial oblique photogrammetry, 3D point clouds have become the primary datasets used in large-scale urban reconstruction [1,2]. In particular, recent advances in structure from motion (SfM) [3] and multi-view stereo (MVS) [4] methods allow detailed coverage of urban scenes, and are particularly suitable for feature matching [5], bundle adjustment [6,7], and dense image matching (DIM) [8,9] in aerial oblique images.

However, as the usability of point clouds or derived triangular meshes is limited by the difficulties in manipulating and managing the datasets, 2D polygons and polygonal building models are still the industrial standard datasets for various applications [10,11], including visualization, spatial analysis, urban planning, and navigation [12–15]. Although the extraction of points on the exterior building boundaries is gracefully handled by a standard convex hull or alpha-shapes for non-convex boundaries, simplification and regularization of the noisy boundaries are still non-trivial tasks and well-established cartographic algorithms do not produce acceptable results when applied to point cloud datasets.

The inherent deficiencies of point clouds data, such as data anisotropy, insufficient sampling, and especially noise, make it challenging to retrieve compact 2D/3D polygons of buildings that have satisfactory geometric quality. For ALS data, which have high altimetric accuracy but relatively low point density [16], the initial boundary points of building plans are often jagged [17] and small structures are not well sampled. Meanwhile, DIM point clouds of aerial oblique images are generally inferior to those created from laser scanning in terms of noise level and the preservation of sharp features [9]. The forward intersected point clouds suffer from inaccurate positioning at the edge of building planes due to disparity discontinuities, and thus sharp features may degenerate at corners.

In the boundary simplification process, traditional edge collapse-based methods are likely to eliminate sharp features, whereas regularization methods that adopt the Manhattan rule based on the dominant orientation tend to be too strict in many real-world applications, leading to large distortions when applied to polygons with multiple orientations.

This paper consists of three major contributions, in order to overcome the above issues: (1) we allow the mutation of boundary points along their normal in the local stage in a least-square manner to resist noise; (2) we relax the hard Manhattan assumptions (either parallel or perpendicular to the dominant orientation) [1] by also considering fidelity terms. Unlike other modified Manhattan-based regularization methods which need to pre-define the number of resulting orientations [18] or difference angle threshold for compulsive orthogonality [19], the proposed method casts the problem as a Markov random field (MRF) and allows multiple and arbitrary orientations with mutual regularities in the same optimization; and (3) we propose a strategy to produce simplified polygons with inter-part regularity in 3D space.

In this study, we extend our preliminary conference paper [20] in the following four directions: (1) improving the local stage regularization algorithm, allowing it to handle the zigzag effect more efficiently; (2) taking regional level relationships into considerations to handle boundary regularities between different planes or buildings; (3) conducting additional experiments on relatively low-density ALS data for building footprint generalization; and (4) supplementing more quantitative comparison with other representative methods. The rest of this paper is organized as follows. Section 2 briefly reviews the existing research on boundary points simplification and regularization. In Section 3, the proposed method is described in detail. The performance of the proposed method is evaluated in two different application scenarios, both qualitatively and quantitatively in Section 4. Discussions are given in Section 5 and the conclusions are presented in Section 6.

2. Related Work

The simplification and regularization of closed boundaries or polygons for discrete points have attracted a lot of interest in the fields of photogrammetry, cartography, and computer graphics. Such procedures can be used in map simplification [21], multi-scale representation [22,23], 3D model simplification [24], building footprint generalization [25], and 3D reconstruction [26]. However, the data redundancy and noise level in building borders from ALS or photogrammetric point clouds are distinctively higher than those in digital maps, existing building footprints, or 3D models. As shown in Figure 1, generally, producing methods for building boundary simplification and regularization can be categorized into three major classes.

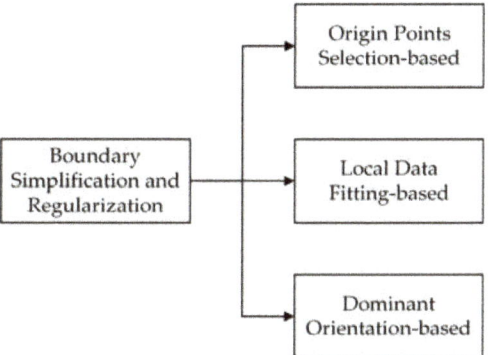

Figure 1. Classification of boundary generalization methods.

The aim of boundary point simplification is to represent an original boundary compactly, which is necessary for efficient computation and visualization. The first class of methods attempted to decrease the number of vertices by discarding original points that met a certain criterion, locally or globally. Among this class of methods, the Douglas–Peucker algorithm, which uses perpendicular distance as a global indicator, is the most common [27]. Some algorithms locally define a line through the first two points of the original polyline, and then calculate the deviation from this line for each successive point. The first point with a deviation that exceeds a pre-defined threshold is kept and treated as the first point of the next line. This deviation could be either a direct perpendicular distance [28] or a sleeve-fitting residual [29]. Sometimes, further constraints are also incorporated into this formulation [30]. More recently, methods that consider both local and global measurements in the process of simplification have been explored. Triangle decimation is used to simplify the curves while enforcing the topology of the curve set [31], resulting in the simplification of curves while maintaining the geometric relationship between curves. This type of simplification algorithm is easy to implement, computationally efficient, and requires only a small amount of memory. However, the resulting polygons depend on the choice of starting points and are easily affected by noise. Furthermore, critical points around sharp features are likely to degenerate during the process of simplification.

To better fit the original boundary points, the second class of methods tries to detect line segments from the original points and then assemble them to form a closed polygon. These methods focus on the preservation of data fidelity. Using the simplified polygon produced by the Douglas–Peucker algorithm, some researchers further strengthen the polylines by fitting the line segments. For example, line segments obtained using the least-squares method can be intersected in a pairwise manner to form a closed polygon [32] or selected for optimal configuration based on the minimum description length (MDL) principle [33,34]. In contrast, some studies fitted line segments through a wide range of approaches, such as detecting critical points for initial boundaries [35], pivot points [36], detecting line segments by Random Sample Consensus (RANSAC) [37], multi-scale line fitting [38], and casting the problem to Gaussian space [39], Hough space [40], or graph space [41]. Adjacent non-parallel segments produce a vertex in the polygon; otherwise, the segments are merged [42]. The above methods tend to fit local smooth line segments and connect them to form a closed polygon with loose (or even without) global constraints. However, when the position of the original boundary points deviates too much from the ground truth location due to data noise or imperfections in sampling during the data collection, the simplified results will not reconstruct the shape very accurately. In addition, the global regularities in the building boundaries are not well preserved.

In the third class of methods, the Manhattan assumption is adopted to produce simplified polygon boundaries [43–45]. It first detects the dominant direction of the polygon and then enforces the rectangular shape constraint on the segments. Many approaches have been used to conduct the first step, including intersecting horizontal lines in 3D space [46], rectangle fitting [43], weighted line-segment lengths analysis [19,45,47], longest fitted line detection in Euclidian space [26,48] or Hough space [40], vanishing points detection in image space [18], and histogram analysis [49–51]. Furthermore, some methods first divide all of the line segments into two orthogonal groups, then refit the dominant orientation by taking all of the original points into consideration [44,52]. For the second step, the two common approaches to reconstructing the simplified polygon are constrained least-squares fitting and graph cut [50,53]. Although the Manhattan assumption successfully characterizes most urban and indoor environments, a strict Manhattan constraint will lead to undesirable distortion in environments that do not conform to this assumption. To address this deficiency, some methods balance data fidelity and shape regularity while simplifying a given point set. By improving the traditional Manhattan assumption, some studies have loosely enforced the rectangular rules by allowing oblique edges [18,19,37,47,54]. Line segments are compared with a pre-defined threshold, and those with less deviation are forced to obey regularity rules; otherwise, their original orientation angles are kept, allowing the reconstruction of non-rectangular shaped buildings. Besides, multiple dominant orientations are assumed [55]. However, the definition of tilt edges is subjective and their potential mutual regularity is ignored. More recently, in related fields, energy functions have been built to explicitly express data fitting residuals and geometric regularity [56–60]. The parameters of an optimal structure can be retrieved by minimizing object functions.

In aforementioned building boundary regularization methods, the noises in ALS and photogrammetric point clouds sometimes make it difficult to regularize building boundaries due to the unstable estimation of edge orientation angles and their mutual regularity. Besides, inter-part relations between different buildings are not considered to produce regular boundaries in regional level. Therefore, in order to produce building boundary with inter-part regularity and a high-degree of data fidelity, in this paper, we propose a hierarchical regularization method, which composed of a local stage and a global stage. In the proposed method, we allow the shifts of the boundary points to resist noise, and in the global stage, inter-part regularities between different polygons are posed as a labeling problem that considers two competitive desires, preserving data fidelity and enforcing the same labels (orthogonal or parallel), and solve the problem with a standard graph cut.

3. Hierarchical Regularization of Building Boundaries from Noisy Point Clouds

3.1. Overview of the Approach

Staring from building point clouds in one region, planar structures are first extracted with simple parallel and orthogonal constraints using the existing RANSAC-based methods [61,62]. Then, the detected 3D planes that share the same normal orientations (parallel or coplanar) are grouped to conduct boundary tracing and boundary regularization, as depicted in the blue box in Figure 2. For each plane group, 3D points are projected to 2D space by translation their centroid to the origin point and rotating the normal of the plane to the positive direction of Z-axis, the consecutive boundary points of each plane, which are the input of the proposed methods, is extracted using alpha-shapes [63]. Besides, non-parallel planes are projected to each plane group to form virtual angles for subsequent global regularization.

As shown in Figure 2, the hierarchical regularization is composed of two main stages: the local stage and the global stage. The objectives of the two stages are complementary. A closed point boundary is shifted and divided into piecewise smooth line segments in the local stage. In the global stage, global parallel and vertical relationships between line segments in the overall area are discovered to further regularize edges, which results in highly regularized polygons.

In the local stage, neighboring relationships between consecutive building boundary points are first built using a forward search strategy; then the initial normal of each point is estimated using principal component analysis (PCA) with a fixed neighbor size. Based on the initial normal orientation and the neighboring relationship, a high-quality 2D normal of each point is reconstructed by minimizing the least-squares function, which decreases the normal differences between neighboring points while ensuring that the final normal does not deviate too much from the initial value. The positions of the initial points are then shifted along the refined normal to harmonize their positions and normal orientations. These points are then grouped according to the reconstructed normal, and line segments with limited endpoints are fitted. These lines have a finite extent that is clipped to a bounding box based on the projection of the points.

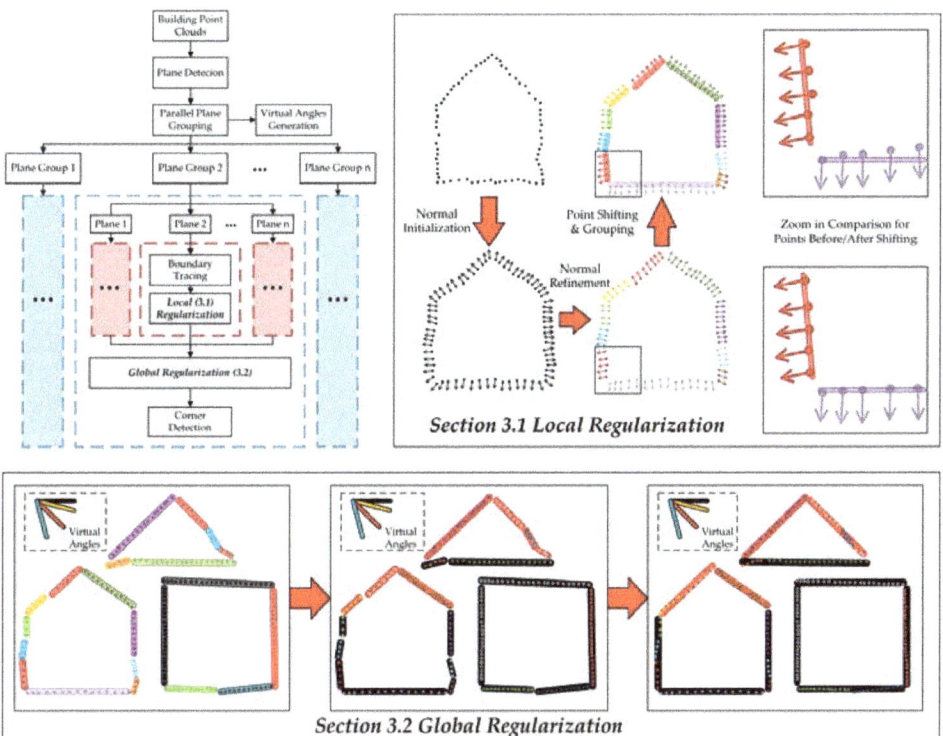

Figure 2. Overview of the proposed regularization method. The red boxes and blue boxes indicate for same types of procedures. In the global regularization, colors of segments and virtual angles are set randomly at first. Then same colors stands for selecting the same orientation angles.

In the global stage, each line segment along with its shifted points is set as an input to generate augmented line segments. The mutual parallel and vertical relations between each segment in the same group are used to form the augmentation. Besides, regularity clues from non-parallel planes are also discovered. Then, we construct an energy function that considers both data fitting cost and overall smoothness using the shifted points as observations. The energy function is minimized via graph cut and the result is highly regularized line segments in the global scene.

After the two stages of refinement, the resulting polygons are generated through a simple corner detection method that merges line segments with the same orientation and intersects. In the following section, we describe the two stages in detail.

3.2. Shiftable Line Fitting for Local Regularization

The goal of this stage is to robustly reconstruct for each boundary a piecewise smooth polygon that approximates the original outline. Since the noises in point clouds may hinder the estimation of point normal, inspired by [64], the refinement of normal vectors are necessary. Therefore, three steps are incorporated to refine the boundaries and produce a piecewise smooth polygon: neighborhood estimation, normal refinement, and shiftable points grouping. A neighbor relationship between points that belong to the same line is first estimated, then the 2D normal is computed and refined based on the assumption that neighboring points are more likely to have a uniform normal (e.g., numerically the included angle is less than 1 degree). After that, piecewise smooth polylines are grouped together in accordance with their 2D normal and the positions of the points are shifted along their 2D normal based on a least-squares function that measures the consistency of points' positions and their normal in a given tolerance which is depicted in the upper right of Figure 2. It should be noted that the shiftable line fitting is different from direct line fitting through RANSAC or other alternatives, in the ways that the proposed method are less sensitive to local noise and will create more local segments for the successive global labeling.

3.2.1. Outlier-Free Neighborhood Estimation

The main purpose of this phase is to identify outlier-free collinear relationships between consecutive points. Therefore, a forward search-based iterative method is used to estimate the neighboring region of a point which consists of collinear consecutive points around this point. The neighboring region of a given point starts from the point (itself). In each iteration, one point (the next consecutive one) is added to this region if the residual of the candidate point to the fitted least-squares linear equation of the last iteration is below a given threshold. To strictly prevent outliers and preserve sharp features, a small threshold is preferred. However, due to the discrete character of point clouds, it is meaningless to set threshold less than the Ground Sample Distance (GSD) and 1.5 GSD is chosen. The iteration stops when a candidate point is rejected, at which point the neighbor region detection of another point starts. Two points are treated as a neighborhood only when they appear in each other's neighbor region. If a point has no neighboring points, then it is treated as a potential outlier or as high-frequency noise and is not used in subsequent processing steps. When the mutual neighbor relationships between all of the points are defined, PCA is used to estimate the initial 2D normal vector of each segment.

3.2.2. Robust Normal Estimation of Boundaries

As 2D normal can be represented by rotation angles in 1D space, we transform the normal vector into angle space, which not only decreases the number of parameters but also promotes computational efficiency. Based on the assumption that neighbor points are more likely to have similar normal, the normal angle of each point is refined in a least-squares adjustment that considers both the initial normal angle and neighbor differences. The cost function is formed based on the neighboring relationships derived in the previous step. Two terms are incorporated, as shown in Equation (1):

$$\min_{\theta} \sum_{(p,q)\in N} \omega_{p,q} |\theta_p - \theta_q|^2 + \lambda \sum_p |\theta_p - \theta_p^0|^2 \tag{1}$$

where N is the neighbor region and p and q are two different points which have neighbor relationships; θ_p and θ_q are the unknown normal angles of points p and q after refinement; θ_p^0 is the initial value of θ_p; and λ controls the balance between data fidelity for initial estimated normal and consistency of normal vectors for neighbor points, a large value of λ would prevent the change of normal angle and it was set as 0.1 in all the experiments. ω is the weight between neighboring points and is computed as follows:

$$\omega_{p,q} = e^{-(\theta_p - \theta_q / \sigma)^4} \tag{2}$$

where σ defines the preferred angle difference, which penalizes big initial normal differences, and fixed at 15° in all the experiments below, because we observed that the deviations of estimated normal at smooth region are lower than this value for most of the cases; θ_p and θ_q are the same as in Equation (1).

The above optimization is a standard regularized least-squares optimization and a standard solver is adopted to solve this problem [65]. The first term of Equation (1) minimizes the angle difference between neighboring points and the second term prevents the normal angles from deviating too much from their initial values. As depicted in Figure 2, after the normal refinement, the normal angle differences of the sharp features are enlarged (see the rooftop and the lower left corner), whereas those in the smooth region are reduced (see points with the same color).

3.2.3. Line Fitting with Shiftable Points

Once the high-quality normal orientations of all of the inlier points have been reconstructed, the consistent positions of boundary points p' are updated by shifting them in the normal direction (represented by n) to smooth the initial boundary points p. The length of the shifting is quantized by a single scalar t, as $p' = p + t \times n$. In order to optimize the local shifting value, similar optimization is approached as defined in Equation (3). It should be noted that, unlike traditional line fitting, this procedure operates on individual point and may involve neighborhood information that does not belong to the same local segment.

$$\min_{t} \sum_{(p,q) \in N} \omega_{pq} \left(\left| (p' - q')^T n_q \right|^2 + \left| (q' - p')^T n_p \right|^2 \right) + \mu \sum_{p} t_p^2 \tag{3}$$

This is also a regularized least-squares problem, in which the weight ω is defined as the same as Equation (2) and gives larger weights to points with similar normal orientations. The left part measures the local deviation in the point's neighboring region N, which allows smoothness across different small segments. Because there are arbitrary many solutions to this free adjustment problem, the other term is also included to ensure data fidelity by constraining the final point position so that it does not shift too much from its initial position. The scale factor μ is used to balance normal smoothness and position deviation, a large value of μ would prevent the points to shift from their initial positions.

Once the smoothed positions are obtained, piecewise smooth polylines are constructed by grouping consecutive points that share similar normal angles, as the normal angles across sharp features have been magnified in the previous step. The shifted positions of the points belonging to each group are used to fit a least-squares line segment, and the corners are detected by intersecting lines with corresponding bounding boxes, based on the projection of points. As shown in Figure 2, after the local stage simplification, the initial polygon boundary points are represented by 10 edges and the sharp features are not lost.

3.3. Constrained Model Selection for Global Regularization

In the stage described above, the consecutive boundary points are integrated as piecewise smooth line segments composed of shifted points with a finite extent and unique orientation angles. These orientation angles may be contaminated by the noise that is inherent in the initial boundary points, leading to irregular polygons.

As man-made buildings always have regular shapes and buildings in nearby regions are likely to confirm some regional level regularity, it is reasonable to infer that all of the edges from the same region of building outlines are mutual related. These kind of inter-polyline relations are mainly parallelism and orthogonality. However, non-uniform edges may also exist in building outlines, and forcing them to become parallel (or perpendicular) to the dominant orientation may induce large distortions. Therefore, in the proposed method, the regularities are only softly enforced. We cast the regularization into the label space, in which the label of two segments are considered the same if they are collinear, parallel or orthogonal. In addition, long segments are preferred in the label inferencing. This problem is, formulated as a standard MRF and solved efficiently with graph cut [66].

3.3.1. Constrained Model Extension

In our hypothesis, the line segments that constitute the same polygon only have a limited number of orientation angles, and these angles are always represented by larger edges. As the orientation of initial angles may be affected by noise, especially short edges, robust detection of inter-polyline relations is a non-trivial problem. Inspired by [57], we extend the potential orientation angle of all of the line segments according to the expected inter-polyline relations. Although only one orientation angle is selected finally, the extended orientation angles help us to recover noisy or under-sampled line segments and allow us to make global optimal orientation angle selections.

In each plane group that contains detected planes that share the same normal vector, a line segment in 2D space is represented by the corresponding point sets as $p \in l = (c,o)^T$, in which c is the center position of the segment and o is the orientation angle of this line segment. In order to enhance the regularities, only a subset of all the possible values for the orientations should be selected. Therefore, all the orientations are first extended to pairs, which also consider the corresponding orthogonal direction as $\varphi = [o, o \pm \pi/2]$. The sign of the extended value is chosen properly so that the two orientations all fit into the range of $[0,\pi)$. Then the label space, Φ, is the union set of two parts Φ_1 and Φ_2. One part is formulated by all the pairs for all the segments as $\Phi_1 = \{\varphi_1{}^i \,|\, i = 1, 2 \dots, M\}$ and M is the total number of segments in this plane group. The other part, nominated as virtual angles in this paper, is formulated by the projections from non-parallel plane groups in the same region as $\Phi_2 = \{\varphi_1{}^i \,|\, i = 1, 2 \dots, N\}$ and N is the total number of plane groups that are not parallel with this plane group. Since building outlines are always mutual appeared in intersected planes and the function of a plane is estimated by a distinct larger amount of points than the function of a line segment, the label space in Φ_2 is more likely to reveal the real orientations.

In the model extension process, the center positions remain unchanged and only the orientation angles for each segment varies according to the selected label. Note that for a given original model and a given candidate angle, only one candidate model can be formed, either parallel or perpendicular. The purpose of the model extension is to ensure that the optimal configuration exists as a subset of the augmented candidate model set and that a regularized polygon can be simultaneously recovered by applying a proper model selection strategy.

3.3.2. Model Selection Using Graph Cut

Next, we need to select the optimal models from the extended candidate models. An initial model can only correspond to one final selected model. An energy function that considers both neighbor smoothness and data fitting degree is built to measure the regularity and data fitting errors, as below:

$$E(\varphi) = \sum_{l \in L} D(l, \varphi_l) + \lambda \sum_{(l,k) \in S} e^{(-|\varphi_l - \varphi_k|/\sigma)}$$
$$D(l, \varphi_l) = \sum_{p \in l} d_\perp^{\varphi_l}(p) \qquad (4)$$

in which, L represents all the line segments in this plane group and l is a line segment which is fitted by several shifted points. The first term measures the refitted error of the line segments, using the selected label orientation and means the distance of a point to the corresponding segments. And the second is the smoothness constraint, which penalizes similar initial angle models that are labeled differently. S defines the pairwise neighborhood of the segments, which includes pairs of models that have similar initial orientation angles; σ is the same as in Equation (2) (15° selected for all the experiments in this paper). λ controls the strength of the regularities, and if the value of λ is set to 0, non-regularity are considered in the optimization process.

We implement the graph cut algorithm [66,67] to minimize the energy function and convert the labeling result to corresponding orientation angles. Successive segments with the same orientations are merged to form a new segment and the corners are identified as the intersection of two non-parallel line segments. As depicted in Figure 2, after the global stage regularization, the piecewise smooth polygons are further merged and intersected to form new polygons with two orientation angles that represent the main structure of the original point boundary while preserving the parallel and perpendicular relationships between the main edges.

4. Experimental Evaluation

4.1. Description of the Test Data and Evaluation Methods

To test our algorithm, we evaluate the performance in two different application contexts: 2D building footprint generalization and 3D building model reconstruction. Furthermore, we use both photogrammetric point clouds generated from aerial oblique images and ALS data to test the robustness to the different type of noises. The datasets all cover urban areas, such as Hong Kong, and Toronto [68].

The generation of photogrammetric point clouds for the Hong Kong datasets is accomplished using off-the-shelves SfM/MVS solutions [69]. And, in the pre-processing, the building point clouds are extracted using existing footprints and post-processing interactions. After that, planes are detected via RANSAC [61,62,70]. In addition, other powerful methods could replace the methods we have chosen in this study. Planes which share the same normal are grouped together to be further regularized and the virtual angle between intersected plane groups are computed. For each plane group, the detected planes are first rotated and transformed into 2D space; then, border points are extracted by applying the 2D alpha-shapes. After regularization, the 2D polygons are transformed back into 3D space. The buildings and corresponding plane segmentations are illustrated in Figure 3, and their basic information is given in Table 1.

Figure 3. Photogrammetric point clouds and detected plans of test areas: (**a**) original building point clouds and (**b**) detected planes in the point clouds.

Table 1. Basic information about the photogrammetric test datasets.

Area Name	Point Number	Point Density (pt/m^2)	Detected Planes (Groups)
Centre	3,005,398	81	381 (29)

This study also tests the performance of the ALS dataset in 2D footprint generalization. The ALS data for two areas are acquired from downtown Toronto. The dataset was released by ISPRS [68] and two areas called Area-4 and Area-5 are used. Building points are extracted using existing building detection methods [71], then manually refined to correct some apparent mistakes (e.g., remove false detection and restore incomplete detection). Then, single buildings are obtained through point clustering and all of the points are projected to the ground plane to conduct boundary extraction and regularization. The ALS point clouds being tested are shown in Figure 4 and their basic information is given in Table 2.

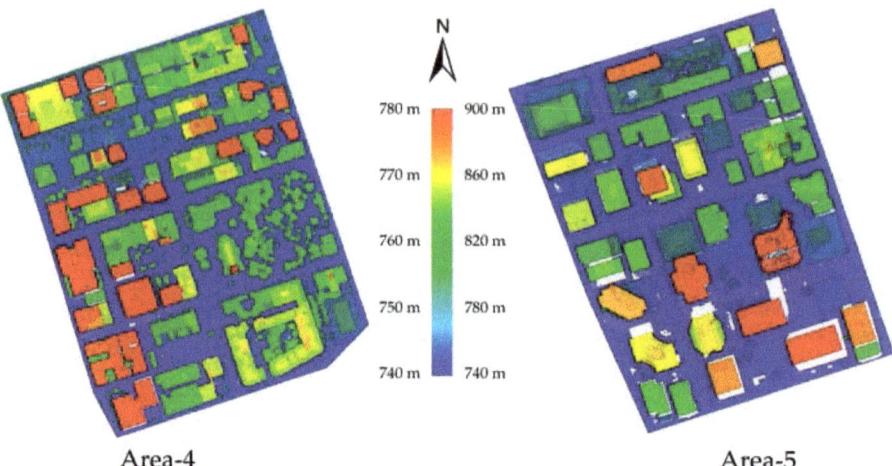

Figure 4. Aerial laser scanning (ALS) point clouds used in the 2D footprint generalization test (color shaded by altitude).

Table 2. Basic information about the ALS test datasets.

Area Name	Point Number	Point Density (pt/m^2)	Detected Planes (Groups)
Area-4	1,291,120	6.15	45 (1)
Area-5	1,138,977	5.42	34 (1)

We compare our algorithm with four representative methods. The first compared method is the well-known Douglas–Peucker (DP) algorithm (with the threshold from 0.2 m to 0.35 m in most cases) [27] and the second compared method is a polygon simplification algorithm published by [31] which we represent as SCS in the following part of this paper. The virtue of these two methods is to maintain of good approximation error and produce simplified polygons with approving data fidelity to the input boundary. Besides, we also compared with regularity-skilled methods, e.g., the dominant orientation based methods [47] (represent as DOB), and a method which balanced between regularity and fidelity [20] (represent as BRF). To give a fair comparison between different polyline simplification and regularization algorithms, the polygons are simplified to similar numbers of vertexes, e.g., changing the parameters in an adaptive manner. Both qualitative and quantitative comparisons are conducted. The output polygons are displayed in their original 3D or 2D space and are compared visually. In order to reveal to what extent do the output polygons approximate the input boundary (fidelity) and how regular do the polygons are (regularity), two indices are calculated to quantitatively measure them respectively. Fidelity is defined by the averages of the distances from the original boundary points to their corresponding output polylines, and regularity is represented by the distribution of the orientation angles that appear in each output polygon. Since the ALS datasets in Toronto are released with ground truth [68], the results are further investigated by calculating the root mean square (RMS) distance and the Hausdorff distance [72] to the reference boundaries. Note that, only footprints with 1:1 correspondence to the reference datasets are incorporated in the comparison.

4.2. Experimental Comparison of the Photogrammetric Point Clouds

As shown in Figure 5, the proposed method could produce neat polygons with a good quality that main structures of buildings are rather recognizable. The detailed comparisons for all the five tested methods are shown in Figures 6 and 7. Although all the five tested algorithms can simplify the polygon outlines to a certain degree, their visual qualities are different in detail. Since regularity constraints are not enforced, the results from SCS and DP could only recover the shape with approving errors, the parallel and the orthogonal relationship between edges are not preserved.

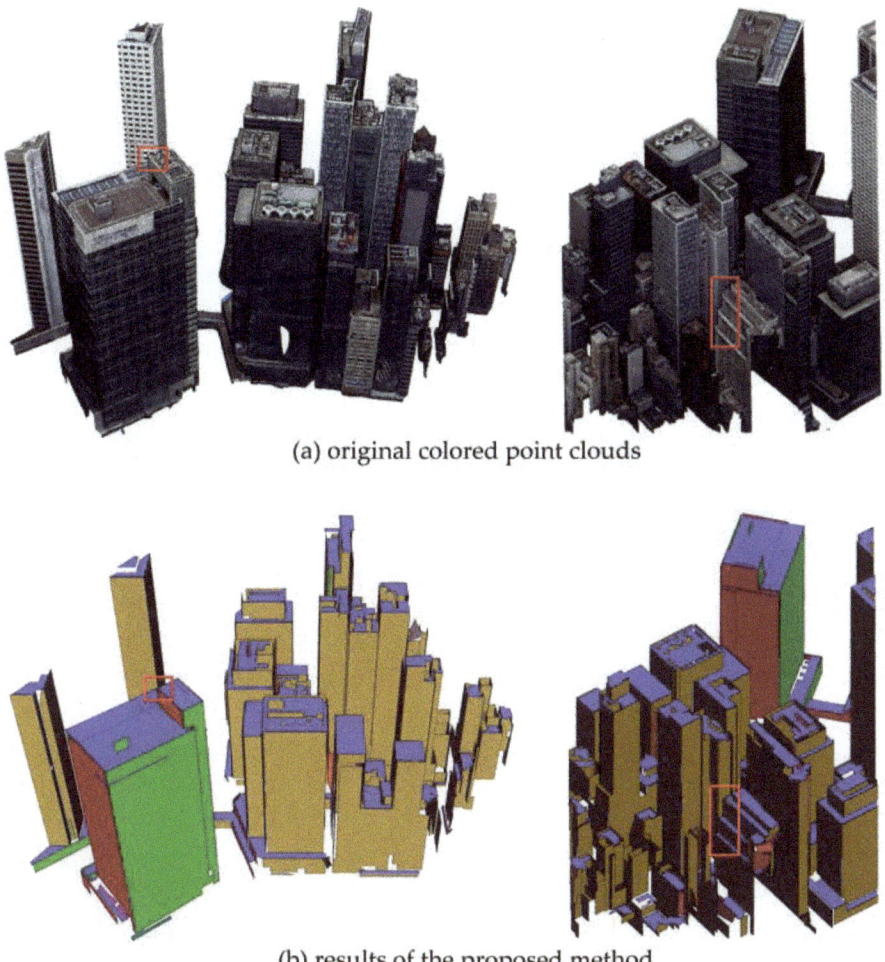

Figure 5. Visual comparison of the regularization results applied to photogrammetric point clouds from two complimentary views: (**a**) original colored point clouds and (**b**) the results obtained by the proposed method.

Meanwhile, for most rectangules the proposed methods and the method in BRF and DOB could correctly recover edges which belong to the major orientations. However, as non-dominant orientations are kept with their original orientations to receive a higher degree of data fidelity in DOB, the output polygons may contain a lot of tilt edges without mutual regularity. In contrast, the proposed methods and BRF could discover regularity constraints between non-dominant orientation edges and softly enforce mutual regularity constraints, resulting in polygons with overall inner regularity. When comparing reconstructed polygons in 3D, it is obvious that corners which shared by three planes are visually recognized from the proposed methods, the mutual orthogonal relationship between edges belong to different planes are also recovered as the intersection angles between different plane groups are served as virtual orientation angles in the global optimization process. This feature is potentially useful for topology reconstruction between polygons. Some polygon-snapping algorithms could use the results of our algorithm as input [56]. As shown in Figures 6 and 7, results from BRF, DOB, SCS, and DP are somehow irregular at building corners due to the inaccurate estimation of dominant orientation in noisy point clouds.

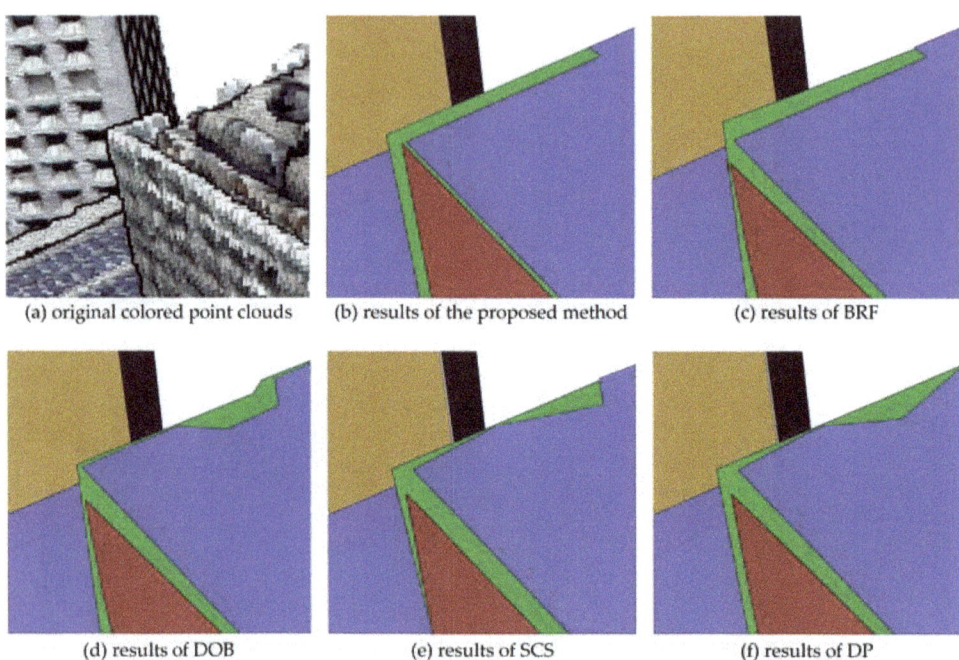

Figure 6. Zoom-in comparison of the regularization methods applied to photogrammetric point clouds in the red box region on the left of Figure 5. (**a**), original colored point clouds; (**b–f**), results of the proposed method, BRF [20], DOB [47], SCS [31], and DP [27] respectively.

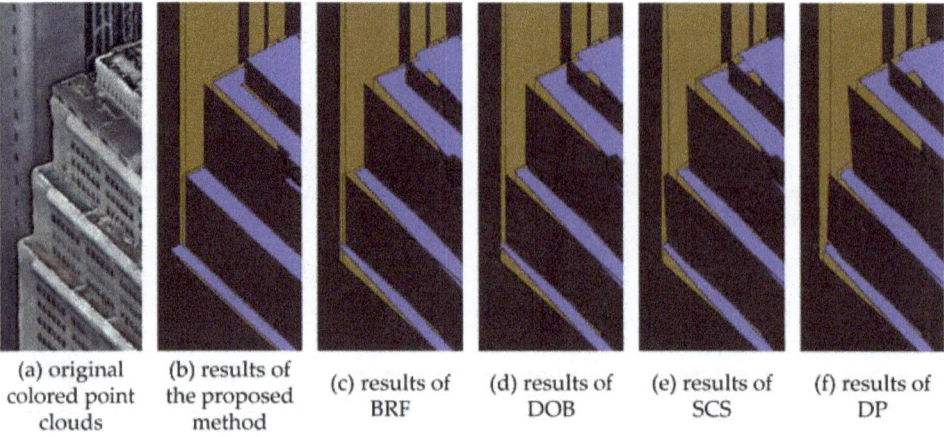

Figure 7. Zoom-in comparison of the regularization methods applied to photogrammetric point clouds in the red box region on the right of Figure 5. (**a**), original colored point clouds; (**b–f**), results of the proposed method, BRF [20], DOB [47], SCS [31], and DP [27] respectively.

The quantified fidelity evaluation is presented in Figure 8. As shown in the graph, for all five tested methods, the average residuals of the output polygon ranges from 0.181 m to 0.253 m while the numbers of edges are between 3049 and 4019. The number of edges for the proposed method is the smallest among all the five methods, stating that the output polygons of our method are quite concise. As for residuals, the minimum average value is 0.181 m which obtained by DOB and the value for the proposed method is slightly larger than that value (3 cm), which indicates a satisfying data fidelity of the output polygons by the proposed method.

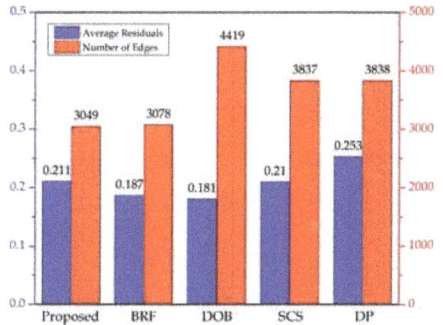

Figure 8. Average data fitting residuals and numbers of edges of the output polygons of the photogrammetric dataset. From left to right are the values for the proposed method, BRF [20], DOB [47], SCS [31], and DP [27], respectively.

In Figure 9, we illustrate the distribution of the orientation angles that appear in the output polygons. Orientation angles with intersection angle less than 0.1 degrees are identified as the same orientation angle. And, for same orientation angles, the length of their line segments are accumulated and divided by the total length of polygons in the scene to get the length percentage for this orientation angle. As shown in Figure 9, the proposed methods produce polygons with the minimum number of orientation angles and 1/3 of the total edges (in length) are shared by two orientation angles which orthogonal with each other. For the other four tested methods, orientation angles also concentred on several orientations but with less affinity. For DOB, edges belong to the dominant orientation are detected and complied, however, the remained edges are still irregular. Meanwhile, for BRF, several distinctive orientations for individual polygons are recovered but lacks inter-polygon regularity. As illustrated in Figure 9d,e, results from SCS and DP which focus on data approximation, regularity is not considered. The statistical values in Figures 8 and 9 indicate that our method produces global regularized polygons that fit the original point boundaries well and maintain data fidelity to a competitive degree, which conforms to the visual observations in Figures 5–7.

4.3. Experimental Comparison of ALS Point Clouds

The original alpha-shapes boundary of the 2D footprints from Area-4 and Area-5, together with the corresponding generalized polygons produced by the proposed method are shown in Figures 10 and 11, respectively. Detailed areas of the two environments are presented in Figures 12 and 13, respectively. Overall, all of the five methods are capable of generalizing building footprints while maintaining the main structure. Long edges are simplified with good data fidelity due to the preferable sampling of the original ALS data. However, the results of the detailed reconstructions of the short edges are different. Shorter edges, which are not as densely sampled as longer ones, are neatly reconstructed in the polygons produced by the proposed method, BRF, and DOB, but are distorted or deformed in the polygons produced by SCS and DP. For rectangular footprints, the output polygons for the Manhattan-based method are concise and regular. However, some tilt edges are not regularized, see Figures 12 and 13. As for the proposed method, nearly all the major structure of original boundaries are well preserved. Potential regularity between edges are discovered to enhance regularity, and non-dominant orientations are preserved to a large degree.

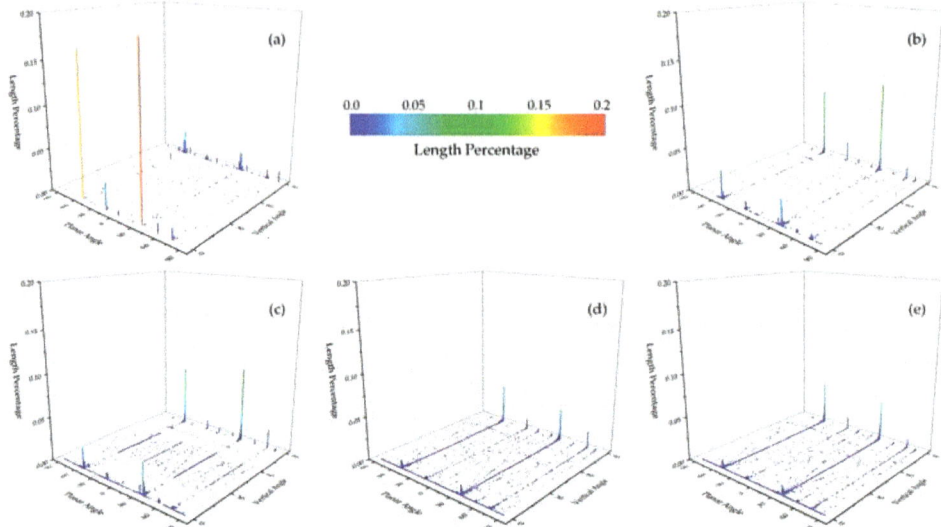

Figure 9. Comparison of the orientation angles that appear in the polygons output by different methods applied to the photogrammetric datasets. In each sub-graph, the dots and the corresponding drop lines mark the appearance of a certain orientation angle in 3D space. (**a**), results for the proposed method; (**b**–**e**), results for BRF [20], DOB [47], SCS [31], and DP [27] respectively.

(a) original alpha-shapes boundary

(b) results of the proposed method

Figure 10. Visual comparison of the 2D building footprint generalization in Area-4. (**a**) original alpha-shapes boundary; (**b**) results of the proposed method.

Figure 11. Visual comparison of the 2D building footprint generalization in Area-5. (**a**) original alpha-shapes boundary; (**b**) results of the proposed method.

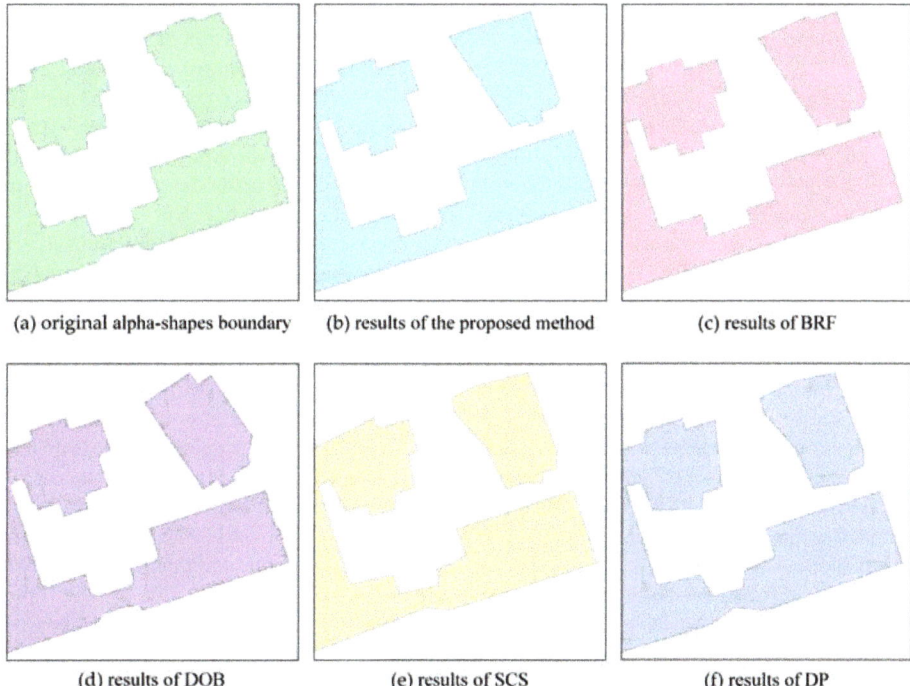

Figure 12. Zoom-in comparison of the regularization methods applied to ALS point clouds in the red box region of Figure 10. (**a**), original alpha-shapes boundary; (**b–f**), results of the proposed method, BRF [20], DOB [47], SCS [31], and DP [27] respectively.

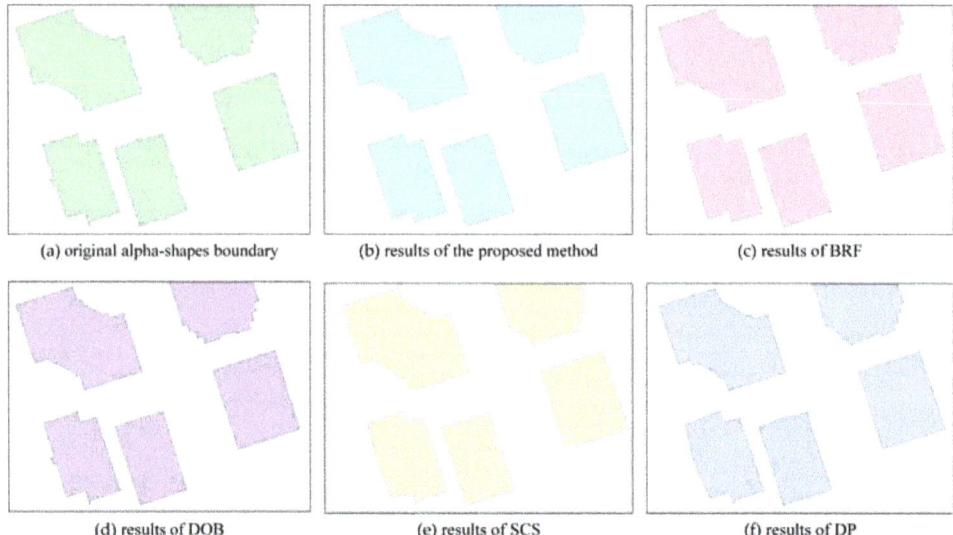

Figure 13. Zoom-in comparison of the regularization methods applied to ALS point clouds in the red box region of Figure 11. (**a**), original alpha-shapes boundary; (**b**–**f**), results of the proposed method, BRF [20], DOB [47], SCS [31], and DP [27] respectively.

The quantified fidelity evaluations of Area-4 and Area-5 are presented in Figure 14. As shown in the bar graph, for the polygons in Area-4 and Area-5, the RMS of the output polygons (setting the original boundary points as references) produced by the five methods are all lower than 0.3 m, and the overall residuals for Area-5 are lower than Area-4. Given the point density in the original ALS data (see Table 2), these results are relatively good. In both the two areas the output edge numbers of the proposed method are the smallest while the residuals for the proposed methods remain a comparable low level among the five test method, proving again the preferable fidelity of boundaries after the two-stage regularization. In the meantime, results for the Manhattan-based method in DOB is low indeed for they keep too many edges with original orientation angles. As a result, the output polygons by DOB contain the largest number of edges. Since the point density is relatively low, SCS and DP may lose sharp features and result in distorted polygons with large data fitting residuals.

In addition, we calculate the RMS and Hausdorff distance of the output polygons by comparing with the ground truth provided by the benchmark sponsors [68]. Apart from evaluating the performance of the five test methods, the RMS values are also compared with other benchmark participants' results reported in their publications [73–75]. As shown in Tables 3 and 4, the RMS for the proposed method is about 0.77 m in Area-4 and 0.68 m in Area-5 while the Hausdorff distances to the ground truth polygons are 1.28 m and 1.13 m, respectively. In all the five tested methods, the RMS and Hausdorff distance of the proposed method is the lowest in both the two test ALS areas, demonstrate that our method can outperform other boundary simplification and regularization methods. When comparing with other results which focus on building detection, the RMS value of the proposed method still shows comparative results. The reason may lie in the fact that the input building point clouds in this paper are extracted with manual post-processing, thus the input boundaries used here are more intact.

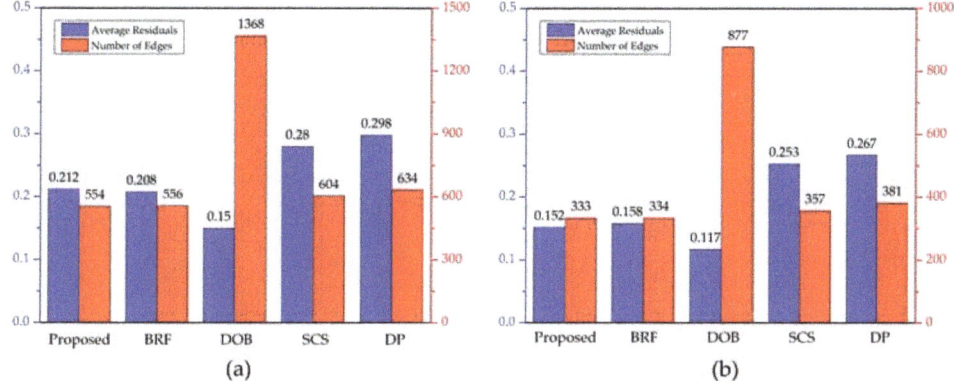

Figure 14. Average data fitting residuals and numbers of edges of the output polygons of the photogrammetric dataset, (**a**) for Area-4 and (**b**) for Area-5. From left to right are the values for the proposed method, BRF [20], DOB [47], SCS [31], and DP [27], respectively.

Table 3. Comparison of the root mean square (RMS) and Hausdorff distance in Area-4.

Method	RMS (m)	Hausdorff Distance (m)
Proposed	0.77	1.28
BRF [20]	0.79	1.35
DOB [47]	1.11	2.01
SCS [31]	0.94	1.56
DP [27]	1.09	1.86
CKU [73]	1.62	–
YOR [74]	0.80	–
MON2 [75]	0.96	–

Table 4. Comparison of the RMS and Hausdorff distance in Area-5.

Method	RMS (m)	Hausdorff Distance (m)
Proposed	0.68	1.13
BRF [20]	0.68	1.14
DOB [47]	1.02	2.00
SCS [31]	0.91	1.31
DP [27]	0.93	1.43
CKU [73]	1.68	–
YOR [74]	0.90	–
MON2 [75]	0.89	–

In Figures 15 and 16, we present the distribution graph of the orientation angles that appear in the output polygons. In Area-4, the output polygons produced by the proposed method are concentrated on 11 orientation angles and two of which, that deviate from each other by 90 degrees, contain more than 90% of edges in length. Meanwhile, in Area-5, our method produced polygons are concentrated on only five orientation angles. The resulting orientation angle histogram of BRF and DOB also show peaks at the two dominant orientations of the scene, but as inter-polygon constraints are not enforced, the regional level regularity is not preserved as well as the proposed method. As for the methods in SCS and DP, due to the low point density, the loss of sharp feature around corners result in irregular boundaries for small edges which are not well-sampled.

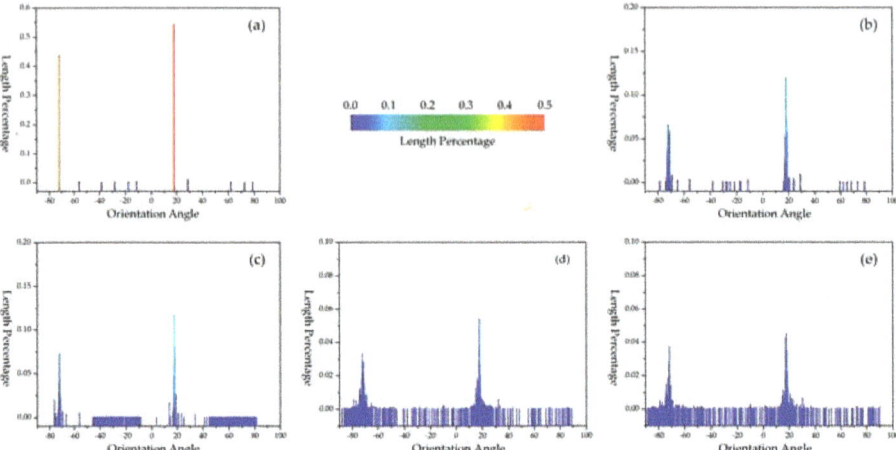

Figure 15. Comparison of the orientation angles that appear in the polygons output by different methods applied to the ALS datasets in Area-4. In each sub-graph, the dots and the corresponding drop lines mark the appearance of a certain orientation angle in horizontal plane. From (**a**–**e**) are the plots for the proposed method, BRF [20], DOB [47], SCS [31], and DP [27], respectively.

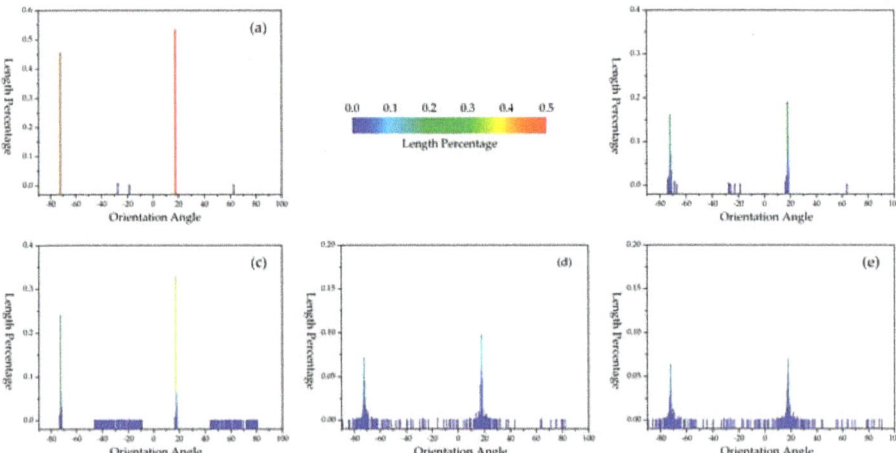

Figure 16. Comparison of the orientation angles that appear in the polygons output by different methods applied to the ALS datasets in Area-5. In each sub-graph, the dots and the corresponding drop lines mark the appearance of a certain orientation angle in horizontal plane. From (**a**–**e**) are the plots for the proposed method, BRF [20], DOB [47], SCS [31], and DP [27], respectively.

5. Discussions

The experiments on photogrammetric and ALS point clouds revealed that the output polygons for the proposed methods have presented better regularities than other methods. Although, in Figures 8 and 14, the average residuals for the proposed methods are larger than those for BRF [20] and DOB [47], it should be noted that, in these two experiments, the original boundary points are set as references to reveal absolute data fidelity; and the data fidelity and regularities are two competed desires. However, if using the ground truth labeled manually as the reference, the RMS values for the proposed methods became the smallest. This suggests that the proposed methods could resist the noise in original boundary and produce regularized polygons with inter-part regularity, see Figure 9, Figure 15, and Figure 16. Meanwhile, the Hausdorff distance which measures shape similarity also confirmed that the results of the proposed methods are the most similar with ground truth in shapes.

Besides, the parameters of the proposed method could be changed regarding the property of input point clouds and surveying area. In our test, the parameters setting strategy used in this paper are quite robust for the three test areas. The reason might be that these parameters stand for some physical meaning which descript the surveyed objects or the input point clouds.

However, there are also some limitations of the proposed methods. Firstly, compared with BRF [20], DOB [47], SCS [31], and DP [27], which regularized each plane of a single building separately, the proposed method works on the whole datasets (in the global stage) and computational complexity increased quadratically with the number of segments. Thus, further incremental processing strategy should be considered to make the method scalable to large datasets. Secondly, the extraction of planes from the point clouds, as a pre-processing stage for the proposed methods, is still an open problem in city scale, especially for the buildings with many fragmented parts.

6. Conclusions

In this study, a method for the hierarchical regularization of noise boundaries of buildings from ALS and photogrammetric large scale point clouds is presented. The method incorporates two stages of regularization that reconstruct piecewise smooth line segments and global regularized polygons in regional level. Qualitative and quantitative comparisons of the proposed method with existing methods have revealed the effectiveness of the proposed method in handling highly noisy point boundaries and producing polygons with both satisfactory regularity and data fidelity. The regularized polygons could fit original boundary points with 0.2 m in average residual while making more than 90% of edges to be regular (either parallel or orthogonal) with each other. The absolute RMS refers to ground truth is about 0.7m and the shape similarity is about 1.2 m in the Hausdorff distance.

In the future, three related problems should be solved. First, rather than just planar primitives, more complex regularities, such as concentric of sphere, cone, and cylinder, may also be investigated. Second, rather than the volumetric segmentation or 3D plane arrangement [55] approaches for building reconstruction, which is hard for interactively intervention when the quality of extracted plane is less desirable, the reconstructed polygons will be explored in an interactive environment, which is more practical in real-world applications. Third, alternative reconstruction paradigms, e.g., 3D edge extraction from point clouds then reconstruct building faces from closed edges, may also be incorporated with the hierarchical regularization methods to produce precision 3D models.

Author Contributions: L.X., Q.Z. and H.H. conceived and designed this study; L.X. and H.H. performed the experimental campaign, the implemented methodology; L.X. wrote the original draft; Q.Z. and B.W. supervised this study; Y.L. helped in software developing; Y.Z. and R.Z. helped in funding acquisition. All authors wrote the manuscript.

Funding: This study was supported by the National Natural Science Foundation of China (No. 41631174, 61602392, 41501421 and 41571392) and Open Fund of Key Laboratory of Urban Land Resource Monitoring and Simulation, Ministry of Land and Resource China (KF-2016-02-022).

Acknowledgments: The authors would like to acknowledge the provision of the Hong Kong data set by Ambit Geospatial Solution Limited and the provision of the Toronto data set by Optech Inc., First Base Solutions Inc.

Conflicts of Interest: The authors declare no conflict of interest.

References

1. Vosselman, G.; Maas, H. *Airborne and Terrestrial Laser Scanning*; Whittles Publishing: Caithness, UK, 2010; p. 336.
2. Berger, M.; Tagliasacchi, A.; Seversky, L.; Alliez, P.; Levine, J.; Sharf, A.; Silva, C. State of the art in surface reconstruction from point clouds. In *EUROGRAPHICS Star Reports*; The Eurographics Association: Geneve, Switzerland, 2014; pp. 161–185.

3. Moulon, P.; Monasse, P.; Marlet, R. Global fusion of relative motions for robust, accurate and scalable structure from motion. In Proceedings of the IEEE International Conference on Computer Vision (ICCV), Sydney, Australia, 1–8 December 2013; pp. 3248–3255.
4. Vu, H.; Labatut, P.; Pons, J.; Keriven, R. High accuracy and visibility-consistent dense multiview stereo. *IEEE Trans. Pattern Anal. Mach. Intell.* **2012**, *34*, 889–901. [CrossRef] [PubMed]
5. Hu, H.; Zhu, Q.; Du, Z.; Zhang, Y.; Ding, Y. Reliable spatial relationship constrained feature point matching of oblique aerial images. *Photogramm. Eng. Remote Sens.* **2015**, *81*, 49–58. [CrossRef]
6. Rupnik, E.; Nex, F.; Toschi, I.; Remondino, F. Aerial multi-camera systems: Accuracy and block triangulation issues. *ISPRS J. Photogramm. Remote Sens.* **2015**, *101*, 233–246. [CrossRef]
7. Xie, L.; Hu, H.; Wang, J.; Zhu, Q.; Chen, M. An asymmetric re-weighting method for the precision combined bundle adjustment of aerial oblique images. *ISPRS J. Photogramm. Remote Sens.* **2016**, *117*, 92–107. [CrossRef]
8. Hirschmuller, H. Stereo processing by semiglobal matching and mutual information. *IEEE Trans. Pattern Anal. Mach. Intell.* **2008**, *30*, 328–341. [CrossRef] [PubMed]
9. Hu, H.; Chen, C.; Wu, B.; Yang, X.; Zhu, Q.; Ding, Y. Texture-aware dense image matching using ternary census transform. *ISPRS Ann. Photogramm. Remote Sens. Spat. Inf. Sci.* **2016**, *3*, 59–66. [CrossRef]
10. Zlatanova, S.; Rahman, A.A.; Shi, W. Topological models and frameworks for 3D spatial objects. *Comput. Geosci.* **2004**, *30*, 419–428. [CrossRef]
11. Biljecki, F.; Ledoux, H.; Stoter, J. An improved LOD specification for 3D building models. *Comput. Environ. Urban Syst.* **2016**, *59*, 25–37. [CrossRef]
12. Singh, S.P.; Jain, K.; Mandla, V.R. Virtual 3D city modeling: Techniques and applications. *ISPRS-Int. Arch. Photogramm. Remote Sens. Spat. Inf. Sci.* **2013**, *2*, 73–91. [CrossRef]
13. Biljecki, F.; Stoter, J.; Ledoux, H.; Zlatanova, S.; Çöltekin, A. Applications of 3D City Models: State of the Art Review. *ISPRS Int. J. Geo-Inf.* **2015**, *4*, 2842–2889. [CrossRef]
14. Poux, F.; Neuville, R.; Nys, G.; Billen, R. 3D Point Cloud Semantic Modelling: Integrated Framework for Indoor Spaces and Furniture. *Remote Sens.* **2018**, *10*, 1412. [CrossRef]
15. Bosche, F.; Haas, C.T.; Akinci, B. Automated recognition of 3D CAD objects in site laser scans for project 3D status visualization and performance control. *J. Comput. Civ. Eng.* **2009**, *23*, 311–318. [CrossRef]
16. Lafarge, F.; Mallet, C. Creating large-scale city models from 3D-point clouds: A Robust Approach with Hybrid Representation. *Int. J. Comput. Vis.* **2012**, *99*, 69–85. [CrossRef]
17. Albers, B.; Kada, M.; Wichmann, A. Automatic extraction and regularization of building outlines from airborne LIDAR point clouds. In Proceedings of the 2016 XXIII ISPRS Congress, Prague, Czech Republic, 12–19 July 2016.
18. Cabral, R.; Furukawa, Y. Piecewise planar and compact floorplan reconstruction from images. In Proceedings of the IEEE Conference on Computer Vision and Pattern Recognition (CVPR), Columbus, OH, USA, 23–28 June 2014; pp. 628–635.
19. Lee, J.; Han, S.; Byun, Y.; Kim, Y. Extraction and regularization of various building boundaries with complex shapes utilizing distribution characteristics of airborne LIDAR points. *ETRI J.* **2011**, *33*, 547–557. [CrossRef]
20. Xie, L.; Hu, H.; Zhu, Q.; Wu, B.; Zhang, Y. Hierarchical regularization of polygons for photogrammetric point clouds of oblique images. *ISPRS-Int. Arch. Photogramm. Remote Sens. Spat. Inf. Sci.* **2017**, *42*, 35–40. [CrossRef]
21. Li, Z.; Openshaw, S. Algorithms for automated line generalization1 based on a natural principle of objective generalization. *Int. J. Geogr. Inf. Syst.* **1992**, *6*, 373–389. [CrossRef]
22. Garrido, A.; Garcia-Silvente, M. Boundary simplification using a multiscale dominant-point detection algorithm. *Pattern Recog.* **1998**, *31*, 791–804. [CrossRef]
23. Sester, M. Optimization approaches for generalization and data abstraction. *Int. J. Geogr. Inf. Sci.* **2005**, *19*, 871–897. [CrossRef]
24. Xie, J.; Feng, C. An integrated simplification approach for 3D buildings with sloped and flat roofs. *ISPRS Int. J. Geo-Inf.* **2016**, *5*, 128. [CrossRef]
25. Kada, M.; Luo, F. Generalisation of building ground plans using half-spaces. *Int. Arch. Photogram. Remote Sens. Spat. Inf. Sci.* **2006**, *36*, 2–5.
26. Xiao, Y.; Wang, C.; Li, J.; Zhang, W.; Xi, X.; Wang, C.; Dong, P. Building segmentation and modeling from airborne LiDAR data. *Int. J. Digit. Earth* **2015**, *8*, 694–709. [CrossRef]

27. Douglas, D.H.; Peucker, T.K. Algorithms for the reduction of the number of points required to represent a digitized line or its caricature. *Cartographica* **1973**, *10*, 112–122. [CrossRef]
28. Reumann, K.; Witkam, A. Optimizing curve segmentation in computer graphics. In Proceedings of the International Computing Symposium, Davos, Switzerland, 4–7 September 1973; pp. 467–472.
29. Zhao, Z.; Saalfeld, A. Linear-time sleeve-fitting polyline simplification algorithms. *Proc. AutoCarto* **1997**, *13*, 214–223.
30. Opheim, H. Fast data reduction of a digitized curve. *Geo-Processing* **1982**, *2*, 33–40.
31. Dyken, C.; Dæhlen, M.; Sevaldrud, T. Simultaneous curve simplification. *J. Geogr. Syst.* **2009**, *11*, 273–289. [CrossRef]
32. Kim, C.; Habib, A.; Mrstik, P. New approach for planar patch segmentation using airborne laser data. In Proceedings of the ASPRS, Tampa, FL, USA, 7–11 May 2007.
33. Weidner, U.; Förstner, W. Towards automatic building extraction from high-resolution digital elevation models. *ISPRS J. Photogramm. Remote Sens.* **1995**, *50*, 38–49. [CrossRef]
34. Jung, J.; Jwa, Y.; Sohn, G. Implicit Regularization for Reconstructing 3D Building Rooftop Models Using Airborne LiDAR Data. *Sensors* **2017**, *17*, 621. [CrossRef]
35. Xu, J.; Wan, Y.; Yao, F. A method of 3D building boundary extraction from airborne LIDAR points cloud. In Proceedings of the IEEE Symposium on Photonics and Optoelectronic (SOPO), Chengdu, China, 19–24 June 2010; pp. 1–4.
36. Li, M.; Nan, L.; Smith, N.; Wonka, P. Reconstructing building mass models from UAV images. *Comput. Graph.* **2016**, *54*, 84–93. [CrossRef]
37. Sester, M.; Neidhart, H. Reconstruction of building ground plans from laser scanner data. In Proceedings of the AGILE08, Girona, Spain, 4–8 August 2008; p. 111.
38. Oesau, S.; Lafarge, F.; Alliez, P. Indoor scene reconstruction using feature sensitive primitive extraction and graph-cut. *ISPRS J. Photogramm. Remote Sens.* **2014**, *90*, 68–82. [CrossRef]
39. Poullis, C. A framework for automatic modeling from point cloud data. *IEEE Trans. Pattern Anal. Mach. Intell.* **2013**, *35*, 2563–2575. [CrossRef]
40. Turker, M.; Koc-San, D. Building extraction from high-resolution optical spaceborne images using the integration of support vector machine (SVM) classification, Hough transformation and perceptual grouping. *Int. J. Appl. Earth Observ. Geoinf.* **2015**, *34*, 58–69. [CrossRef]
41. Ley, A.; Hänsch, R.; Hellwich, O. Automatic building abstraction from aerial photogrammetry. *ISPRS Ann. Photogramm. Remote Sens. Spat. Inf. Sci.* **2017**, *4*, 243–250. [CrossRef]
42. Cao, R.; Zhang, Y.; Liu, X.; Zhao, Z. 3D building roof reconstruction from airborne LiDAR point clouds: A framework based on a spatial database. *Int. J. Geogr. Inf. Sci.* **2017**, *31*, 1359–1380. [CrossRef]
43. Gross, H.; Thoennessen, U.; Hansen, W.V. 3D modeling of urban structures. *Int. Arch. Photogram. Remote Sens.* **2005**, *36*, W24.
44. Ma, R. Building Model Reconstruction from Lidar Data and Aerial Photographs. Ph.D. Thesis, The Ohio State University, Columbus, OH, USA, 2005.
45. Gamba, P.; Dell' Acqua, F.; Lisini, G.; Trianni, G. Improved VHR urban area mapping exploiting object boundaries. *IEEE Trans. Geosci. Remote Sens.* **2007**, *45*, 2676–2682. [CrossRef]
46. Vosselman, G. Building reconstruction using planar faces in very high density height data. *Int. Arch. Photogram. Remote Sens.* **1999**, *32*, 87–94.
47. Zhao, Z.; Duan, Y.; Zhang, Y.; Cao, R. Extracting buildings from and regularizing boundaries in airborne lidar data using connected operators. *Int. J. Remote Sens.* **2016**, *37*, 889–912. [CrossRef]
48. Awrangjeb, M. Using point cloud data to identify, trace, and regularize the outlines of buildings. *Int. J. Remote Sens.* **2016**, *37*, 551–579. [CrossRef]
49. Alharthy, A.; Bethel, J. Heuristic filtering and 3D feature extraction from LIDAR data. *Int. Arch. Photogram. Remote Sens. Spat. Inf. Sci.* **2002**, *34*, 29–34.
50. Li, M.; Wonka, P.; Nan, L. Manhattan-world urban reconstruction from point clouds. In *European Conference on Computer Vision*; Springer: Cham, Switzerland, 2016; pp. 54–69.
51. Pohl, M.; Meidow, J.; Bulatov, D. Simplification of polygonal chains by enforcing few distinctive edge directions. In *Scandinavian Conference on Image Analysis*; Springer: Cham, Switzerland, 2017; pp. 3–14.
52. Sampath, A.; Shan, J. Building boundary tracing and regularization from airborne LiDAR point clouds. *Photogramm. Eng. Remote Sens.* **2007**, *73*, 805–812. [CrossRef]

53. Guercke, R.; Sester, M. Building footprint simplification based on hough transform and least squares adjustment. In Proceedings of the 14th Workshop of the ICA Commission on Generalisation and Multiple Representation, Paris, France, 3–8 July 2011.
54. Ikehata, S.; Yang, H.; Furukawa, Y. Structured indoor modeling. In Proceedings of the IEEE International Conference on Computer Vision, Santiago, Chile, 11–18 December 2015; pp. 1323–1331.
55. Gilani, S.; Awrangjeb, M.; Lu, G. An Automatic Building Extraction and Regularisation Technique Using LiDAR Point Cloud Data and Orthoimage. *Remote Sens.* **2016**, *8*, 258. [CrossRef]
56. Arikan, M.; Schwärzler, M.; Flöry, S.; Wimmer, M.; Maierhofer, S. O-snap: Optimization-based snapping for modeling architecture. *ACM Trans. Graph.* **2013**, *32*, 6. [CrossRef]
57. Monszpart, A.; Mellado, N.; Brostow, G.J.; Mitra, N.J. RAPter: Rebuilding man-made scenes with regular arrangements of planes. *ACM Trans. Graph.* **2015**, *34*, 1–12. [CrossRef]
58. Nan, L.; Jiang, C.; Ghanem, B.; Wonka, P. Template assembly for detailed urban reconstruction. *Comput. Graph. Forum* **2015**, *34*, 217–228. [CrossRef]
59. Favreau, J.; Lafarge, F.; Bousseau, A. Fidelity vs. simplicity: A global approach to line drawing vectorization. *ACM Trans. Graph.* **2016**, *35*, 120. [CrossRef]
60. Wang, J.; Fang, T.; Su, Q.; Zhu, S.; Liu, J.; Cai, S.; Tai, C.; Quan, L. Image-based building regularization using structural linear features. *IEEE Trans. Visual. Comput. Graph.* **2016**, *22*, 1760–1772. [CrossRef]
61. Schnabel, R.; Wahl, R.; Klein, R. Efficient RANSAC for point-cloud shape detection. *Comput. Graph. Forum* **2007**, *26*, 214–226. [CrossRef]
62. Verdie, Y.; Lafarge, F.; Alliez, P. LOD generation for urban scenes. *ACM Trans. Graph.* **2015**, *34*. [CrossRef]
63. Guo, B.; Menon, J.; Willette, B. Surface reconstruction using alpha shapes. In *Computer Graphics Forum*; Wiley Online Library: Hoboken, NJ, USA, 1997; pp. 177–190.
64. Avron, H.; Sharf, A.; Greif, C.; Cohen-Or, D. ℓ 1-Sparse reconstruction of sharp point set surfaces. *ACM Trans. Graph.* **2010**, *29*, 135. [CrossRef]
65. Agarwal, S.; Mierle, K. Ceres Solver. 2016. Available online: http://ceres-solver.org/ (accessed on 25 October 2018).
66. Kolmogorov, V.; Zabih, R. What energy functions can be minimized via graph cuts? *IEEE Trans. Pattern Anal. Mach. Intell.* **2004**, *26*, 147–159. [CrossRef]
67. Boykov, Y.; Veksler, O.; Zabih, R. Fast approximate energy minimization via graph cuts. Pattern Analysis and Machine Intelligence. *IEEE Trans. Pattern Anal. Mach. Intell.* **2001**, *23*, 1222–1239. [CrossRef]
68. Rottensteiner, F.; Sohn, G.; Gerke, M.; Wegner, J.D. ISPRS test project on urban classification and 3D building reconstruction. In *Commission III-Photogrammetric Computer Vision and Image Analysis*; Working Group III/4-3D Scene Analysis; 2013; pp. 1–17. Available online: http://www2.isprs.org/tl_files/isprs/wg34/docs/ComplexScenes_revision_v4.pdf (accessed on 9 December 2018).
69. Bentley. Context Capture Create 3D Models from Simple Photographs. 2018. Available online: https://www.bentley.com/en/products/brands/contextcapture (accessed on 25 October 2018).
70. Rusu, R.B.; Cousins, S. 3D is here: Point cloud library (PCL). In Proceedings of the 2011 IEEE International Conference on Robotics and automation (ICRA), Shanghai, China, 9–13 May 2011; pp. 1–4.
71. Zhu, Q.; Li, Y.; Hu, H.; Wu, B. Robust point cloud classification based on multi-level semantic relationships for urban scenes. *ISPRS J. Photogramm. Remote Sens.* **2017**, *129*, 86–102. [CrossRef]
72. Huttenlocher, D.P.; Klanderman, G.A.; Rucklidge, W.J. Comparing images using the Hausdorff distance. *IEEE Trans. Pattern Anal. Mach. Intell.* **1993**, *15*, 850–863. [CrossRef]
73. Rau, J.Y. A Line-based 3D Roof Model Reconstruction Algorithm: Tin-Merging and Reshaping (TMR). *ISPRS Ann. Photogramm. Remote Sens. Spat. Inf. Sci.* **2012**, *3*, 287–292. [CrossRef]
74. Sohn, G.; Jwa, Y.; Jung, J.; Kim, H. An Implicit Regularization for 3D Building Rooftop Modeling Using Airborne LIDAR Data. *ISPRS Ann. Photogramm. Remote Sens. Spat. Inf. Sci.* **2012**, *1*, 305–310. [CrossRef]
75. Awrangjeb, M.; Lu, G.; Fraser, C. Automatic Building Extraction from LIDAR Data Covering Complex Urban Scenes. *ISPRS Ann. Photogramm. Remote Sens. Spat. Inf. Sci.* **2014**, *40*, 25–32. [CrossRef]

© 2018 by the authors. Licensee MDPI, Basel, Switzerland. This article is an open access article distributed under the terms and conditions of the Creative Commons Attribution (CC BY) license (http://creativecommons.org/licenses/by/4.0/).

Article

Extraction of Buildings from Multiple-View Aerial Images Using a Feature-Level-Fusion Strategy

Youqiang Dong [1,2], Li Zhang [2], Ximin Cui [1,*], Haibin Ai [2] and Biao Xu [2,*]

1. College of Geoscience and Surveying Engineering, China University of Mining and Technology (Beijing), Xueyuan Road DING No. 11, Beijing 100083, China; dyqstruggling@126.com
2. Chinese Academy of Surveying and Mapping, Lianhuachixi Road No. 28, Beijing 100830, China; zhangl@casm.ac.cn (L.Z.); ahb32@163.com (H.A.)
* Correspondence: cxm@cumtb.edu.cn (X.C.); biaoxv@casm.ac.cn (B.X.); Tel.: +86-10-6233-9305 (X.C.); +86-10-6388-0531 (B.X.)

Received: 27 September 2018; Accepted: 28 November 2018; Published: 4 December 2018

Abstract: Aerial images are widely used for building detection. However, the performance of building detection methods based on aerial images alone is typically poorer than that of building detection methods using both LiDAR and image data. To overcome these limitations, we present a framework for detecting and regularizing the boundary of individual buildings using a feature-level-fusion strategy based on features from dense image matching (DIM) point clouds, orthophoto and original aerial images. The proposed framework is divided into three stages. In the first stage, the features from the original aerial image and DIM points are fused to detect buildings and obtain the so-called blob of an individual building. Then, a feature-level fusion strategy is applied to match the straight-line segments from original aerial images so that the matched straight-line segment can be used in the later stage. Finally, a new footprint generation algorithm is proposed to generate the building footprint by combining the matched straight-line segments and the boundary of the blob of the individual building. The performance of our framework is evaluated on a vertical aerial image dataset (Vaihingen) and two oblique aerial image datasets (Potsdam and Lunen). The experimental results reveal 89% to 96% per-area completeness with accuracy above almost 93%. Relative to six existing methods, our proposed method not only is more robust but also can obtain a similar performance to the methods based on LiDAR and images.

Keywords: building detection; aerial images; feature-level-fusion; straight-line segment matching; occlusion; building regularization technique

1. Introduction

Buildings, as key urban objects, play an important role in city planning [1,2], disaster management [3–5], emergency response [6], and many other application fields [7]. Due to the rapid development of cities and the requirement for up-to-date geospatial information, automatic building detection from high-resolution remote sensing images remains a primary research topic in the communities of computer vision and geomatics. Over the past two decades, a variety of methods have been proposed for automatic building detection. Based on the types of input data sources, the existing automatic building detection methods can be divided into two categories [8]:

- Single-source-data-based methods, where the data include Airborne Laser Scanning (ALS) point clouds [9,10], ALS-based Digital Surface Model (DSM) grids [11,12], and images [13,14].
- Multisource-data-based methods, where the methods include fusion of ALS-based digital DSM grid data and orthophoto data [15,16] and fusion of ALS data and the images [17].

Many researchers have reported that multisource-data-based methods perform better than single-source-data-based methods [18–20]. This is mainly because multisource-data-based methods use not only the spectral features provided by images but also the height information for the building detection. However, the cost of multisource-data-based methods may be higher than that of single-source-data-based methods. Moreover, with the fast development of multi-camera aerial platforms and dense matching techniques, reliable and accurate Dense Image Matching (DIM) point clouds can be generated from the overlapping aerial images [21]. Under this condition, instead of using multisource remote sensing data, the approach of solely employing aerial images to extract buildings in complex urban scenes is feasible. The building extraction mainly contains building detection and the regularization of the building boundary in this paper. Hence, we first discuss the related works in building detection paradigm only using aerial images and then cover the relevant literature on boundary regularization in the following work.

From the perspective of the photogrammetric processing, the DIM point cloud, DSM and orthophoto data can be generated from original aerial images, and all these data can provide various features for building extraction. According to the sources of the features used in the process of building detection, we distinguish aerial-image-based methods into four groups: methods using features from images (including original aerial images or orthophoto), methods using features from DIM point clouds, methods fusing the features from orthophoto and DSM and methods fusing the features from aerial images and DIM point clouds.

The first two methods mainly make use of the features from either the images or the DIM point clouds to detect buildings. The images mainly contain the spectral features (such as the color, tone, and texture) and spatial features (such as the area and shape). Based on these features, both pixel-based and segment-/object-oriented classification methods are proposed [22,23] for building detection. Because the pixel-based methods have a salt-and-pepper effect, researchers prefer the latter. The methods in [24–26] are several examples. However, objects of the same type may appear to have different spectral signatures, whereas different objects may appear to have similar spectral signatures under various background conditions. Therefore, the methods using features from images alone cannot obtain satisfactory performance, particularly in a complex urban scene.

The DIM point clouds mainly provide the height information for buildings detection. Although the cues from DIM point clouds are more robust than the spectral features from images, the 3-Dimensional (3D) shape features of buildings and trees are similar, as shown in Figure 1a–d, which increases the challenges of building detection using DIM point clouds in an urban scenario. In addition, the noise of the DIM point clouds is higher than the ALS data, which causes that the methods based on ALS data for building detection may be not suitable for DIM point clouds. Therefore, compared with the first two methods, the latter two methods are more popular in building detection.

The methods in [27–29] are some examples of the methods fusing the features from orthophoto and DSM. Compared with the methods solely using either the spectral information from images or height information from DIM point clouds, the detection results of the methods fusing the features from orthophoto and DSM were more robust and had greater accuracy. However, some disadvantages should not be ignored. On the one hand, compared with the original images, the orthophoto introduces the wrapping phenomenon, as shown in Figure 1e,f. This results that the features from orthophoto, such as texture, cannot reflect the true objects. On the other hand, the DSM, which represent the elevation of the tallest surfaces at that point, is unable to provide the information about the occluded objects such as low buildings. Hence, in terms of the features, the methods fusing the DIM point clouds and original aerial images have more potential for building detection. Our proposed building detection method falls under this category.

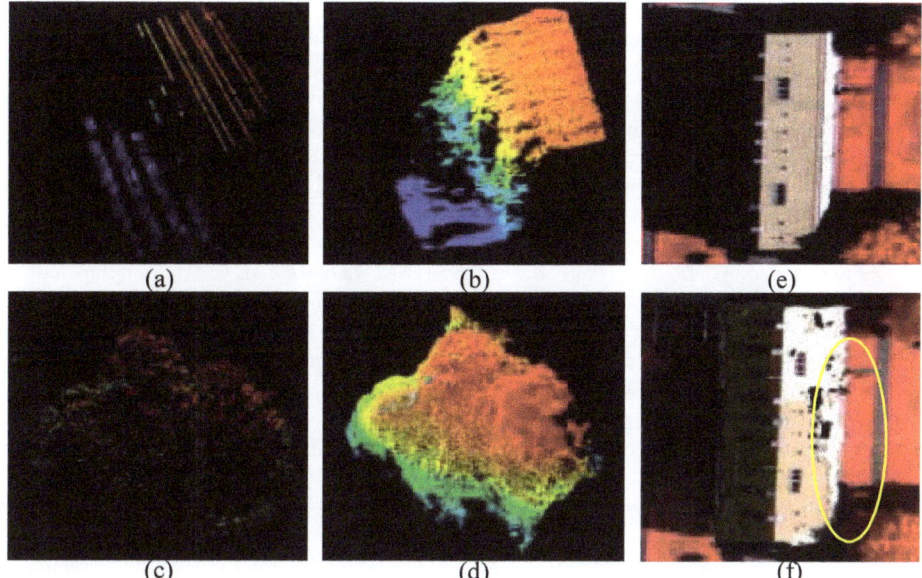

Figure 1. (**a**,**b**) the building from Airborne Laser Scanning (ALS) data and Digital Surface Model (DSM) point clouds, respectively; (**c**,**d**) the tree from ALS data and dense image matching (DIM) point clouds, respectively; (**e**,**f**) the building from original aerial image and orthophoto, respectively.

Some approaches based on the fusion of DIM point clouds and original aerial images for building detection have been proposed. Xiao et al. [30] makes use of the façade features that are detected from oblique aerial images using edge and height information to extract the buildings first; then, the DIM point cloud is employed to verify the detected buildings. In this method, façades are the most important features, which leads to the situation where the small or low buildings are removed due to the loss of windows on the façades. Rau et al. [31] proposed a method based on a rule hierarchical semantic network. First, this method makes use of a multi-resolution segmentation technique to segment the images into patches and calculate the features of each patch. Subsequently, the object height and gradient features from DIM point clouds and the spectral features from the patch are combined to classify the objects into correct categories. Finally, the DIM point clouds are classified into the correct classes with the aid of back projection. However, this method relies on experience to set the classification thresholds, which severely affected the accuracy of the classification. Second, the method makes use of only nine features to classify the DIM point clouds into five objective categories, resulting in an accuracy of 81% for buildings in National Cheng Kung University (NCKU), Taiwan campus, which is a flat area.

Based on the above analysis, a new building detection method using a feature-level-fusion strategy is proposed in this paper. Specifically,

- Filter the DIM point clouds to get the non-ground points.
- Apply the object-oriented classification method to detect vegetation with the aid of the features from the original aerial images.
- Make use of the back-projection to remove the tree points from the non-ground points so that we can obtain the building DIM points according to classified original aerial images.
- Create the building mask using the building DIM points.

The second task in this paper is the regularization of the building boundary after the building detection. In terms of the boundary regularization, multiple methods [32–38] have been proposed.

Most of these methods are aimed at the LiDAR point clouds. However, the accuracy of the detected boundaries is often compromised due to the point cloud sparsity. In fact, the nature of building boundary regularization is to refine and delineate the boundary of a building mask. If the rough building edges of a building mask are replaced by straight-line segments that present the true building edges, then the building boundary can be regularized. A previous study [39] made use of the straight-line segments from orthophoto to regularize the building footprint, and the results showed that a larger number of straight-line segments corresponds to higher performance. Therefore, we aim to extract robust lines as much as possible in our proposed method to assist the building footprint regularization.

Besides the orthophoto, all the original aerial images, DIM point clouds and the building mask can provide straight-line features. Compared with the photogrammetric products, original aerial images can provide more line segments. Hence, we choose the straight-line segments from original aerial images to replace the rough edge of building mask so that we can regularize the building boundary. Furthermore, these segments have higher accuracy.

However, the straight-line segments from original aerial images are located in 2D image space. Hence, the line matching is necessary so that these 2D straight-line segments can be converted into 3D straight-line segments. In fact, line matching is a challenging task [40,41]. In [41], Habib et al. projected the two-dimensional lines from left and right images onto the roof extracted from the ALS data. If the two projected lines onto the planar coordinates satisfy the given thresholds, the two lines can be regarded as a pair of corresponding lines. In this method, the LiDAR point clouds need to be segmented into the planar. For DIM point clouds, the segmentation is a problem because of the high noise of DIM point clouds. Therefore, a new strategy of the straight-line segments matching is proposed with the aid of the straight-line features from the building mask, orthophoto and DIM point clouds. The details of this strategy are described as follows:

- Extract the coarse building edges from the DIM points, orthophoto and the blob of the individual building.
- Extract the building edge from the original aerial images and match these straight-line segments with the help of the coarse building edges.

Theoretically, the boundary of building mask cannot be replaced with the matched lines completely. Therefore, the straight-line segments from the blob of the individual building are still essential. We can integrate these two kinds of lines to generate the closed building footprint. Sections 2.3 and 2.4 show the details. In this paper, our main contributions include the following:

- Make use of the aerial images alone to detect the buildings by the combination of the features from original aerial images and DIM point clouds.
- A new straight-line segment matching strategy based on three images is proposed with the help of the coarse building edge from the DIM points, orthophoto and the blob of the individual building.
- In the regularization stage, a new strategy is proposed for the generation of a building footprint.

The rest of this paper is organized as follows: Section 2 details the proposed building detection and the boundary regularization techniques. Section 3 presents the performance and discusses the experimental results using three test datasets followed by a comparative analysis. Finally, Section 4 concludes this paper.

2. Our Proposed Method

Our proposed approach for building detection and boundary regularization consists of four stages: (1) The generation of DIM point clouds and orthophoto from original aerial images; (2) individual building detection from DIM point clouds with the aid of original aerial images; (3) building edge detection using a feature-level fusion strategy; and (4) regularization of building boundaries by the fusion of matched lines and boundary lines. The entire workflow is shown in Figure 2.

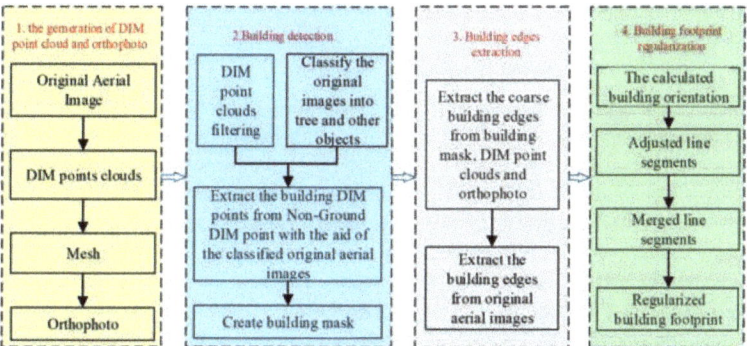

Figure 2. Entire workflow of building detection and boundary regularization as described in this paper.

2.1. Generation of DIM Point Clouds and Orthophoto from Original Aerial Images

Our proposed method takes the aerial images as an input. The features from aerial images and DIM points will be employed in the process of detecting buildings and the straight-line features from orthophoto will be used to help building edge detection. Therefore, the generation of DIM point clouds and true orthophoto from original aerial images is the top priority of our proposed method.

In this paper, we applied the commercial package Agisoft PhotoScan [42] which is able to automatically orient and match large datasets of images to generate the DIM point clouds and orthophoto. Due to commercial considerations, little information is available concerning the internal algorithms employed. Hence, we will describe the process of the generation of the DIM point clouds and orthophoto based on the workflow of Agisoft PhotoScan. Specially,

- Add the original images, the positioning and orientation system (POS) data and the camera calibration parameters into this software.
- Align photos. In this step, the point features are detected and matched; and the accuracy camera locations of each aerial image are estimated.
- Build the dense points clouds. According to a previous report [21], a stereo semi-global matching like (SGM-like) method is used to generate the DIM point clouds.
- Make use of the DIM points to generate a mesh. In nature, the mesh is the DSM of the survey area.
- Build the orthophoto using the generated mesh and the original aerial images.

Figure 3a,b shows the generated orthophoto and the DIM point clouds of the test area 3 of the Vaihingen dataset.

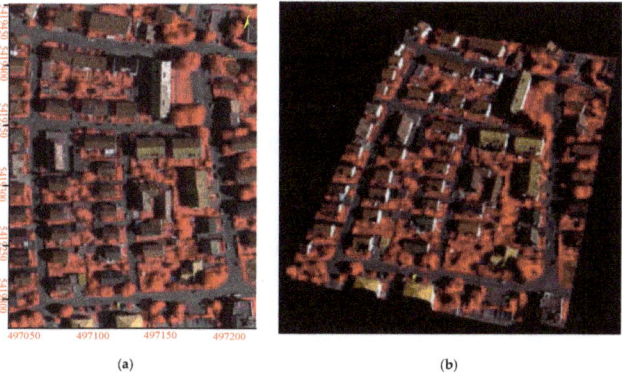

Figure 3. Derived photogrammetric products (**a**) the generated orthophoto; (**b**) DIM point clouds.

2.2. Building Detection from DIM Point Clouds with the Aid of the Original Aerial Image

2.2.1. Filtering of DIM Point Clouds

Filtering of the point clouds is the process of separation of ground points from the non-ground points and is the first step to detect buildings. Multiple filtering methods [43,44] have been proposed. Among these methods, the progressive triangular irregular network (TIN) densification (PTD) algorithm [45] is one of the most popular algorithms in engineering applications. However, this method may fail to extract the ground points because the density and the noise of DIM point clouds are higher than that of the ALS point cloud. Hence, an improved PTD algorithm [46] is selected

The improved PTD is divided into two steps. The first step is selecting seed points and constructing the initial TIN. The second step is an iterative densification of the TIN. The largest differences between the PTD and the improved PTD is the procedure for calculating densification thresholds. In the improved PTD, the angle threshold changes from high to low, and with the increase of the density of points added into the TIN, the angle threshold becomes large. Using this method, we can remove the ground points and low objectives points. The remaining points only contain building and tree points.

2.2.2. Object-Oriented Classification of Original Aerial Images

Generally, it is easier to detect trees from aerial images than to detect buildings. Based on this, we classify the aerial images into two categories: trees and other objects. Here, the commercial classification software eCognition 9.0.2 is used to detect trees from each aerial images. In this section, the original aerial images are classified, so that the classification results of each original aerial image can be employed in the later stage. The process of object-oriented classification is divided into three steps:

- Multi-resolution segmentation of the aerial image. This technique is used to extract reasonable image objects. In the segmentation stage, several parameters, such as layer weight, compactness, shape, and scale, must be determined in advance. These algorithms and related parameters are described in detail in [47]. Generally, the parameters were determined through visual assessment as well as trial and error. We set the scale factor as 90 and set the weights for red, green, blue and straight-line layers are 1, 1, 1, and 2, respectively. The shape and compactness parameters are 0.3 and 0.7, respectively.
- Feature selection. The normalized difference vegetation index (NDVI) has been used extensively to detect vegetation [48]. However, relying on the NDVI alone to detect the vegetation is not accurate due to influence of shadows and colored buildings on aerial images [16]. Therefore, besides the NDVI, the texture information in the form of entropy [49] is also used on the basis of the observation that trees are rich in texture and have higher surface roughness than building roofs [16]. Moreover, R, G, B, and brightness are also selected as the features in our proposed method. Notably, if the near-infrared band is not available, color vegetation indices can be calculated from color aerial images. In this paper, we applied the green leaf index (GLI) [50,51] to replace NDVI. The formula of the GLI is expressed in the formula (1). In this formula, R, G and B represents the value of the red, green and blue bands of each pixel from original aerial image, respectively.

$$GLI = (2 \times G - R - B)/(2 \times G + R + B) \qquad (1)$$

- Supervised classification of segments using a Random Forest (RF) [52]. The reference labels are created by an operator. The computed feature vector per segment is fed into the RF learning scheme. To monitor the quality of learning, the training and prediction is performed several times.

2.2.3. Removal of Tree Points from Non-Ground DIM Points

In this stage, the tree DIM points will be removed from non-ground DIM points with the aid of the classified original aerial images. The process of removing tree points from non-ground DIM points is described as follows:

- At first, define a vector L. The size of L is equal to the number of non-ground DIM points and each element within this vector is used to mark the category of each corresponding non-ground DIM point. In the initial stage of this process, we set the value of each element within this vector to 0, which indicates that the corresponding DIM point is unclassified.
- Then, select a classified original aerial image and make use of this classified original aerial image to label the category of each element within the vector L. The fundamental of this step is back projection. If the calculated projected point of a non-ground DIM point falls within the region which is labelled as tree in the selected image, the corresponding element within L plus 1; otherwise, the corresponding element within L minus 1.
- Continue the second step until all classified original aerial images are traversed.
- If the value of an element within the vector L is greater than 0, the corresponding non-ground DIM point is regarded as a tree point; otherwise, the DIM point is a building point.

In the second step of the above process, the visibility analysis and occlusion detection is necessary during the back projection. The purpose of visibility analysis is to determine whether the selected DIM point is within the selected image's field of view (FOV) and face the image's direction without occlusion from other objects. In this paper, we make use of the collinear equations and an image's interior/exterior orientation parameters to determine whether the selected DIM point is within an image FOV and applied the method in [53] for occlusion detection. If a non-ground DIM point is invisible or occluded in the selected classified original aerial image, the value of the corresponding element within L remains unchanged.

Here, we take a DIM point as an example to further describe the process of removing tree points from non-ground DIM points. As is shown in Figure 4, the projected point of the selected DIM point is marked as A and is invisible at the viewpoint of image IMG_147000509. Therefore, the value of the corresponding element within L remains unchanged when the image IMG_147000509 is applied in this process. In image IMG_147000345, due to the impact of illumination and texture, the region where the projected point A falls is incorrectly divided into the tree category. Based on the description of the above process, the value of the corresponding element plus 1. In the image IMG_147000413 and IMG_147000449, the projected point A falls in the region marked as other objects. Hence, the value of the corresponding element within the vector L minus 1, respectively. After 4 images are traversed, the corresponding element value is less than 0. Obviously, the DIM point A will be regarded as a building point.

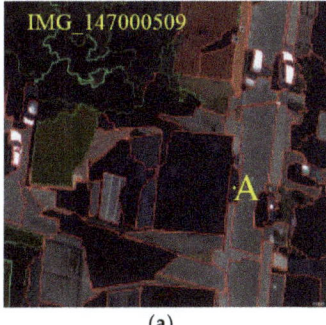 

(a)   (b)

Figure 4. *Cont.*

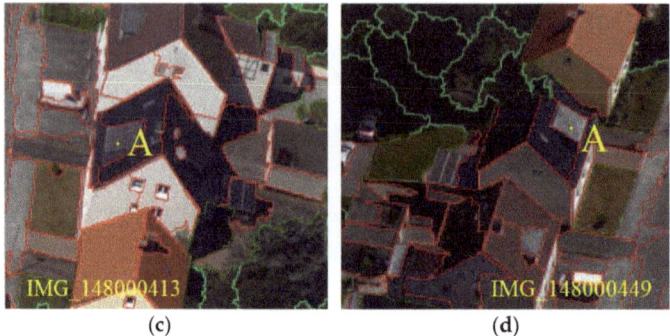

Figure 4. Process of removing tree points from non-ground DIM points: (**a**) the DIM point A which can't be seen at the viewpoint of this image; (**b**) the DIM point A which is incorrectly classified as trees; (**c,d**) the DIM point A which is correctly classified as buildings.

The classification results are shown in Figure 5b. Figure 5c,d show the advantages of our proposed method. The low building marked as B in Figure 5c is occluded at the nadir viewpoint, which results in this building remaining undetected from the orthophoto. However, in our proposed method, this low building can be detected from DIM points as is shown in Figure 5d.

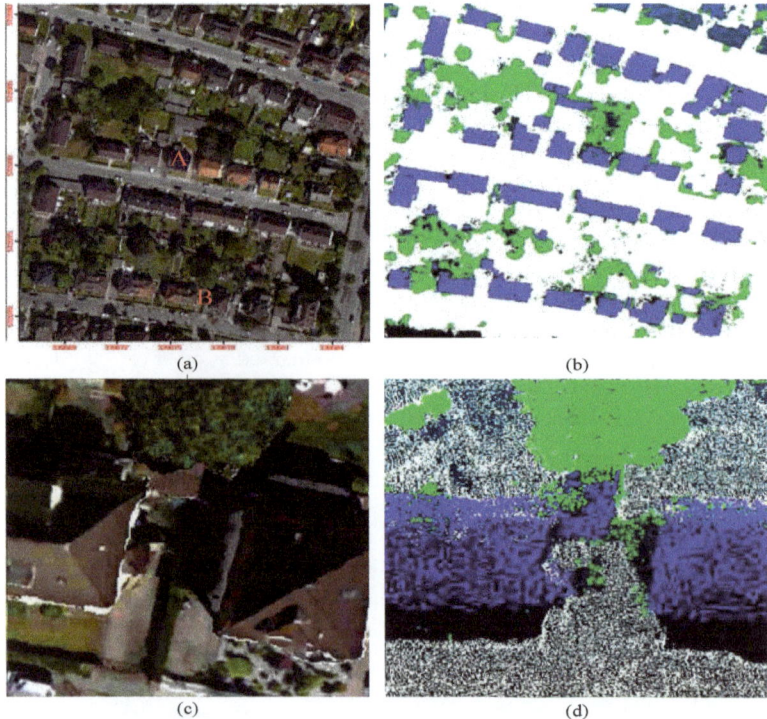

Figure 5. Intermediate result illustrations of building detection: (**a**) generated orthophoto; (**b**) classification results of DIM points: white—ground points, green—tree points, blue—building points; (**c**) occluded building example in the orthophoto; (**d**) detected occluded building (marked as B in (**a**)): blue—building points.

2.2.4. Extraction of Individual Building from the DIM Points

Some scattered points remain distributed in 3D space because of the wrong category of DIM points. To remove these interference points, the octree method is used to divide the point cloud into 3D grids. If the number of the DIM points in a user-defined 3D grid is beyond a certain threshold, the grid is preserved. This threshold is determined by the size of the grid and the density of the DIM points. Generally speaking, the smaller the grid size is, the smaller this threshold is; and the higher the density is, the larger the threshold is.

Project the 3D points onto the XY planar coordinates with the resolution of the orthophoto; and a morphological close operator with a 3×3 square structuring element is used to generate an initial binary image. This structuring element is estimated by the density of DIM points and the resolution of planar coordinates. Then, a two-pass algorithm [54] is used to extract the connected components. Each connected component represents an individual building blob.

2.3. Building Edge Detection Using a Feature-Level Fusion Strategy

The boundaries of individual buildings obtained in Section 2.2 are irregular. In this section, the straight-line segments are extracted from the DIM points, orthophoto and building blob. Subsequently, the coarse building edges from the DIM points, orthophoto, building mask, and height information from DIM points are fused to help match extracted straight-line segments from aerial images so that the matched line segments can be used for regularization of building boundaries.

2.3.1. Detection of Coarse Building Edges from the DIM Points, Orthophoto and Building Mask

In this section, three kinds of building edges are extracted: the building edges from DIM points, the building edges from orthophoto and the building edges from extracted building blob. Figure 6a–c shows the extracted results of the three kinds of building edges. All these building edges are mainly used to assist the straight-line segments from original aerial images to match in the later stage. The process of detecting the three kinds of building edges is described as follows:

- Detection of building edges from DIM points. In terms of DIM point clouds, the building facades are the building edges. Hence, how to extract the building edges from DIM points is converted to how to detect the building facades from DIM points. The density of DIM points at the building façades is larger than that at other locations. Based on this, a method [55] named as the density of projected points (DoPP) is used to obtain the building façades. If the number of the points located in a grid cell is beyond the threshold $DoPP_{thr}$, the grid is labelled 255. After these steps, the generated façade outlines still have a width of 2–3 pixels. Subsequently, a skeletonization algorithm [56] is performed to thin the façade outlines. Finally, a straight-line detector based on the freeman chain code [57] is used to generate building edges.
- Detection of building edges from orthophoto. The process of extracting building edges from orthophoto is divided into three steps. First, a straight-line segment detector [58] is used to extract straight-line segments from the orthophoto. Second, a buffer region is defined by a specified individual building blob obtained in Section 2.2. If the extracted straight-line segment intersects the buffer region, this straight-line segment is considered as a building candidate edge. Finally, the candidate edge is discretized into Num_t points. The number of points located in buffer region is Num_i. Num_i/Num_t represents the length of the candidate edge falling into the buffer area. The larger this ratio is, the greater the probability that this candidate edge is the building edge is. In this paper, if this ratio is greater than 0.6, the candidate edge is considered a building edge.
- Detection of building edges from the building blob. The boundaries of buildings are estimated using the Moore Neighborhood Tracing algorithm [59], which provides an organized list of points for an object in a binary image. To convert the raster images of individual buildings into a vector, the Douglas-Peucker algorithm [60] is used to generate the building edges.

Figure 6. The extracted building edges: (**a**) the extracted building edges from DIM points; (**b**) the extracted building edges from orthophoto; (**c**) the extracted building edges from building blob; (**d**) the generated matched lines from original aerial images.

2.3.2. Detection of the Building Boundary Line Segments from the Original Aerial Images by Matching Line Segments

In this section, the basic process unit is an individual building. Before the line matching, both the straight-line segments from original aerial images and the DIM points of the selected individual building are obtained; then, the height information from DIM points is used to obtain an alternative matching line pool of a selected coarse building edge; finally, a line matching algorithm based on the three-vision images is applied to obtain the matched building edges from the alternative matching line pool. Specifically,

- Obtain the associated straight-line segments from multiple original aerial images and DIM points of an individual building

    - Select an individual building detected in Section 2.2.4.
    - According to the planar coordinate values of the individual building, obtain the corresponding DIM points.
    - Project the DIM points onto the original aerial images, and obtain the corresponding regions of interest (ROIs) of selected individual building from multiple aerial images.

Employ the straight-line segments detector [58] to extract the straight-line segments from corresponding ROIs. The extracted straight-line segments from ROIs are shown in Figure 7.

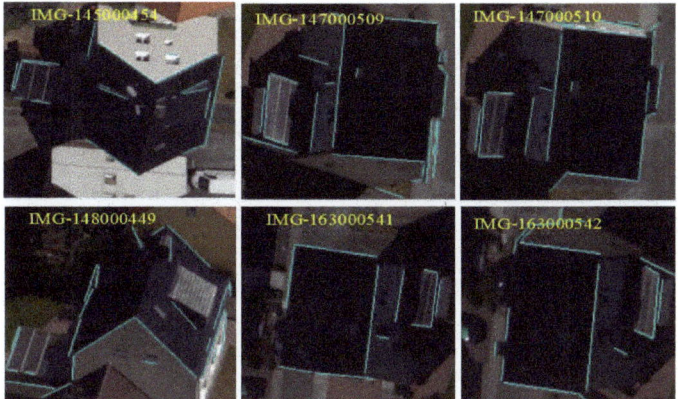

Figure 7. The detected straight-line segments from multiple images.

- The generation of an alternative matching straight-line segments pool of a coarse building edge

Figure 8 shows the process of the generation of alternative matching straight-line segments of a coarse building edge. We divide the process into three steps; and the details are described as follows.

■ Choose a coarse building edge, and discretize this straight-line segment into 2D points according to the given interval 0.1 m. Use the nearest neighbor interpolation algorithm [61] to convert the discretized 2D points into 3D points by fusing the height information from the DIM points. Notably, the points at the building facades which have been detected by the method DoPP should be removed before interpolation so that we can obtain the exact coordinate values of each 3D point.
■ Project the 3D points onto the selected original image according to the collinear equation. In this step, occlusion detection is necessary.
■ Fit the projected 2D points into a straight-line segment on the selected aerial image. A buffer region with the given size is created. Check whether the associated line segments from aerial image intersect the buffer region. If a straight-line segment from aerial image intersects within the buffer, the line is labelled as an alternative line.

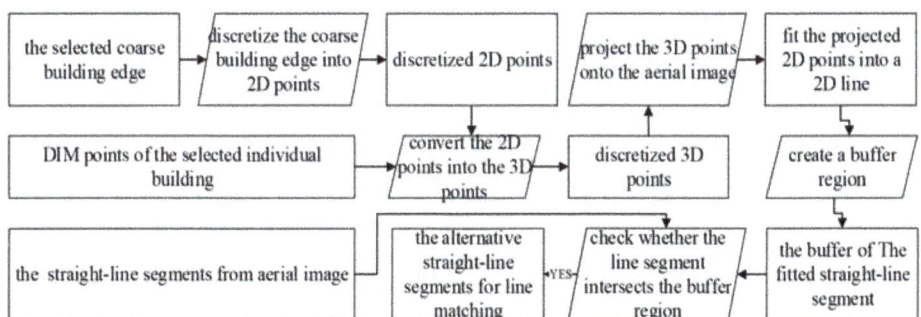

Figure 8. The process of generating alternative matching lines of a coarse building edge from an original aerial image.

The number of line segments that intersect the buffer may be more than one in an image. To reduce the complexity of line matching, only the longest line is selected. After all the associated images are processed, an alternative matching line pool of the selected coarse building edge is created. Figure 9 shows the generated alternative matching line pool. In Figure 9, the IMG-n represents the n_{th} image in the generated alternative matching line pool and the red line is the selected alternative matching line.

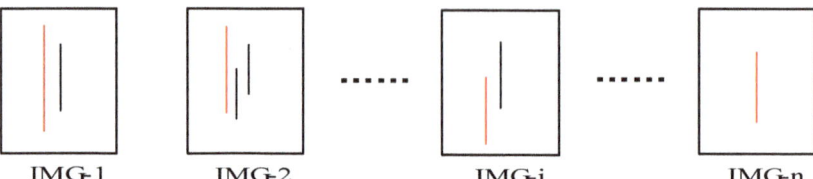

Figure 9. The generated alternative matching line pool of a coarse building edge.

- Straight-line segments matching based on three-vision images

After the generation of alternative matching straight-line segments pool of a selected coarse building edge, the straight-line segments matching process starts.

■ Select two longest line segments from the alternative matching line pool. As is shown in Figure 10a, it is obvious that the camera C1 is not on the straight-line segment Line 1. Similarly, both the camera C1 and C2 are not on the straight-line segment Line 2 and Line 3, respectively. Hence, a 3D plane is generated by the camera C1 and the straight-line segment Line 1. Another plane is created by camera C2 and the line segment Line 2. Two planes intersect into a 3D line segment.

■ Project the 3D line segment onto the IMG-3. If the projected line and Line3 overlap each other, calculate the angle α and normal distance d between Line 3 and the projected line; otherwise, return to the first step. In the process, the occlusion detection is necessary. The normal distance d is expressed as Equation (2), and $d_1, d_2, d_3, and\ d_4$ are shown in Figure 10b.

$$d = (d_1 + d_2 + d_3 + d_4)/4 \qquad (2)$$

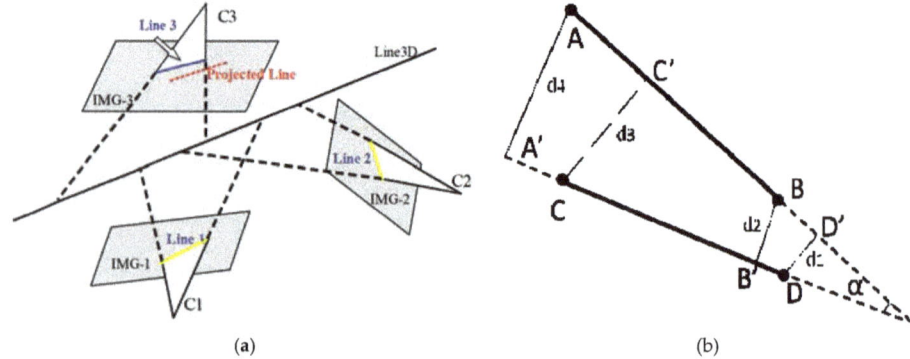

Figure 10. The process of the straight-line segment matching (a) line 3D matching; (b) definition of normal distance.

If the angle is less than $6°$ and the normal distance is less than 0.5 m, the three straight-line segments can be regarded as homonymous lines; otherwise, return to the first step.

- To ensure the robustness and accuracy of line matching, the three homonymous lines should be checked. In accordance with the method described in the previous step, use the camera C1 and the line segment Line 1 to generate a 3D plane, and use the camera C3 and the line segment Line 3 to generate another 3D plane. Two planes intersect into a 3D line. Project the 3D line onto IMG-2. Check whether the normal distance and the angle between Line 2 and the projected line satisfy the given thresholds. Similarly, check the normal distance and angle between Line 1 and the projected line, which is generated from Line 2 and Line 3. If the normal distance and angle still satisfy the given thresholds, the three lines can be considered homonymous lines; otherwise, return to the first step.
- Make use of the three homonymous lines to create a 3D line. Project the 3D line onto the other aerial images. According to the given thresholds, search the homonymous line segments. The results are shown in Figure 11a. Make use of the homonymous lines to create a 3D line, and project the 3D line onto the XY plane. A new building edge can be created.

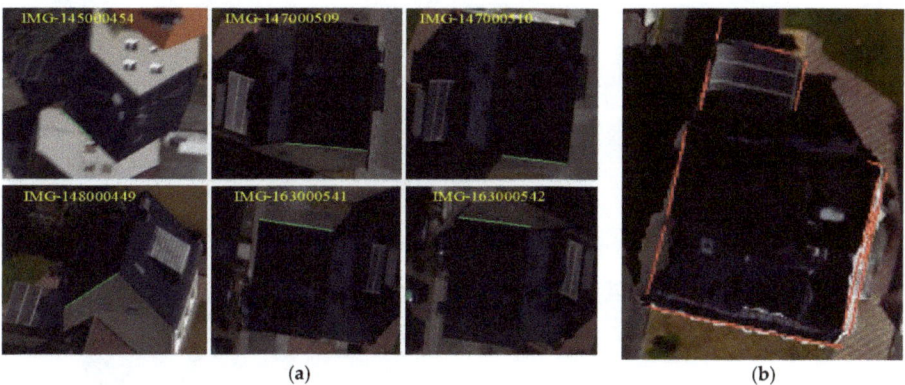

Figure 11. The results of line matching: (a) homonymous lines of a selected coarse building edge; (b) the generated matching straight-line segments of an individual building.

- Iterate the second and third steps until all the coarse building edges are processed. The matched lines of the selected individual building are shown in Figure 11b.

2.4. Regularization of Building Boundaries by the Fusion of Matched Lines and Boundary Lines

In this section, the matched straight-line segments from original aerial images and the straight-line segments from building blob are fused to generate the building footprint. First, the boundary line is adjusted to the specified angle. Then, the parallel line segments are merged. Finally, a new method for building boundary regularization is proposed.

2.4.1. Adjustment of Straight-Line Segments

Buildings typically have one or more principal directions. The direction of the true building edge should be in consonance with the building's direction. Under this principle, the lines should be adjusted to the building's main direction. The matched straight-line segments from original aerial images can be used to calculate the building's direction. The selected straight-line segments are divided into nine intervals according to the line angles. Given that there are **n** lines in an interval, the total length of n straight-line segments can be calculated and labelled as $Length_i$. A histogram is generated using $Length_i$. The peak of the histogram represents the building's direction. The equation can be expressed as follows.

$$Domi = \sum_{i=1}^{n} dis_i \times angle_i / \sum_{i=1}^{n} dis_i \qquad (3)$$

Buildings are diverse; some buildings have complex structures and contain relatively more directions. If the total value of an interval is more than 0.3 times the maximum value, the angle value can be regarded as another direction of the building. After the estimation of the dominant directions, the lines from the blob of individual buildings are adjusted slightly according to the dominant directions of buildings. The line adjustment process can be described as follows. First, a line is selected, and the adjustment angle is calculated; then, the line is rotated around the midpoint of the processing line. The adjustment results are shown in Figure 12a.

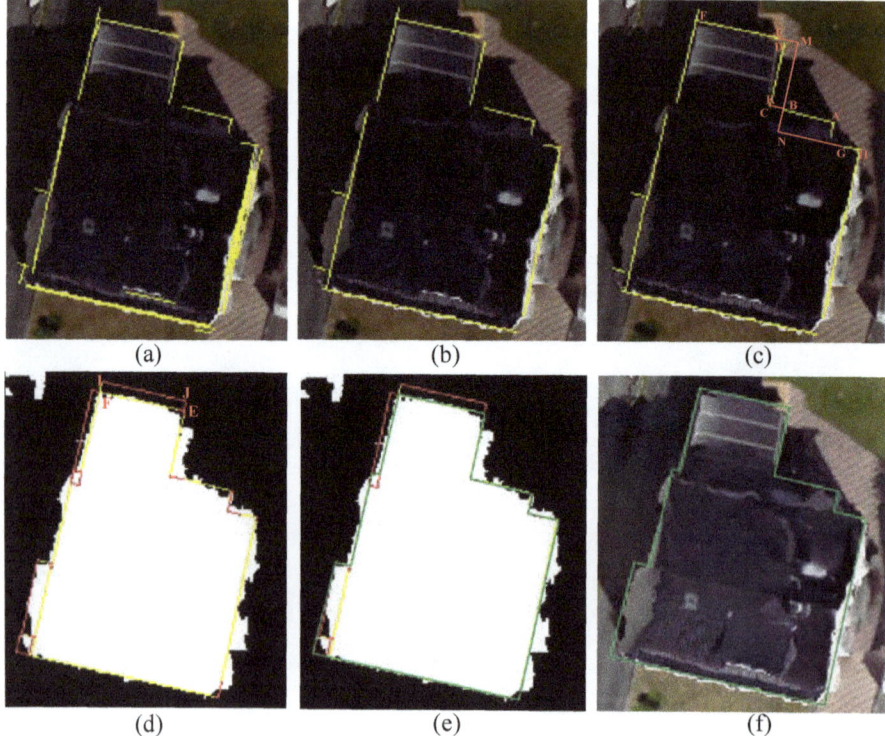

Figure 12. The process of building regularization: (**a**) results of adjusted line segments; (**b**) results of merged line segments; (**c**) process of filling a gap between two line segments; (**d**) results of filling gaps; (**e**) preserved polygons; (**f**) generated building footprint.

2.4.2. Merging of Straight-Line Segments

Figure 13 shows the process of five parallel line segments that are merged to one average line segment. First, the longest line segment L2 is selected, and a buffer area is created by a given threshold of 0.7 m, as shown in Figure 13a. Then, each line segment is checked for parallelism to L2. If a line is parallel to L2 and intersects the buffer area, the line is added to the line segments set L_m. When all line segments are processed, line segments L1, L3, L4 and L5 are selected. Given that the equation of the straight-line L_i is $y = ax + b_i$, we conclude that the length of the line segment in L_m affects the location of the merged line. A longer line segment corresponds to a greater weight. The equation of the merged straight-line is expressed as follows.

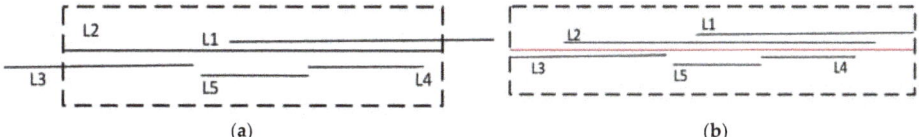

Figure 13. Merging of 5 straight-line segments: (**a**) five parallel line segments; (**b**) the merged line segment of five five parallel line segments.

$$ax - y - \frac{\sum_0^n w_i^2 b_i}{\sum_0^n w_i^2} = 0 \qquad (4)$$

In this equation $w_i = l_i / \sum_0^n l_i$, and l_i represents the length of the line segment L_i. To obtain the merged line segment, the endpoint of each line segment is projected to the generated straight-line segments. All the projected points are added to the point set P. The longest line segment within two projected points is the merged line segment as shown in Figure 13b. The results of merged lines are shown in Figure 12b.

2.4.3. Building Footprint Generation

Gaps exist between the two merged line segments. To obtain a complete building profile, these gaps should be filled. As shown in Figure 12c, point B is an endpoint of the line segment AB, and a gap exists between line segment AB and the line segment CD. To fill this gap, a new strategy is proposed.

- First, calculate the distances from the point B to the other line segments in three directions—along the direction and two vertical directions of line segment AB. As shown in Figure 12c, the distance from the point B to the line segment CD is BK; the distance from the point B to the line segment EF is BM + ME; and the distance from the point B to the line segment HG is BN + NG.
- Search the minimum distance among the calculated distances from the point B to the other line segments. The line segment BK is the gap between the line segment AB and CD.
- Continue the above two steps until all the gaps are filled. The results are shown in Figure 12d.

After the gaps are filled, the invalid polygons that contain a small amount of areas of the detected building mask, such as the rectangle EFIJ in Figure 12d, should be removed. Before the process, we need to search from the non-overlapping single polygon from the entire polygons. Figure 14 shows the process of searching the non-overlapping single polygon.

- First, the endpoints are coded again, and the line segments are labelled. If the line segment is located on the external contour of the entire polygon, the straight-line segment is labelled as 1; otherwise, the line segment is labelled as 2, as shown in Figure 14a.
- Choose a line arbitrarily from the lines dataset as the starting edge. Here, the line segment EF is selected.
- Search the line segments that share the same endpoint F. If only a straight-line segment is searched, then the line segment is the next line segment. If two or more straight-line segments are found, calculate the angle clockwise between each alternative line segment and this line segment EF. The angle between the line FT and FE is 270°, and the angle between the line FG and FE is 90°. Choose the straight-line segment FG as the target line segment based on the smaller calculated angle value.
- Continue until the starting edge EF. For all the straight-line segments on the search path, subtract 1, as shown in Figure 14b, and save the single polygon ABCDEFGHKLMNOPRA. If a line segment is labelled as 0, the straight-line segment is removed from the polygons, as shown in Figure 14c.
- Search until all the straight-line segments are removed. The result is shown in Figure 14f.

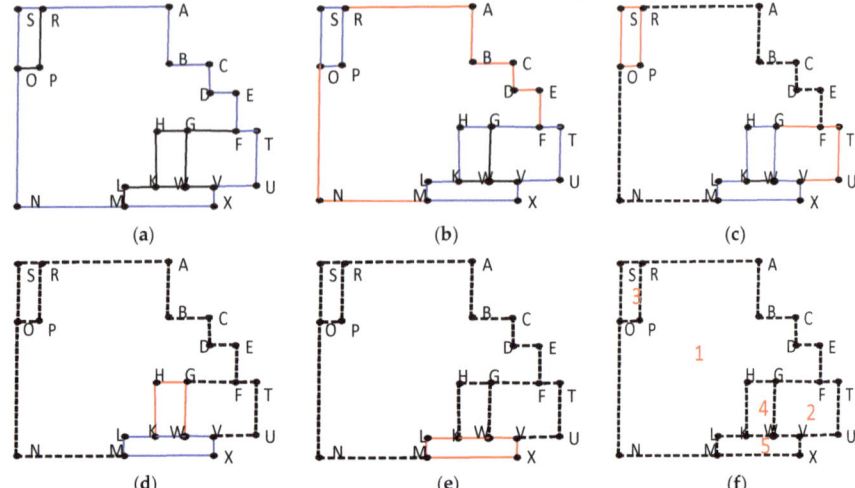

Figure 14. Process of searching the non-overlapping single polygon: (**a**) the initial state of line segments of the polygon; (**b**–**e**) the state of line segments of the polygon in the search process; (**f**) the preserved non-overlapping single polygons. Black line—line labelled as 2; blue line—line labelled as 1; red line—line labelled as 0; the dashed line—removed line.

Within each preserved polygon, the total number of pixels is Num_1, and the number of pixels present in the building is Num_2. If the Num_2/Num_1 is larger than the given threshold 0.5, the polygon is preserved; otherwise, the polygon is removed. The preserved polygon is shown in Figure 12e. Merge the preserved single polygon; the outline of the merged polygon is the building footprint, as shown in Figure 12f.

3. Experiments and Performance Evaluation

3.1. Data Description and Study Area

The performance of the proposed approach is tested on three datasets: Vaihingen (VH), Potsdam and Lunen (LN). The VH dataset consists of vertical aerial images; the Potsdam and LN datasets are composed of oblique aerial images. These datasets have different complexities and surrounding conditions. Each dataset is listed as follows:

- VH. The VH dataset is published by the International Society for Photogrammetry and Remote Sensing (ISPRS), and has the real ground reference value. The object coordinate system of this dataset is the system of the Land Survey of the German federal state of Baden Württemberg, based on a Transverse Mercator projection. There are three test areas in this dataset and have been presented in Figure 15a–c. The orientation parameters are used to produce DIM points, and the derived DIM points of three areas have almost the same point density of approximately 30 points/m².

  - ■ VH 1. This test area is situated in the center of the city of Vaihingen. It is characterized by dense development consisting of historic buildings having rather complex shapes, but also has some trees. There are 37 buildings in this area and the buildings are located on the hillsides.
  - ■ VH 2. This area is flat and is characterized by a few high-rising residential buildings that are surrounded by trees. 14 buildings are in this test area.

- VH 3. This is a purely residential area with small detached houses and contain 56 buildings. The surface morphology of this area is relatively flat.

• Potsdam. The dataset including 210 images was collected by TrimbleAOS on 5 May 2008 in Potsdam, Germany and has a GSD of about 10 cm. The reference coordinate system of this dataset is the WGS84 coordinate system with UTM zone 33N. Figure 15d shows the selected test area, which contains 4 patches according to the website (http://www2.isprs.org/commissions/comm3/wg4/2d-sem-label-potsdam.html): 6_11, 6_12, 7_11 and 7_12. The selected test area contains 54 buildings and is characterized by a typical historic city with large building blocks, narrow streets and a dense settlement structure. Because the collection times between the ISPRS VH benchmark dataset and the Potsdam dataset differ, the reference data are slightly modified by an operator.

• LN. This dataset including 170 images was collected by the Quattro DigiCAM Oblique system on 1 May 2011 in Lunen, Germany and has a GSD of 8–2 cm. The object coordinate system of this dataset is the European Terrestrial Reference System 1989. Three patches with flat surface morphology were selected as the test areas. The reference data are obtained by a trained operator. Figure 15e–f shows the selected areas.

- LN 1. This is a purely residential area. In this area, there are 57 buildings and some vegetation.
- LN 2. In this area, there are 36 buildings and several of these buildings are the occluded by tress.
- LN 3. This area is characterized by a few high-rising buildings with complex structures. In this area, 47 buildings exist.

Figure 15. Datasets. Vaihingen: (**a**) VH 1, (**b**) VH 2, and (**c**) VH 3; (**d**) Potsdam; Lunen: (**e**) LN 1, (**f**) LN 2, and (**g**) LN 3.

3.2. Evaluation Criterion

The evaluation index system adopted by ISPRS [62] is applied for a quality assessment of our proposed approach. In this evaluation system, three categories of evaluations are performed: object-based, pixel-based, and geometric, where each category uses several metrics. The object-based metrics (completeness, accuracy, quality, under- and over-segmentation errors, and reference cross-lap rates) evaluate the performance by counting the number of buildings, while the pixel-based metrics (completeness, accuracy, quality, area-omission, and area-commission errors) measure the detection accuracy by counting the number of pixels. The geometric metric (root mean square error, i.e., RMSE) indicates the planimetric distance in meters accuracy from extracted outlines to a reference outline. The RMSE, correctness, completeness, and quality equations are shown in Equations (5).

$$
\begin{aligned}
RMSE &= \sqrt{\tfrac{\sum d^2}{N}} \\
Correctness &= TP/(TP+FP) \\
Completness &= TP/(TP+FN) \\
Quality &= TP/(TP+FP+FN)
\end{aligned}
\tag{5}
$$

where N represents the number of points for which a correspondence has been found within a predefined search buffer, d is the distance between the corresponding points, TP represents the number of true positives, FP represents the number of false positives, and FN represents the number of false negatives. In the evaluation process of our proposed approach, the building area is taken into consideration. The minimum areas for large and small buildings are set to 50 m^2 and 2.5 m^2, respectively. The symbols used in the assessment are described in Table 1.

Table 1. Symbol descriptions in the performance evaluation.

Symbols	Description
C_m, $C_{m,50}$	Completeness for all the buildings, over 50 m^2 and over 2.5 m^2 object-based detection
C_r, $C_{r,50}$	Correctness for all the buildings, over 50 m^2 and over 2.5 m^2 object-based detection
Q_l, $Q_{l,50}$	Quality for all the buildings, over 50 m^2 and over 2.5 m^2 object-based detection
C_{mp}, $C_{mp,50}$	Completeness all the buildings, over 50 m^2 and over 2.5 m^2 pixel-based detection
C_{rp}, $C_{rp,50}$	Correctness all the buildings, over 50 m^2 and over 2.5 m^2 pixel-based detection
Q_{lp}	Quality of all building pixel-based detection
RMSE	Planimetric accuracy in meters
1 : M	M detected buildings correspond to one building in the reference (over-segmented)
N : 1	A detected building corresponds to N buildings in the reference (under-segmented)
N : M	Both over- and under-segmentation in the number of buildings

3.3. Results and Discussion

Sections 3.3.1 and 3.3.2 describe and discuss the evaluation results of our proposed technique on the VH, Potsdam and LN datasets.

3.3.1. Vaihingen Results

Tables 2 and 3 show the official per-object and per-area level evaluation results for the three test areas of the VH dataset. Figure 16 shows the per-pixel level visual evaluation of all the test areas (column 1) for the building delineation technique (column 3) and their corresponding regularization outcome (column 2).

Table 2 shows that the overall object-based completeness and accuracy are 83.83% and 98.41%, respectively; the buildings over 50 m^2 are extracted with 100% objective completeness and accuracy before regularization. The missing buildings (marked "green circle" in Figure 16a,d,g) are eliminated in the process of building detection. Building detection consists of two steps: DIM point clouds filtering and object-oriented classification of original aerial images. Figure 17 shows the results of

DIM point cloud filtering on the VH dataset. Only two small buildings (marked "green circle" in Figure 17a,c) are removed due to the improper filtering thresholds. Most of the missing buildings (marked "yellow circle" in Figure 17a–c) are removed in the process of object-oriented classification of original aerial images, which indicates that the errors of building detection mainly occur at this stage. To increase the detection rate, the accuracy of object-oriented classification of original aerial images should be improved.

Certain close buildings were combined unexpectedly (marked as P, Q and R), as shown in Figure 16i because the quality of DIM point clouds is lower than that of LiDAR point clouds, particularly in the regions between two buildings. In addition to the many-to-1 (N : 1) segmentation errors, many-to-many (M : N) segmentation errors marked as O in Figure 15b exist. These many-to-many (M : N) segmentation errors are caused by two factors. On the one hand, the filter threshold removes the low part of the building, as shown in Figure 16a, and the buildings are segmented into several parts; on the other hand, a building and a segmented part are merged into a building at the same time. Moreover, there are two incorrectly detected buildings (objects under a tree and shuttle bus), as shown in Figure 16h marked as X, Y. After regularization, the object-based results are the same as those before regularization.

After regularization, the per-area completeness, accuracy, and quality are improved. The pixel-based completeness and accuracy are 91.15% and 94.91%, respectively. This increase might be substantial if the straight-line segments from the DIM points, orthophoto and original images could replace the line segments from the boundary of the detected building mask. After the boundary is regularized efficiently, the planimetric accuracy is improved from 0.731 m to 0.69 m.

Table 2. Object-based building detection results on the VH dataset before and after the regularization stage.

Method	Areas	C_m	C_r	Q_l	$C_{m,50}$	$C_{r,50}$	$Q_{l,50}$	1:M	N:1	N:M
Before regularization	VH 1	81.00	100.00	81.00	100.00	100.00	100.00	0	2	1
	VH 2	83.30	100.00	83.30	100.00	100.00	100.00	0	0	0
	VH 3	87.20	95.23	81.60	100.00	100.00	100.00	0	2	0
	Average	83.83	98.41	81.97	100.00	100.00	100.00	0	1.33	0.33
After regularization	VH 1	81.00	100.00	81	100.00	100.00	100.00	0	2	1
	VH 2	83.30	100.00	83.3	100.00	100.00	100.00	0	0	0
	VH 3	87.20	95.23	81.6	100.00	100.00	100.00	0	2	0
	Average	83.83	98.41	81.97	100.00	100.00	100.00	0	1.33	0.33

Table 3. Pixel-based building detection results on the VH dataset before and after the regularization stage.

Proposed Detection	Areas	C_{mp}	C_{rp}	Q_{lp}	RMSE
Before regularization	VH 1	90.7	91.2	83.45	0.853
	VH 2	89.14	95.47	88.55	0.612
	VH 3	90.42	95.22	86.5	0.727
	Average	90.09	93.96	86.17	0.731
After regularization	VH 1	90.38	92.29	84.04	0.844
	VH 2	92.37	96.75	89.59	0.591
	VH 3	90.7	95.68	87.13	0.632
	Average	91.15	94.91	89.92	0.69

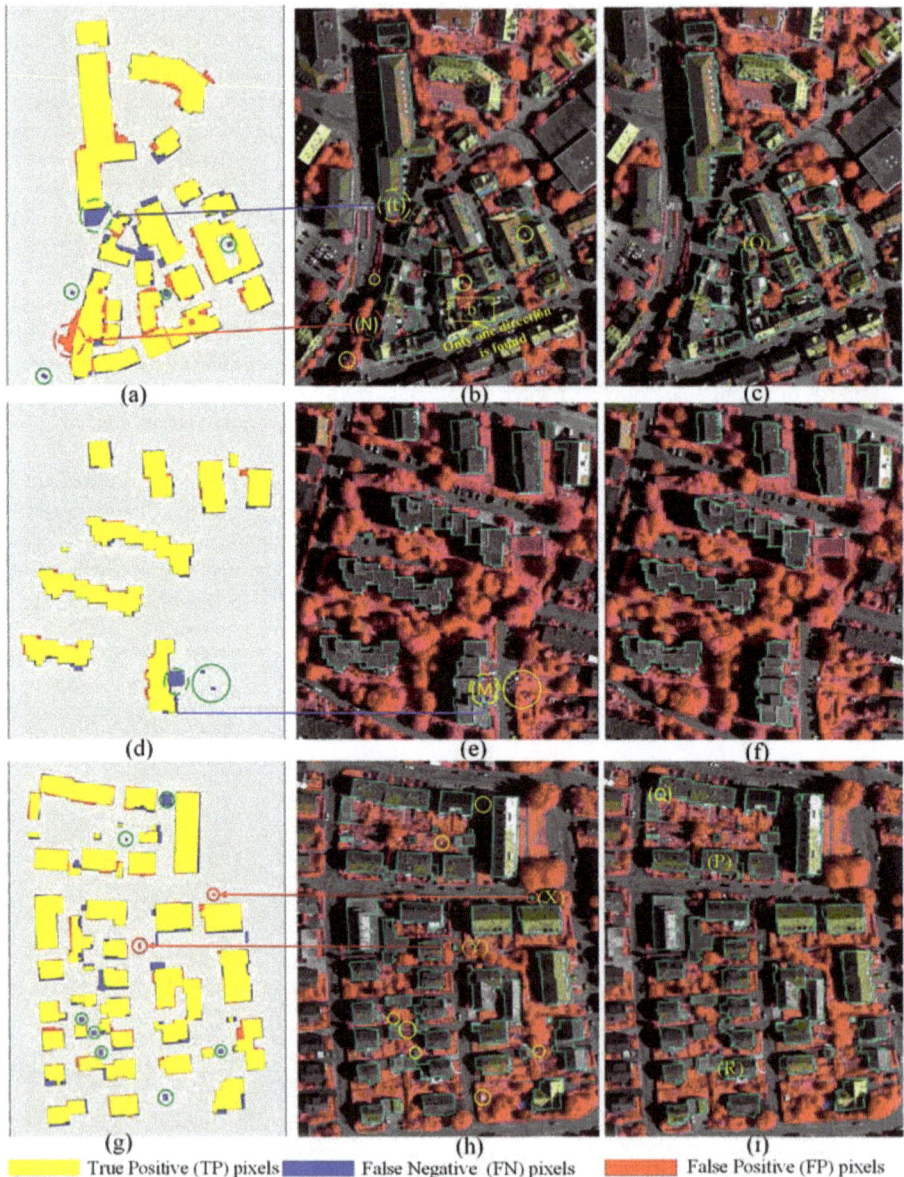

Figure 16. Building detection on the VH data set: (**a**–**c**) Area 1, (**d**–**f**) Area 2, and (**g**–**i**) Area 3. Column 1: pixel-based evaluation, Column 2: regularized boundary, and Column 3: boundary before regularization.

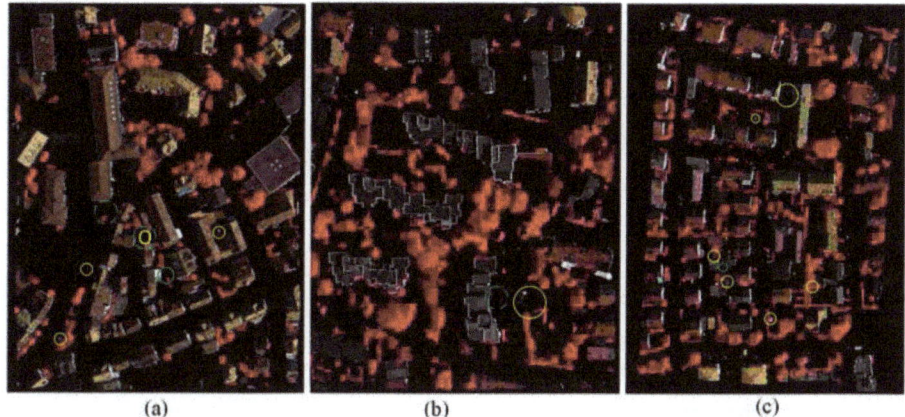

Figure 17. Results of DIM point cloud filtering on the VH German dataset: (**a**) VH 1, (**b**)VH 2, (**c**) VH 3.

3.3.2. Potsdam and LN Results

Tables 4 and 5 present the object- and pixel-based evaluation of the detection technique before and after the regularization. Figures 18 and 19 show the extracted buildings and their corresponding building footprints for Potsdam, LN1, LN2, and LN3. The proposed building detection method extracted 49, 52, 35, and 39 buildings out of 54, 57, 36, and 47 reference buildings in the Potsdam, LN1, LN2 and LN3, respectively.

Table 4. Object-based building detection results on the Potsdam and Lunen (LN) datasets before and after the regularization stage.

Method	Areas	C_m	C_r	Q_l	$C_{m,50}$	$C_{r,50}$	$Q_{l,50}$	1:M	N:1	N:M
Before regularization	Potsdam	90.74	100	90.74	100	100	100	1	4	0
	LN1	87.71	96.15	84.75	97.06	100	97.06	0	8	0
	LN2	94.4	97.14	91.89	100	100	100	0	3	0
	LN3	86.05	94.59	82.22	100	100	100	0	2	0
	Average	89.73	96.97	87.4	99.27	100	99.27	0.25	4.25	0.25
After regularization	Potsdam	90.74	100	90.74	100	100	100	1	3	0
	LN1	87.71	96.15	84.75	97.06	100	97.06	0	8	0
	LN2	94.4	97.14	91.89	100	100	100	0	2	1
	LN3	86.05	94.59	82.22	100	100	100	0	2	0
	Average	89.73	96.97	87.4	99.27	100	99.27	0.25	4.25	0.25

Table 5. Pixel-based building detection results on the Potsdam and LN datasets before and after the regularization stage.

Proposed Detection	Areas	C_{mp}	C_{rp}	Q_{lp}	RMSE
Before regularization	Potsdam	94.9	95.1	90.5	0.913
	LN1	93.15	89.33	83.82	0.882
	LN2	91.81	93.05	85.92	0.816
	LN3	95.72	93.78	90	0.754
	Average	93.9	92.82	87.56	0.841
After regularization	Potsdam	94.61	95.7	90.76	0.843
	LN1	93.32	90.02	84.56	0.813
	LN2	92.3	94.12	87.27	0.742
	LN3	96.4	93.69	90.52	0.715
	Average	94.16	93.38	88.28	0.778

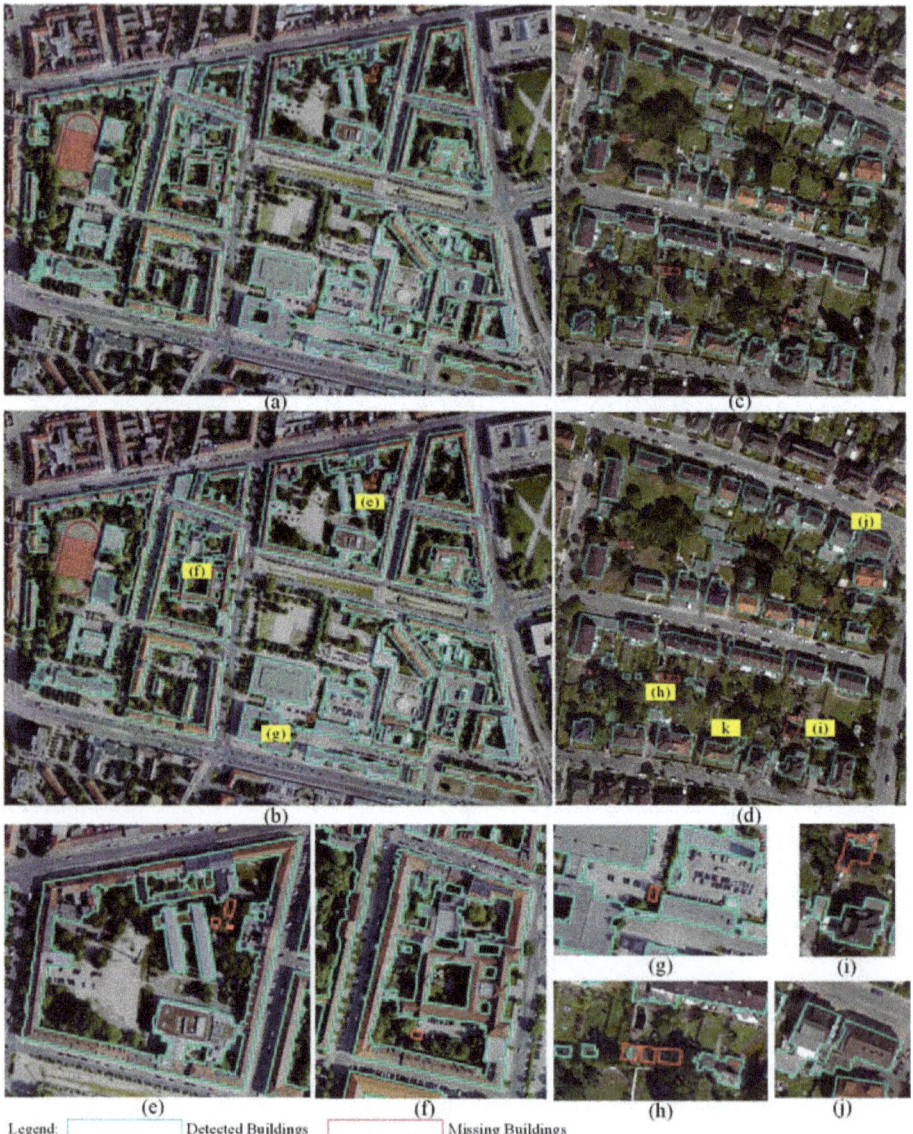

Figure 18. Building detection on the Potsdam and LN datasets: (**a,b**) building detection and regularization on the Potsdam dataset; (**c,d**) building detection and regularization on LN1; (**e–g**) building regularization example in Potsdam; (**d,e**); and (**f–h**) building regularization examples in LN1. Areas marked in (**b**) and (**d**) are magnified in (**e–g**) and (**h–i**), respectively.

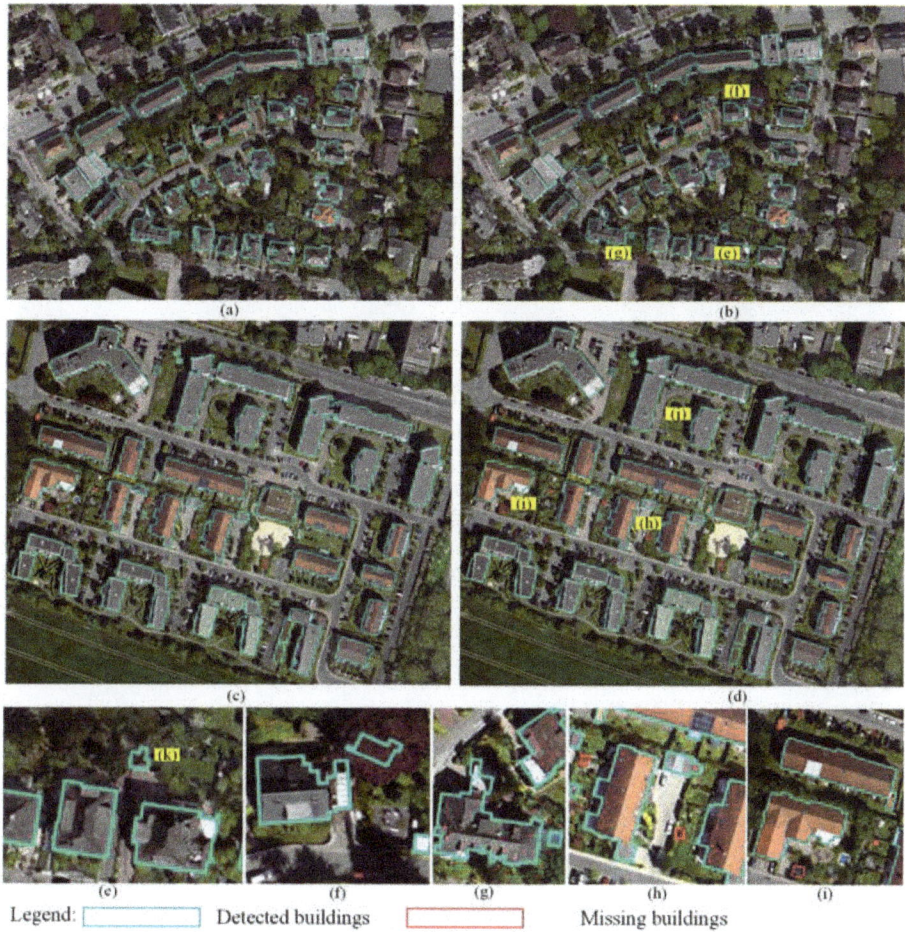

Figure 19. (**a,b**) Building detection and regularization on LN 2; (**c,d**) Building detection and regularization on LN 3; (**e–g**) Building regularization example in LN 2; and (**g–i**) Building regularization examples in LN 3.

The object-based completeness and accuracy on all the buildings are 89.73% and 96.97%, respectively, as shown in Table 4. A building over 50 m² marked as i in Figure 18d is missed due to the influence of shadows in the process of tree removal. Table 5 shows the total per-area evaluation (completeness and accuracy 93.9% and 92.82%, respectively) of the proposed building detection method for Potsdam and Lunen before regularization. The per-area completeness in LN2 is lower than that of the three remaining areas (Potsdam, LN1, and LN3), mainly because the roof of the buildings is not detected completely because of the misclassification of the tree and building, as shown in Figure 19f. Moreover, there are some sundries surround the building; these objects are regarded as the parts of the true building, which causes the per-area accuracy in LN1 to be lower (89.33%) than that in the other three areas. Figure 18j shows this example. The average pixel-based completeness and accuracy increase significantly from 93.9% and 92.82% to 94.16% and 93.38% after the regularization, respectively. Similarly, three factors hinder the detection of buildings in the process of building detection.

- Initially, the filtering threshold may be larger than the height of low buildings. In our test arrangement, some low buildings (< the given thresholds) are excluded in the filtering process.

Fortunately, the height of most buildings is higher than the given thresholds in the Potsdam and Lunen test areas. Only the building marked as **e** in Figure 18b is removed.

- Second, the misclassification of trees and other objects in the process of object-oriented classification of original aerial images is the main reason that affects the accuracy of building detection. Some buildings are removed due to the influence of similar spectral features and shadow. Figure 18f,g,i and Figure 19h,i shows examples of missed buildings. Similarly, some trees are classified as buildings, as shown in Figure 18f.
- Third, the noise of DIM points degrades the accuracy of building detection. Uncoupled to the over-segmentation error, under- and many-to-many segmentation errors exist in Table 4. Figure 19g shows the many-to-many segmentation error example. The shadow on the surface of the building leads to the over-segmentation of the building, and the adjacent buildings are combined with the building at the same time due to the small connecting regions between these two buildings.

3.4. Comparative Analysis

In our proposed method (PM), we solely employed aerial images as the data source to extract buildings in complex urban scenes. To compare our proposed method with other methods, we select those which use aerial images alone and the supervised classification strategy. According to the ISPRS portal and the classified methods in [8], three methods—DLR, RAM and Hand—are selected for the comparative analysis. In addition, we hope that our proposed method is superior to other methods combining the images and LiDAR data. Hence, in addition to these three methods just mentioned, Fed_2, ITCR and MON4 are also chosen. In Fed_2, LiDAR data are used to detect buildings, and the footprint of buildings are generated by the straight-line segment from the orthoimagery. In ITCR, the LiDAR data and the original images are fused, and a supervisory strategy is used for building detection. MON4 make use of a method named as Gradient-based Building Extraction (GBE) to extract the building planes and their neighboring ground planes from images and Lidar; then, analyzes the height difference and connectivity between the extracted building planes and their neighboring ground planes to extract low buildings. These chosen methods can be seen in Table 6.

Table 6. Six existing methods compared with our proposed technique.

Benchmark Data Set	Methods	Data Type	Processing Strategy	Reference
VH	DLR	Image	supervised	[8]
	RMA	Image	data-driven	[8]
	Hand	Image	Dempster-Shafer	[8]
	Fed_2	LiDAR + image	data-driven	[39]
	ITCR	LiDAR + image	supervised	[17]
	MON4	LiDAR + image	Data-driven	[28]

Table 7 presents a comparison between our proposed method and six other methods. From the comparison results of the different methods in the VH dataset, several conclusions can be obtained as follows:

- Relative to RMA and Hand, our proposed method can obtain similar object-based completeness and accuracy in VH1 and VH3. The object-based accuracy of RMA and Hand in VH2 is 52% and 78%, respectively, and is significantly lower than that of the proposed method because NDVI is the main feature in RMA and Hand. The wrong NDVI estimate decreases the object-based accuracy in VH2 due to the influence of shadow pixels. The results show that our proposed method is more robust than RMA and Hand due to the use of additional features.
- The DLR not only divides the objects into buildings and vegetation but also takes into account vegetation shadowed in the separation of buildings and other objects. Therefore, the pixel-based completeness and accuracy of DLR is slightly higher than that of our proposed method in VH1,

VH2 and VH3. Moreover, the pixel-based completeness and accuracy of DLR is also higher than that of other 4 methods. The object-based completeness of DLR is lower than that of our proposed method in VH2 and VH3, mainly because the features from original aerial images are used to detect buildings and because the small buildings can be easily detected in the two areas. Actually, there are seldom buildings that are occluded in the three test areas of Vaihingen. The advantages of our proposed method are not thoroughly demonstrated. The buildings labelled as k in Figures 18 and 19 are partially or completely occluded, and DLR cannot detect the occluded buildings. In our proposed method, the occluded building can be detected. Notably, because the noise of DIM points is higher than that of LiDAR point cloud, the object-based accuracy of the proposed method is slightly lower than that of DLR in VH3. Large buildings (50 m^2) were extracted with 100% accuracy and completeness.

- The proposed method offers better in object-based completeness and accuracy than ITCR, mainly because the LiDAR point clouds need to be segmented into 3D segments that are regarded as the processed unit in ITCR. The process produces segmented errors that damage the building detection results. Furthermore, the segmentation of DIM points is more challenges because the quality of DIM points is lower than that of the LiDAR point cloud. Therefore, we make use of the aerial images to detect buildings from non-ground points to replace the segmentations of the DIM point cloud in our proposed method.
- Fed_2 and our proposed method obtain better performance than ITCR and IIST because the segmented errors are avoided in the process of building detection. The performance of Fed_2, MON4 and our proposed method is approximately the same.

Table 7. Comparison of the results of different methods in Vaihingen.

Area	Method	C_m	C_r	$C_{m,50}$	$C_{r,50}$	C_{mp}	C_{rp}	RMSE	1:M	N:1	N:M
VH1	DLR	83.8	96.9	100	100	91.9	95.4	0.9	-	-	-
	RMA	83.8	96.9	100	100	91.6	92.4	1.0	-	-	-
	Hand	83.8	93.9	100	100	93.8	90.5	0.9	-	-	-
	Fed_2	83.8	100	100	100	85.4	86.6	1.0	0	6	0
	ITCR	86.5	91.4	100	100	91.2	90.3	1.1	-	-	-
	MON4	89.2	93.9	100	100	92.1	83.9	1.3	-	-	-
	PM	81	100	100	100	90.38	92.29	0.8	0	2	1
VH2	DLR	78.6	100	100	100	94.3	97	0.6	-	-	-
	RMA	85.7	52.2	100	100	95.4	85.9	0.9	-	-	-
	Hand	78.6	78.6	100	100	95.1	89.8	0.8	-	-	-
	Fed_2	85.7	100	100	100	88.8	84.5	0.9	0	2	0
	ITCR	78.6	42.3	100	100	94	89	0.8	-	-	-
	MON4	85.7	91.7	100	100	97.2	83.5	1.1	-	-	-
	PM	83.8	100	100	100	92.37	96.75	0.6	0	0	0
VH3	DLR	78.6	100	100	100	93.7	95.5	0.7	-	-	-
	RMA	78.6	93.9	100	100	91.3	92.4	0.8	-	-	-
	Hand	78.6	92.8	92.1	100	91.9	90.6	0.8	-	-	-
	Fed_2	82.1	95.7	100	100	89.9	84.7	1.1.	0	5	0
	ITCR	75	78.2	94.7	100	89.1	92.5	0.8	-	-	-
	MON4	76.8	95.7	97.4	100	93.7	81.3	1.0	-	-	-
	PM	85.11	95.23	100	100	90.7	95.68	0.6	0	2	0

The above analysis shows that our proposed algorithm is robust and can obtain similar performance of building detection methods that fuse LiDAR data and images. Moreover, the RMSE is decreased due to the boundary lines being replaced with the matched lines from the original aerial images.

3.5. Performances of Our Proposed Building Regularization

Section 3.4 shows that the directions of buildings are the precondition for the building regularization. To evaluate the performance of the regularization technique, we divide the buildings into three categories: single-direction buildings, multi-direction buildings and complex structure buildings. The regularization results of single-direction buildings and multi-direction buildings are shown in Figure 20a–n.

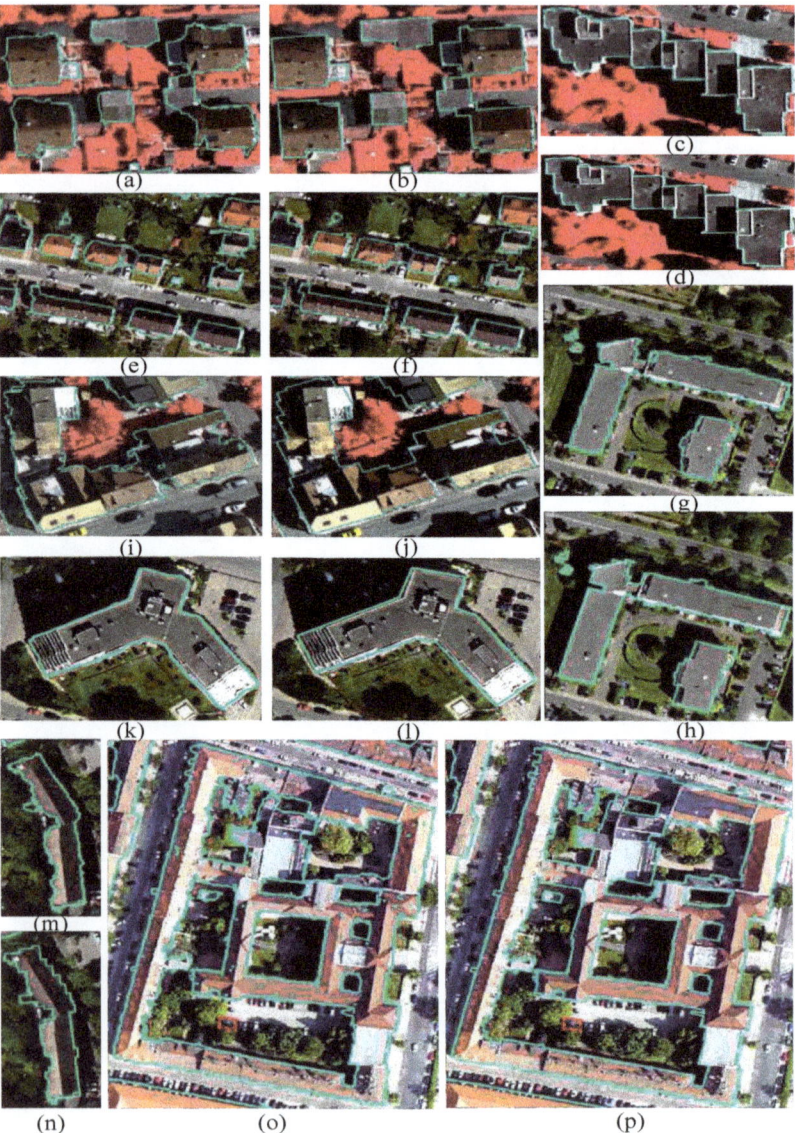

Figure 20. Building examples from the Vaihingen (VH), Potsdam and LN datasets. Detected and regularized buildings on the (**a,b**) VH 3; (**c,d**) VH2; (**e,f**) LN1; (**i,j**) VH1; (**g–l**) LN 3; (**m,n**) LN2; and (**o,p**) Potsdam.

In our proposed method, the matched lines are used in the process of searching the direction of building. According to the line segments from the mask of the detected buildings, the usage of adjusted lines increases the RMSE. Figure 20a–f,k–l shows relevant examples. Nevertheless, incorrectly adjusted line segments remain, particularly involving the multi-direction buildings, as shown in Figure 20g–j,m,n, mainly because building's directions cannot be computed completely due to the absence of line segments that provide the directions of buildings. As shown in Figure 20g–j,m–p, these buildings have two directions. However, only one direction is calculated correctly; the other direction is ignored in the examples. As a result, some lines deviate from the true direction.

In terms of the complex structure building, holes exist. To obtain the accurate footprint, both the external contours and inner contours from the mask of detected building are calculated, respectively. According to the contours, we group the lines and regularize the contours. These regular contours construct the footprint of the complex structure building. Figure 20o,p shows the examples of this building. Notably, if the size of the hole is smaller than the given thresholds, we can ignore the inner contour.

4. Conclusions

This paper focuses on both building detection and footprint regularization using solely aerial images. A framework for building detection and footprint regularization using a feature-level-fusion strategy is proposed. Following a comparison with six other methods, the experiment results show that the proposed method can not only provide comparable results to building detection methods using LiDAR and aerial images but also generate building footprints in complex environments. However, several limitations for our proposed method cannot be ignored:

- In the building detection stage, the results of building detection rely on the results of object-oriented classification of original aerial images. Shadows are an important factor influencing the classification of original aerial images. In the process of classification, we categorize the objects into only two classes (trees and other objects) and do not train and classify shadows as separate objects. As a result, the classification results cannot perform perfectly. In the further work, we will make use of the intensity and chromatic to detect the shadow so that we can get better performance.
- The noise of the DIM point cloud produced by dense matching is higher than that of the LiDAR point clouds. The buildings detection results and footprint regularization are affected by noise.
- A threshold exists in the process of DIM point cloud filtering. The filtering threshold cannot ensure that all the buildings can be preserved. In fact, the low buildings may be removed, while some objects that are higher than the threshold are preserved. The improper threshold causes detection errors.
- At the stage of building footprint regularization, only straight lines are used. Therefore, for a building with a circular boundary, our proposed method cannot provide satisfactory performance.
- Finally, the morphology of buildings also influences the accuracy of building detection results. In fact, the more complex the building morphology is, the more difficult the building boundary regularization is.

Author Contributions: Y.D. has developed, implemented and conducted the tests. In addition, he has written the paper. L.Z. and X.C. wrote part of the manuscript, and performed the experiments and experimental analysis. H.A. proposed the original idea. B.X. made the contribution on the programming, and revised the manuscript.

Funding: This research was funded: (1) National Natural Science Foundation of China, grant number 51474217; (2) National Key Research and Development Program of China, grant number 2017YFB0503004; (3) the Basic Research Fund of Chinese Academy of Surveying and Mapping, grant number 7771801, respectively.

Acknowledgments: The authors would like to thank the anonymous reviewers for their constructive comments, which greatly improved the quality of our manuscript. The Vaihingen data set was provided by the German Society for Photogrammetry, Remote Sensing and Geoinformation (DGPF). The Lunen data set was provided by Integrated Geospatial Innovations (IGI). The Potsdam data set was provided by Trimble. Moreover, we thank X.L. for revising the language.

Conflicts of Interest: The authors declare no conflict of interest.

References

1. Krüger, A.; Kolbe, T.H. Building Analysis for Urban Energy Planning Using Key Indicators on Virtual 3d City Models—The Energy Atlas of Berlin. *ISPRS Int. Arch. Photogramm. Remote Sens. Spat. Inf. Sci.* **2012**, *XXXIX-B2*, 145–150.
2. Arinah, R.; Yunos, M.Y.; Mydin, M.O.; Isa, N.K.; Ariffin, N.F.; Ismail, N.A. Building the Safe City Planning Concept: An Analysis of Preceding Studies. *J. Teknol.* **2015**, *75*, 95–100.
3. Murtiyoso, A.; Remondino, F.; Rupnik, E.; Nex, F.; Grussenmeyer, P. Oblique Aerial Photography Tool for Building Inspection and Damage Assessment. *ISPRS Int. Arch. Photogramm. Remote Sens. Spat. Inf. Sci.* **2014**, *XL-1*, 309–313. [CrossRef]
4. Vetrivel, A.; Gerke, M.; Kerle, N.; Vosselman, G. Identification of damage in buildings based on gaps in 3D point clouds from very high resolution oblique airborne images. *ISPRS J. Photogramm. Remote Sens.* **2015**, *105*, 61–78. [CrossRef]
5. Karimzadeh, S.; Mastuoka, M. Building Damage Assessment Using Multisensor Dual-Polarized Synthetic Aperture Radar Data for the 2016 M 6.2 Amatrice Earthquake, Italy. *Remote Sens.* **2017**, *9*, 330. [CrossRef]
6. Stefanov, W.L.; Ramsey, M.S.; Christensen, P.R. Monitoring urban land cover change: An expert system approach to land cover classification of semiarid to arid urban centers. *Remote Sens. Environ.* **2001**, *77*, 173–185. [CrossRef]
7. Antonarakis, A.S.; Richards, K.S.; Brasington, J. Object-based land cover classification using airborne LiDAR. *Remote Sens. Environ.* **2008**, *112*, 2988–2998. [CrossRef]
8. Rottensteiner, F.; Sohn, G.; Gerke, M.; Wegner, J.D.; Breitkopf, U.; Jung, J. Results of the ISPRS benchmark on urban object detection and 3D building reconstruction. *ISPRS J. Photogramm. Remote Sens.* **2014**, *93*, 256–271. [CrossRef]
9. Poullis, C. A Framework for Automatic Modeling from Point Cloud Data. *IEEE Trans. Pattern Anal. Mach. Intell.* **2013**, *35*, 2563–2575. [CrossRef]
10. Sun, S.; Salvaggio, C. Aerial 3D Building Detection and Modeling from Airborne LiDAR Point Clouds. *IEEE J. Sel. Top. Appl. Earth Obs. Remote Sens.* **2013**, *6*, 1440–1449. [CrossRef]
11. Mongus, D.; Lukač, N.; Žalik, B. Ground and building extraction from LiDAR data based on differential morphological profiles and locally fitted surfaces. *ISPRS J. Photogramm. Remote Sens.* **2014**, *93*, 145–156. [CrossRef]
12. Liua, C.; Shia, B.; Yanga, X.; Lia, N. Legion Segmentation for Building Extraction from LIDAR Based Dsm Data. *ISPRS Int. Arch. Photogramm. Remote Sens. Spat. Inf. Sci.* **2012**, *XXXIX-B3*, 291–296. [CrossRef]
13. Dahlke, D.; Linkiewicz, M.; Meissner, H. True 3D building reconstruction: Façade, roof and overhang modelling from oblique and vertical aerial imagery. *Int. J. Image Data Fusion* **2015**, *6*, 314–329. [CrossRef]
14. Nex, F.; Rupnik, E.; Remondino, F. Building Footprints Extraction from Oblique Imagery. *ISPRS Ann. Photogramm. Remote Sens. Spat. Inf. Sci.* **2013**, *II-3/W3*, 61–66. [CrossRef]
15. Arefi, H.; Reinartz, P. Building Reconstruction Using DSM and Orthorectified Images. *Remote Sens.* **2013**, *5*, 1681–1703. [CrossRef]
16. Grigillo, D.; Kanjir, U. Urban object extraction from digital surface model and digital aerial images. *ISPRS Ann. Photogramm. Remote Sens. Spat. Inf. Sci.* **2012**, *I-3*, 215–220. [CrossRef]
17. Gerke, M.; Xiao, J. Fusion of airborne laser scanning point clouds and images for supervised and unsupervised scene classification. *ISPRS J. Photogramm. Remote Sens.* **2014**, *87*, 78–92. [CrossRef]
18. Awrangjeb, M.; Zhang, C.; Fraser, C.S. Automatic Reconstruction of Building Roofs through Effective Integration of Lidar and Multispectral Imagery. *ISPRS Ann. Photogramm. Remote Sens. Spat. Inf. Sci.* **2012**, *I-3*, 203–208. [CrossRef]
19. Rottensteiner, F.; Trinder, J.; Clode, S.; Kubik, K. Using the Dempster-Shafer method for the fusion of LiDAR data and multi-spectral images for building detection. *Inf. Fusion* **2005**, *6*, 283–300. [CrossRef]

20. Chen, L.; Zhao, S.; Han, W.; Li, Y. Building detection in an urban area using LiDAR data and QuickBird imagery. *Int. J. Remote Sens.* **2012**, *33*, 5135–5148. [CrossRef]
21. Remondino, F.; Spera, M.G.; Nocerino, E.; Menna, F.; Nex, F. State of the art in high density image matching. *Photogramm. Rec.* **2014**, *29*, 144–166. [CrossRef]
22. Myint, S.W.; Gober, P.; Brazel, A.; Grossman-Clarke, S.; Weng, Q. Per-pixel vs. object-based classification of urban land cover extraction using high spatial resolution imagery. *Remote Sens. Environ.* **2011**, *115*, 1145–1161. [CrossRef]
23. Blaschke, T. Object based image analysis for remote sensing. *ISPRS J. Photogramm. Remote Sens.* **2010**, *65*, 2–16. [CrossRef]
24. Salehi, B.; Zhang, Y.; Zhong, M.; Dey, V. Object-Based Classification of Urban Areas Using VHR Imagery and Height Points Ancillary Data. *Remote Sens.* **2012**, *4*, 2256–2276. [CrossRef]
25. Pu, R.; Landry, S.; Yu, Q. Object-Based Urban Detailed Land Cover Classification with High Spatial Resolution IKONOS Imagery. *Int. J. Remote. Sens.* **2011**, *32*, 3285–3308.
26. Burnett, C.; Blaschke, T. A multi-scale segmentation/object relationship modeling methodology for landscape analysis. *Ecol. Model.* **2003**, *168*, 233–249. [CrossRef]
27. Rottensteiner, F.; Trinder, J.; Clode, S.; Kubik, K. Building detection by fusion of airborne laserscanner data and multi-spectral images: Performance evaluation and sensitivity analysis. *ISPRS J. Photogramm. Remote Sens.* **2007**, *62*, 135–149. [CrossRef]
28. ISPRS. ISPRS Test Project on Urban Classification, 3D Building Reconstruction and Semantic Labeling. 2013. Available online: http://www2.isprs.org/commissions/comm3/wg4/tests.html (accessed on 20 May 2018).
29. Tarantino, E.; Figorito, B. Extracting Buildings from True Color Stereo Aerial Images Using a Decision Making Strategy. *Remote Sens.* **2011**, *3*, 1553–1567. [CrossRef]
30. Xiao, J.; Gerke, M.; Vosselman, G. Building extraction from oblique airborne imagery based on robust façade detection. *ISPRS J. Photogramm. Remote Sens.* **2012**, *68*, 56–68. [CrossRef]
31. Rau, J.Y.; Jhan, J.P.; Hsu, Y.C. Analysis of Oblique Aerial Images for Land Cover and Point Cloud Classification in an Urban Environment. *IEEE Trans. Geosci. Remote Sens.* **2015**, *53*, 1304–1319. [CrossRef]
32. Jwa, Y.; Sohn, G.; Tao, V.; Cho, W. An implicit geometric regularization of 3D building shape using airborne LiDAR data. *Int. Arch. Photogramm. Remote Sens. Spat. Inf. Sci.* **2008**, *XXXVII*, 69–76.
33. Brunn, A.; Weidner, U. Model-Based 2D-Shape Recovery. In Proceedings of the DAGM-Symposium, Bielefeld, Germany, 13–15 September 1995; Springer: Berlin, Germany, 1995; pp. 260–268.
34. Ameri, B. Feature based model verification (FBMV): A new concept for hypothesis validation in building reconstruction. *Int. Arch. Photogramm. Remote Sens.* **2000**, *33*, 24–35.
35. Sampath, A.; Shan, J. Building Boundary Tracing and Regularization from Airborne LiDAR Point Clouds. *Photogramm. Eng. Remote Sens.* **2007**, *73*, 805–812. [CrossRef]
36. Jarzabekrychard, M. Reconstruction of Building Outlines in Dense Urban Areas Based on LIDAR Data and Address Points. *ISPRS Int. Arch. Photogramm. Remote Sens. Spat. Inf. Sci.* **2012**, *39*, 121–126. [CrossRef]
37. Rottensteiner, F. Automatic generation of high-quality building models from LiDAR data. *Comput. Graph. Appl. IEEE* **2003**, *23*, 42–50. [CrossRef]
38. Sohn, G.D.I. Data fusion of high-resolution satellite imagery and LiDAR data for automatic building extraction. *ISPRS J. Photogramm. Remote Sens.* **2007**, *62*, 43–63. [CrossRef]
39. Gilani, S.; Awrangjeb, M.; Lu, G. An Automatic Building Extraction and Regularization Technique Using LiDAR Point Cloud Data and Ortho-image. *Remote Sens.* **2016**, *8*, 258. [CrossRef]
40. Jakobsson, M.; Rosenberg, N.A. CLUMPP: A cluster matching and permutation program for dealing with label switching and multimodality in analysis of population structure. *Bioinformatics* **2007**, *23*, 1801–1806. [CrossRef]
41. Habib, A.F.; Zhai, R.; Kim, C. Generation of Complex Polyhedral Building Models by Integrating Stereo-Aerial Imagery and LiDAR Data. *Photogramm. Eng. Remote Sens.* **2010**, *76*, 609–623. [CrossRef]
42. Agisoft. Agisoft PhotoScan. 2018. Available online: http://www.agisoft.ru/products/photoscan/professional/ (accessed on 25 May 2018).
43. Sithole, G.; Vosselman, G. Experimental comparison of filter algorithms for bare-Earth extraction from airborne laser scanning point clouds. *ISPRS J. Photogramm. Remote Sens.* **2004**, *59*, 85–101. [CrossRef]
44. Zhang, W.; Qi, J.; Wan, P.; Wang, H.; Xie, D.; Wang, X.; Yan, G. An Easy-to-Use Airborne LiDAR Data Filtering Method Based on Cloth Simulation. *Remote Sens.* **2016**, *8*, 501. [CrossRef]

45. Axelsson, P. DEM generation from laser scanner data using adaptive TIN models. *Int. Arch. Photogramm. Remote Sens.* **2000**, *33 Pt B4*, 110–117.
46. Dong, Y.; Cui, X.; Zhang, L.; Ai, H. An Improved Progressive TIN Densification Filtering Method Considering the Density and Standard Variance of Point Clouds. *ISPRS Int. J. Geo-Inf.* **2018**, *7*, 409. [CrossRef]
47. Devereux, B.J.; Amable, G.S.; Posada, C.C. An efficient image segmentation algorithm for landscape analysis. *Int. J. Appl. Earth Obs. Geo-Inf.* **2004**, *6*, 47–61. [CrossRef]
48. Defries, R.S.; Townshend, J.R.G. NDVI-derived land cover classifications at a global scale. *Int. J. Remote Sens.* **1994**, *15*, 3567–3586. [CrossRef]
49. Awrangjeb, M.; Ravanbakhsh, M.; Fraser, C.S. Automatic detection of residential buildings using LIDAR data and multispectral imagery. *ISPRS J. Photogramm. Remote Sens.* **2010**, *65*, 457–467. [CrossRef]
50. Tomimatsu, H.; Itano, S.; Tsutsumi, M.; Nakamura, T.; Maeda, S. Seasonal change in statistics for the floristic composition of Miscanthus- and Zoysia-dominated grasslands. *Jpn. J. Grassl. Sci.* **2009**, *55*, 48–53.
51. Meyer, G.E.; Neto, J.C. Verification of color vegetation indices for automated crop imaging applications. *Comput. Electron. Agric.* **2008**, *63*, 282–293. [CrossRef]
52. Pal, M. Random forest classifier for remote sensing classification. *Int. J. Remote Sens.* **2005**, *26*, 217–222. [CrossRef]
53. Habib, A.F.; Kim, E.M.; Kim, C.J. New Methodologies for True Orthophoto Generation. *Photogramm. Eng. Remote Sens.* **2007**, *73*, 25–36. [CrossRef]
54. Wu, K.; Otoo, E.; Suzuki, K. Optimizing two-pass connected-component labeling algorithms. *Pattern Anal. Appl.* **2009**, *12*, 117–135. [CrossRef]
55. Zhong, S.W.; Bi, J.L.; Quan, L.Q. A Method for Segmentation of Range Image Captured by Vehicle-borne Laser scanning Based on the Density of Projected Points. *Acta Geod. Cartogr. Sin.* **2005**, *34*, 95–100.
56. Iwanowski, M.; Soille, P. Fast Algorithm for Order Independent Binary Homotopic Thinning. In Proceedings of the International Conference on Adaptive and Natural Computing Algorithms, Warsaw, Poland, 11–14 April 2007; Springer: Berlin, Germany, 2007; pp. 606–615.
57. Li, C.; Wang, Z.; Li, L. An Improved HT Algorithm on Straight Line Detection Based on Freeman Chain Code. In Proceedings of the 2009 2nd International Congress on Image and Signal Processing, Tianjin, China, 17–19 October 2009; pp. 1–4.
58. Lu, X.; Yao, J.; Li, K.; Li, L. CannyLines: A parameter-free line segment detector. In Proceedings of the 2015 IEEE International Conference on Image Processing (ICIP), Quebec City, QC, Canada, 27–30 September 2015; pp. 507–511.
59. Cen, A.; Wang, C.; Hama, H. A fast algorithm of neighbourhood coding and operations in neighbourhood coding image. *Mem. Fac. Eng. Osaka City Univ.* **1995**, *36*, 77–84.
60. Visvalingam, M.; Whyatt, J.D. The Douglas-Peucker Algorithm for Line Simplification: Re-evaluation through Visualization. *Comput. Graph. Forum* **1990**, *9*, 213–225. [CrossRef]
61. Maeland, E. On the comparison of interpolation methods. *IEEE Trans Med Imaging* **1988**, *7*, 213–217. [CrossRef] [PubMed]
62. Rutzinger, M.; Rottensteiner, F.; Pfeifer, N. A comparison of evaluation techniques for building extraction from airborne laser scanning. *IEEE J. Sel. Top. Appl. Earth Obs. Remote Sens.* **2009**, *2*, 11–20. [CrossRef]

© 2018 by the authors. Licensee MDPI, Basel, Switzerland. This article is an open access article distributed under the terms and conditions of the Creative Commons Attribution (CC BY) license (http://creativecommons.org/licenses/by/4.0/).

Article

Building Extraction in Very High Resolution Imagery by Dense-Attention Networks

Hui Yang [1,2], Penghai Wu [2,3,4,*], Xuedong Yao [2], Yanlan Wu [2,3,4,*], Biao Wang [2,4] and Yongyang Xu [5]

1. School of Resource and Environmental Science, Wuhan University, Wuhan 430079, China; yanghui@whu.edu.cn
2. School of Resources and Environmental Engineering, Anhui University, Hefei 230601, China; yaoxd9501@163.com (X.Y.); wangbiao-rs@ahu.edu.cn (B.W.)
3. Institute of Physical Science and Information Technology, Anhui University, Hefei 230601, China
4. Anhui Engineering Research Center for Geographical Information Intelligent Technology, Hefei 230601, China
5. Department of Information Engineering, China University of Geosciences, Wuhan 430074, China; yongyangxu@cug.edu.cn
* Correspondence: wuph@ahu.edu.cn (P.W.); wylmq@sina.com (Y.W.)

Received: 6 September 2018; Accepted: 6 November 2018; Published: 8 November 2018

Abstract: Building extraction from very high resolution (VHR) imagery plays an important role in urban planning, disaster management, navigation, updating geographic databases, and several other geospatial applications. Compared with the traditional building extraction approaches, deep learning networks have recently shown outstanding performance in this task by using both high-level and low-level feature maps. However, it is difficult to utilize different level features rationally with the present deep learning networks. To tackle this problem, a novel network based on DenseNets and the attention mechanism was proposed, called the dense-attention network (DAN). The DAN contains an encoder part and a decoder part which are separately composed of lightweight DenseNets and a spatial attention fusion module. The proposed encoder–decoder architecture can strengthen feature propagation and effectively bring higher-level feature information to suppress the low-level feature and noises. Experimental results based on public international society for photogrammetry and remote sensing (ISPRS) datasets with only red–green–blue (RGB) images demonstrated that the proposed DAN achieved a higher score (96.16% overall accuracy (*OA*), 92.56% *F*1 score, 90.56% mean intersection over union (*MIOU*), less training and response time and higher-quality value) when compared with other deep learning methods.

Keywords: building extraction; deep learning; attention mechanism; very high resolution; imagery

---

1. Introduction

Extracting 2D (two-dimensional) buildings footprints in very high resolution (VHR) imagery has many applications in navigation, urban planning, disaster management, and population estimation [1]. However, many complicated factors such as various scales, complex background (shadow, vegetation, water, and man-made non-building features), heterogeneity of roof, and rich topological appearances [2] make 2D building extraction from VHR images quite a challenging task.

Over the past decade, some methods have tried to extract buildings through VHR imagery, which applied different strategies such as new frameworks [3], new parameters [4], new indices [5], other related information [6], and some hybrid algorithms [7]. Based on the used data, building extraction methods can generally be divided into three categories: 2D (two-dimensional) information based, fused 2D–3D information based and 3D information based [8–10]. 2D information is mainly derived from images, including aerial images and space-borne images, while 3D information is mainly derived from airborne laser scanning technology, such as light detection and ranging (LiDAR) data [11–13]. To extract buildings from 2D information or/and 3D information, some feature extraction technologies have been developed, such as the handcrafted features-based traditional technologies and deep learning-based technologies. The traditional technologies use handcrafted features as a key feature for building extraction, which may contain spectral information or/and spatial information or/and geometrical information [14]. The performance of these technologies relies on the extraction of low-level hand-engineered local features. This limits the representative ability and restricts their performance. Therefore, the extraction of more representative high-level features is desirable, which plays a dominant role in building extraction. The deep learning technologies, as a new framework, have the ability to learn high-level hierarchical features from both 2D/3D information corresponding to the different levels of abstraction, making it dominant in the field of building extraction [15,16]. For extracting buildings, some promising convolutional neural network (CNN) approaches [17–19] and fully convolutional network (FCN) approaches [20,21] have been proposed. However, CNN and FCN only use high-level feature maps to perform pixel-classification; low-level feature maps with rich detailed information are discarded. As a result, CNN and FCN have very limited capacity to deal with small and complex buildings. In order to address this issue, reusing low-level feature maps has become a popular solution as these maps possess rich spatial information and fine-grained details. Some supervised semantic segmentation procedures based on excellent networks such as U-Net [22], DeconvNet [20], Segnet [23], and RefineNet [24] have also appeared.

Recently, an interesting network, called the Dense Networks (DenseNets), has been very popular, which was awarded the best paper in the IEEE conference on computer vision and pattern recognition (CVPR) 2017 [25]. The DenseNets are built from dense blocks and pooling operations, where each dense block is an iterative concatenation of previous feature maps. Several compelling advantages have been proven: they alleviate the vanishing-gradient problem, strengthen feature propagation, encourage feature reuse, and substantially reduce the number of parameters. Therefore, the advantages of DenseNets make them a very good fit for semantic segmentation as they naturally induce skip connections and multi-scale supervision. DenseNets are extended to fully convolutional DenseNets (FC-DenseNets) for semantic segmentation [26], which can improve the state-of-the-art performance in challenging urban scene understanding datasets, without additional post-processing, pretraining, or including temporal information. For instance, Li et al. extended the FC-DenseNets called multiple-feature reuse network (MFRN) to extract buildings from remote sensing data with a high accuracy [27].

However, over-using low-level features may introduce redundant information into the network and result in over-segmentation when the model tends to receive more information from lower layers [28]. How to rationally utilize different level feature remains an open research question. In this study, a novel network was proposed to effectively utilize both high-/low-level feature maps, based on DenseNets and an attention mechanism, called the dense-attention network (DAN). The visual attention refers to the fact that when human vision deals with images, people tend to select the most pertinent piece of information rather than using all available information. The nature of the attention mechanism is to pick the information that contributes a lot to the target from the source. The attention mechanism usually uses the higher-level semantic information to re-weight the low-level information to suppress the background and noises [29]. In the DAN, a spatial attention fusion module was designed to enhance useful low-level feature information and remove noise to avoid over-using low-level features. Therefore, when building multi-scale features in the skip-connection operations, higher-level feature information was used to suppress the low-level features and noises.

The rest of this paper is organized as follows. Related work is presented in Section 2. A detailed description of the proposed method is given in Section 3. The results of the experiments are listed in Section 4. Finally, the discussion and concluding remarks are in Sections 5 and 6.

2. Related Works

2.1. Semantic Segmentation of Remote-Sensing Images

In essence, semantic segmentation algorithms assign a label to every pixel in an image. Semantic segmentation is the term more commonly used in computer vision and is becoming increasingly used in remote sensing. Semantic segmentation of remote-sensing images has numerous applications, such as land-cover classification, urban planning, natural hazard detection, and environment monitoring [30]. Building extraction from remote-sensing images is essentially a problem of segmenting semantic objects. Compared with ordinary digital images, remote-sensing images, especially very high resolution, have different characteristics, which bring challenges for semantic segmentation purposes, such as complex backgrounds, intricate spatial details and limited spectral resolution. Hence, an effective feature representation and mining is a matter of great importance to a semantic segmentation system for very high resolution remote-sensing images.

There has been a vast literature focusing on segmenting remotely sensed images into desired objects. Traditional methods, such as watershed, mean shift, clustering method, active contours, and Markov random field model, have been widely used to produce segments for remotely sensed images [31]. One of the common drawbacks is that their performance heavily relies on handcrafted feature selection, which is hard to optimize. More recently, deep learning approaches have achieved great success in semantic segmentation on both remotely sensed images and other images [31]. State-of-the-art approaches for semantic image segmentation are built on convolutional networks [27,31]. The convolution network is usually a pre-trained deep convolutional neural network (DCNN) designed to classify images from, VGG-16 [32], ResNet [33], Deeplab-V3 [34] and DenseNet [25]. For each network, features at different levels need to be extracted and jointly combined to fulfill the segmentation task. High-level and abstract features are more suitable for the semantic segmentation of large and confused objects, while small objects benefit from low-level and raw features. Basically, successful networks should have the ability to integrate low- and high-level features for semantic segmentation.

2.2. Attention Mechanism

An attention mechanism is an effective tool to extract the most useful information of the input signal [35]. An attention mechanism is achieved by using the filter function (e.g., a softmax or sigmoid) and sequential techniques. The attention mechanism has recently been widely used in image captioning [36,37], image classification [34] and visual question answering [38,39], image recognition [40], and other fields [41,42]. In these applications, they used the filter function to activate the gathered top information to weight the activations channel-wisely or spatially and introduce feedback connections. For example, Wang et al. used a soft mask structure to generate attention-aware features [33], where attention-awareness is to use image segmentation to capture the visual attention focus area. Hu et al. designed a squeeze-and-excitation block to recalibrate channel-wise features [43]. Wang et al. built an entropy control module to select low-feature maps for semantic segmentation [28]. Li et al. introduced a global attention upsample module to guide the integration low- and high-level features in semantic segmentation [44]. Studies have shown the attention mechanism can strengthen some neurons that featured by the target, and improve their performance. Therefore, when designing networks of the building extraction of remote-sensing images, an attention mechanism module was built to integrate low- and high-level features for semantic segmentation, and avoid over-using low-level features.

3. Methods

The encoder–decoder architecture [22–24,45,46] is widely used in semantic segmentation based on deep learning. The encoder part is mainly used to extract multi-scale features of the input data. The decoder part aims to recovers the spatial resolution of feature maps and to extract target objects using these feature maps. Inspired by the architecture, building the semantic segmentation model (named DAN) proposed in this paper also adopted an encoder–decoder architecture, and its overall architecture is shown in Figure 1. In the encoder part, lightweight DenseNets are used to extract the feature maps from inputs; while in the decoder part, a spatial attention fusion module is used to guide the low-feature maps to help high-level features recover the detail of images.

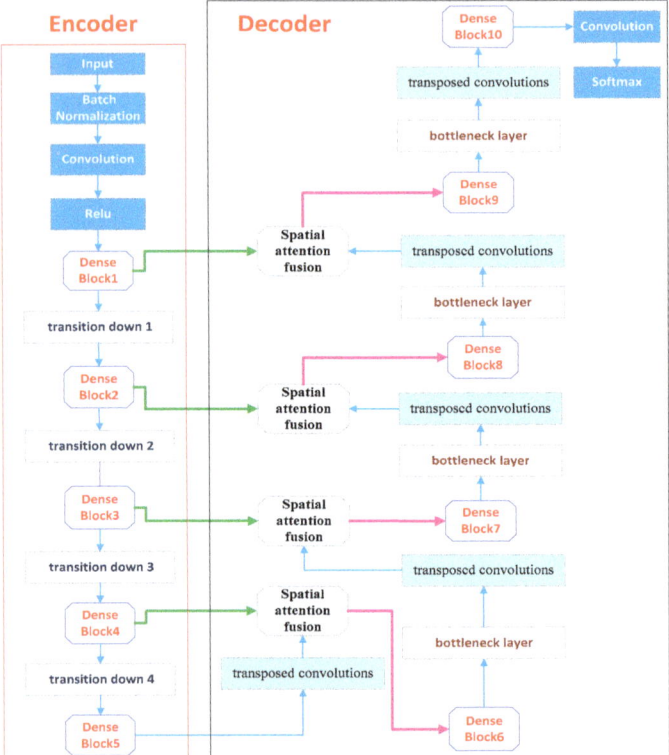

Figure 1. Diagram of overall architecture of dense-attention networks.

3.1. Lightweight DenseNets

DenseNets have a high demand for graphics processing unit (GPU) memory [47]. In order to reduce the GPU memory consumption, a lightweight DenseNets was designed. In DenseNets, to encourage the reuse of features and facilitate gradient propagation, there is a direct connection between any two layers. In other words, the feature maps of all preceding layers will be directly passed to all the behind layers as the input of the layer. Suppose the transformation function of L layer is $H_L(.)$ and the output is X_L, the transformation of each layer of DenseNets is as shown in Equation (1):

$$X_L = H_L([X_0, X_1, \ldots, X_{L-1}]) \tag{1}$$

where $H_L(.)$ is a combination of three operations: Batch Normalization–ReLU–Convolution (3 × 3). [...] is the concatenation of the feature maps, and the output dimension of $H_L(.)$ is K. K is called the growth rate, which controls the number of channels of the feature map of the networks. These layers with the same spatial resolution are called dense blocks. The transition down is located between dense blocks, which is used for down-sampling. It consists of a 1 × 1 convolutional layer followed by a dropout layer and a 2 × 2 average pooling layer. The dense block is an iterative concatenation of a previous feature map. Therefore, the lightweight DenseNets architecture was built from one input convolution layer, five dense blocks, and four transition downs. Figure 2 shows the lightweight DenseNets architecture.

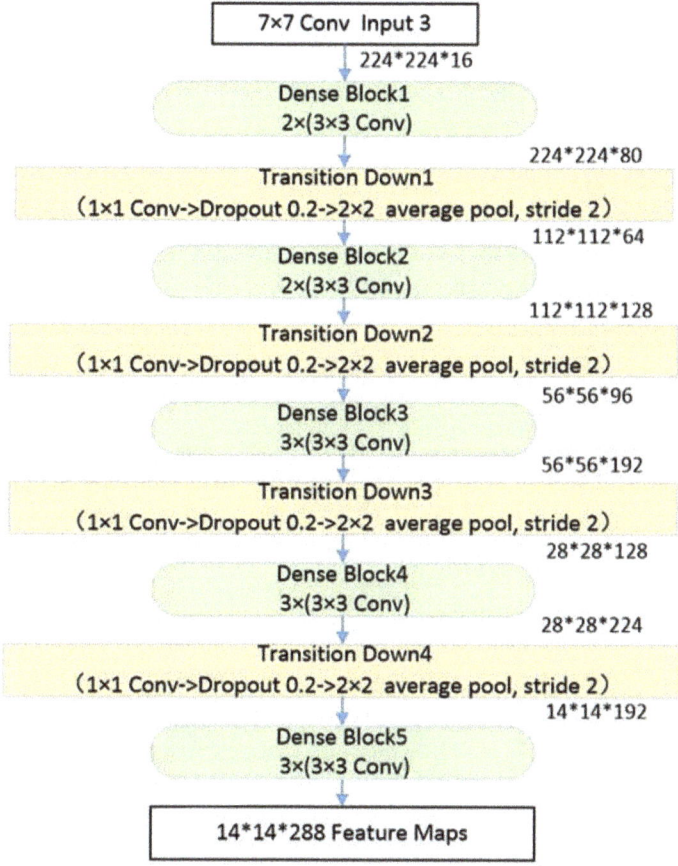

Figure 2. The lightweight DenseNets architectures. The growth rate for all networks is $k = 32$. Note that each "conv" layer shown in the figure corresponds to the sequence Batch Normalization–ReLU–Convolution.

3.2. Spatial Attention Fusion Module

The common encoder–decoder networks mainly use different scales of feature maps to help decoders gradually recover the object details information. The skip connection of U-Nets is a common way to help decoders recover object details information from the encoder path by reusing feature maps. However, this way will result in over-using low-level features and cause over-segmentation [28]. The attention mechanism can weight lower-level information using higher-level visual information. Inspired by the attention mechanism, a spatial attention fusion module was designed to enhance useful low-level feature information and remove noise to avoid over-using low-level features. In the spatial attention fusion module, first, the high-level features are activated by a sigmoid layer, the output of the activation normalizes to [0,1] and is used as the weight of low-level features. Then, the high-level features activation output is multiplied by low-level features to obtain the weighted low-level feature. Finally, the high-level features and weighted low-level information are added as the input of the dense block to gradually recover the object details information. The architecture of the spatial attention features fusion module is shown in Figure 3.

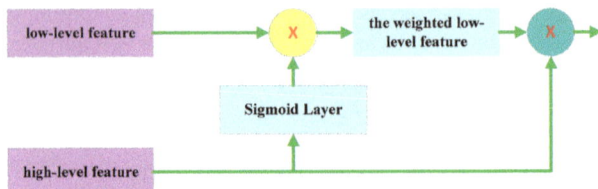

Figure 3. The architecture of the spatial attention features fusion module.

3.3. The Architecture's Decoder Part

As can be seen from Figure 1, in addition to the spatial attention features fusion modules and the transposed convolutions, the decoder path also contained five dense blocks and four bottleneck layers. Dense Block 6–8 contain three "conv" layers, respectively. Dense Block 9–10 contained two "conv" layers, respectively. The growth rate for these dense blocks was $k = 32$. Bottleneck layers were used to reduce the number of input feature-maps, which was built from a 1×1 convolutional layer followed by a dropout layer (drop rate: 0.2), In the four bottleneck layers, the output channels of bottleneck layers 1–2 were 256, and the output channels of bottleneck layers 3–4 were 128. The strides of transposed convolutions were equal to 2.

4. Experiments

In this section, the effectiveness of the proposed scheme for building extraction in very high resolution remote-sensing images was investigated. All networks were trained and tested with Tensorflow on GPU (TITAN X).

4.1. Training Details

4.1.1. Dataset

The proposed method was evaluated on the ISPRS 2D semantic labeling contest (Potsdam), which is an open benchmark dataset, which can be downloaded from the ISPRS official website (http://www2.isprs.org/commissions/comm3/wg4/2d-sem-label-potsdam.html). This dataset contains 38 very high-resolution true orthophoto (TOP) tiles extracted from a large TOP mosaic. ISPRS only provides 24 labeled images for training, while the remaining 14 tiles are unreleased. In this study, five of the labeled images were randomly used as the validation set and the remaining labeled images as the training models. Each tile contained around 6000×6000 pixels with a resolution of 5 cm, which made small details visible. The ground truth contained six of the most common land cover classes including impervious surfaces, buildings, low vegetation, trees, cars, and clutter/background.

4.1.2. Dataset Preprocessing

Given the limited memory of the GPU and obtaining more training samples, images of an average size of 6000×6000 were split into smaller patches in sizes of 224×224, 448×448, and 512×512. Thus, 14,339 patches of three sizes for training the networks can be obtained. However, to aid in testing, the size of the evaluation data was set as 3000×3000. So, a total of 20 images can be obtained for evaluation. Moreover, according to the defined red–green–blue (RGB) values of the six land cover classes, required objects can be extracted such as (0, 0, 255), which means the building type.

4.1.3. Implementation Details

For an individual network, the network was trained with an initial learning rate of 0.01. To ensure an outstanding learning result, a manual adjustment of the learning rate was made according to the speed of the training convergence, and will be about 0.00001 at last. There were 500 epochs during the training and each epoch had 1000 samples. As Adam is an adaptive optimizer with implementation simple, high computational efficiency and low memory requirement, which is used as the optimizer to optimize the network when adjusting parameters like weights, biases, and so on. To contribute to the geoscience community, the implementation code, the trained network, labeled images, and test images will release in open-source format and can be publicly accessed via GitHub (https://github.com/shenhuqiji/DAN).

4.1.4. Evaluation

Pixel- and object-based metrics are used in this study. While the latter quantifies the number of buildings and offers a quick assessment, the former is based on the number of pixels within the extracted buildings and provides a more rigorous evaluation [48]. In pixel-based evaluation, overall accuracy (*OA*), *F*1 score, and mean intersection over union (*MIOU*) were used to assess the quantitative performance. The *F*1 score is calculated by:

$$F1 = 2 \times \frac{precision \times recall}{precision + recall} \quad (2)$$

where,

$$precision = \frac{tp}{tp+fp}, \quad recall = \frac{tp}{tp+fn} \quad (3)$$

where *tp*, *fp*, and *fn* are true positive, false positive, and false negative, respectively. These values can be calculated by the pixel-based confusion matrices per tile, or an accumulated confusion matrix. Overall accuracy is the normalization of the trace from the confusion matrix, and *IoU* is an average value of the intersection of the prediction and ground truth regions over their union, as follows. Then the *MIOU* can be computed by averaging the *IoU* of all classes.

$$IoU = \frac{precision \times recall}{precision + recall - precision \times recall} \quad (4)$$

In object-based evaluation, completeness, correctness, and quality values were used to assess the quantitative performance. Completeness is known as detection rate or producer's accuracy, and correctness is known as user's accuracy [48]. The quality values is calculated by:

$$Quality = \frac{\|tp\|}{\|tp\| + \|fp\| + \|fn\|} \quad (5)$$

4.2. Extraction Results

After 500,000 iterations, our best model achieved state-of-the-art results on the datasets (Table 1). Note that all of the results are listed based on the RGB images without any pre-processing and post-processing. The changing accuracies and losses of the datasets with the increasing epochs are shown in Figure 4. The architecture reached high scores (96.16% *OA*, 92.56% *F*1 score, 90.56% *MIOU*, 0.9521 *Precision* and 0.9066 *recall*) for all five validation datasets, which indicated the proposed dense-attention network performed well on the buildings. Furthermore, dataset 2 and dataset 5 obtained the highest accuracy and the lowest accuracy for the *OA*, respectively (97.21% vs. 94.39%). Visual inspection and comparison of building extraction maps were performed for the two datasets. The original images, ground truth, and prediction results of the dataset 2 and dataset 5 are listed in Figure 5. Although the extraction result of dataset 5 was the worst of the five validation datasets, the prediction of Figure 5e was also close to the ground truth of Figure 5f.

Table 1. Pixel-based evaluation results of the average accuracy for the overall accuracy (*OA*), F1 score, mean intersection over union (*MIOU*), *precision* and *recall* for buildings on all validation datasets and on individual datasets, respectively.

Validation Datasets	OA (%)	F1 Score (%)	MIOU (%)	Precision	Recall
All five datasets	96.16	92.56	90.56	0.9521	0.9066
Only dataset 1	96.63	89.34	88.08	0.8573	0.9327
Only dataset 2	97.21	95.05	93.38	0.9786	0.9240
Only dataset 3	97.10	95.54	93.64	0.9878	0.9251
Only dataset 4	95.92	92.45	90.26	0.9690	0.8839
Only dataset 5	94.39	90.39	87.43	0.9342	0.8755

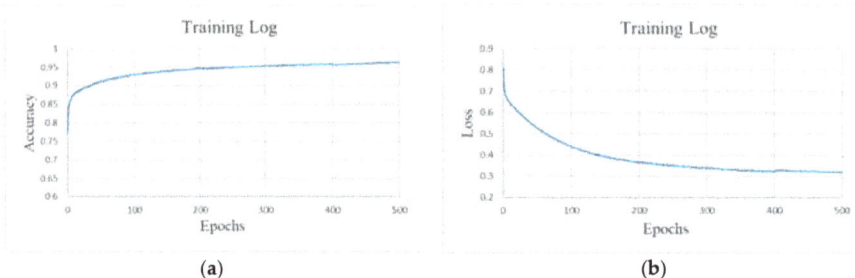

Figure 4. Plots showing the accuracy and loss of the dense-attention network (DAN) for training the datasets. The training accuracy (**a**) and the loss (**b**) changed with the increasing epochs.

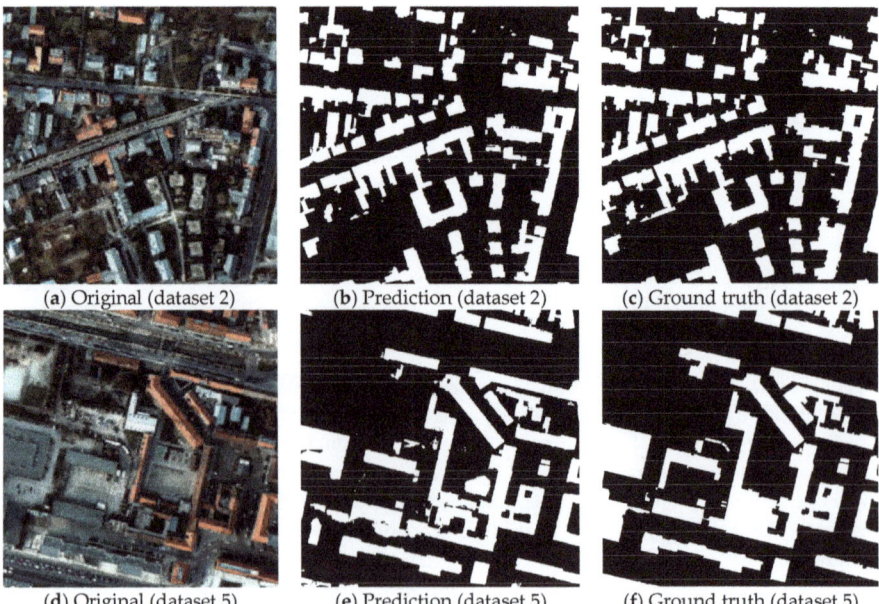

Figure 5. The worst results and the best results of the building extraction using the DAN. (**a**) represents the original RGB image of dataset 2; (**b**) represents the corresponding prediction from deep learning; (**c**) represents the corresponding ground truth. (**d–f**) are the original RGB image of dataset 5, prediction from deep learning, and the corresponding ground truth, respectively.

Moreover, the object-based evaluation results were also given in Table 2. Overall, these values from object-based evaluation are lower than the pixel-based results. Furthermore, dataset 2 and dataset 1 obtained the highest accuracy and the lowest accuracy for the quality value, respectively (0.8267 vs. 0.55), which indicate there are some differences in pixel- and object-based metrics.

Table 2. Object-based evaluation results on all validation datasets and on individual datasets using the completeness, C_m; correctness, C_r; quality metrics, Q; ($C_{m,2.5}$, $C_{r,2.5}$, and $Q_{2.5}$ are for buildings over 2.5 m^2), respectively.

Validate Data	tp	fn	fp	$C_{m,2.5}$	$C_{r,2.5}$	$Q_{2.5}$
All datasets	159	30	32	0.8413	0.8325	0.7195
Only dataset 1	22	4	14	0.8462	0.6111	0.55
Only dataset 2	62	11	2	0.8493	0.9688	0.8267
Only dataset 3	32	7	2	0.8205	0.9412	0.7805
Only dataset 4	18	5	6	0.7826	0.75	0.6207
Only dataset 5	25	3	8	0.8929	0.7576	0.6944

4.3. Comparisons with Related Networks

To show the effectiveness of the proposed network, comparisons were performed against two recent state-of-the-art building extraction methods, as showed in Figure 6. Note that the two methods were implemented and tested on the same experimental datasets (RGB images) of the ISPRS 2D semantic-labeling contest (Potsdam). In order to compare the test results, the same training datasets with corresponding stable loss and test datasets were used for the Deeplab-V3 [34], MFRN [27], and the proposed DAN. Due to lack of spatial attention fusion module, some low-level features are over-used and result in over-segmentation, which bring trivial and fragmentary buildings for the five validation datasets from the Deeplab-V3 and MFRN networks (see red boxes in Figure 6a,b). The results have improved markedly from the proposed DAN network in red boxes of Figure 6c. This finding suggests that the spatial attention mechanism can significantly improve the performance of a DCNN-based method. In order to quantitatively compare the proposed DAN network with the Deeplab-V3 and MFRN networks, the average *OA*, *F*1 score, *MIOU*, training time (*TT*) and recognition time (*RT*) of the three networks are listed in Table 3. The MFRN and the proposed DAN clearly outperforms the Deeplab-V3 by about 5% in the *OA*, about 8% in the *F*1 score and about 10% in the *MIOU*, respectively. In addition, although the proposed DAN network produced a minor improvement compared with the very new MFRN network, the DAN network used less layer fully convolutional DenseNet. Actually, a 56-layer fully convolutional DenseNet is concluded in MFRN, while the DAN use only a 24-layer fully convolutional DenseNet. More layers means that MFRN will take more time to train and test. From Table 3, the training time (*TT*) and recognition time (*RT*) of the three network are listed. The proposed DAN outperforms again the MFRN and Deeplab-V3 with less time (*TT*: 42.1 h < 51.4 h < 86.7 h, *RT*: 77.6 s < 85 s < 88.8 s). Therefore, compared with the very new MFRN network, another advantage of DAN is its high efficiency. Overall, although there exist a litter false classified buildings (see yellow boxes in Figure 6c,d), the proposed DAN network can achieve a better performance to extract buildings from VHR remote-sensing images without any other processing.

Similarly, object-based evaluation results of the proposed DAN network with the Deeplab-V3 and MFRN networks on Potsdam datasets using the completeness, correctness and quality metrics are listed in Table 4. From the metrics of completeness, there are no distinct differences for the three networks, while the scores of correctness and quality metrics from DAN are better than that from the other networks. Although the scores from object-based evaluation are lower than that from the pixel-based results for the three networks, the DAN still obtain the highest scores. Therefore, the results from both the pixel-based and the object-based evaluation system again shows that the proposed DAN network can achieve a better performance.

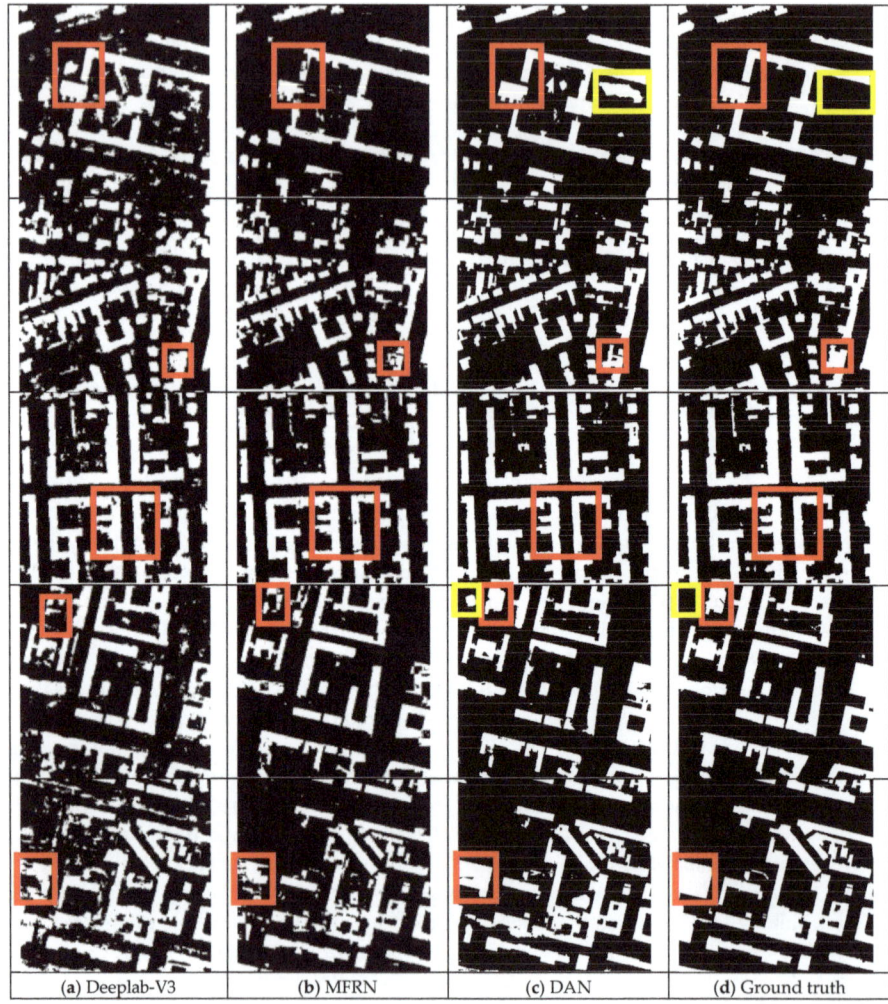

Figure 6. The results of the building extraction from Deeplab-V3, multiple-feature reuse network (MFRN) and the proposed DAN.

Table 3. Compared with the results of the proposed network with other networks on Potsdam datasets using the average *OA*, *F*1 score, MIOU, training time (*TT*) and recognition time (*RT*). The *RT* is tested in datasets which include 20 images with a size in 3000 × 3000.

	OA (%)	F1 Score (%)	MIOU (%)	TT (h)	RT (s)
Deeplab-V3	90.25	83.36	79.37	86.7	88.8
MFRN	95.61	91.80	89.74	51.4	85.0
DAN	96.16	92.56	90.56	42.1	77.6

Table 4. Object-based evaluation results of the proposed network with other networks on Potsdam datasets using the completeness, C_m; correctness, C_r; quality metrics, Q; ($C_{m,2.5}$, $C_{r,2.5}$, and $Q_{2.5}$ are for buildings over 2.5 m²).

	tp	fn	fp	$C_{m,2.5}$	$C_{r,2.5}$	$Q_{2.5}$
Deeplab-V3	168	21	392	0.8889	0.3	0.2892
MFRN	163	26	56	0.8624	0.7443	0.6653
DAN	159	30	32	0.8413	0.8325	0.7195

5. Discussion

Extracting 2D buildings footprints in VHR imagery has wide applications in navigation, urban planning, disaster management, and population estimation. It is necessary to develop techniques to extract 2D buildings information. Considering the limitations of the existing extraction methods, this study proposes a dense-attention network (DAN) to extract 2D building in VHR images. This study has examined in detail the theoretical basis of the proposed method and compared it with other deep learning-based approaches using ISPRS 2D semantic labeling contest datasets. In general, some groups or blocks of 2D buildings can be extracted in their entirety from the proposed DAN network, while the Deeplab-V3 and MFRN often bring trivial and fragmentary buildings. The extracted 2D buildings footprints were evaluated against the ground truth (label data). The experiments indicated that the proposed DAN achieved a higher score than Deeplab-V3 and MFRN on both accuracy and efficiency.

The primary reason for the superior performance of DAN is that different-level features are rationally utilized by combining the DenseNets and a spatial attention fusion module. The DenseNets had been proven the superiority in extracting the feature maps from inputs. The DAN network based on DenseNet also has these capability. In addition, the common encoder–decoder networks mainly use skip connection to help decoders gradually recover the object details. Then, the acquired low-level feature maps in the early layer are less discriminating. Such feature maps are fused with the higher-level feature maps by skip connection may increase the ambiguity of the final result. This will result in over-using low-level features and cause over-segmentation. The attention mechanism can weight lower level information using higher-level visual information to suppress the background and noises. Therefore, a spatial attention fusion module based on the attention mechanism can better guide the low-feature maps to help high-level features recover details of the images and reduce the over-segmentation.

However, it should be noted that the accuracy of building extraction in VHR images could be affected by some factors, as can be seen from Figure 7:

- Complex background. Although water, bare, and sparse vegetation are in the minority in some test samples, they were also detected as buildings because of the similar hue to the foreground object (building), see Figure 6c,d (yellow boxes). The complex background may cause precision to be lower than recall, see the evaluation result of dataset 1 in Table 1. In addition, the water was not included into the above six land cover classes, which makes it difficult to fully learn the characteristics of the complex background, see Figure 7a–c. The misclassification may be a main limitation of the proposed DAN.
- Special buildings (SB). In some training samples, the characteristics (such as color, texture, and material) of a few buildings' roofs were quite different from most buildings. Moreover, the shape of some buildings that were covered by trees could not be detected precisely, and some blurry and irregular boundaries were hardly classified. Therefore, it was hard to detect these buildings, see Figure 7d–f.
- Unremarkable buildings (UB). In most training samples, when compared to the background, the foreground objects were very distinct. However, in some of the test samples, a few images were covered with large amounts of bare and sparse vegetation, and small-sized buildings. These small-sized buildings were displayed in patchy distributions and were even hard to detect with the naked eye, which added to the difficulty of detection, see Figure 7g–i.

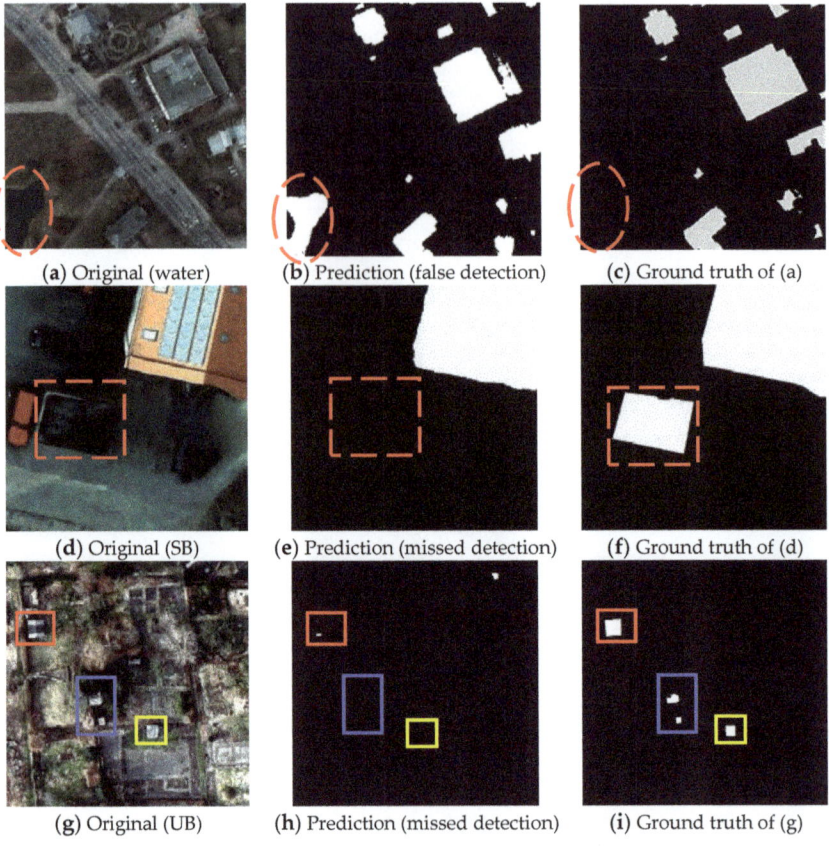

Figure 7. Some factors caused false detection and missed detection in building extraction. (**a**) represents the original red–green–blue (RGB) image with water; (**b**) represents the corresponding prediction with false detection; (**c**) represents the corresponding ground truth. (**d**–**f**) and (**g**–**i**) are the original RGB image with special buildings and unremarkable buildings, corresponding prediction with missed detection, and the corresponding ground truth, respectively.

To weaken the effects of the above factors, some more optimized deep learning networks were required to improve the efficiency and accuracy of building extraction for special buildings and unremarkable buildings. Furthermore, other attempts such as pre-processing or post-processing methods also play an important role in distinguishing complex backgrounds in building extraction. For pre-processing, edge-enhancing techniques may be introduced to increase the differences among objects, which leads to better performance during classification. For example, the water may not be detected as a building if the digital surface model (DSM), normalized difference vegetation index (NDVI), or normalized difference water index (NDWI) are used to enhance the edge. Post-processing methods are often used at the end of the chosen deep learning network classifier for further reducing the noise (false classified buildings) and to sharpen the boundary of the buildings. For example, the conditional random field and guided filters have proven to be a very effective post-processing way to optimize the classification results and further improve the efficiency and accuracy of building extraction [49]. Moreover, recent work on data fusion of multi-modal remote-sensing data also might help improve the accuracy of building extraction, as proposed in Audebert et al. [50]. However, the objective of this study was to propose a novel network for 2D building extraction in VHR (only RGB) images without any other processing or data fusion strategies. The pre-processing or post-processing methods can improve the efficiency and accuracy of building extraction, which will be further addressed in our future work.

6. Conclusions

In this paper, a dense-attention network (DAN) was proposed for 2D building extraction in VHR images. The dense-attention network contained an encoder part and a decoder part, which can guide message passing between high- and low-feature maps. As the encoder part, lightweight DenseNets were used to extract the feature maps from inputs; while the decoder part, named the spatial attention fusion module, was used to guide the low-feature maps to help high-level features recover details of the images. Therefore, the DAN can effectively help in integrating useful features, and weakening the noises and background. Experiments were carried out on VHR imageries from the ISPRS dataset. Six land classes could be extracted successfully using the DAN and the results showed the effectiveness and feasibility of the proposed network in improving the performance of the building extraction. The DAN was compared with two recent networks such as the Deeplab-V3 and MFRN, which had the potential to perform better in terms of *OA*, *MIOU*, *F*1 score, *TT* and *RT* (pixel-based) and quality metrics (object-based). However, the extraction accuracy was affected by complex background, special buildings, and unremarkable buildings. These problems may be overcome by combining more optimized deep neural networks and pre-processing or post-processing methods, which can reduce errors and omissions. Instead of separately treating the network and pre-/post-processing methods, both of them will be considered simultaneously for higher accuracy and efficiency.

Author Contributions: P.W. and H.Y. designed the experiments; Y.W. and B.W. contributed analysis tools; X.Y. and Y.X. performed the experiments; P.W. and H.Y. wrote the paper. All authors have read and approved the final manuscript.

Funding: This research was funded by the National Natural Science Foundation of China (grant number 41501376, 41571400), Natural Science Foundation of Anhui Province (grant number 1608085MD83), the Key Laboratory of Earth Observation and Geospatial Information Science of NASG (grant number 201805), the Science Research Project of Anhui Education Department (grant number KJ2018A0007) and the open fund for Discipline Construction, Institute of Physical Science and Information Technology, Anhui University.

Acknowledgments: The authors thank the ISPRS for making the Potsdam datasets available and organizing the semantic labeling challenge. The authors also thank Jian Liang (Wuhan University) and Xinxin Zhou (Anhui University) for providing the code of MFRN and data pre-processing, respectively, and the anonymous reviewers and the external editor for providing valuable suggestion and comments, which have greatly improved this paper.

Conflicts of Interest: The authors declare no conflict of interest.

References

1. Ghanea, M.; Moallem, P.; Momeni, M. Building extraction from high-resolution satellite images in urban areas: Recent methods and strategies against significant challenges. *Int. J. Remote Sens.* **2016**, *37*, 5234–5248. [CrossRef]
2. Ok, A.O. Automated detection of buildings from single vhr multispectral images using shadow information and graph cuts. *ISPRS J. Photogramm. Remote Sens.* **2013**, *86*, 21–40. [CrossRef]
3. Zhang, L.; Zhang, L.; Du, B. Deep learning for remote sensing data: A technical tutorial on the state of the art. *IEEE Geosci. Remote Sens. Mag.* **2016**, *4*, 22–40. [CrossRef]
4. Ahmadi, S.; Zoej, M.J.V.; Ebadi, H.; Moghaddam, H.A.; Mohammadzadeh, A. Automatic urban building boundary extraction from high resolution aerial images using an innovative model of active contours. *Int. J. Appl. Earth Obs. Geoinf.* **2010**, *12*, 150–157. [CrossRef]
5. Huang, X.; Zhang, L. A multidirectional and multiscale morphological index for automatic building extraction from multispectral geoeye-1 imagery. *Photogramm. Eng. Remote Sens.* **2011**, *77*, 721–732. [CrossRef]
6. Jin, X.; Davis, C.H. Automated building extraction from high-resolution satellite imagery in urban areas using structural, contextual, and spectral information. *EURASIP J. Appl. Signal Process.* **2005**, *14*, 745309. [CrossRef]

7. Ghanea, M.; Moallem, P.; Momeni, M. Automatic building extraction in dense urban areas through geoeye multispectral imagery. *Int. J. Remote Sens.* **2014**, *35*, 5094–5119. [CrossRef]
8. Du, S.; Zhang, Y.; Zou, Z.; Xu, S.; He, X.; Chen, S. Automatic building extraction from LiDAR data fusion of point and grid-based features. *ISPRS J. Photogramm. Remote Sens.* **2017**, *130*, 294–307. [CrossRef]
9. Campos-Taberner, M.; Romero-Soriano, A.; Gatta, C.; Camps-Valls, G.; Lagrange, A.; Saux, B.L.; Beaupere, A.; Boulch, A.; Chan-Hon-Tong, A.; Herbin, S.; et al. Processing of extremely high-resolution lidar and rgb data: Outcome of the 2015 ieee grss data fusion contest—Part A: 2-D contest. *IEEE J. Sel. Top. Appl. Earth Obs. Remote Sens.* **2017**, *9*, 5547–5559. [CrossRef]
10. Vo, A.V.; Truong-Hong, L.; Laefer, D.F.; Tiede, D.; d'Oleire-Oltmanns, S.; Baraldi, A.; Shimoni, M.; Moser, G.; Tuia, D. Processing of extremely high resolution LiDAR and RGB data: Outcome of the 2015 IEEE GRSS data fusion contest—Part B: 3-D contest. *IEEE J. Sel. Top. Appl. Earth Obs. Remote Sens.* **2016**, *9*, 5560–5575. [CrossRef]
11. Gilani, S.A.N.; Awrangjeb, M.; Lu, G. An automatic building extraction and regularisation technique using lidar point cloud data and orthoimage. *Remote Sens.* **2016**, *8*, 258. [CrossRef]
12. Niemeyer, J.; Rottensteiner, F.; Soergel, U. Contextual classification of lidar data and building object detection in urban areas. *ISPRS J. Photogramm. Remote Sens.* **2014**, *87*, 152–165. [CrossRef]
13. Aljumaily, H.; Laefer, D.F.; Cuadra, D. Urban point cloud mining based on density clustering and MapReduce. *J. Comput. Civ. Eng.* **2017**, *31*, 04017021. [CrossRef]
14. Li, E.; Xu, S.; Meng, W.; Zhang, X. Building extraction from remotely sensed images by integrating saliency cue. *IEEE J. Sel. Top. Appl. Earth Obs. Remote Sens.* **2017**, *10*, 906–919. [CrossRef]
15. Vakalopoulou, M.; Karantzalos, K.; Komodakis, N.; Paragios, N. In Building detection in very high resolution multispectral data with deep learning features. In Proceedings of the Geoscience and Remote Sensing Symposium, Milan, Italy, 26–31 July 2015; pp. 1873–1876.
16. Sun, Y.; Zhang, X.; Zhao, X.; Xin, Q. Extracting Building Boundaries from High Resolution Optical Images and LiDAR Data by Integrating the Convolutional Neural Network and the Active Contour Model. *Remote Sens.* **2018**, *10*, 1459. [CrossRef]
17. Alshehhi, R.; Marpu, P.R.; Wei, L.W.; Mura, M.D. Simultaneous extraction of roads and buildings in remote sensing imagery with convolutional neural networks. *ISPRS J. Photogramm. Remote Sens.* **2017**, *130*, 139–149. [CrossRef]
18. Yang, H.L.; Yuan, J.; Lunga, D.; Laverdiere, M.; Rose, A.; Bhaduri, B. Building extraction at scale using convolutional neural network: Mapping of the united states. *IEEE J. Sel. Top. Appl. Earth Obs. Remote Sens.* **2018**. [CrossRef]
19. Saito, S.; Yamashita, T.; Aoki, Y. Multiple object extraction from aerial imagery with convolutional neural networks. *Electron. Imaging* **2016**, *60*. [CrossRef]
20. Huang, Z.; Cheng, G.; Wang, H.; Li, H.; Shi, L.; Pan, C. Building extraction from multi-source remote sensing images via deep deconvolution neural networks. In Proceedings of the 2016 IEEE International Geoscience and Remote Sensing Symposium (IGARSS), Beijing, China, 10–15 July 2016; pp. 1835–1838.
21. Bittner, K.; Cui, S.; Reinartz, P. Building extraction from remote sensing data using fully convolutional networks. In Proceedings of the International Archives of the Photogrammetry, Remote Sensing and Spatial Information Sciences, ISPRS Hannover Workshop, Hannover, Germany, 6–9 June 2017; pp. 481–486.
22. Ronneberger, O.; Fischer, P.; Brox, T. U-Net: Convolutional networks for biomedical image segmentation. In Proceedings of the International Conference on Medical Image Computing and Computer-Assisted Intervention, Munich, Germany, 5–9 October 2015; pp. 234–241.
23. Badrinarayanan, V.; Kendall, A.; Cipolla, R. Segnet: A deep convolutional encoder-decoder architecture for scene segmentation. *arXiv* **2016**, arXiv:1511.00561v3.
24. Lin, G.; Milan, A.; Shen, C.; Reid, I. Refinenet: Multi-path refinement networks for high-resolution semantic segmentation. *arXiv* **2016**, arXiv:1611.06612.
25. Huang, G.; Liu, Z.; Laurens, V.D.M.; Weinberger, K.Q. Densely connected convolutional networks. In Proceedings of the IEEE Conference on Computer Vision and Pattern Recognition (CVPR), Honolulu, HI, USA, 21–26 July 2017; pp. 2261–2269.
26. Jégou, S.; Drozdzal, M.; Vazquez, D.; Romero, A.; Bengio, Y. The one hundred layers tiramisu: Fully convolutional densenets for semantic segmentation. In Proceedings of the IEEE Conference on Computer Vision and Pattern Recognition (CVPR), Honolulu, HI, USA, 21–26 July 2017; pp. 1175–1183.

27. Li, L.; Liang, J.; Weng, M.; Zhu, H. A Multiple-Feature Reuse Network to Extract Buildings from Remote Sensing Imagery. *Remote Sens.* **2018**, *10*, 1350. [CrossRef]
28. Wang, H.; Wang, Y.; Zhang, Q.; Xiang, S.; Pan, C. Gated convolutional neural network for semantic segmentation in high-resolution images. *Remote Sens.* **2017**, *9*, 446. [CrossRef]
29. Yang, Y.; Zhong, Z.; Shen, T.; Lin, Z. Convolutional neural networks with alternately updated clique. In Proceedings of the IEEE Conference on Computer Vision and Pattern Recognition, Salt Lake City, UT, USA, 18–22 June 2018; pp. 2413–2422.
30. Yu, B.; Yang, L.; Chen, F. Semantic segmentation for high spatial resolution remote sensing images based on convolution neural network and pyramid pooling module. *IEEE J. Sel. Top. Appl. Earth Obs. Remote Sens.* **2018**, *11*, 3252–3261. [CrossRef]
31. Kemker, R.; Salvaggio, C.; Kanan, C. Algorithms for semantic segmentation of multispectral remote sensing imagery using deep learning. *ISPRS J. Photogramm. Remote Sens.* **2018**. [CrossRef]
32. Simonyan, K.; Zisserman, A. Very deep convolutional networks for large-scale image recognition. *arXiv* **2014**, arXiv:1409.1556.
33. Wang, F.; Jiang, M.; Qian, C.; Yang, S.; Li, C.; Zhang, H.; Wang, X.; Tang, X. Residual attention network for image classification. In Proceedings of the IEEE Conference on Computer Vision and Pattern Recognition (CVPR), Honolulu, HI, USA, 21–26 July 2017; pp. 6450–6458.
34. Chen, L.C.; Papandreou, G.; Schroff, F.; Adam, H. Rethinking atrous convolution for semantic image segmentation. *arXiv* **2017**, arXiv:1706.05587v3.
35. Itti, L.; Koch, C. Computational modelling of visual attention. *Nat. Rev. Neurosci.* **2001**, *2*, 194–203. [CrossRef] [PubMed]
36. Xu, K.; Ba, J.; Kiros, R.; Cho, K.; Courville, A.; Salakhutdinov, R.; Zemel, R.; Bengio, Y. Show, attend and tell: Neural image caption generation with visual attention. *Cmputer Sci.* **2015**.
37. Chen, L.; Zhang, H.; Xiao, J.; Nie, L.; Shao, J.; Liu, W.; Chua, T.S. Sca-cnn: Spatial and channel-wise attention in convolutional networks for image captioning. In Proceedings of the IEEE Conference on Computer Vision and Pattern Recognition (CVPR), Honolulu, HI, USA, 21–26 July 2017; pp. 6298–6306.
38. Yang, Z.; He, X.; Gao, J.; Deng, L.; Smola, A. Stacked attention networks for image question answering. In Proceedings of the IEEE Conference on Computer Vision and Pattern Recognition (CVPR), Las Vegas, NV, USA, 26 June–1 July 2016; pp. 21–29.
39. Chen, K.; Wang, J.; Chen, L.C.; Gao, H.; Xu, W.; Nevatia, R. Abc-cnn: An attention based convolutional neural network for visual question answering. *arXiv* **2015**, arXiv:1511.05960v2.
40. Fu, J.; Zheng, H.; Mei, T. Look closer to see better: Recurrent attention convolutional neural network for fine-grained image recognition. In Proceedings of the IEEE Conference on Computer Vision and Pattern Recognition (CVPR), Honolulu, HI, USA, 21–26 July 2017; pp. 4476–4484.
41. Yao, L.; Torabi, A.; Cho, K.; Ballas, N.; Pal, C.; Larochelle, H.; Courville, A. Describing videos by exploiting temporal structure. In Proceedings of the 2015 IEEE International Conference on Computer Vision (ICCV), Santiago, Chile, 7–13 December 2015; pp. 4507–4515.
42. Kuen, J.; Wang, Z.; Wang, G. Recurrent attentional networks for saliency detection. In Proceedings of the IEEE Conference on Computer Vision and Pattern Recognition (CVPR), Las Vegas, NV, USA, 26 June–1 July 2016; pp. 3668–3677.
43. Hu, J.; Shen, L.; Sun, G. Squeeze-and-excitation networks. *arXiv* **2018**, arXiv:1709.01507v2.
44. Li, H.; Xiong, P.; An, J.; Wang, L. Pyramid attention network for semantic segmentation. *arXiv* **2018**, arXiv:1805.10180v2.
45. Yu, C.; Wang, J.; Peng, C.; Gao, C.; Yu, G.; Sang, N. Learning a discriminative feature network for semantic segmentation. In Proceedings of the IEEE Conference on Computer Vision and Pattern Recognition, Salt Lake City, UT, USA, 18–22 June 2018; pp. 1857–1866.
46. Golnaz, G.; Fowlkes, C.C. Laplacian pyramid reconstruction and refinement for semantic segmentation. In Proceedings of the European Conference on Computer Vision (ECCV), Amsterdam, The Netherlands, 8–16 October 2016; pp. 519–534.
47. Pleiss, G.; Chen, D.; Huang, G.; Li, T.; Laurens, V.D.M.; Weinberger, K.Q. Memory-efficient implementation of densenets. *arXiv* 2017, arXiv:1707.06990v1.

48. Awrangjeb, M.; Fraser, C.S. An Automatic and Threshold-Free Performance Evaluation System for Building Extraction Techniques from Airborne LIDAR Data. *IEEE J. Sel. Top. Appl. Earth Obs. Remote Sens.* **2014**, *7*, 4184–4198. [CrossRef]
49. Xu, Y.; Wu, L.; Xie, Z.; Chen, Z. Building extraction in very high resolution remote sensing imagery using deep learning and guided filters. *Remote Sens.* **2018**, *10*, 144. [CrossRef]
50. Audebert, N.; Saux, B.; Lefèvre, S. Beyond RGB: Very high resolution urban remote sensing with multimodal deep networks. *ISPRS J. Photogramm. Remote Sens.* **2018**, *140*, 20–32. [CrossRef]

© 2018 by the authors. Licensee MDPI, Basel, Switzerland. This article is an open access article distributed under the terms and conditions of the Creative Commons Attribution (CC BY) license (http://creativecommons.org/licenses/by/4.0/).

Article

An Effective Data-Driven Method for 3-D Building Roof Reconstruction and Robust Change Detection

Mohammad Awrangjeb [1,*,†], Syed Ali Naqi Gilani [1,2,†] and Fasahat Ullah Siddiqui [1,2,†]

1. School of Information and Communication Technology, Griffith University, Nathan, QLD 4111, Australia; alinaqig@gmail.com (S.A.N.G.); fasahat02@gmail.com (F.U.S.)
2. Faculty of Information Technology, Monash University, Clayton, VIC 3800, Australia
* Correspondence: m.awrangjeb@griffith.edu.au; Tel.: +61-7-373-55032
† Authors contributed equally to this work.

Received: 13 June 2018; Accepted: 19 September 2018; Published: 21 September 2018

Abstract: Three-dimensional (3-D) reconstruction of building roofs can be an essential prerequisite for 3-D building change detection, which is important for detection of informal buildings or extensions and for update of 3-D map database. However, automatic 3-D roof reconstruction from the remote sensing data is still in its development stage for a number of reasons. For instance, there are difficulties in determining the neighbourhood relationships among the planes on a complex building roof, locating the step edges from point cloud data often requires additional information or may impose constraints, and missing roof planes attract human interaction and often produces high reconstruction errors. This research introduces a new 3-D roof reconstruction technique that constructs an adjacency matrix to define the topological relationships among the roof planes. It identifies any missing planes through an analysis using the 3-D plane intersection lines between adjacent planes. Then, it generates new planes to fill gaps of missing planes. Finally, it obtains complete building models through insertion of approximate wall planes and building floor. The reported research in this paper then uses the generated building models to detect 3-D changes in buildings. Plane connections between neighbouring planes are first defined to establish relationships between neighbouring planes. Then, each building in the reference and test model sets is represented using a graph data structure. Finally, the height intensity images, and if required the graph representations, of the reference and test models are directly compared to find and categorise 3-D changes into five groups: *new*, *unchanged*, *demolished*, *modified* and *partially-modified* planes. Experimental results on two Australian datasets show high object- and pixel-based accuracy in terms of completeness, correctness, and quality for both 3-D roof reconstruction and change detection techniques. The proposed change detection technique is robust to various changes including addition of a new veranda to or removal of an existing veranda from a building and increase of the height of a building.

Keywords: building; modelling; reconstruction; change detection; LiDAR; point cloud; 3-D

1. Introduction

The fundamental task of building reconstruction is the transformation of low-level building primitives (e.g., lines and planes) to a high-level model description. In 3-D change detection, it is inspected whether there are changes in buildings over a period in terms of new, modified, and/or demolished buildings and/or building-parts. The reconstruction step can be considered as an essential prerequisite for 3-D change detection, particularly for detection of informal buildings or extensions and for update of 3-D map database. In a 3-D map database, there are buildings along with other important man-made objects such as roads and electric power lines. A direct comparison of 3-D building models generated from a recent dataset to the models in the (old) map database will not only identify the changes in buildings but also help an effective and efficient update of the database.

In practice, there will be only a small number of buildings being changed in an area for a given period of time, unless this is a newly built-up area or an area hit by a calamity (e.g., bushfire or earthquake). Therefore, automatic modelling and change detection steps will be helpful in indicating the potential changed areas of buildings in a user interface, where a human operator can quickly accept and/or reject the indications and update the database accordingly [1]. Moreover, the state and local government officials can check if the indicative changes were previously authorised or not. Thus, they can send inspectors to the unauthorised (informal) areas only, instead of all areas, saving both money and time.

Many scientists have developed building reconstruction and change detection techniques utilising image information only, others have utilised LiDAR (Light Detection And Ranging) point cloud, and some have attempted to integrate LiDAR and aerial images for several Geographic Information System (GIS) applications, including city mapping, map database updating, disaster estimation, and city infrastructure planning. However, among spatial data researchers and mapping professionals, LiDAR data have gained popularity because of fast pulsation, precision, and accuracy in capturing 3-D geo-referenced spatial information about buildings, roads, vegetation, and other objects expediently at a high point density. These characteristics make these data feasible to examine natural and built environments across a wide range of scales for automatic reconstruction and change detection in buildings and their distinct features. Recent improvements in the automation of building reconstruction and change detection methods are reducing the labour and time consumption in these applications.

The 3-D reconstruction of building models and change detection include several non-trivial processes, such as segmentation, classification, structuring, hypothesis generation, and geometric modelling. A seamless integration of these in a conventional way would be not only unrealistic but also labourious. This paper, therefore, presents a workflow that uses building roof planes for 3-D model generation and subsequently uses these models for change detection. To achieve our goals and address the particular challenges, this research presents two techniques aiming at 3-D reconstruction of building roof models and building change detection separately. The proposed techniques are entirely data-driven using LiDAR point cloud data only. The first technique, 3-D building roof modelling, reconstructs buildings represented at lower levels with coarse boundaries (3-D roof planes) to higher levels (3-D building models). The second technique, building change detection, subsequently uses the constructed 3-D building models and LiDAR data for identification of changes in buildings.

In Section 2, the related works are discussed. Section 3 provides the detail on challenges for 3-D building modelling and change detection methods, along with the contributions of this paper. The proposed 3-D building modelling and 3-D change detection methods are presented in Sections 4 and 5, respectively. In Section 6, the dataset, parameter settings, experimental setup, and results are discussed. Finally, conclusions are presented in Section 7.

2. Related Works

In recent years, studies on building extraction, reconstruction and change detection have made significant advances and a wide range of methods have been proposed on façade segmentation and opening area detection [2], building extraction [3–5], change detection and map database update [1,6–8], roof plane extraction [9] and 3-D reconstruction [10]. A number of techniques [11,12] have also been proposed for evaluation of these methods. In addition, since different methods were evaluated using different datasets and evaluation techniques, there have been several attempts to benchmark them on common platforms [13,14].

In 3-D building roof reconstruction and change detection most early methods were manual, with the involvement of a trained human operator who performed accurate measurements. However, human intervention is not only expensive but also reduces the speed of execution in achieving high productivity and in processing large datasets. Recently developed reconstruction and change detection methods [1,15] aim to reduce these limitations in a semiautomatic manner.

2.1. 3-D Building Roof Modelling

The 3-D building reconstruction methods can broadly be classified into three categories on the basis of their processing strategies: model-driven, data-driven, and hybrid approaches [16–18]. A model-driven method uses a set of predefined building models (shape) and fits into the input data for the extraction purposes, in contrast to a data-driven method that uses the input data and extracts one or more features (e.g., lines and planes) for the detection and reconstruction of buildings. A hybrid method, on the other hand, exhibits the characteristics of both model- and data-driven approaches.

Among the model-driven methods, Oude Elberink and Vosselman [19] proposed a graph matching approach to handle both complete and incomplete laser data. While a complete matching of data with a target model allows an automated 3-D reconstruction of a building roof, an incomplete match leads to a manual interaction for a correct model. To reduce human interaction, Xiong et al. [20] proposed a graph-based error correction for roof topology. A graph edit dictionary that stores representative erroneous subgraphs was used to automatically identify and correct errors.

Kim and Shan [10] proposed a novel data-driven roof plane segmentation and building reconstruction technique using airborne LiDAR data. Although this technique shows good results, it suffers from over-segmentation and neglects the effect of vegetation in the segmentation process. Jung et al. [21] presented a reconstruction technique to develop 3-D rooftop models at city-scale from LiDAR data. Although the experimental results showed a good performance for large buildings, some small roof planes were not detected, and were therefore not reconstructed. Moreover, this method suffers from under-segmentation issues since many roof planes were merged into their adjacent clusters. Wu et al. [22] offered a graph-based technique to reconstruct urban building models from airborne LiDAR data. Building models were reconstructed by gluing all individual parts of the building models obtained from bipartite graph matching into a complete model. Although this technique provides detailed models, it fails to capture the sides of buildings and produces high geometric distortion resulting in low completeness and high modelling errors.

Rottensteiner et al. [14] reported comparative research results for urban object detection and 3-D building reconstruction using the ISPRS (International Society for Photogrammetry and Remote Sensing) benchmark datasets. They selected fourteen different building reconstruction methods and evaluated their performances on the basis of different quality metrics.

2.2. 3-D Building Change Detection

Based on the input data sources, the building change detection methods can be categorised into three groups: image only, LiDAR only or integration of LiDAR and image-based methods. The image-based methods are mainly for 2-D change detection and are mostly unable to differentiate between partially-modified buildings and planes. For example, both Gu et al. [8] and Leichtle et al. [7] divided the image region into changed and unchanged buildings only and were unable to find out the changes in individual roof planes.

Raw LiDAR data or LiDAR-derived Digital Surface Model (DSM) is also used as the only source of information to detect the 3-D building changes [6,23]. Tran et al. [6] proposed a method where, in addition to the ground and tree, they classified buildings into new, demolished and unchanged types. Teo and Shih [23] extracted changed building regions by applying the height threshold and morphological filter to nDSM (normalised DSM) of two different days. However, this method missed many small planes because of inadequate selection of window size of the filter. In addition, this method was unable to detect partially-modified building planes.

There are also methods that use a segmentation technique for building change detection from both LiDAR and images [1,24]. For instance, Awrangjeb [1] generated a building mask by extracting the building boundaries from non-ground LiDAR data. Later, the building boundaries were refined manually using visual analysis of the aerial image. This method could detect all possible changes in a building such as new, demolished, unchanged, modified, and partially-modified buildings.

In addition, there are also methods that detect the changes in a building using 3-D building models. These methods compare the height information of individual buildings and detect the height changes in the buildings for change detection. A 3-D building model can be generated from LiDAR data by using thresholding method [25] or from stereo-images by using a least square matching process [26,27]. For example, Chen et al. [25] used LiDAR-based building models at two different dates to detect building changes. The non-building portion of the models was removed by using a height threshold and NDVI (Normalized Difference Vegetation Index) analysis. However, small changes in buildings and occluded buildings were also removed from the models and were not detected. Therefore, this method had up to 96% of pixel-based accuracy. Stal et al. [26] and Qin [27] methods compare the stereo-images- and LiDAR-based building models to detect building changes. In the method of Stal et al. [26], a morphological filter and NDVI analysis were applied to remove the unwanted regions, while Qin [27] used three more filters, such as shadow index, noise filter, and irregular structure filter. However, the methods that use 3-D building models need several pre-defined parameters for accurate detection of building changes. In addition, modified and partially-modified building planes were not detected by these methods.

3. Challenges and Contributions

The existing 3-D building modelling techniques differ significantly, based on the primitive shapes used and the input data sources [28]. They often impose constraints on the minimum footprint size and positional accuracy values for reconstruction of specific models at a certain level of detail [29]. Although many of the existing approaches have demonstrated promising results in building reconstruction, there are still a number of issues to be improved. For instance, building roofs are mostly disconnected after segmentation process, causing difficulty in determining the neighbourhood relationships among the roof planes. Furthermore, locating step edges from LiDAR data only is also hard and often requires additional information or constraints. In addition, the approximation of roof patches, which are generally missed because of the low density of the LiDAR data, requires the operator to make assumptions and often produces high reconstruction errors.

To resolve the above issues, this research introduces a data-driven 3-D reconstruction technique that constructs buildings represented at lower levels with coarse boundaries (3-D roof planes) to the higher levels (3-D building models). The proposed reconstruction technique receives building roof planes (extracted from LiDAR point could data by any roof plane extraction method such as Awrangjeb and Fraser [9] and Mongus et al. [4]) as inputs and offers the following contributions:

- Insertion of missing planes: A missing plane can be a small plane from where the number of reflected laser points is limited, possibly due to a low point density. It can also be due to a height jump between planes. A slanted plane is grown if there is an existence of unsegmented LiDAR points between any two planes. Otherwise, a vertical plane is inserted between the planes.
- Reconstruction of complete building models: When there are missing planes, the topological relationship among the roof planes is incorrect. Thus, an adjacency matrix that defines the topological relationships among the input roof planes of a building is first constructed. Then, the matrix is updated (i.e., the topological relationship is corrected) based on the inserted missing planes. Finally, the building model is generated using the correct topological relationship and the revised intersection lines among the inserted missing planes and the input roof planes.

Generally, the building change detection step is applied after the building detection or 3-D building modelling stage to update a map database [1]. The update to the map database is largely dependent on visual interpretation, estimation of numerous parameters, and human interaction. Consequently, it is a time-consuming procedure to analyse the changes. In addition, most of the existing methods mainly focus on detecting the 2-D changes in a building and, thus, are unable to distinguish the specific height change in individual planes. Moreover, majority of the existing 3-D change detection methods, which are based on height difference in the DSM, classify the detected changes into three groups: new, unchanged and demolished planes [23,25,30]. Consequently, these methods are unable to distinguish the partially-modified planes from the unchanged and new planes, and modified planes from the new planes. Therefore, the contribution of the proposed building change detection technique can be summarised as follows:

- Change detection: Automatic detection of 3-D (2-D space and height) changes in buildings and their planes into five groups: *new, unchanged, demolished, modified,* and *partially-modified* planes. Unlike the existing methods, the proposed method is capable of detecting changes on a per-plane basis. A newly proposed graphical representation of 3-D building models helps in identification of *modified* and *partially-modified* planes.

Since in reality only a small number of buildings are changed in an area, the proposed change detection technique first uses height difference values between the reference and test models to identify new, completely demolished and unchanged buildings. The planes of these buildings are denoted as *new*, *demolished* and *unchanged*, respectively. Then, only the modified building regions are compared using a graph-based representation of the reference and test models to obtain other changes such as *modified* and *partially-modified* planes.

4. Proposed 3-D Building Modelling Technique

Figure 1 shows the workflow of the proposed 3-D building modelling technique. Awrangjeb and Fraser's [9] technique is used for the extraction of building roof planes using LiDAR data and the corresponding Digital Terrain Model (DTM) as the input. The used technique offers high detection performance but has low accuracy in extracting small roof planes and tiny structures since it uses only the LiDAR data. This section presents the different steps in Figure 1 in detail for generation of 3-D building models.

4.1. Adjacency of Roof Planes and Their Intersection Lines

The primary elements for the generation of building models are the roof planes which are input to the proposed technique. A neighbourhood relation matrix is established among the roof planes to determine the topological relations among these 3-D roof primitives, which is originally an adjacency matrix and stores the records of the neighbours of each roof plane.

Let $S_p = \{P_1, P_2, ..., P_n\}$ be a set of n input roof planes and an adjacency matrix M of the same size is instantiated, i.e., $M_{n \times n}$. The roof planes that remain within the Euclidean distance $dist_p$ of a source plane P_i are considered its neighbours and the corresponding rows and columns of M are updated accordingly with the roof plane's ID. The value of $dist_p$ is chosen as twice the maximum distance (d_{max}) of a point to its nearest point in the input LiDAR point cloud [9]. To speed up the generation of M, the planes within an appropriate rectangular region (e.g., a bit larger than the bounding box) around the source plane P_i are first determined, rather than computing the distances of a plane P_i to the rest of the input roof planes. Next, for the planes which lie within $dist_p$ of the boundary points of P_i, their particular records against the ith row and column of M are updated. The procedure continues and all the input roof planes are processed iteratively to establish the interrelations among them.

Then, to find the intersection line between two adjacent planes P_1 and P_2, their plane equations are used to assess whether these planes mutually intersect in 3-D space. If they do, a point called the intersection point I_{pnt} and a direction vector $\hat{n}$ in 3-D space are obtained. The two end points of the intersection line are not known from I_{pnt} and $\hat{n}$ alone. Subsequently, 2-D straight lines are first approximated using the roof boundary points which face each other. The MATLAB built-in function `polyfit` is used for the approximation of line segments. Following the concepts of 3-D coordinate geometry and using the approximated 2-D lines, I_{pnt}, and $\hat{n}$ together, the 3-D intersection lines between the adjacent roof planes are estimated.

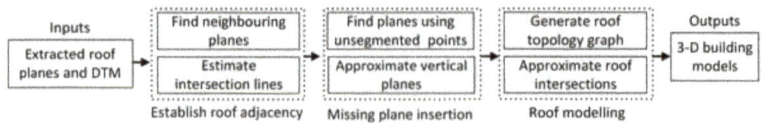

Figure 1. The workflow of the proposed 3-D building modelling technique.

The position of I_{pnt} from the two corresponding plane boundaries is considered, entirely according to the resolution of the input LiDAR data, to be $dist_p = 2d_{max}$. The plane insertion procedure first assesses the nearest distances of I_{pnt} from two intersecting plane boundaries. If any of the two nearest distances exceeds $2d_{max}$, it will potentially indicate either of the two general possibilities: (1) a missing LiDAR-based plane; or (2) a missing vertical plane, between these adjacent roof planes. Both o possible scenarios are described using diagrams in the following sections.

4.2. Detection and Insertion of Missing Roof Planes

The proposed modelling technique now iteratively takes a plane and its neighbours to approximate their intersection lines. However, if the position of I_{pnt} is away (more than $2d_{max}$) from the two intersecting plane boundaries, the plane insertion process attempts to search for any unsegmented LiDAR points between the participating roof planes. If such points are found (at least 4), the process infers the presence of a plane between these roof planes. Figure 2a,b shows an example building and a small missing plane among already extracted planes P_1, P_2, and P_3.

The process invokes the region-growing segmentation technique in Awrangjeb [9] to extract a planar region P_n using the unsegmented LiDAR points (see the green points in Figure 2c). In addition to the available points, the segmentation process uses points of the neighbouring planes (P_1, P_2, and P_3) for the extraction of a new plane P_n assuming these points might have been added wrongly to the neighbouring planes because P_n was missed earlier. Therefore, each iteration of the region-growing technique computes a plane-fitting error between new and neighbouring planes. If the new plane results in a height error smaller than those of the neighbouring planes' errors, the LiDAR points of the neighbouring planes are removed from their respective regions and added to P_n.

The segmentation process continues growing the region until it finds no points complying with the above height error criterion. After the segmentation process stops, the proposed technique estimates the boundary of the new plane and updates the boundary information of the neighbouring planes, as shown in Figure 2c. The process also updates the neighbourhood matrix M with the information on the new plane and new neighbouring relations. Subsequently, the intersection lines between the identified plane and the participating planes are estimated using their boundary points, as explained in Section 4.1. These intersection lines are also recorded against their adjacent roof planes. Figure 2d shows all the roof planes of the sample building and the intersection lines between them.

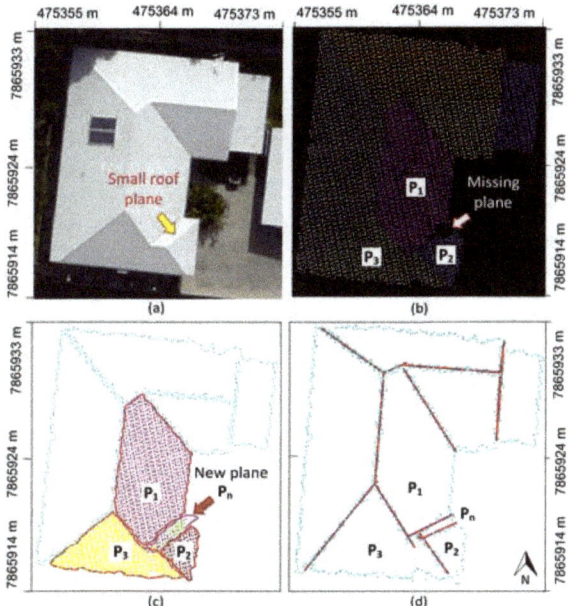

Figure 2. Insertion of a missing plane: (**a**) test building image for demonstration; (**b**) input roof planes for corresponding building and the location of a missing plane; (**c**) new roof plane using unsegmented LiDAR points (green) and points from neighbours; and (**d**) 3-D intersection lines between roof planes.

4.3. Insertion of Vertical Roof Planes

Figure 3a shows a building, where a small slanted plane is located between two adjacent roof planes labelled P_1 and P_2. This plane was not detected by the involved segmentation technique due to the unavailability or absence of enough LiDAR points, as shown in Figure 3b. Therefore, a new plane P_4 between P_1 and P_2 is inserted as follows. The neighbouring boundary points of P_1 and P_2 are first used to form a vertical plane P_4. Then, the intersection lines of P_4 with P_1 and P_2, respectively, are determined. Figure 3c shows P_4 and its intersection lines with P_1 and P_2. The procedure not only keeps track of new planes, but also maintains the correct neighbourhood information in M. Therefore, even after the insertion of P_4, it is found through the neighbourhood selection method (described in Section 4.1) that P_4 has a new neighbouring plane, i.e., P_3, as can be seen visually in Figure 3b,d. There is a dire need to determine the intersection between P_3 and P_4 to precisely establish the topological relationships among the building roofs and reconstruct a model with a good level of detail.

As shown in Figure 3, P_4 is actually a thin slanted (nearly vertical) plane. However, due to shortage of points on this plane, it could not be inserted using LiDAR data following the procedure in Section 4.2. Since a vertical plane is inserted instead, the intersection line between P_3 and P_4 cannot be found as expected. To solve this, the plane insertion procedure is executed following the similar steps above and a new vertical plane P_5 is inserted between P_3 and P_4 using their boundary points. The procedure further approximates the intersection lines between the participating roof planes, between P_3 and P_5 and between P_5 and P_4, as shown in Figure 3e. The procedure stops once all the roof planes are processed. Figure 3f shows all the building roof planes and their intersection lines in 3-D space for better visualisation.

Thereafter, the boundary of each building is extracted using the boundary tracing procedure in Ali et al. [31] and regularised using the technique proposed by Awrangjeb [32] to form an appropriate building footprint, which is a polygon consisting of 3-D corner points.

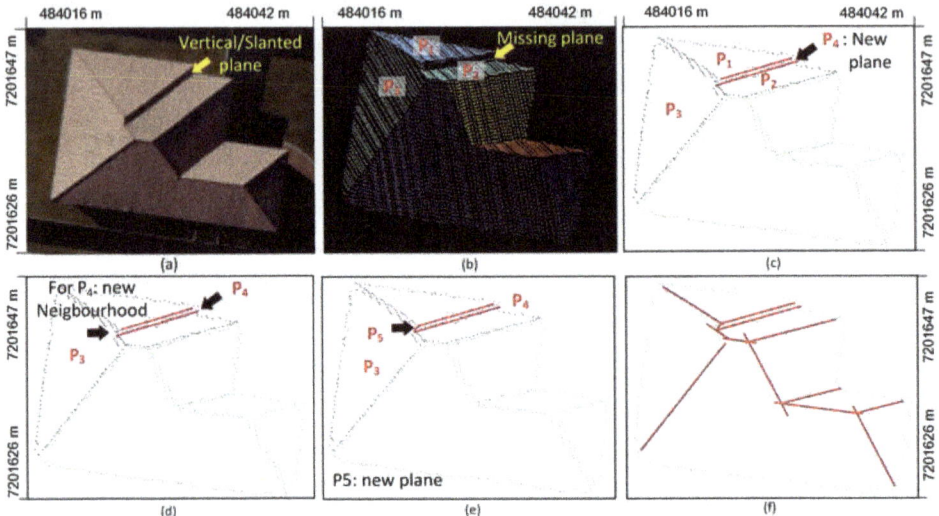

Figure 3. Insertion of a real vertical plane: (**a**) test building image for demonstration; (**b**) input roof planes for corresponding building and location of missing plane; (**c**) insertion of a new vertical plane P_4; (**d**) assessing adjacency between existing plane P_3 and new plane P_4; (**e**) insertion of a new vertical plane P_5 between P_3 and P_4; and (**f**) 3-D view of building roof planes and intersection lines for construction of interrelation between roof planes.

4.4. Rooftop Topology and Modelling

At this juncture, the information about the buildings, all their possible roof planes, intersection lines, and the adjacency relationship is available. For 3-D roof modelling, it is important to obtain 3-D ridge (intersection of ridge lines) and edge (intersection of ridge line and building boundary) points. However, as shown in Figure 4a, there is a small gap between each end point of a ridge line and its corresponding (actual) ridge or edge points. To fill the gaps and establish connectivity among the adjacent roof planes, the adjacency matrix M is used along with the Roof Topology Graph (RTG) following the principles proposed by Verma et al. [33]. As shown in Figure 4b, each roof plane is represented as a vertex in RTG and two adjacent planes are connected through an edge. These roof planes are labelled with their vertex numbers in Figure 4a.

In the context of RTG, a basic cycle indicates a ridge point that belongs to several ridge lines [34]. For instance, the roof planes P_1, P_4, and P_5 form a basic cycle and the intersection of the corresponding ridge lines determines a ridge intersection point. In addition, the corresponding vertices of the ridge lines participating in the intersection determination process are updated. These points can also be referred as ridge points, and will be used at the later stage to approximate the model shape. An RTG can also be represented as a composition of several basic cycles, as shown in Figure 4c. The building rooftop shown in Figure 4a has six basic cycles and so do the ridge intersection points. The least squares approach is applied to approximate the intersection among the ridge lines of the participating roof planes, as shown in Figure 4d. Thereafter, 3-D edge points are found by intersecting ridge lines with the building boundary. Figure 5a,b shows the edge points in two different perspective views. Note that, in a real scenario, two or more ridge lines intersect at the same point on the building boundary. However, the intersection of these ridge lines and the building boundary in the proposed modelling method may generate two or more individual points on the boundary. A 3-D single intersection point is not estimated for these points, since such an estimated 3-D single point may not be on the building boundary. Hence, individual edge points are considered.

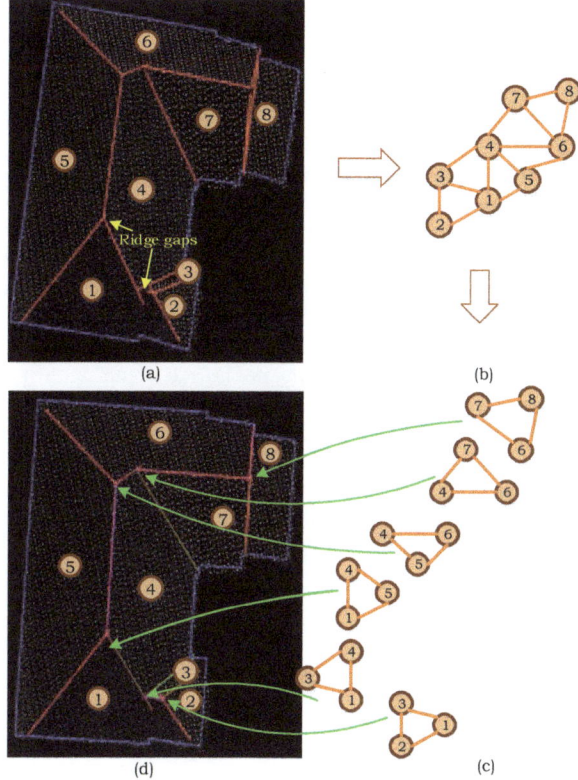

Figure 4. Determination of ridge intersection points: (**a**) roof planes and ridge (intersection) lines; (**b**) Roof topology graph; (**c**) closed cycles; and (**d**) corresponding ridge intersection points.

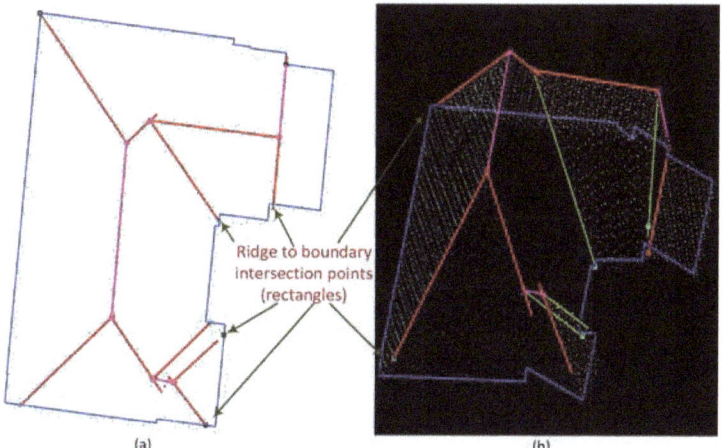

Figure 5. Determination of edge intersection points: (**a**) edge points (ridge to building boundary intersection points); and (**b**) 3-D view of building showing edge points.

4.5. Complete 3-D Building Models

For roof modelling, each building is processed separately, and the procedure first finds the 3-D points around each plane boundary and constructs each roof segment. To do this, 3-D intersection lines, whose junction points (red ovals in Figure 6) have been updated, are recalled. Then, the junction points (edge or ridge points) of 3-D intersection lines are re-ordered in succession around the plane using the information on the corresponding LiDAR-based building boundary points. This is shown in Figure 6a, where the junction points are labelled as N_1–N_6 to represent a roof segment. All the roof planes of each building are processed iteratively and the corresponding roof model is generated that has regularised plane boundaries, as shown in Figure 6b. Note that during the above modelling process (e.g., using least squares to approximate the intersection among the ridge lines), the planarity of the roof segment, say using the 3-D junction points N_1–N_6, may be slightly changed with respect to the original plane equation generated from the segmented LiDAR points. However, the plane equation is not updated since it is estimated by a large number of LiDAR points on the plane.

For a complete 3-D building model, it is necessary to generate walls from the periphery of the roof model to its floor. In this regard, the edge points are used to generate the approximate building floor first. The ground height of each edge point is determined from the DTM so that the model seems to be a replica of its respective real building. All the consecutive ground points are connected to obtain the building floor. Finally, the building walls are determined by extruding the edge points to their corresponding floor points. Figure 7 presents the real building and its 3-D reconstructed model, where the walls are represented in a transparent grey colour.

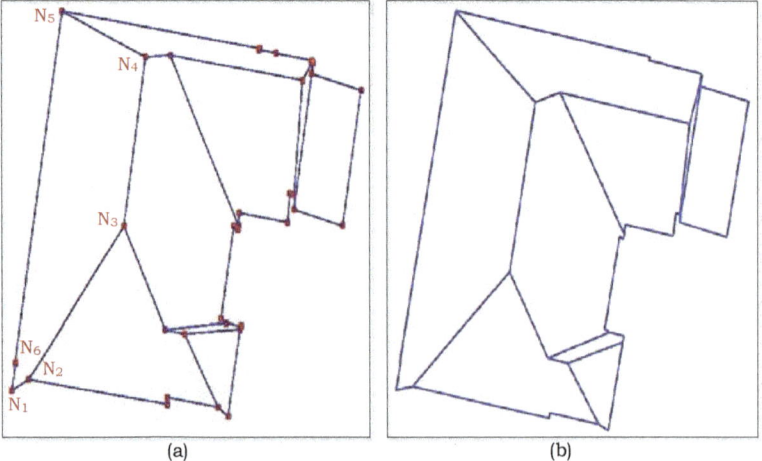

Figure 6. Individual roof segments and building model: (**a**) a roof segment is shown with a sequence of ridge and edge points (N_1 to N_6); and (**b**) roof model of the sample building.

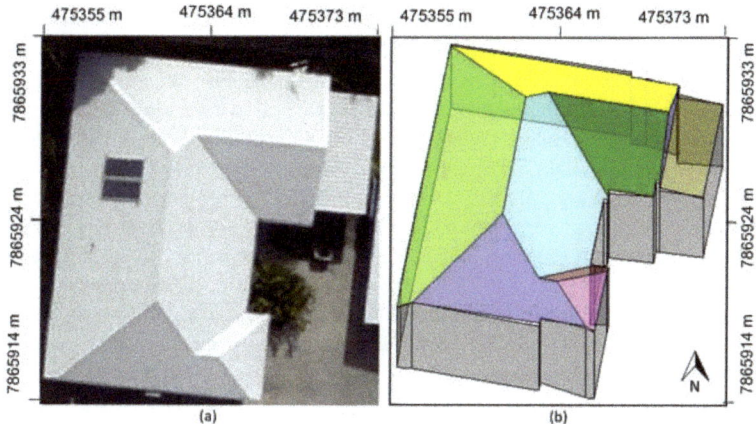

Figure 7. Complete building model for the sample building: (**a**) aerial image; and (**b**) 3-D model.

5. Proposed 3-D Building Change Detection Method

Any building change detection method requires two sets of building models: a reference set and a test model set. The reference dataset is collected at an earlier date than that of the test dataset. For our investigation, however, datasets from two different dates were not available. Therefore, the 3-D models generated from the available dataset are considered to be in the reference model set and a model modification step is carried out to generate the test model set.

Figure 8 shows the flow diagram of the proposed building change detection method. The inputs to the proposed method are aerial images, LiDAR, DTM, and extracted 3-D building (reference) models. The proposed method has three major steps: (i) test model generation; (ii) creation of building data structure; and (iii) automatic change detection. The test model generation step is a manual step. If the test model set is generated from an available dataset captured at a later date than that of the reference dataset, then this step is not necessary.

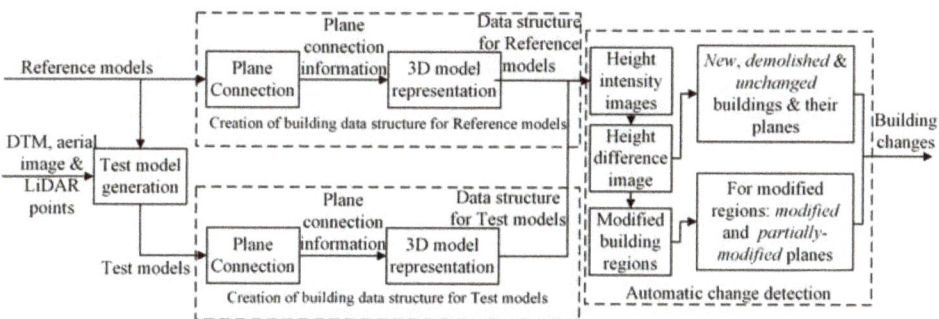

Figure 8. The flow diagram of the proposed building change detection method.

To identify the changes, the reference and test models are represented in a graph-based data structure. Thereafter, they can be compared in the automatic change detection step. However, since the number of actual changes in buildings is small in practice, such a model-by-model comparison will be unnecessarily time consuming. Therefore, the height information of the test and reference models are first used to identify potential change locations. Then, the models are compared, if necessary, using the 3-D representations. Consequently, the planes from both models are classified into five groups: *unchanged*, *new*, *modified*, *demolished* and *partially-modified* planes.

5.1. Test Model Generation

The building models generated by the proposed 3-D building modelling technique are considered as reference models. To obtain test models, the size and height changes are introduced to the reference models. Among the input data of the proposed change detection technique (see Figure 8), while the aerial image contains sharp boundaries of objects including the buildings, the DTM and plane equation contain the bare earth and height information of buildings, respectively. Therefore, the DTM, plane equation, and aerial images are collectively used to produce changes in the reference data.

The changes in the reference data are performed by selecting points around a plane's boundary in the aerial image and import X (Easting) and Y (Northing) coordinates of that location. This procedure is also shown in Figure 9. Based on X and Y coordinates of a point, its Z (Height) value is approximated by using the respective plane equation. For the wall planes, Z value at a ground point of the wall plane is directly extracted from the DTM. The extracted information of planes is later used to modify the building model of the reference data. For example in Figure 10, five different changes are made to the reference models: addition of height to two buildings, addition of a veranda to a building, removal of verandas from two buildings, addition of a new building, and relocation of all buildings in the scene.

5.2. 3-D Model Representation

Unlike the existing methods, the proposed 3-D change detection method not only obtains building changes on a per-plane basis, but also detects *modified* and *partially-modified* planes. A graph-based 3-D data structure, where individual planes and their relationships are represented, is proposed to detect *modified* and *partially-modified* planes. The relationship between planes helps identification of different types of planes (e.g., roof and wall planes) and relative position of a plane with respect to its neighbouring planes. In addition, this ensures the detection of wall planes that are usually undetected by analysing only height change.

The 3-D building model consists of roof and wall planes. The relationship between the connected roof planes (indicated by the adjacency matrix M) can either be a parent-to-child, a parent-to-parent, or a child-to-child connection. The parent-to-parent and child-to-child are also called sibling connections. All connected roof planes are initially labelled as siblings. Then, the parent-to-child connection of two connected roof planes is labelled by verifying the two conditions: (1) the child and parent planes intersect each other; and (2) the child plane has a height value lower than the parent. Whereas the relationship between a roof plane and a wall plane is a parent-to-child connection, the relationship between two neighbouring wall planes is a child-to-child connection. Parents that intersect each other are in a parent-to-parent connection. For instance, in Figure 11, the parent-to-child, parent-to-parent, and child-to-child connections are marked by black, red and magenta arrows, respectively, for the building shown in the orange coloured rectangle in Figure 10a.

Figure 9. Extracting X and Y coordinates of a point of the cyan plane, where X and Y values are highlighted by a purple box.

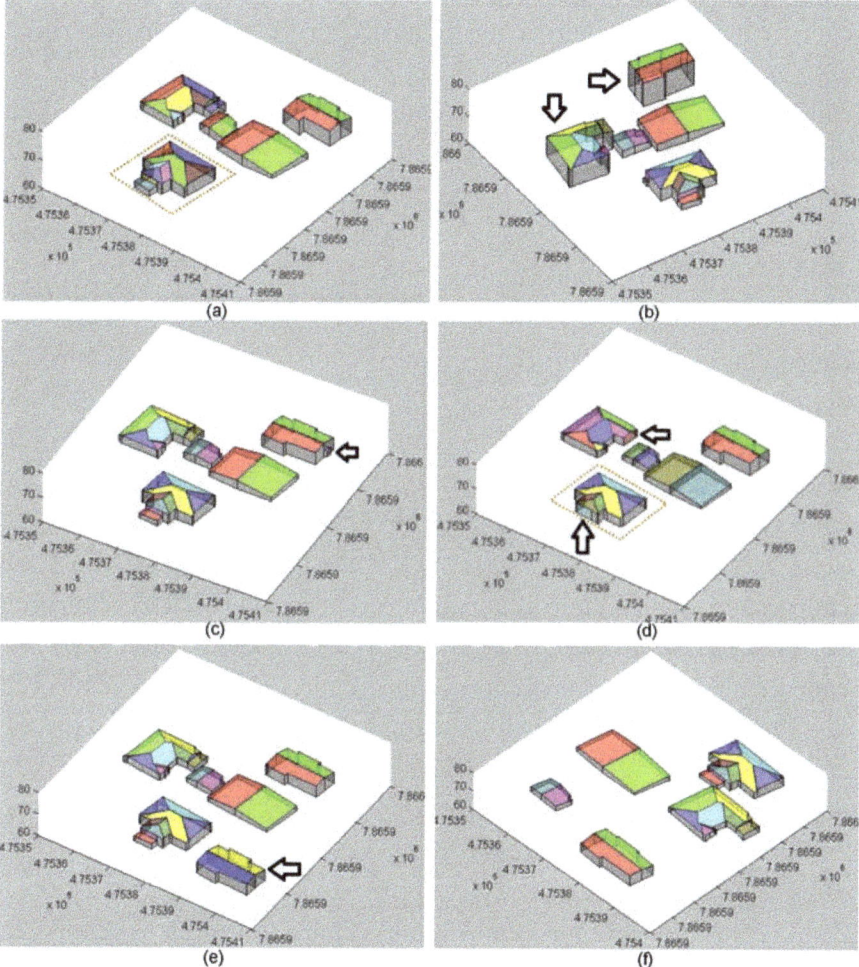

Figure 10. Introduction of five different changes: (**a**) reference data; (**b**) addition of height to buildings; (**c**) addition of a veranda; (**d**) removal of verandas; (**e**) addition of a new building; and (**f**) relocation of building positions.

Thereafter, a 3-D building model is represented in a graphical structure, where each plane of a building is considered as a separate node. A node contains the complete information of a plane, e.g., plane ID, plane equation, 3-D plane points (i.e., polygon), plane type (i.e., roof or wall plane), and its connected plane information. Figure 12 shows an example to illustrate the data structure of a 3-D building model shown within the orange rectangle in Figure 10a. The roof and wall planes are represented by blue and red nodes, respectively. As complete information of a node is not possible to mention in the data tree, the plane ID is shown at each node. The connections of nodes can either be sibling (parent-to-parent and child-to-child connections) or parent-to-child, which are highlighted by black and green coloured edges, respectively. The arrowhead of a green edge indicates the child in the parent-to-child relationship.

5.3. Automatic Building Change Detection

The proposed change detection method measures the change between the 3-D models of the reference and test scenes by comparing their structural information. The full structural comparison using the graphical representation, shown in Figure 12, between the two corresponding building models B_r and B_t, from the reference and test model sets, respectively, is unnecessary for the following reasons. Firstly, the full comparison of all corresponding building pairs (B_r, B_t) is computationally expensive. Secondly, in practice, there will be no or only a small number of existing buildings being changed in a given test scene. Therefore, the building structure of both models are first compared based on the height difference. Then, if necessary, only the related parts of the graph data structures are compared to obtain a more specific types and details of the involved changes. By using these two tests, the proposed method classifies the building planes into five groups: *unchanged*, *demolished*, *new*, *modified*, and *partially-modified* planes.

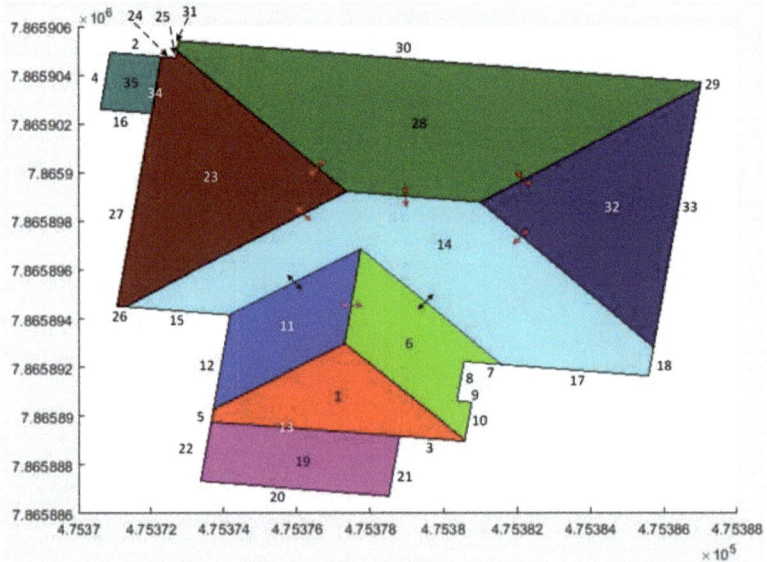

Figure 11. Relationship between the connected building planes; black, red, and magenta arrows represent the parent-to-child, parent-to-parent, and child-to-child connections, respectively. This building is the one within the orange rectangle in Figure 10a and its wall plane numbers are shown outside the building boundary.

5.3.1. Height Test

For the height test, two intensity images are generated: one for the reference scene and the other for the test scene. Each intensity image has a resolution of 0.25 m and represents heights from individual roof planes with respect to the ground. Thus, a zero height represents no buildings. The input LiDAR points within the segmented planes from [9] are used to generate the height image. If the LiDAR points are not available, then the individual plane equations can be used to estimate heights on each pixel of the height image. In the height test, the height intensity image (say, I_r) of the reference scene is subtracted from that (say, I_t) of the test scene. Figure 13c shows the absolute (pixel-to-pixel) height difference image ($I_d = I_t - I_r$) for I_r and I_t in Figure 13a,b. In I_d, there can be the following cases for buildings in the scene (see Figure 13). Case A: For a completely new building, all height differences are positive in I_d and there is only zero height value in I_r. All planes within this new building are marked as *new*. Case B: Likewise, for a completely demolished building, all height differences are negative in I_d and there is only zero height value in I_t. All planes within this demolished building are marked as *demolished*. Case C: If there are no or negligible height differences everywhere within a building region, then all of its planes are marked as *unchanged*. In reality, however, if the two models of an unchanged building are generated from two datasets obtained at different times, the final models can be slightly different. Therefore, a height tolerance threshold (0.5 m) is used. In addition, there may be non-overlapping thin parts (due to high height difference) along the building boundaries, this should not be more than 1 m in width for an unchanged building. Otherwise, the building is considered modified and Cases D and E below are considered.

If the above three cases (Cases A–C) from I_d are excluded, the remaining regions are all modified building regions in the scene. Figure 13d shows the image I_m containing only modified regions after exclusion of completely new, demolished and unchanged buildings from I_d in Figure 13c. In I_m, there can be the following cases for modified buildings in the scene. Case D: A building is modified through removing one or more parts, e.g., a veranda is removed, which is shown within the orange rectangles in Figure 10a,d. This case is observed by the same absolute height values within the corresponding modified region in I_m and I_r, but zero height value in I_t. Case E: A building is modified through extending one or more parts, e.g., a veranda is added or extended. This case is observed by the same absolute height values within the corresponding modified region in I_m and I_t, but zero height value in I_r (see Figure 13d). Case F: A building is modified in height direction, e.g., a one-storey building is modified to a two-storey building, or vice versa (see Figure 13d). For these three cases, all the unchanged areas, if any, are identified and planes within these areas are marked *unchanged* (the same procedure is followed as in Case C above). Consequently, only the planes in each modified region in I_m are now subject to the plane test below exploiting the graphical representation in Figure 12.

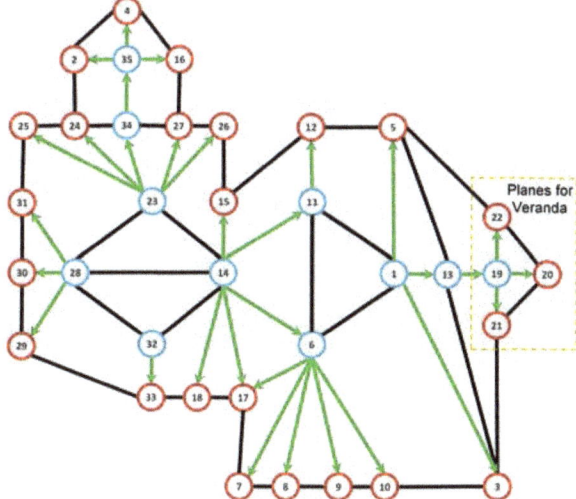

Figure 12. A graph (data structure) for a building model shown within the orange rectangle in Figure 10a: roof planes are blue nodes; wall planes are red nodes; parent-to-child connections are in green edges (where an arrowhead indicates a child side); and sibling connections are in black edges.

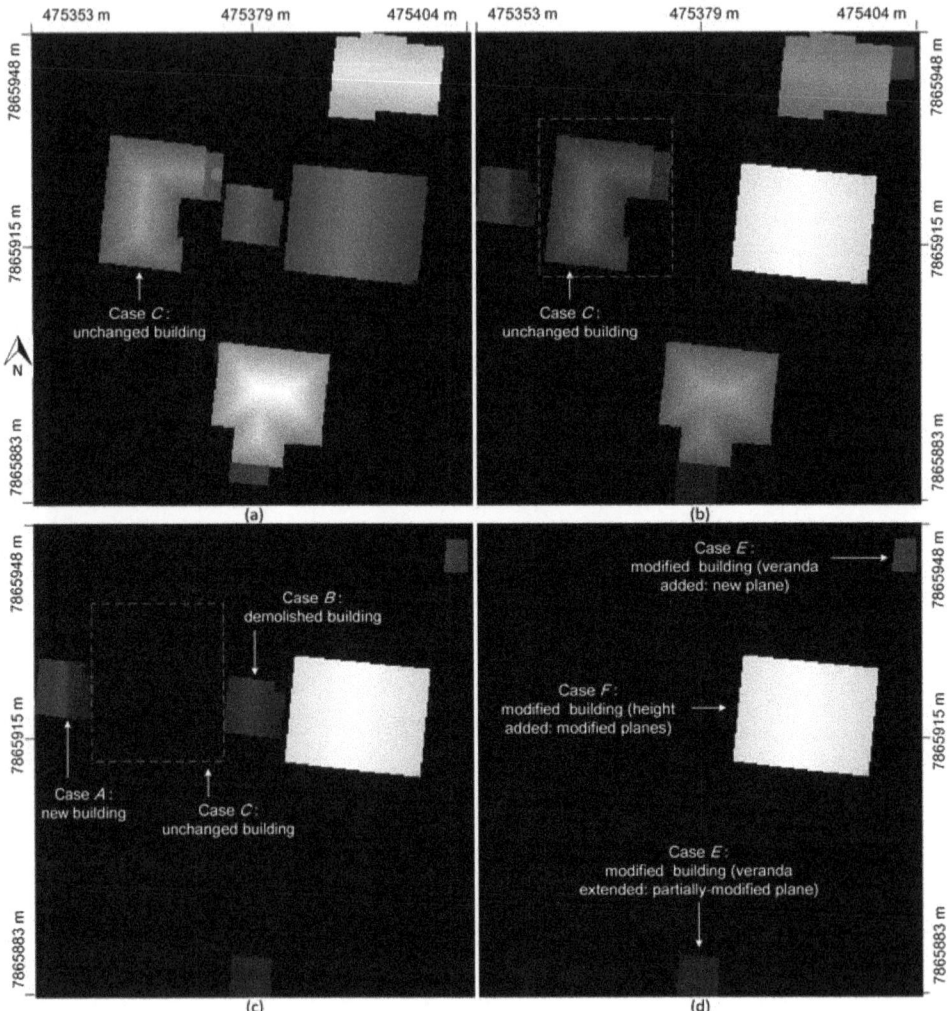

Figure 13. Finding the 3-D plane changes between the reference and test models: Height intensity images for: (**a**) reference models; (**b**) test models; (**c**) absolute height difference image (test minus reference); and (**d**) modified building regions.

5.3.2. Plane Test

Let the two graphical representations of B_r and B_t be G_r and G_t, respectively. A removal of a building part (Case D) is a result of a full and/or partially demolishment of one or more existing building planes. In this situation, the partially-demolished planes are present in both G_r and G_t, but fully demolished planes are only present in G_r. In contrast, an extension to an existing building (Case E) consists of an addition of one or more completely new planes and/or an extension of one or more existing planes. In such a case, the extended planes are present in both G_r and G_t, but completely new planes are only present in G_t. When a building is modified in height direction (Case F), in addition to the height change, the modification can also include Cases D and E. Therefore, Cases D and E can be considered as minor modifications in buildings and Case F is a major modification. Each of the planes within a modified region can be classified as a new, demolished, modified or partially-modified plane.

The proposed plane test tries to establish correspondences between the planes (from G_r and G_t) by applying the point-in-polygon (PIP) test [12]. A plane in G_t is marked as *new* if a corresponding plane is not found in G_r (see the top-right corner of Figure 13d). A plane in G_r is marked as *demolished* if a corresponding plane is not found in G_t. For instance, for removal of the veranda in Figure 10, planes within the orange coloured rectangle in Figure 12 will be absent in G_t. Thus, no correspondences are established for new and demolished planes. Nevertheless, for a fully or partially modified plane, a correspondence can be found through the PIP test.

To differentiate between a fully modified and partially-modified reference planes, the height differences in I_m are again used. A reference plane is fully modified when there are height changes everywhere in the plane. This reference plane and its corresponding plane in G_t are marked as *modified* (see the mid-right side building in Figure 13d). Otherwise, the reference plane is a partially-modified plane when there are height changes in some areas and no height changes in rest of the plane. Both the reference plane and its corresponding plane in G_t are marked as *partially-modified* (see the building at the bottom in Figure 13d).

All these groups of planes in the reference and the test building models are shown in Figure 14, where the unchanged, demolished, new, modified, and partially-modified planes are marked by yellow, green, red, cyan, and blue colours, respectively.

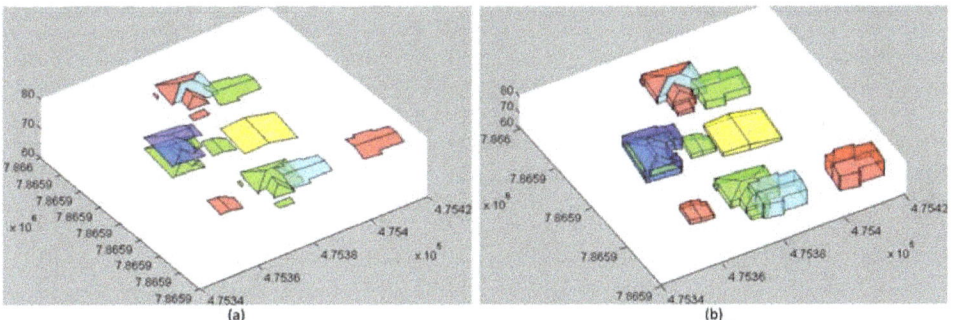

Figure 14. Classification of building planes into five groups, i.e., unchanged (yellow), new (red), modified (blue), partially-modified (cyan) and demolished (green) planes: (**a**) roof planes only; and (**b**) roof and wall planes.

6. Performance Study

The proposed 3-D building modelling technique requires only point cloud data for individual planes and the corresponding DTM. The proposed change detection technique needs datasets, which are from the same area but collected at two well separated dates, reflecting some real building changes as illustrated in Section 5.1. However, it is hard to obtain such datasets publicly. As a result, two Australian datasets, which have high density point cloud data, high resolution multi-spectral orthophotos, and DTMs, were used. Since these datasets are available for one date only, for verification of the 3-D change detection technique, tests models were manually generated. In this section, results for 3-D building modelling and change detection techniques are separately presented.

6.1. Datasets

Two datasets namely, Aitkenvale (AV) and Hervey Bay (HB), as shown in Figures 15a and 16a, were selected to evaluate the performance of our proposed building modelling and change detection methods. The AV site has a point density of 29.3 points per m^2 and it covers an area of 108×80 m^2. This dataset has five buildings and comparatively high vegetation. The HB site covers an area of 108×104 m^2 and has a point density of 12 points per m^2. It contains 26 single-storey residential buildings of different sizes that are surrounded by vegetation cut into different shapes. For both datasets, multi-spectral orthophotos of resolution 5 cm and 20 cm, respectively, were available. In addition, DTMs with 1 m resolution were used for both the sites.

Since the data were not available from two dates, for verification of the proposed change detection method, the above available datasets were exploited to generate the reference building models by using the proposed 3-D modelling technique. However, the test models were manually generated (see Section 5.1). In total, 62 changes (test sites) were made to the AV and HB reference models by introducing one or more of the following operations: addition of a new building, addition of a veranda, increasing height of a building, removal of a veranda, removal of a roof plane, removal of a building, rotation of a building, and change of the positions of three or five buildings (for details, see Table 1).

6.2. Evaluation System

To verify the performance of the proposed building modelling and change detection method, a previously proposed automatic evaluation system [12] was employed. For given two sets of input data (i.e., reference and test objects), this evaluation system estimates a set of evaluation metrics without any human interaction.

For evaluation of the generated building models, there was an absence of the 3-D reference data (building models). In the literature, there is also a lack of appropriate evaluation metrics for 3-D models. Thus, it is hard to make a proper evaluation for the generated 3-D models. Earlier, the extracted roof planes and building boundaries were evaluated against the 2-D reference data that were collected through monoscopic image measurement [9]. The proposed building modelling method uses those extracted planes for generation of complete 3-D models. These input (extracted) planes consist of raw LiDAR points, thus are incomplete and have zigzag boundaries. The proposed modelling technique inserts possible missing planes on the roof and finds the plane boundaries using plane intersection lines. Consequently, the 2-D reference data from Awrangjeb and Fraser [9] were used to evaluate the planes in the generated 3-D models. In addition, since the main contribution of the proposed 3-D modelling method is to reconstruct missing planes, it is also shown how many of missing planes the proposed method successfully inserted.

The same limitation, i.e., the absence of actual 3-D reference data from two dates and the lack of appropriate evaluation metrics for 3-D changes, has been observed for evaluation of 3-D change detection performance. As a result, the reference (generated from the data of an earlier date) and test (generated from the data of a later date) models were directly compared to evaluate the change detection performance. Changes between these two sets of models were exploited to verify the changes detected by the proposed automatic change detection technique.

Mainly two categories of evaluation metrics, i.e., object-based and pixel-based, are used for 2-D evaluation of building models and changes. In object-based evaluation, completeness (C_m), correctness (C_r), and quality (Q_l) metrics are estimated by counting the number of objects, whereas in pixel-based evaluation completeness (C_{mp}), correctness (C_{rp}) and quality (Q_{lp}) are calculated by counting the number of pixels in the objects. In building model evaluation, the root-mean-square-error (RMSE) is also used in both planimetric (2-D space) and height directions to evaluate the geometric accuracy. In addition, reference cross-lap, detection cross-lap, area commission and omission errors are used to indicate segmentation errors.

The detail about the above evaluation metrics can be found in Awrangjeb and Fraser [12]. In addition to quantitative results, qualitative analysis is also presented via visualisation.

Table 1. Operations performed on reference models from Aitkenvale (AV) and Hervey Bay (HB) datasets to generate test models. Tick symbol shows a particular operation applied on a corresponding dataset. Ah, addition of height; Ab, addition of building; Av, addition of veranda; Rv, removal of veranda; Bp, building position change; Rob, rotation of building; Rp, removal of plane.

Test Sites		Ah	Ab	Av	Rv	Interchange 3 Bp	5 Bp	Rob	Rp
AV(1)	HB(1)				✓				
AV(2)	HB(2)	✓							
AV(3)	HB(3)			✓					
AV(4)	HB(4)		✓						
AV(5)	HB(5)	✓				✓			
AV(6)	HB(6)	✓						✓	
AV(7)	HB(7)		✓			✓			
AV(8)	HB(8)		✓					✓	
AV(9)	HB(9)	✓	✓			✓			
AV(10)	HB(10)	✓	✓					✓	
AV(11)	HB(11)							✓	
AV(12)	HB(12)			✓		✓			
AV(13)	HB(13)			✓				✓	
AV(14)	HB(14)	✓		✓					
AV(15)	HB(15)		✓	✓					
AV(16)	HB(16)		✓	✓		✓			
AV(17)	HB(17)	✓	✓	✓		✓			
AV(18)	HB(18)		✓	✓				✓	
AV(19)	HB(19)	✓	✓	✓				✓	
AV(20)	HB(20)	✓			✓				
AV(21)	HB(21)	✓			✓	✓			
AV(22)	HB(22)	✓			✓			✓	
AV(23)	HB(23)			✓	✓				
AV(24)	HB(24)		✓		✓	✓			
AV(25)	HB(25)		✓		✓			✓	
AV(26)	HB(26)	✓	✓		✓	✓			
AV(27)	HB(27)	✓	✓		✓			✓	
AV(28)	HB(28)					✓			
AV(29)	HB(29)			✓					
AV(30)	HB(30)							✓	
AV(31)	HB(31)								✓

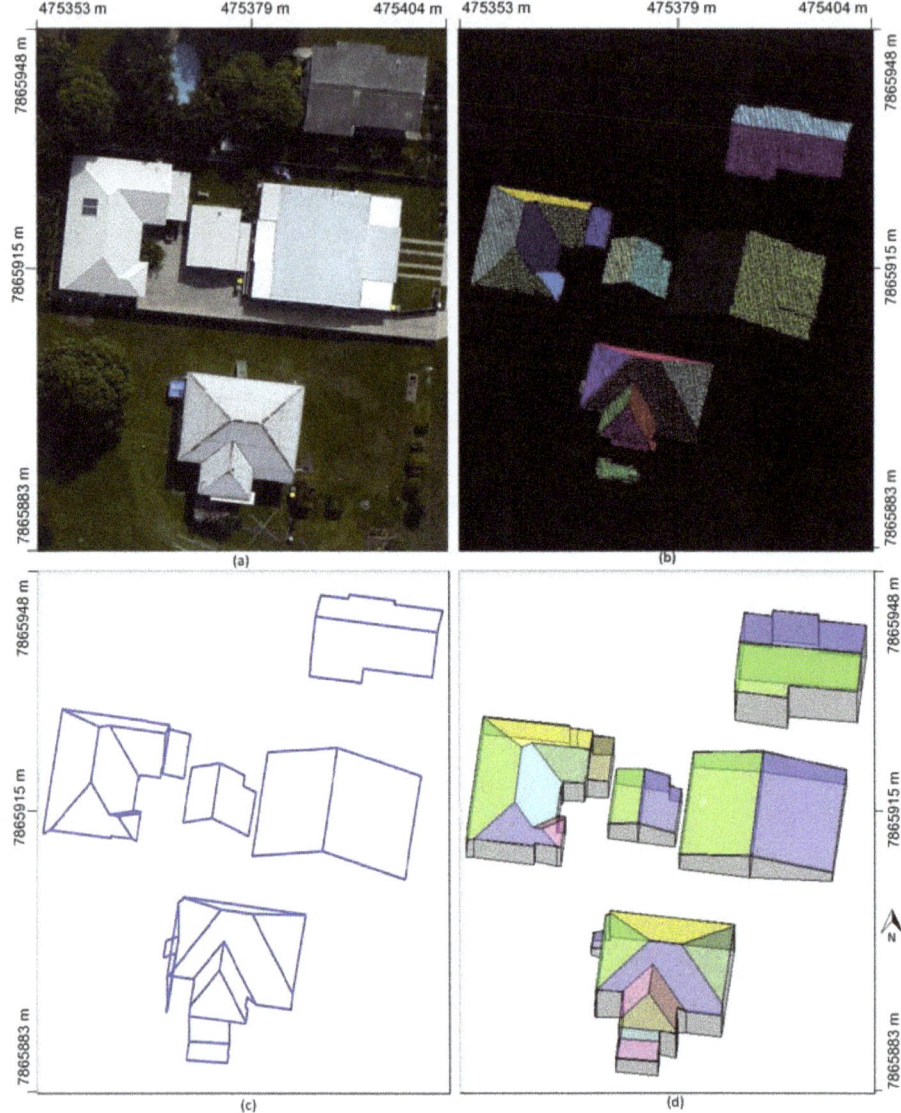

Figure 15. 3-D reconstructed models from Aitkenvale dataset: (**a**) aerial image; (**b**) LiDAR points of input roof planes; (**c**) building roof models; and (**d**) 3-D building models.

6.3. Parameter Setting

There are limited parameters used by the proposed building modelling and change detection techniques. Most of the parameter values were chosen from the existing literature. For example, the distance to find neighbouring planes or neighbouring LiDAR points ($dist_p = 2d_{max}$), height image resolution (0.25 m), minimum width of a thin (unchanged) region (1 m), plane fitting error (0.10 m) and plane height error (0.15 m) are from Siddiqui et al. [35], Awrangjeb [1] and Awrangjeb and Fraser [9], and the Gaussian smoothing scale ($\sigma = 3$) is from Awrangjeb et al. [36].

In the proposed change detection technique, a height tolerance and a distance thresholds are also used. Since reference models have been used to generate test models, there was no height difference for unchanged planes. However, due to error in the LiDAR data (which are collected using the same or different systems on two different dates), there may still be some height differences for unchanged planes and buildings. The value of the height threshold could be set at 0.5 m allowing the maximum error in the LiDAR data. The value of the distance threshold is set at $dist_p = 2d_{max}$, which is the minimum length and width of an overlap, changed or unchanged area.

6.4. 3-D Building Modelling Results

The proposed geometric modelling technique relies entirely on LiDAR data, while images are used in this article for visualisation. The performance in terms of insertion of missing planes is presented in Table 2. The 3-D model generation results on the two test datasets AV and HB are presented quantitatively in Table 3 and qualitatively in Figures 15 and 16, respectively.

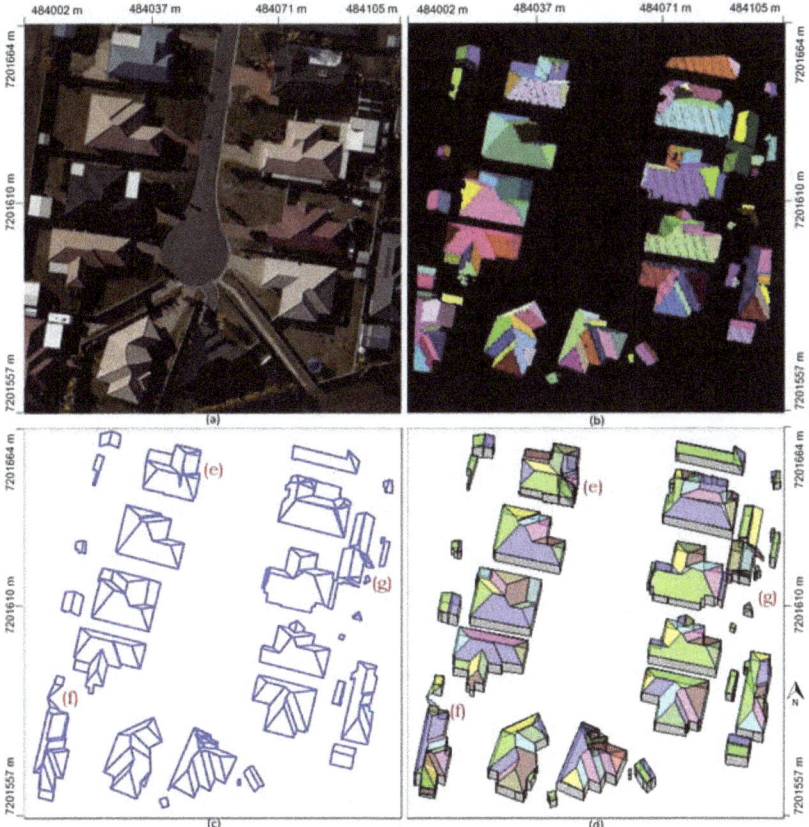

Figure 16. 3-D reconstructed models from Hervey Bay dataset: (**a**) aerial image; (**b**) LiDAR points of input roof planes; (**c**) building roof models; and (**d**) 3-D building models.

6.4.1. Quantitative Results for 3-D Reconstruction

Table 2 shows the performance of the proposed 3-D modelling method in insertion of missing planes. In both test cases, since the reference and input plane sets do not include the wall planes, they are not counted here. In addition, the reference and input plane sets do not include vertical planes in real height jumps (e.g., between a main building and a connected veranda). However, these vertical planes in the height jumps are required to reconstruct the building models. Therefore, the vertical planes inserted for height jumps as well as for missing small slanted planes (e.g., see Figure 3) are counted here.

Table 2. Performance in insertion of missing planes. For all buildings in a scene, N_{ref} is the number of reference planes, N_{input} is the number of input planes from [9] and $N_{missing}$ is the number of missing planes in input plane sets. For the proposed method, N_{recon} is the number of planes in reconstructed models, N_{lidar} is the number of inserted LiDAR-based planes, $N_{vertical}$ is the number of inserted vertical planes, N_{total} is the number of total inserted planes, and $N_{stillmiss}$ is the number of still missing planes.

Test-Case	Total Planes				Inserted Planes			
	N_{ref}	N_{input}	$N_{missing}$	N_{recon}	N_{lidar}	$N_{vertical}$	N_{total}	$N_{stillmiss}$
AV	25	24	1	29	1	4	5	0
HB	167	147	20	158	7	4	11	9

For the AV dataset, the proposed modelling method successfully extracted the only missing slanted roof plane using the unsegmented LiDAR points (see Figure 2). In addition, four vertical planes were inserted in height jumps between the main buildings and verandas. For the HB dataset, to fill the 20 missing planes, it inserted seven LiDAR-based planes and four vertical planes, including one shown in Figure 3. Therefore, while there are no more missing planes in the AV dataset, there are still nine missing planes in the HB dataset. All these still missing planes are small in size, sometimes less than 1 m² in area, and, therefore, could not be recognised and inserted (see Section 6.4.3 below for further discussion).

6.4.2. Comparative Results

It is hard to compare the results of different 3-D building modelling methods. Firstly, the approaches that the 3-D building modelling methods adopt are different, for instance, model-driven and data-driven approaches. Secondly, different methods are evaluated using different datasets, which vary in input point density and complexity of buildings. Therefore, it is hard to find an appropriate method for comparisons. For example, Xiong et al. [20] presented a model-driven method that uses a set of pre-defined building models. It is evaluated using two datasets: the ISPRS [14] and Enschede. While the point density in the ISPRS dataset is 4–7 points/m², in the Enschede dataset it is 20 points/m². However, the Enschede dataset includes complex buildings with non-planar surfaces. Since the method proposed in this paper is a data-driven method that works on high density point cloud data comprising buildings with planar surfaces only, it may not be fair to compare it with Xiong et al. [20]. In the experimentation, the proposed method was tested against the ISPRS dataset, but it did not work well. Consequently, the proposed method is compared with Awrangjeb and Fraser [9].

For the AV dataset, per-plane statistics in Table 3 show that the proposed modelling technique achieved 100% object-based completeness (C_m), correctness (C_r), and quality (Q_l), indicating 4.35% increase in object-based accuracy from that of input roof planes. This is primarily because of the insertion of the missing roof planes. The proposed modelling technique has no detection cross-lap (under-segmentation) rate for the AV dataset because of the insertion of the new roof planes, where the statistics of the input roof planes showed under-segmentation errors with a detection cross-lap rate of $C_{rd} = 4.5$. In terms of pixel-based accuracy of the AV dataset, the evaluation results show a gradual increase in per-plane completeness (C_{mp}), correctness (C_{rp}), and quality (Q_{lp}).

Table 3. Roof planes evaluation results using threshold-free reference classification of Australian datasets. C_m, completeness; C_r, correctness; Q_l, quality in percentage; C_{mp}, pixel completeness; C_{rp}, pixel correctness; Q_{lp}, pixel quality; C_{rd}, detection cross-lap (under-segmentation); C_{rr}, reference cross-lap (over-segmentation) rates; O_e, area omission error; C_e, area commission error; RMS_{XY}, planimetric accuracy (metres); $RMSE_z$, height accuracy (metres).

Test-Case	Per-Plane Object			Segmentation		Per-Plane Pixel			Error in Area		RMS_{XY}	$RMSE_z$
	C_m	C_r	Q_l	C_{rd}	C_{rr}	C_{mp}	C_{rp}	Q_{lp}	O_e	C_e		
Input roof planes [9]												
AV	95.65	100.0	95.65	4.5	0	88.96	93.63	83.89	11.0	5.9	0.02	0.03
HB	85.62	95.33	82.18	8.7	0.5	73.44	82.13	63.32	26.55	17.86	0.39	0.03
Proposed 3-D reconstructed roof planes												
AV	100.0	100.0	100.0	0	0	90.95	94.02	85.14	9.0	5.1	0.02	0.03
HB	88.02	98.0	86.47	7.9	0.8	76.38	85.43	72.42	23.61	14.86	0.34	0.02

Table 3 shows that the proposed modelling technique has achieved 3–4% better C_m, C_r, and Q_l for the HB dataset, and has a subsequent impact on the detection cross-lap rate, which is indicated by a decrease value of C_{rd} from 8.7% to 7.9%. In contrast, C_{rr} shows a slight increase in reference cross-lap rate, which is due to the insertion of (missing) vertical roof planes. Two area indices (O_e and C_e errors) show better accuracy in terms of non-detected (omitted) area and incorrectly detected (committed) area between the input and reconstructed roof planes for both the datasets. In addition, Table 3 further indicates that the reconstructed roof planes have high planimetric and height accuracies.

6.4.3. Qualitative Analysis for 3-D Models

Visual inspection of Figures 15 and 16 not only indicates the ability of the proposed modelling technique to reconstruct variably-shaped buildings but also validates its application for the development of complex building models. However, there were some modelling errors mainly as a result of missing small roof planes in the input planes. Figure 17 shows some of these errors in the magnified versions of buildings (labelled e–g in Figure 16d). These small planes were missed mainly because of under-segmentation errors by the involved segmentation technique [9] and lack of available LiDAR points, especially in the HB dataset, where the point density was low. The rectangles in Figure 17 show the buildings where the height discontinuities (step edges) were not extracted properly due to sparsity of the data points and under-detected sides of the roof planes. The ovals, however, show the locations where the proposed technique was unable to recover the intersection points because of small missing planes. These shortcomings can be overcome by using spectral features from the corresponding aerial imagery. For instance, information of lines extracted from images can be used with the LiDAR-approximated intersection lines to obtain accurate building models at low reconstruction errors.

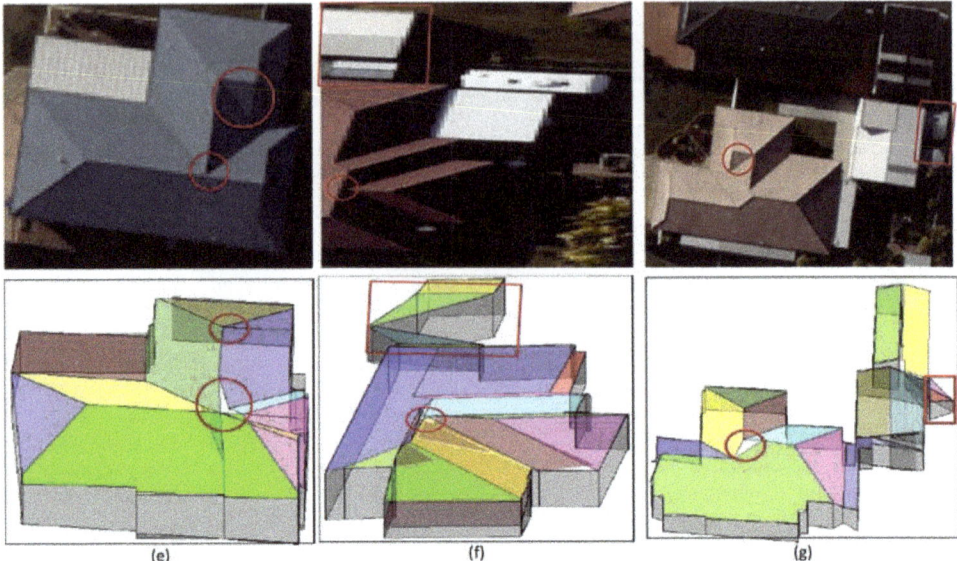

Figure 17. Issues in building modelling: The first row shows the building images and the second row illustrates the building models: (**e**–**g**) are indicated in Figure 16d.

6.5. 3-D Building Change Detection Results

The proposed change detection method classifies building planes into five groups: *new*, *unchanged*, *changed*, *modified*, and *partially-modified* planes. In this paper, there are 62 changes introduced to the two reference sites to test the performance of the proposed building change detection method. Both qualitative analysis and quantitative results are presented to show its performance.

6.5.1. Quantitative Results

Table 4 shows the change detection results for the thirty one AV test sites (see Table 1) in terms of the object- and pixel-based completeness, correctness, and quality. The proposed building change detection method achieved 100% object-based completeness, correctness, and quality for building planes larger than 10 m^2. In addition, the proposed change detection method achieved 100% object-based completeness and over 95% object-based correctness and quality for all sizes of planes. Similarly, the pixel-based metrics for all the planes in the AV test sites are mostly greater than 90%.

Table 4. Building change detection results for 2-D (size) changes in the AV site. Object-based: C_m, completeness; C_r, correctness; Q_l, quality (C_{m10}, C_{r10} Q_{l10} and C_{m50}, C_{r50} Q_{l50} are for building planes over 10 m^2 and 50 m^2, respectively). Pixel-based: C_{mp}, completeness; C_{rp}, correctness; Q_{lp}, quality are in percentage.

Modified Sites	Object-Based									Pixel-Based		
	C_m	C_r	Q_l	C_{m10}	C_{r10}	Q_{l10}	C_{m50}	C_{r50}	Q_{l50}	C_{mp}	C_{rp}	Q_{lp}
AV(1)	100	95.2	95.2	100	100	100	100	100	100	93.6	97.4	91.4
AV(2)	100	95.8	95.8	100	100	100	100	100	100	95	97.4	92.6
AV(3)	100	96.0	96.0	100	100	100	100	100	100	95.1	97.4	92.7
AV(4)	100	96.1	96.1	100	100	100	100	100	100	94.0	86.8	82.2
AV(5)	100	95.8	95.8	100	100	100	100	100	100	95	97.4	92.6
AV(6)	100	95.8	95.8	100	100	100	100	100	100	94.9	97.4	92.6
AV(7)	100	96.1	96.1	100	100	100	100	100	100	94.4	96.5	91.3
AV(8)	100	96.1	96.1	100	100	100	100	100	100	94.3	96.5	91.3
AV(9)	100	96.1	96.1	100	100	100	100	100	100	94.4	96.6	91.3
AV(10)	100	96.1	96.1	100	100	100	100	100	100	94.4	96.6	91.3
AV(11)	100	95.8	95.8	100	100	100	100	100	100	94.9	97.3	92.6
AV(12)	100	96.0	96.0	100	100	100	100	100	100	95.1	97.4	92.7
AV(13)	100	96.0	96.0	100	100	100	100	100	100	95.0	97.4	92.7
AV(14)	100	96	96	100	100	100	100	100	100	95.1	97.4	92.7
AV(15)	100	96.4	96.4	100	100	100	100	100	100	94.4	96.5	91.4
AV(16)	100	96.4	96.4	100	100	100	100	100	100	94.4	96.5	91.4
AV(17)	100	96.4	96.4	100	100	100	100	100	100	94.4	96.5	91.4
AV(18)	100	96.4	96.4	100	100	100	100	100	100	94.4	96.5	91.3
AV(19)	100	96.4	96.4	100	100	100	100	100	100	94.4	96.5	91.4
AV(20)	100	95.2	95.2	100	100	100	100	100	100	93.6	97.4	91.4
AV(21)	100	95.2	95.2	100	100	100	100	100	100	93.6	97.5	91.4
AV(22)	100	95.2	95.2	100	100	100	100	100	100	93.6	97.5	91.4
AV(23)	100	95.6	95.6	100	100	100	100	100	100	92.7	86.51	81.0
AV(24)	100	95.6	95.6	100	100	100	100	100	100	93.2	96.6	90.2
AV(25)	100	95.6	95.6	100	100	100	100	100	100	93.1	96.6	90.2
AV(26)	100	95.6	95.6	100	100	100	100	100	100	93.2	96.6	90.3
AV(27)	100	95.6	95.6	100	100	100	100	100	100	93.2	96.6	90.3
AV(28)	100	95.8	95.8	100	100	100	100	100	100	95.0	97.4	92.6
AV(29)	100	96.1	96.1	100	100	100	100	100	100	83.0	97.2	81.1
AV(30)	100	95.8	95.8	100	100	100	100	100	100	95.1	97.3	92.7
AV(31)	100	95.2	95.2	100	100	100	100	100	100	93.4	96.6	90.5
Average	100	95.8	95.8	100	100	100	100	100	100	93.9	96.4	90.7

The results on the thirty one HB test sites (see Table 1) are tabulated in Table 5. The proposed building change detection method achieved 100% object-based completeness, correctness, and quality for building planes larger than 50 m^2. For all sizes of planes, our proposed method achieved more than 90% object-based correctness, 98% object-based completeness and 95% pixel-based correctness.

Table 5. Building change detection results for changes in the HB site. Object-based: C_m, completeness; C_r, correctness; Q_l, quality (C_m, C_{r10} Q_{l10} and C_{m50}, C_{r50} Q_{l50} are for building planes over 10 m² and 50 m², respectively). Pixel-based: C_{mp}, completeness; C_{rp}, correctness; Q_{lp}, quality are in percentage.

Modified Sites	Object-Based									Pixel-Based		
	C_m	C_r	Q_l	C_{m10}	C_{r10}	Q_{l10}	C_{m50}	C_{r50}	Q_{l50}	C_{mp}	C_{rp}	Q_{lp}
HB(1)	90.9	99.2	88.5	98.0	100	96.5	100	100	100	84.4	96.9	82.2
HB(2)	90.5	99.3	88.5	97.4	100	96.5	100	100	100	84.7	99.6	84.4
HB(3)	90.5	99.3	88.6	97.4	100	96.5	100	100	100	85.0	99.6	84.8
HB(4)	90.9	99.3	89.0	97.5	100	96.7	100	100	100	85.4	99.6	85.1
HB(5)	90.5	99.3	88.5	97.4	100	96.5	100	100	100	84.7	99.6	84.4
HB(6)	90.5	99.3	88.5	97.4	100	96.5	100	100	100	84.7	99.6	84.4
HB(7)	90.9	99.3	89.0	97.5	100	96.7	100	100	100	85.4	99.6	85.1
HB(8)	90.9	99.3	89.0	97.5	100	96.7	100	100	100	85.4	99.6	85.1
HB(9)	90.9	99.3	89.0	97.6	100	96.7	100	100	100	85.4	99.6	85.1
HB(10)	90.9	99.3	89.0	97.5	100	96.7	100	100	100	85.4	99.6	85.1
HB(11)	90.4	99.3	88.5	97.4	100	96.5	100	100	100	84.7	99.6	84.4
HB(12)	90.5	99.3	88.6	97.4	100	96.5	100	100	100	85.0	99.6	84.8
HB(13)	90.5	99.3	88.6	97.4	100	96.5	100	100	100	85.0	99.6	84.8
HB(14)	90.5	99.3	88.6	97.4	100	96.5	100	100	100	85.0	99.6	84.8
HB(15)	90.5	93.9	84.3	97.4	95.2	92.1	100	100	100	85.0	91.8	79.0
HB(16)	90.9	99.3	89.1	97.5	100	96.7	100	100	100	85.7	99.6	85.5
HB(17)	90.9	99.3	89.1	97.6	100	96.7	100	100	100	85.7	99.6	85.5
HB(18)	90.9	99.3	89.1	97.5	100	96.7	100	100	100	85.7	99.6	85.5
HB(19)	90.9	99.3	89.1	97.6	100	96.7	100	100	100	85.8	99.6	85.5
HB(20)	90.9	99.3	88.5	98.0	100	96.5	100	100	100	84.5	96.9	82.3
HB(21)	90.9	99.3	88.5	98.0	100	96.5	100	100	100	84.5	96.9	82.3
HB(22)	90.9	99.3	88.5	98.0	100	96.5	100	100	100	84.5	96.9	82.3
HB(23)	91.3	99.3	89.0	98.1	100	96.7	100	100	100	85.2	97.1	83.1
HB(24)	91.3	99.3	89.0	98.1	100	96.7	100	100	100	85.2	97.1	83.1
HB(25)	91.3	99.3	89.0	98.1	100	96.7	100	100	100	85.2	97.1	83.1
HB(26)	91.3	99.3	89.0	98.1	100	96.7	100	100	100	85.3	97.1	83.1
HB(27)	91.3	99.3	89.0	98.1	100	96.7	100	100	100	85.3	97.1	83.1
HB(28)	90.4	99.2	88.5	97.4	100	96.5	100	100	100	84.7	99.6	84.4
HB(29)	90.2	99.2	88.2	97.3	100	96.4	100	100	100	85.1	99.6	84.4
HB(30)	90.4	99.2	88.5	97.4	100	96.5	100	100	100	84.7	99.6	84.4
HB(31)	90.1	98.5	87.4	97.3	99.2	95.6	100	100	100	79.4	95.7	76.7
Average	90.8	99.1	88.6	97.6	99.8	96.4	100	100	100	84.9	98.5	83.8

The high object- and pixel-based correctness and quality values for all modified sites of the dataset indicate that the proposed change detection method detects all kinds of changes in building roof planes. The proposed change detection method achieved a high pixel-based correctness, but it achieved low pixel-based completeness. As compared to the best-obtained results by existing change detection methods, i.e., 95.7% of overall completeness for all size planes [37] and 76.1% of overall correctness for all size planes [38], the proposed change detection method achieved 89.4% and 97.45% of overall completeness and correctness for all sizes of planes. However, this comparison may be unfair as different datasets were used in evaluation of these methods.

6.5.2. Qualitative Analysis

In this paper, 62 sets of changes have been made in the reference sites, AV(1)–(31) and HB(1)–(31), to evaluate the performance of the proposed building change detection method. Five test sites are used for visual demonstration, where two test sites of each reference site have the height changes in 2-D space, and the other three test sites of each reference site have changes in 2-D and/or 3-D spaces. In the first two test sites of the AV and HB reference sites, AV(17), AV(26), HB(17), and HB(26), the changes in building model are made in height, removal or addition of new verandas, addition of a new building, and relocation of three buildings. The other three test sites, i.e., AV(29)–(31) and HB(29)–(31), are obtained by rotation of a building, destruction of a plane, and introduction of a new building in the building model.

As shown in Figures 18 and 19, the changes in the reference sites are accurately detected by the proposed change detection method. For example, in Figures 18b,c, and 19b,c, the height change (in blue modified planes) are successfully detected by the proposed method. In addition, new veranda and new building planes (in red), demolished planes (in green), and unchanged planes (in yellow) are successfully detected from the building models of AV(17), AV(26), HB(17), and AV(26) sites. In the case of rotation of a building in Figures 18d and 19d and addition of a new building at the demolished building in Figures 18e and 19e, the modified planes (in blue), partially-modified planes (in cyan), demolished planes (in green), and unchanged planes (in yellow) are successfully detected by the proposed change detection method. In Figures 18f and 19f, a few building planes are removed from the building models of AV(31) and HB(31) sites, but these demolished planes (in green) and unchanged planes (in yellow) are also accurately detected by proposed change detection method.

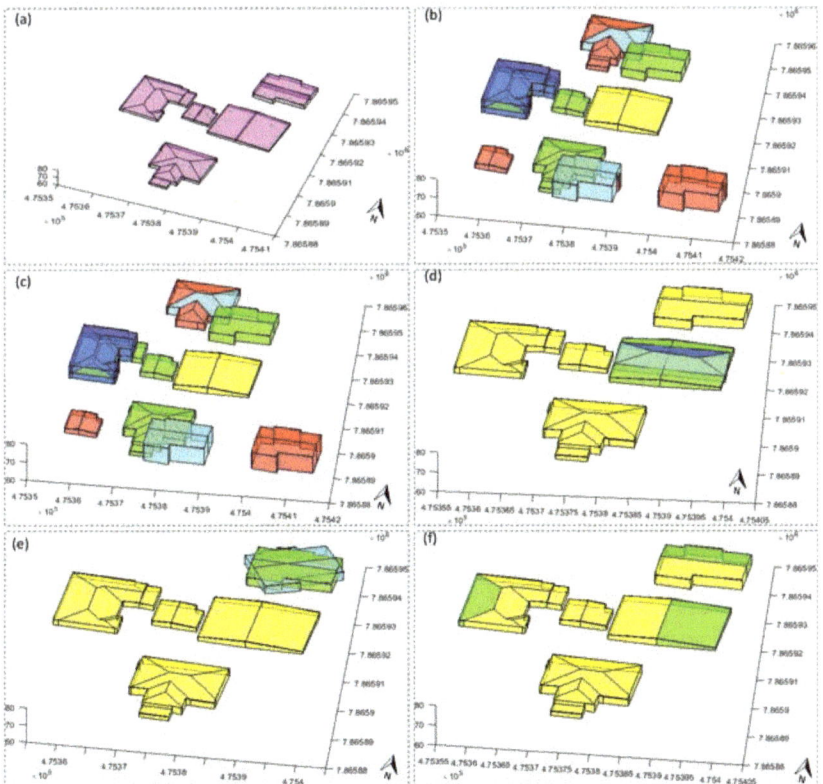

Figure 18. Selected qualitative results of Aitkenvale (AV) reference site after applying the proposed change detection method: (**a**) AV reference models; and change detections in test sites: (**b**) AV(17); (**c**) AV(26); (**d**) AV(29); (**e**) AV(30); and (**f**) AV(31) (Table 1). The reference models in (**a**) are shown in pink and grey colours. In the change detection results (**b**–**f**), the unchanged, new, modified, partially-modified and demolished planes are marked by yellow, red, blue, cyan and green colours, respectively.

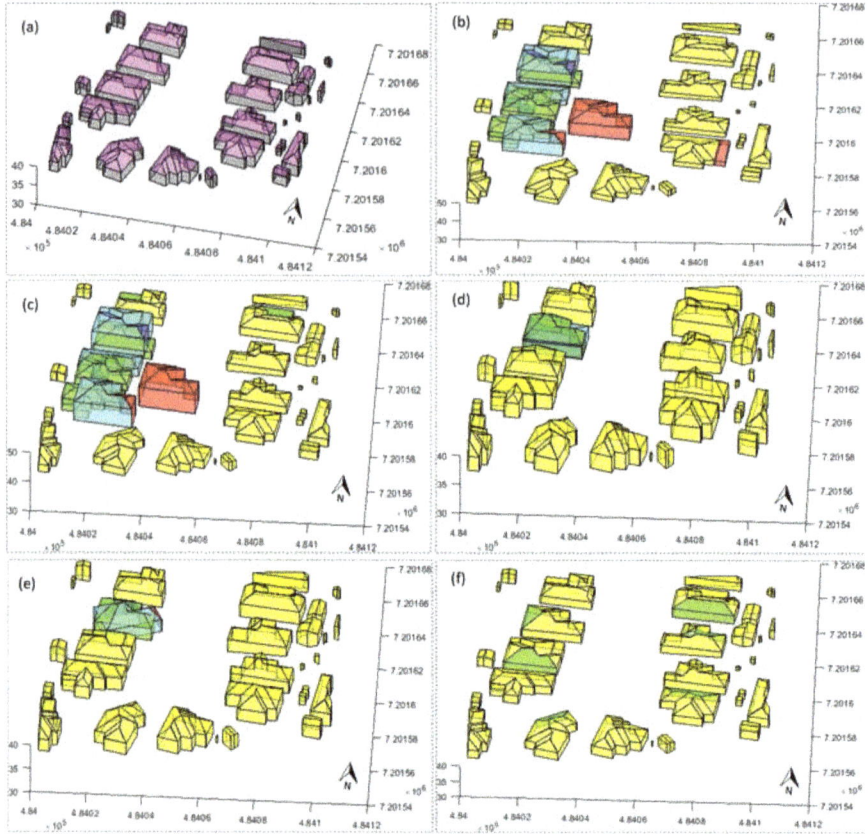

Figure 19. Selected qualitative results of Hervey Bay (HB) reference site after applying the proposed change detection method: (**a**) HB reference models; and change detections in test sites: (**b**) HB(17); (**c**) HB(26); (**d**) HB(29); (**e**) HB(30); and (**f**) HB(31) (Table 1). The reference models in (**a**) are shown in pink and grey colours. In the change detection results (**b–f**), the unchanged, new, modified, partially-modified and demolished planes are marked by yellow, red, blue, cyan and green colours, respectively.

7. Conclusions

Here, the building modelling task is performed in an unsupervised and data-driven fashion. Unlike the model-driven techniques, the roof types are not restricted to a pre-existing model catalogue. The roof planes, which are not extracted due to low point density, noise, and/or the vertical nature of the structures, are hypothesised using the roof topology assumption. As part of the modelling process, interrelations and interconnections among the building roof planes are used for the reconstruction of building models. It was demonstrated that the buildings at higher levels of detail are reconstructed by using individual roof planes and their interconnections based on their spatial adjacency.

The proposed 3-D change detection technique first defines the plane connections into three types of relations: parent-to-child, parent-to-parent and child-to-child. Then, it represents each generated building model into a graph-based data structure. Since, in practice, there are only a small number of buildings being changed in a period of time, the height difference values between the reference and the test models are initially used to find new, completely demolished and unchanged buildings. The corresponding building planes of these buildings are marked as *new*, *demolished* and *unchanged*. Thereafter, for only the modified building regions, the reference and building models are compared using the graph data structure. The planes in the modified building regions are classified as *new, unchanged, demolished, modified* and *partially-modified* planes.

The performance study using two Australian datasets shows the high effectiveness for both the reconstruction and change detection techniques. The proposed reconstruction method successfully inserted 57% (12 out of 21) of missing roof planes. The remaining nine planes that were still missing are from the HB dataset. This dataset has a low input point density (12 point/m^2) and contains buildings with complex roof structures. In addition, the missing planes are mainly small in size (i.e., less than 1 m^2 in area). Moreover, compared with Awrangjeb and Fraser [9], the proposed reconstruction method has shown 3–5% better performance in terms of object-based completeness, correctness and quality. The proposed change detection method has shown 100% completeness, correctness and quality values on both datasets for planes more than 50 m^2 in area. However, when the minimum plane size was set to 10 m^2, these values drop to between 95% and 100% for the HB dataset. When all planes were considered, the correctness value drops to 95% for the AV dataset and the completeness value to 90% for the HB dataset. This indicates that it is harder to detect changes for planes smaller in size.

The output of the proposed building modelling and 3-D change detection techniques can be exploited to semiautomatically create a new or update an existing map database. A graphical user interface (GUI) similar to the one presented by Awrangjeb [1] can be used to quickly rectify the building models, if there are errors, and then store as a new building database. The same GUI can be used to indicate the changes in buildings by overlaying a new building database over an old one. A user can quickly accept or reject the indicative changes to update the database. As in reality the number of changes is small, such an semiautomatic update of the 3-D map database is cost-effective and can be scaled up to a large geographic area.

However, the proposed building reconstruction technique was found ineffective when the input point cloud data to the roof plane extraction technique (e.g., Awrangjeb and Fraser [9]) were low in density (less than 10 points/m^2). Therefore, when the proposed 3-D reconstruction method was applied to the ISPRS benchmark dataset [14], it was not found to work well. When there are small planes or multiple missing planes which are neighbours of one another, the proposed technique fails to generate and insert the appropriate missing planes. In addition, in some particular situations, for example, when buildings have a non-planar roof component or a pyramid-hip roof, the proposed reconstruction procedure will not work well. Future work will include the investigation of a new 3-D roof reconstruction technique that can generate accurate high level building models with complex building roof structures and even using low density point cloud data.

Author Contributions: All three authors carried out the experiments and wrote the paper; S.A.N.G. mainly contributed in building reconstruction part; F.U.S. worked in change detection part; M.A. worked in both parts and oversaw all the activities.

Funding: This work was supported by the Australian Research Council under Grant DE120101778.

Acknowledgments: The Aitkenvale and Hervey Bay datasets were provided by Ergon Energy (www.ergon.com.au) in QLD, Australia.

Conflicts of Interest: The authors declare no conflict of interest.

References

1. Awrangjeb, M. Effective generation and update of a building map database through automatic building change detection from LiDAR point cloud data. *Remote Sens.* **2015**, *7*, 14119–14150. [CrossRef]
2. Zolanvari, S.M.I.; Laefer, D.F.; Natanzi, A.S. Three-dimensional building façade segmentation and opening area detection from point clouds. *ISPRS J. Photogramm. Remote Sens.* **2018**, in press. [CrossRef]
3. Aljumaily, H.; Laefer, D.F.; Cuadra, D. Big-Data Approach for Three-Dimensional Building Extraction from Aerial Laser Scanning. *J. Comput. Civil Eng.* **2016**, *30*, 04015049. [CrossRef]
4. Mongus, D.; Lukač, N.; Žalik, B. Ground and building extraction from LiDAR data based on differential morphological profiles and locally fitted surfaces. *ISPRS J. Photogramm. Remote Sens.* **2014**, *93*, 145–156. [CrossRef]

5. Bizjak, M. The segmentation of a point cloud using locally fitted surfaces. In Proceedings of the 18th Mediterranean Electrotechnical Conference (MELECON), Lemesos, Cyprus, 18–20 April 2016; pp. 1–6.
6. Tran, T.H.G.; Ressl, C.; Pfeifer, N. Integrated Change Detection and Classification in Urban Areas Based on Airborne Laser Scanning Point Clouds. *Sensors* **2018**, *18*, 448. [CrossRef] [PubMed]
7. Leichtle, T.; Geiß, C.; Wurm, M.; Lakes, T.; Taubenböck, H. Unsupervised change detection in VHR remote sensing imagery—An object-based clustering approach in a dynamic urban environment. *Int. J. Appl. Earth Obs. Geoinf.* **2017**, *54*, 15–27. [CrossRef]
8. Gu, W.; Lv, Z.; Hao, M. Change detection method for remote sensing images based on an improved Markov random field. *Multimedia Tools Appl.* **2017**, *76*, 17719–17734. [CrossRef]
9. Awrangjeb, M.; Fraser, C.S. Automatic Segmentation of Raw LiDAR Data for Extraction of Building Roofs. *Remote Sens.* **2014**, *6*, 3716–3751. [CrossRef]
10. Kim, K.; Shan, J. Building roof modeling from airborne laser scanning data based on level set approach. *ISPRS J. Photogramm. Remote Sens.* **2011**, *66*, 484–497. [CrossRef]
11. Truong-Hong, L.; Laefer, D.F. Quantitative evaluation strategies for urban 3D model generation from remote sensing data. *Comput. Graph.* **2015**, *49*, 82–91. [CrossRef]
12. Awrangjeb, M.; Fraser, C.S. An automatic and threshold-free performance evaluation system for building extraction techniques from airborne LIDAR data. *IEEE J. Sel. Top. Appl. Earth Obs. Remote Sens.* **2014**, *7*, 4184–4198. [CrossRef]
13. Vo, A.V.; Truong-Hong, L.; Laefer, D.F.; Tiede, D.; d'Oleire Oltmanns, S.; Baraldi, A.; Shimoni, M.; Moser, G.; Tuia, D. Processing of Extremely High Resolution LiDAR and RGB Data: Outcome of the 2015 IEEE GRSS Data Fusion Contest—Part B: 3-D Contest. *IEEE J. Sel. Top. Appl. Earth Obs. Remote Sens.* **2016**, *9*, 5560–5575. [CrossRef]
14. Rottensteiner, F.; Sohn, G.; Gerke, M.; Wegner, J.D.; Breitkopf, U.; Jung, J. Results of the ISPRS benchmark on urban object detection and 3D building reconstruction. *ISPRS J. Photogramm. Remote Sens.* **2014**, *93*, 256–271. [CrossRef]
15. Li, Y.; Wu, H. An improved building boundary extraction algorithm based on fusion of optical imagery and LiDAR Data. *Int. J. Light Electron Opt.* **2013**, *124*, 5357–5362. [CrossRef]
16. Awrangjeb, M.; Zhang, C.; Fraser, C.S. Automatic extraction of building roofs using LiDAR data and multispectral imagery. *ISPRS J. Photogramm. Remote Sens.* **2013**, *83*, 1–18. [CrossRef]
17. Habib, A.; Kwak, E.; Al-Durgham, M. Model-based automatic 3D building model generation by integrating lidar and aerial images. *Arch. Photogramm. Cartogr. Remote Sens.* **2011**, *22*, 187–200.
18. Cao, R.; Zhang, Y.; Liu, X.; Zhao, Z. 3D building roof reconstruction from airborne LiDAR point clouds: A framework based on a spatial database. *Int. J. Geograph. Inf. Sci.* **2017**, *31*, 1359–1380. [CrossRef]
19. Oude Elberink, S.; Vosselman, G. Building Reconstruction by Target Based Graph Matching on Incomplete Laser Data: Analysis and Limitations. *Remote Sens.* **2009**, *9*, 6101–6118. [CrossRef] [PubMed]
20. Xiong, B.; Oude Elberink, S.; Vosselman, G. A graph edit dictionary for correcting errors in roof topology graphs reconstructed from point clouds. *ISPRS J. Photogramm. Remote Sens.* **2014**, *93*, 227–242. [CrossRef]
21. Jung, J.; Jwa, Y.; Sohn, G. Implicit Regularization for Reconstructing 3D Building Rooftop Models Using Airborne LiDAR Data. *Sensors* **2017**, *17*, 621. [CrossRef] [PubMed]
22. Wu, B.; Yu, B.; Wu, Q.; Yao, S.; Zhao, F.; Mao, W.; Wu, J. A graph-based approach for 3D building model reconstruction from airborne LiDAR point clouds. *Remote Sens.* **2017**, *9*, 92. [CrossRef]
23. Teo, T.A.; Shih, T.Y. Lidar-based change detection and change-type determination in urban areas. *Int. J. Remote Sens.* **2013**, *34*, 968–981. [CrossRef]
24. Tian, J.; Cui, S.; Reinartz, P. Building Change Detection Based on Satellite Stereo Imagery and Digital Surface Models. *IEEE Trans. Geosci. Remote Sens.* **2014**, *52*, 406–417. [CrossRef]
25. Chen, L.C.; Huang, C.Y.; Teo, T.A. Multi-type change detection of building models by integrating spatial and spectral information. *Int. J. Remote Sens.* **2012**, *33*, 1655–1681. [CrossRef]
26. Stal, C.; Tack, F.; De Maeyer, P.; De Wulf, A.; Goossens, R. Airborne photogrammetry and lidar for DSM extraction and 3D change detection over an urban area—A comparative study. *Int. J. Remote Sens.* **2013**, *34*, 1087–1110. [CrossRef]
27. Qin, R. Change detection on LOD 2 building models with very high resolution spaceborne stereo imagery. *ISPRS J. Photogramm. Remote Sens.* **2014**, *96*, 179–192. [CrossRef]

28. Sohn, G.; Huang, X.; Tao, V. A data-driven method for modeling 3D building objects using a binary space partitioning tree. In *Topographic Laser Ranging and Scanning: Principles and Processing*; CRC Press: Boca Raton, FL, USA, 2009; pp. 479–509.
29. Kolbe, T.H.; Gröger, G.; Plümer, L. CityGML—Interoperable access to 3D city models. In *Geo-Information for Disaster Management*; Springer: Berlin/Heidelberg, Germany, 2005; pp. 883–899.
30. Butkiewicz, T.; Chang, R.; Wartell, Z.; Ribarsky, W. Visual analysis and semantic exploration of urban lidar change detection. *Comput. Graph. Forum* **2008**, *27*, 903–910. [CrossRef]
31. Gilani, S.A.N.; Awrangjeb, M.; Lu, G. Segmentation of Airborne Point Cloud Data for Automatic Building Roof Extraction. *GISci. Remote Sens.* **2018**, *55*, 63–89. [CrossRef]
32. Awrangjeb, M. Using point cloud data to identify, trace, and regularize the outlines of buildings. *Int. J. Remote Sens.* **2016**, *37*, 551–579. [CrossRef]
33. Verma, V.; Kumar, R.; Hsu, S. 3D Building Detection and Modeling from Aerial LIDAR Data. In Proceedings of the Conference on Computer Vision and Pattern Recognition, New York, NY, USA, 17–22 June 2006; Volume 2, pp. 2213–2220.
34. Perera, G.S.N.; Maas, H.G. Cycle graph analysis for 3D roof structure modelling: Concepts and performance. *ISPRS J. Photogramm. Remote Sens.* **2014**, *93*, 213–226. [CrossRef]
35. Siddiqui, F.U.; Teng, S.W.; Awrangjeb, M.; Lu, G. A Robust Gradient Based Method for Building Extraction from LiDAR and Photogrammetric Imagery. *Sensors* **2016**, *16*, 1110. [CrossRef] [PubMed]
36. Awrangjeb, M.; Lu, G.; Fraser, C.S. Performance comparisons of contour-based corner detectors. *IEEE Trans. Image Process.* **2012**, *21*, 4167–4179. [CrossRef] [PubMed]
37. Olsen, B.P.; Knudsen, T. Automated change detection for validation and update of geodata. In Proceedings of the 6th Geomatic Week, Barcelona, Spain, 8–10 February 2005.
38. Matikainen, L.; Hyyppä, J.; Ahokas, E.; Markelin, L.; Kaartinen, H. Automatic detection of buildings and changes in buildings for updating of maps. *Remote Sens.* **2010**, *2*, 1217–1248. [CrossRef]

© 2018 by the authors. Licensee MDPI, Basel, Switzerland. This article is an open access article distributed under the terms and conditions of the Creative Commons Attribution (CC BY) license (http://creativecommons.org/licenses/by/4.0/).

Detecting Building Edges from High Spatial Resolution Remote Sensing Imagery Using Richer Convolution Features Network

Tingting Lu [1], Dongping Ming [1,*], Xiangguo Lin [2], Zhaoli Hong [1], Xueding Bai [1] and Ju Fang [1]

[1] School of Information Engineering, China University of Geosciences (Beijing), Beijing 10083, China; lutingtingGIS@163.com (T.L.); hongzhaoli9237@163.com (Z.H.); dztintin@163.com (X.B.); fangju_nanxia@163.com (J.F.)

[2] Institute of Photogrammetry and Remote Sensing, Chinese Academy of Surveying and Mapping, Beijing 100830, China; linxiangguo@casm.ac.cn

* Correspondence: mingdp@cugb.edu.cn

Received: 21 August 2018; Accepted: 18 September 2018; Published: 19 September 2018

Abstract: As the basic feature of building, building edges play an important role in many fields such as urbanization monitoring, city planning, surveying and mapping. Building edges detection from high spatial resolution remote sensing (HSRRS) imagery has always been a long-standing problem. Inspired by the recent success of deep-learning-based edge detection, a building edge detection model using a richer convolutional features (RCF) network is employed in this paper to detect building edges. Firstly, a dataset for building edges detection is constructed by the proposed most peripheral constraint conversion algorithm. Then, based on this dataset the RCF network is retrained. Finally, the edge probability map is obtained by RCF-building model, and this paper involves a geomorphological concept to refine edge probability map according to geometric morphological analysis of topographic surface. The experimental results suggest that RCF-building model can detect building edges accurately and completely, and that this model has an edge detection F-measure that is at least 5% higher than that of other three typical building extraction methods. In addition, the ablation experiment result proves that using the most peripheral constraint conversion algorithm can generate more superior dataset, and the involved refinement algorithm shows a higher F-measure and better visual effect contrasted with the non-maximal suppression algorithm.

Keywords: richer convolution features; building edges detection; high spatial resolution remote sensing imagery

---

1. Introduction

Buildings are one of the most important and most frequently updated parts of urban geographic databases [1]. As an important and fundamental feature for building description, the building edges detection plays a key role during building extraction [2,3]. Building edges detection has extensive applications in real estate registration, disaster monitoring, urban mapping and regional planning [4–6]. With the rapid development of remote sensing imaging technology, the number of high spatial resolution remote sensing (HSRRS) imagery has increased dramatically. HSRRS imagery have improved the spectral features of objects and highlighted information on the structure, texture, and other details of the objects. At the same time, they also brought severe image noise, "different objects with similar spectrum" and other problems [7]. In addition, due to the diversity of the structure of the buildings themselves and the complexity of the surroundings, the detection of building edges from HSRRS imagery is a challenge in the field of computer vision and remote sensing urban application.

In the rich history of edge detection, typically, the early edge detectors were designed by gradient and intensity. Later, researchers began to use artificial design features to detect edges. But these traditional edge detection algorithms mainly rely on handcrafted low-level features to detect edges, whose accuracy are difficult to guarantee and cannot adapt to application. However, with the rapid progress of artificial intelligence, deep learning has excellent performance in the field of natural image edge detection. N4-fields [8], DeepContour [9], DeepEdge [10], HFL [11], HED [12], and richer convolution features (RCF) network [13] were successively proposed. The accuracy of their test results on the BSDS500 [14] dataset has been continuously improved, while the accuracy of the newly proposed RCF network has even exceeded the human performance.

Lots of studies have shown that the deep-learning-based edge detection model can not only detect the edge of the image effectively, but also generate a higher accuracy than the traditional edge detection algorithm. However, it is not applicable to directly extract building edges from HSRRS imagery by using pre-trained deep learning network. The reasons come as follows:

- The dataset used in network training is natural image rather than remote sensing imagery. Remote sensing imagery has some features that natural images do not possess such as resolution information [15] and spatial autocorrelation.
- The remote sensing imagery has other superfluous objects in addition to the building. The network trained by the natural image cannot identify the edges of a certain object, so it is difficult to obtain the building edges directly through the pre-trained deep learning network.

Although it is difficult to acquire a high quality building edges dataset for deep learning, the limitation of the data can be overcome by modifying the existing datasets. Due to the special architecture of RCF and its excellent performance in the deep-learning-based edge detection, this paper presents a new method to detect building edges. Using the most peripheral constraint conversion algorithm, a high-quality HSRRS imagery building edges dataset for deep learning is built for the first time. This paper constructs a building edges detection model by fine-tuning the pre-trained RCF network with this self-build dataset, and the generated RCF-building model can exclusively detect the building edges. In the post-processing stage, this paper involves a geomorphological concept to refine the edge probability map generated by the RCF-building model and obtains accurate building edges. In particular, the advantage of the RCF network special architecture is exploited, which can make full use of all the convolution layers to improve the edge detection accuracy.

The rest of this paper is organized as follows. In Section 2, we briefly present the related work. RCF-based building edges detection model is described in Section 3. Section 4 presents the experiment and contrast results and analyzes the performance of the proposed methods. Finally, the discussion and conclusions are drawn in Sections 5 and 6, respectively.

2. Related Work

Although there are various edge detection algorithms and theories, but there is a great gap between theory and application, only considering the edge detection algorithm cannot directly extract buildings from imagery. As the edge detection algorithm does not have the function to distinguish what kind of object the edge belongs to, it is difficult to obtain the building edges directly by the edge detection. The previous building edges detection methods can be grouped into the following 3 categories:

- Edge-driven methods. This category usually extracts line segments by low-level edge detection algorithm first, then, groups the building edges from the line segments based on various rules [16–26]. Those rules, for example, can be perceptual grouping [16–19], Graph structure theory [20,21], Markov random field models [22], geometry theory [23], circle detection [24], heuristic approach [25], and dense matching [26]. Additionally, a series of models [27–30] have been set up to directly detect the building edges. This kind of method, in comparison to the classical methods, can detect building edges more accurately, and avoid the boundaries of features in the building neighborhood such as streets and trees. Snake model [31], also called active contour model, was widely applied in fields of building edges detection [27–29]. The research in Garcin et al. [30] built a shape-model using Markov Object processes and a MCMC Algorithm, and this model used perspective of the whole building to detect the building.

- Region-driven methods. The building region feature and the edge feature are the important elements of the building description. Under certain circumstances, building edges can be converted from building region. Various classification strategies were utilized to extract building region, here are only a few classification strategies for HSRRS imagery:

    ◆ Object-based image analysis (OBIA) extraction method has gradually been accepted as an efficient method for extracting detailed information from HSRRS imagery [7,32–39]. For example, references [7,32–37] comprehensively used object-based image segmentation and various features of objects such as spectrum, texture, shape, and spatial relation to detect buildings. Due to the scale parameter has an important influence on OBIA, Guo et al. [38] proposed a parameter mining approach to mine parameter information for building extraction. In addition, Liu et al. [39] adopted the probabilistic Hough transform to delineate building region which extract by multi-scale object oriented classification, and result showed that with the boundary constraint, most rectangular building roofs can be correctly detected, extracted, and reconfigured.

    ◆ Extraction method based on deep learning is a long-standing problem in recent years [40–48]. References [40–45] designed an image segmentation using convolutional neural network, full convolutional network or other network, to effectively extract building region from imagery. The above research is still pixel-level-based, references [46–48] proposed superpixel-based convolution neural network (SML-CNN) model in hyperspectral image classification in which superpixels are taken as the basic analysis unit instead of pixels. Compared to other deep-learning-based methods, superpixel-based method gain promising classification results. Gao et al. [49] combined counter map with fully convolutional neural networks to offer a higher level of detection capabilities on image, which provided a new idea for building detection. In addition, constantly proposed theories, such as transfer hashing [50] and structured autoencoders [51] can also be introduced into this application field to solve problems, such as data sparsity and data mining.

    ◆ Extraction method based on mathematical morphology [52–58]. Huang et al. and Rongming et al. [52,53] used morphological building index by differential morphological profile to extract buildings and optimized methods are proposed in references [54–58].

- Auxiliary-information-based methods. Due to the complexity of the structure and surrounding environment of the building, many scholars have proposed the method of extracting the building by the shadow, stereoscopic aerial image or digital elevation model (DEM) data to assist the building extraction. Liow et al. [59] pioneering proposed a new idea of using shadow to extract buildings. Later, research in [59–62] proposed to identify and extract buildings based on the shadow features and graph based segmentation in high-resolution remote sensing imagery. In addition, local contrast in the image where shadow and building interdepend will be increase. Based on this principle, references [63,64] proposed PanTex method with gray level co-occurrence matrix contrast features, which is practically used to identify buildings and build-up areas. Hu et al. [65] used the shadow, shape, color features, similarity of angle between shade lines and so on multiple cues to extract buildings. In addition, stereo information can provide great convenience for the extraction of buildings information [5,66–78].

Among the methods mentioned above, the first category normally used semantic analysis to grouping lines segments, and they have shown relatively good performance on moderate and low spatial resolution remote sensing imagery because of its high signal noise ratio (SNR). However, for

HSRRS imagery, the high spatial resolution and low SNR substantially increases the difficulties of locating and identifying the accurate building edges [39]. For the second category, they have many advantages, such as a comprehensive consideration of prior knowledge, image features, pattern recognition theory and other factors. However, the related methods still have the problems of cumbersome workflow, which requires more prior knowledge, and unable to meet the practical requirements of buildings extraction from high spatial resolution images with high scene complexity. The applicability is also limited by buildings type, density, and size. Moreover, the edge of extraction results is not ideal, so it is difficult to ensure the edge integrity of complex objects. For the last category, although the accuracy of building extraction can be improved based on stereo information, it is greatly limited by multiple data sources scarcity and data misalignment.

Therefore, to overcome these limitations of single data, building structure, surrounding complexity and prior knowledge, this paper tries to detect building edges using state-of-the-art method of edge detection with deep learning, which is only based on two-dimensional HSRRS imagery, also needs no prior knowledge once the deep supervision based dataset is perfectly built.

3. Methodology

As shown in Figure 1, the workflow of proposed method is mainly divided into three stages. In the dataset construction stage, the initial dataset is processed by conversion, clipping, rotation, and selection into a special dataset which can be dedicated to deep-learning-based edge detection. The second stage is network training. Based on the training set, the RCF network is retrained to generate a RCF-building edges detection model. The third stage is detecting and post-processing. The edge probability map is obtained by using RCF-building model. Subsequently, the edge probability map is refined by the involved algorithm, so that the building edges are obtained.

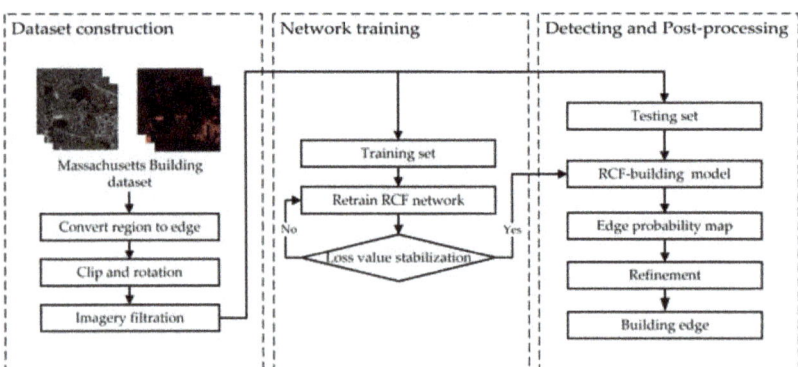

Figure 1. Workflow of fine-tuning RCF network.

3.1. Dataset Construction

As mentioned previously, in the field of deep learning, there is no experimental HSRRS imagery dataset available to building edges detection. Therefore, this paper builds an edge based sample dataset that satisfies the training and testing requirements of the RCF network by pre-processing Massachusetts Building Dataset [79]. The Massachusetts Building dataset is constructed by Mnih and publicly available at http://www.cs.toronto.edu/vmnih/data/. This dataset has a resolution of 1 m and sizes of 1500 × 1500 pixels. It contains 137 training images, 10 testing images, and four validation images between which has no intersection. Each set of data includes an original remote sensing image and a manually traced building region map, as shown in Figure 2a,b. Since the output of RCF network is based on the fusion of multi layers, RCF network is tolerable to slight overfitting. Thus, RCF network does not need validation sets.

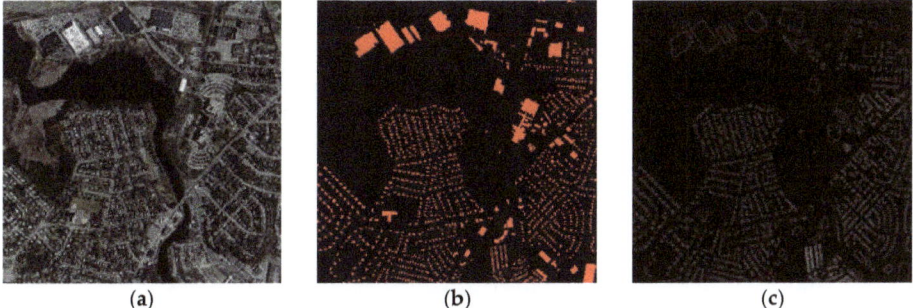

Figure 2. Dataset sample. (**a**) Original image; (**b**) building region map; and (**c**) building edges ground truth map.

Edge detection is different from region extraction, and the location shift of only one pixel may cause the model fail to extract features and reduce the overall precision. To ensure that there is no error occurred when convert building region to building edges, this paper proposes most peripheral constraint algorithm. With constraint of "most peripheral", it emphasizes on only extracting the outermost pixels of the building region features as building edges, and the width of edge is only one pixel. Figure 3 shows the diagram of this conversion algorithm. The steps come as follows:

(1) Binarization of the building region map. Supposing the building pixel value is 1, and the non-building pixel is 0;

(2) Generating an image with the same size as the original image, and all the pixel values are 0. Scanning the building region map row by row to find all pixels (marked as P_r) satisfying two conditions: the pixel value is 1, and the pixel value shifts from 1 to 0 or from 0 to 1. In the newly generated image, setting the pixel values at the same locations with P_r as 1. Thus, building edge pixels on each row are detected;

(3) Generating an image with the same size as the original image, and all the pixel values are 0. Repeating step 2 to detect all building edge pixels on each column;

(4) All building edge pixels on each row and each column are combined. Thus, the building edge is finally detected.

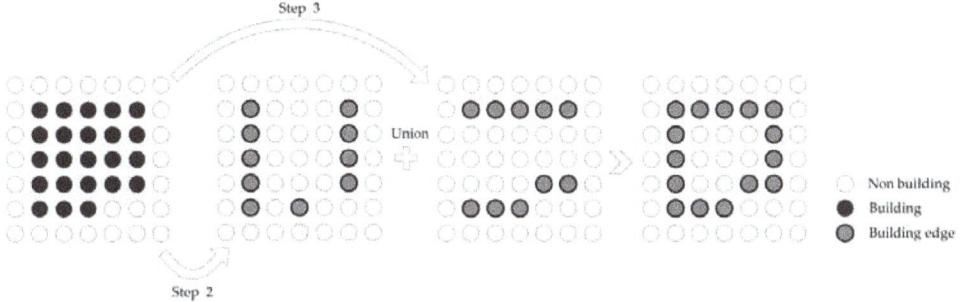

Figure 3. Diagram of conversion from building region into building edges.

Figure 2c shows the conversion result of Figure 2b. After conversion, in order to improve the accuracy of the training network, we augment the data by rotating the imagery by 90, 180, and 270 degrees. Meanwhile, to avoid memory overflow and invalid imagery, this paper ultimately constructs the dataset after image clipping and choosing. The final dataset contains 1856 training images with size of 750 × 750 pixels and 56 testing images with size of 750 × 750 pixels, named Massachusetts Building-edge dataset.

3.2. RCF Network

The RCF network was originally proposed by Liu in 2017 [13]. It was optimized on the basis of VGG16 [80] network. The input of the RCF network is an RGB image with unlimited size, and the output is the edge probability map with the same size. Figure 4 shows the architecture of RCF network when the input image size is 224 × 224 pixels. The main convolutional layers in RCF (as shown in the red dashed rectangle) are divided into five stages and the adjacent two stages are connected through the pooling layer. After the down sampling of the pooling layer, different scales of features can be extracted, and useful information can be obtained while reducing the amount of data. Different from VGG16 network, the RCF network discards all the fully connected layers as well as the fifth pooling layer, and each main convolution layer is connected to a convolution layer with kernel size 1 × 1 and channel depth 21. Then, RCF network sets an element_wise layer for accumulation after each stage. Afterwards, each element_wise layer is connected to a convolution layer with kernel size 1 × 1 and channel depth 1. The difference between the RCF network and the traditional neural network lies in: for the boundary extraction, the previous neural networks only use the last layer as the output, and lose many feature details, while the RCF network fuses the convoluted element_wise layers of each stage (convoluted element_wise layers of 2, 3, 4, and 5 stages need to be restored to its original image size by deconvolution) with the same weights to get a fusion output. This special network architecture allows the RCF network to make full use of semantic information and detailed information for edge detection.

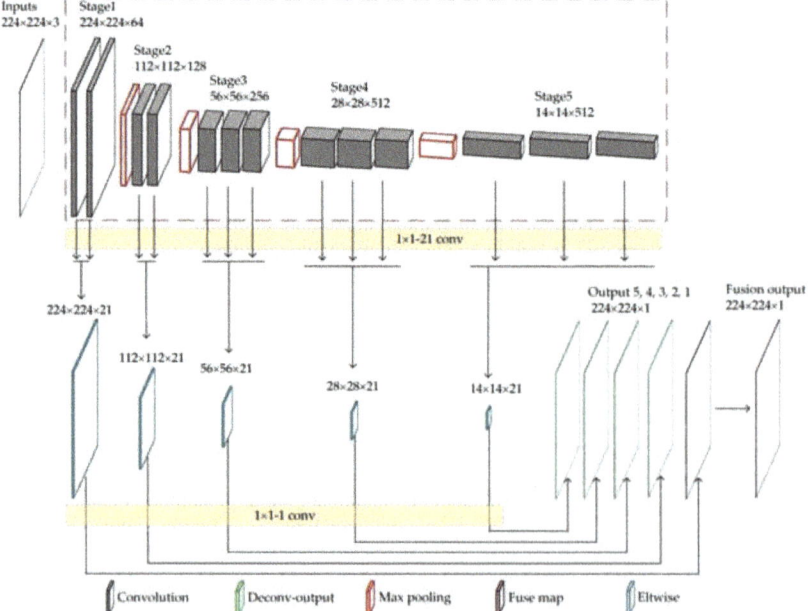

Figure 4. Overview of the RCF network architecture.

3.3. Refinement of Edge Probability Map

The test results of the RCF network are gray-scale edge probability map, on which the greater the gray value is, the higher the probability that the pixel is on an edge. To accurately detect the building edges, it is necessary to refine edge probability map. In computer vision filed, non-maximal suppression (NMS) algorithm is a commonly used refining method. However, as observed in Figure 5, the results show that using NMS algorithm to refine building edges has the following problems: broken outliers, isolated points and flocculent noises.

 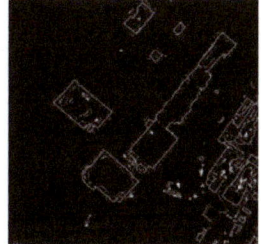 

Figure 5. The refinement results by NMS algorithm.

Therefore, this paper involves a geomorphological concept to refine edge probability map according to geometric morphological analysis of topographic surface. As illustrated in Figure 6, our basic idea is to regard the edge probability value as elevation, according to the principles of geometric morphology, and the points with maxima elevation (i.e., the watershed point) on the topographic profile curve are extracted as accurate edges.

Figure 6. Diagram of edge refinement algorithm.

As described in Figure 7, the procedures of this refinement algorithm are as follows:

(1) Scanning from four directions (vertical, horizontal, left diagonal, and right diagonal) to find the local maxima points as candidate points;
(2) Setting a threshold to discard the candidate points whose probability is less than 0.5 (After many experiments, the highest accuracy is obtained under this threshold. For gray image, the threshold value is 120.);
(3) Calculating the times that each candidate point is detected out. When a candidate point is detected at least twice, it is classified as an edge point;
(4) Checking the edge points got by step (3) one by one. When there is no other edge point in an eight neighborhood, this point is determined as an isolated point and deleted;
(5) Generating edge mask map based on the edge point map got by step (4) to refine the edge probability map and obtain the final edge refinement map.

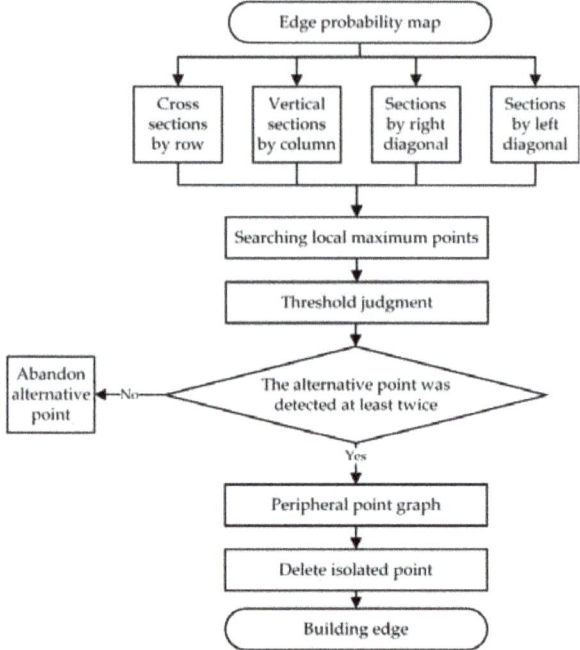

Figure 7. Workflow of edge probability map refinement.

4. Experiments and Analysis

The experimental environment for RCF network re-training and testing is the Caffe framework [81] in Linux system with support of NVIDIA GTX1080 GPU. The learning rate refers to the rate of descent to the local minimum of the cost function, and the initial learning rate is 1×10^{-7}. Every ten thousand iterations, the learning rate will be divided by 10 in training process. The experimental data are the self-processed Massachusetts Building-edge dataset which has been introduced in Section 3.1.

4.1. Experimental Results

In this paper, a trained model generated by 40,000 iterations is selected to extract the building edges. Some example of the building edges detection results are shown in Figure 8(e1–e3). From the visual perspective, the RCF-based building edges detection method adapts to the background very well. As can be seen from the third line of data (Figure 8(a3,b3,c3,d3,e3), which are highlighted by red rectangle, the fine-tuned RCF-building model can not only detect building edges correctly, but also extract building edges that the human unrecognized. Additionally, the refinement results of involved refinement algorithm (Figure 8(e1–e3) are experimentally compared with the results of NMS algorithm (Figure 8(d1–d3). There are less isolated points and flocculent noises in the building edges detection results by the involved refinement algorithm.

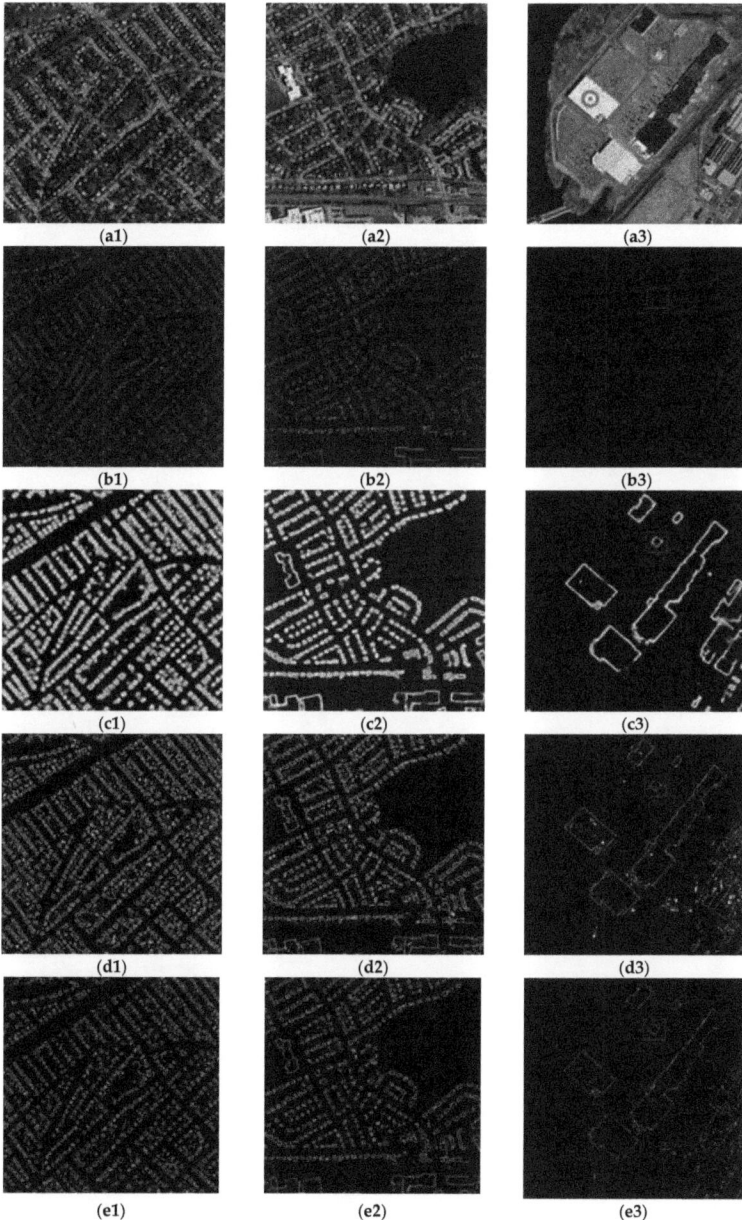

Figure 8. RCF based building edges detection results. (**a1–a3**) Original imagery; (**b1–b3**) building edges ground truth map; (**c1–c3**) building edges probability map generated by RCF-building; (**d1–d3**) building edges refinement map generated by NMS algorithm; and (**e1–e3**) building edges refinement map generated by involved algorithm.

4.2. Precision and Recall Evaluation

In this paper, inspired by references [82–84], we used recall, precision, and F-measure as the criteria for RCF-building model. The evaluation indices can be descripted by Equations (1)–(3):

$$\text{Recall} = \frac{TP}{TP + FP} \quad (1)$$

$$\text{Precision} = \frac{TP}{TP + FN} \quad (2)$$

$$\text{F-measure} = \frac{2 \times \text{Precision} \times \text{Recall}}{\text{Precison} + \text{Recall}} \quad (3)$$

where true positive (TP) represents the number of coincident pixel between detected edges and referenced building edges of ground truth. False positive (FP) represents the number of non-coincident pixel between detected edges and referenced building edges of ground truth. False negative (FN) represents the number of non-coincident pixel between detected non-building objects and non-building edges in the referenced ground truth. F-measure is a synthetic measurement of precision and recall. Actually, the precision and the recall are two contradictory measurements. Generally, they are negatively correlated [85,86]. Based on recall and precision, the precision-recall curve (P-R curve) can be drawn.

As shown in Figure 9, it can be noted that our RCF-building model has an F-measure of 0.89 on the test set, which is higher than the 0.51 from the original RCF network. In addition, compared with the original RCF network, the precision of RCF-building model increases at least 45%. It means that the retraining RCF network has the function of recognizing the edges of buildings. The generated RCF-building model can exclusively detect the building edges, and effectively avoid the superfluous objects edges.

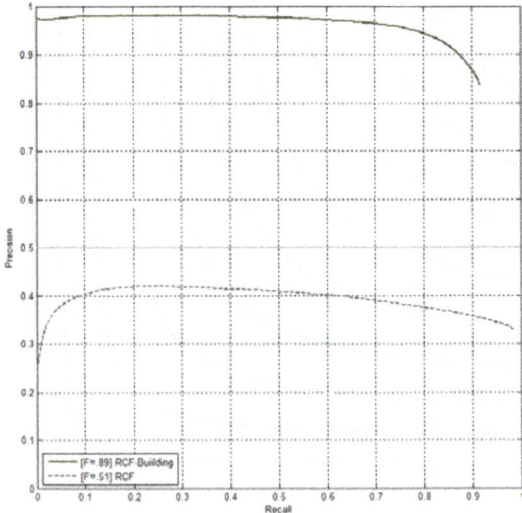

Figure 9. The P-R curves. The solid curve is the result of proposed RCF-building on the test set and the dotted one is the original RCF network.

4.3. Comparison with Other Building Extraction Methods

In this paper, four remote sensing images with different characteristics from the testing set are selected to compare the performance of our method with other three representative building detection methods. Figure 10 illustrates the visual results of our method, OBIA-based ENVI

Feature Extraction [87], Superpixel-based SML-CNN [47] and CNN-based Saito's Method [43]. The ENVI Feature Extraction was implemented through ENVI's Example-Based Classification [88]. The segmentation scale and merge level parameters were set respectively as 40, 30, 30, 40 and 50, 40, 50, 80. Classification was accomplished by training the nearest neighbor classifier with the selected samples of building and non-building objects point. The scale parameters of the SML-CNN are set to 15. In addition, image1 has similar characteristics to the image3, so the model generated by image1 was selected to classify image3. The sample sizes of the data are shown in Table 1. The results of the Saito's Method are derived from the experimental results in the reference [43] which uses the same dataset as this paper. To ensure fair comparison, this paper cuts the related image data into the same size with those used in this paper.

Table 1. Number of sample points marked on the Figure 10 original images.

Sample Category	Image1	Image2	Image3	Image4
Building	1200	194	1200	1954
Non building	2088	1298	2088	1774

It can be clearly seen from Figure 10 that the method used in this paper has better visual effects than ENVI Feature Extraction and SML-CNN. Compared with the overall view results of Saito's Method, although our results have more broken line segments inside the building, as can be seen from the last row in Figure 10, the detailed image shows that the method we used can maximize the integrity of the building edges. In the corner part of the building, the angular characteristics are preserved better by our method.

Table 2 shows the evaluation results of building edges detected by ENVI Feature Extraction [87], SML-CNN [47], Saito's Method [43] and the proposed method in four images. It can be seen from the comparison of the F-measure values that the RCF network has the best performance regardless of whether the building group is high-density (image1) or low-density (image2), or the structure of the building is simple (image3) or complicated (image4). ENVI Feature Extraction is a traditional module for extracting information from high-resolution panchromatic imagery by spatial, spectral, and texture characteristics. Although we manage to cover all types of buildings in the selection of samples, the classification results of buildings still have serious noises and misclassifications, and the building edges extracted by this method would be mixed with more non-architectural edges and closed noise lines inside the building. Compared to traditional building edge detection methods based on image processing, RCF-building is more robust and it is applicable in complicated environment because this model depends on not only image but also supervised dataset. Manually labeled building samples implement deep supervision of each layer of network to achieve optimal fitting of building edge information at different scales, and enhance the saliency-guided building feature learning. Thus the method of ENVI Feature Extraction has similar Recall as the proposed method but much lower Precision. SML-CNN first divides the image into superpixels, and then uses CNN network in classification. Therefore, SML-CNN can extract building edges completely, but at the same time, it might have misclassification. This method has a slightly higher recall and much lower precision than the method we proposed. Saito's Method is a CNN network which simultaneously extracts multiple kinds of objects. Due to the limitation of network architecture, only region features are emphasized while line features are ignored. Although this method can roughly locate buildings in the imagery, the boundaries between the buildings and the non-buildings are not accurate, and present lower Recall value and higher precision value. The method proposed in this paper has a good performance on both precision and recall. Compared to deep-learning-based building extraction methods, RCF-building could better retain building edges angular characteristics.

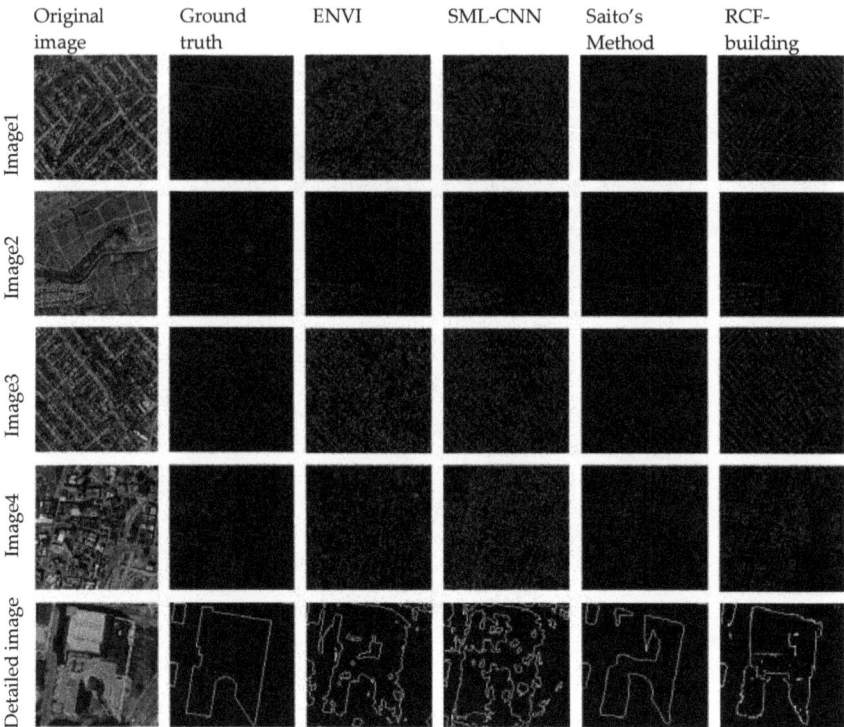

Figure 10. Building edges detection results on Massachusetts Building Dataset. The last row shows the details image of one building.

Table 2. Evaluation results of four different methods on four typical images.

Approach	Index	Image1	Image2	Image3	Image4	Mean
ENVI Feature Extraction	Precision	0.35	0.71	0.44	0.45	0.49
	Recall	0.97	0.90	0.96	0.87	0.93
	F-measure	**0.52**	**0.80**	**0.61**	**0.60**	**0.63**
SLIC-CNN	Precision	0.51	0.54	0.57	0.35	0.49
	Recall	0.99	0.97	0.97	0.96	0.97
	F-measure	**0.68**	**0.70**	**0.72**	**0.52**	**0.65**
Saito's Method	Precision	0.99	1.00	0.99	0.78	0.94
	Recall	0.55	0.72	0.50	0.75	0.63
	F-measure	**0.70**	**0.84**	**0.67**	**0.77**	**0.74**
RCF-building	Precision	0.85	0.96	0.88	0.74	0.86
	Recall	0.94	0.82	0.93	0.94	0.91
	F-measure	**0.89**	**0.89**	**0.91**	**0.82**	**0.88**

5. Discussion

5.1. Ablation Experiment

To verify the effectiveness of different steps of the proposed method, this paper compares the performance of RCF model trained by the self-processing dataset (Massachusetts Building-edge dataset) with the RCF model trained by Canny algorithm [89] converted dataset on all testing set. We also quantitatively compare the performance of the involved edge refining algorithm with NMS edge refining algorithm. Table 3 lists the evaluation results of different pre-processing and post-processing methods. Our methods present the best performance in the Precision, Recall and F-measure. The experimental results verify the effectiveness of proposed conversion algorithm for dataset pre-processing, which proves that the superior dataset has positive influence on RCF network. Furthermore, comparison results also reveal that the good performance of our approach takes the advantage of the involved refinement algorithm. For all testing set, the refining algorithm presented in this paper has better performance.

Table 3. The performance of training set generated by different conversion methods and performance comparison of different refinement algorithms.

Conversion Algorithm	Refinement Algorithm	Precision	Recall	F-Measure
Canny algorithm	NMS	0.46	0.99	**0.63**
Canny algorithm	Our refinement algorithm	0.70	0.94	**0.80**
Our conversion algorithm	NMS	0.60	0.98	**0.75**
Our conversion algorithm	Our refinement algorithm	0.85	0.89	**0.87**

5.2. Influence of the RCF Fusion Output

To explore why RCF-building can recognize the edge of building, this paper compares the average Precision, Recall and F-measure values of all testing set imagery at each stage of network. As shown in Figure 11, with the deepening of the network, the precision and recall value rises gradually during the first three stages, and then the precision and recall value descend (or roughly descend) during the fourth and fifth stage. During the first three stages, the network gradually learns the characteristics of the building edge, so the precision and recall of the detected building edges increase gradually. However, during the fourth and fifth stages, the network is overfitting and regards the characteristics of one training sample as the general nature of all the potential samples. This phenomenon of reduced generalization performance eventually leads to the failure of detecting some parts of the building edges. On the other hand, the overfitting of edge detection is different from the overfitting in other fields, which means after overfitting, if one pixel is judged as edge, the probability of actually being edges is higher. Above all, to make full use of the information generated at each stage, the RCF network utilizes a special architecture that the traditional neural networks do not have: the fusion output layer. The fusion output layer fuses all the output of each stage with the same weight, so that it can perfectly inherit the advantages of each stage and suppress the useless information at first two stages. Thus, the fusion output guarantees the highest precision and recall value.

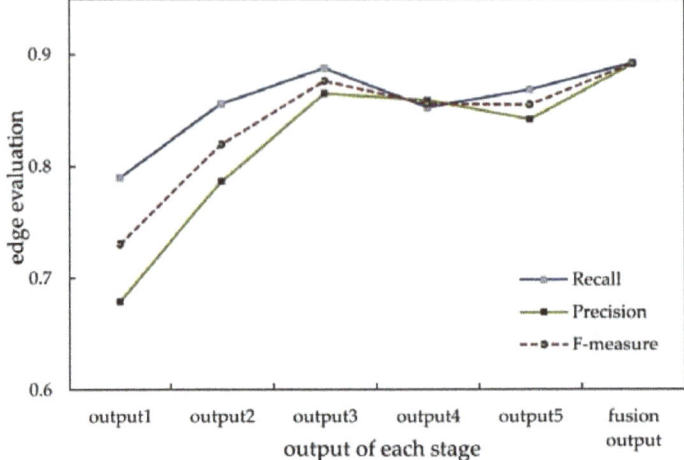

Figure 11. Comparison of precision, recall and F-measure of the output maps at different stages.

Take a test image as an example, the output of the each stage and fusion output images are shown in Figure 12. It is clear that with the deepening of the network stages, the model can gradually extract the edge of the building and eliminate the edges of other superfluous objects, but in the fourth or fifth stage, the edge of the building cannot be completely extracted in the image. The visual result of the fusion output image has the best performance, and the edge of the building can be extracted completely and accurately compared with other stages output. Therefore, RCF's special fusion output architecture makes it suitable for building edges extraction from high resolution remote sensing images.

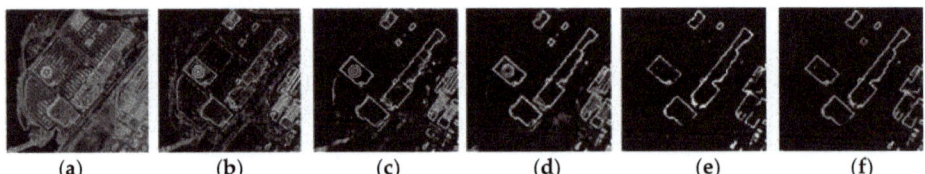

Figure 12. Output images of each stage and fusion output image. From (**a**–**f**): stage1 output, stage2 output, stage3 output, stage4 output, stage5 output, and fusion output.

6. Conclusions

This paper proposes a method for detecting building edges from HSRRS imagery based on the RCF network. The highlights of this work are listed as follows:

- The RCF network is firstly combined with HSRRS imagery to detect building edges and then an RCF-building model that can accurately and comprehensively detect the building edges is built. Compared to the traditional building edge extraction method, the method used in this paper can make use of high-level semantic information and can get a higher accuracy evaluation value and better visual effects. Compared to deep-learning-based building extraction methods, RCF-building could better retain the corner part building edges. In addition, this paper also analyzes the influence of the RCF fusion output architecture on the building edges detection accuracy, and the precision and recall lines affirm that this unique architecture of RCF can perfectly inherit the advantages of each stage and has a strong applicability to the detection of building edges.

- In the preprocessing stage, on the basis of Massachusetts Building dataset, we proposed the most peripheral constraint edge conversion algorithm and created the Massachusetts Building-edge dataset specifically for deep-learning-based building edges detection. The comparison result shows that the dataset produced by the most peripheral constraint algorithm can effectively improve the performance of RCF-building model, and affirms the positive impact of accurately labeled data on network training. This Massachusetts Building-edge dataset makes the foundation for future research on deep-learning-based building edges detection.

- In the post-processing stage, this paper involves a geomorphological concept to refine edge probability map according to geometric morphological analysis of topographic surface. Compared to the NMS algorithm, the involved refinement algorithm could balance the precision and recall value, and get a higher F-measure. It can preserve the integrity of the building edges to the greatest extent and reduce noise points. However, there are still some broken lines, as well as some discontinuities in the detected building edges results after the post-processing.

Additionally, it is worth noting that building edges detection is not the terminal goal of building extraction from HSRRS imagery. The future work will include: (1) connection of the broken edges of the building; (2) vectorization of building edges features; (3) the improvement of RCF network architecture; and (4) using various strategies to ensure that large images can be processed in memory [90].

Author Contributions: T.L. and D.M. conceived and designed the experiments; T.L., D.M., and X.L. performed the experiments; Z.H. contributed dataset construction; T.L. wrote the paper, and X.B. and J.F. contributed to the manuscript.

Funding: This research was supported by the National Natural Science Foundation of China (41671369), the National Key Research and Development Program (2017YFB0503600) and the Fundamental Research Funds for the Central Universities.

Acknowledgments: The authors would like to thank Volodymyr Mnih, from University of Toronto, Canada, for providing the Massachusetts Building Dataset used in the experiments

Conflicts of Interest: The authors declare no conflict of interest.

References

1. Du, S.; Luo, L.; Cao, K.; Shu, M. Extracting building patterns with multilevel graph partition and building grouping. *ISPRS J. Photogramm. Remote Sens.* **2016**, *122*, 81–96. [CrossRef]
2. Li, Y.; Wu, H. Adaptive building edge detection by combining lidar data and aerial images. *Int. Arch. Photogramm. Remote Sens. Spat. Inf. Sci.* **2008**, *37*, 197–202.
3. Hu, X.; Shen, J.; Shan, J.; Pan, L. Local edge distributions for detection of salient structure textures and objects. *IEEE Geosci. Remote Sens. Lett.* **2013**, *10*, 4664–4670. [CrossRef]
4. Yang, H.-C.; Deng, K.-Z.; Zhang, S. Semi-automated extraction from aerial image using improved hough transformation. *Sci. Surv. Mapp.* **2006**, *6*, 32.
5. Siddiqui, F.U.; Teng, S.W.; Awrangjeb, M.; Lu, G. A robust gradient based method for building extraction from lidar and photogrammetric imagery. *Sensors* **2016**, *16*, 1110. [CrossRef] [PubMed]
6. Wu, G.; Guo, Z.; Shi, X.; Chen, Q.; Xu, Y.; Shibasaki, R.; Shao, X. A boundary regulated network for accurate roof segmentation and outline extraction. *Remote Sens.* **2018**, *10*, 1195. [CrossRef]
7. Ming, D.-P.; Luo, J.-C.; Shen, Z.-F.; Wang, M.; Sheng, H. Research on information extraction and target recognition from high resolution remote sensing image. *Sci. Surv. Mapp.* **2005**, *30*, 18–20.
8. Ganin, Y.; Lempitsky, V. N 4-fields: Neural network nearest neighbor fields for image transforms. In Proceedings of the Asian Conference on Computer Vision, Singapore, 1–5 November 2014; Springer: Berlin, Germany, 2014; pp. 536–551.
9. Shen, W.; Wang, X.; Wang, Y.; Bai, X.; Zhang, Z. Deepcontour: A deep convolutional feature learned by positive-sharing loss for contour detection. In Proceedings of the IEEE Conference on Computer Vision and Pattern Recognition, Boston, MA, USA, 7–12 June 2015; pp. 3982–3991.
10. Bertasius, G.; Shi, J.; Torresani, L. Deepedge: A multi-scale bifurcated deep network for top-down contour detection. In Proceedings of the 2015 IEEE Conference on Computer Vision and Pattern Recognition (CVPR), Boston, MA, USA, 7–12 June 2015; pp. 4380–4389.

11. Bertasius, G.; Shi, J.; Torresani, L. High-for-low and low-for-high: Efficient boundary detection from deep object features and its applications to high-level vision. In Proceedings of the IEEE International Conference on Computer Vision, Santiago, Chile, 7–13 December 2015; pp. 504–512.
12. Xie, S.; Tu, Z. Holistically-nested edge detection. In Proceedings of the IEEE International Conference on Computer Vision, Santiago, Chile, 7–13 December 2015; pp. 1395–1403.
13. Liu, Y.; Cheng, M.-M.; Hu, X.; Wang, K.; Bai, X. Richer convolutional features for edge detection. In Proceedings of the 2017 IEEE Conference on Computer Vision and Pattern Recognition (CVPR), Honolulu, HI, USA, 21–26 July 2017; pp. 5872–5881.
14. Martin, D.R.; Fowlkes, C.C.; Malik, J. Learning to detect natural image boundaries using brightness and texture. Advances in Neural Information Processing Systems, Vancouver, BC, Canada, 8–13 December 2003; pp. 1279–1286.
15. Chen, Z.; Zhang, T.; Ouyang, C. End-to-end airplane detection using transfer learning in remote sensing images. *Remote Sens.* **2018**, *10*, 139. [CrossRef]
16. Lin; Huertas; Nevatia. Detection of buildings using perceptual grouping and shadows. In Proceedings of the IEEE Computer Vision & Pattern Recognition, Seattle, WA, USA, 21–23 June 1994.
17. Jaynes, C.O.; Stolle, F.; Collins, R.T. Task driven perceptual organization for extraction of rooftop polygons. In Proceedings of the Second IEEE Workshop on Applications of Computer Vision, Sarasota, FL, USA, 5–7 December 1994; pp. 152–159.
18. Mohan, R.; Nevatia, R. Using perceptual organization to extract 3d structures. *IEEE Trans. Pattern Anal. Mach. Intell.* **1989**, *11*, 1121–1139. [CrossRef]
19. Turker, M.; Koc-San, D. Building extraction from high-resolution optical spaceborne images using the integration of support vector machine (SVM) classification, hough transformation and perceptual grouping. *Int. J. Appl. Earth Obs. Geoinf.* **2015**, *34*, 586–589. [CrossRef]
20. Kim, T.; Muller, J.P. Development of a graph-based approach for building detection. *Image Vis. Comput.* **1999**, *17*, 31–34. [CrossRef]
21. Tao, W.B.; Tian, J.W.; Liu, J. A new approach to extract rectangle building from aerial urban images. In Proceedings of the 2002 6th International Conference on Signal Processing, Beijing, China, 26–30 August 2002; Volume 141, pp. 143–146.
22. Krishnamachari, S.; Chellappa, R. Delineating buildings by grouping lines with mrfs. *IEEE Trans. Image Process.* **2002**, *5*, 1641–1668. [CrossRef] [PubMed]
23. Croitoru, A.; Doytsher, Y. Right-angle rooftop polygon extraction in regularised urban areas: Cutting the corners. *Photogramm. Rec.* **2010**, *19*, 3113–3141. [CrossRef]
24. Cui, S.; Yan, Q.; Reinartz, P. Complex building description and extraction based on hough transformation and cycle detection. *Remote Sens. Lett.* **2012**, *3*, 1511–1559. [CrossRef]
25. Partovi, T.; Bahmanyar, R.; Krauß, T.; Reinartz, P. Building outline extraction using a heuristic approach based on generalization of line segments. *IEEE J. Sel. Top. Appl. Earth Obs. Remote Sens.* **2017**, *10*, 9339–9347. [CrossRef]
26. Su, N.; Yan, Y.; Qiu, M.; Zhao, C.; Wang, L. Object-based dense matching method for maintaining structure characteristics of linear buildings. *Sensors* **2018**, *18*, 1035. [CrossRef] [PubMed]
27. Rüther, H.; Martine, H.M.; Mtalo, E.G. Application of snakes and dynamic programming optimisation technique in modeling of buildings in informal settlement areas. *ISPRS J. Photogramm. Remote Sens.* **2002**, *56*, 269–282. [CrossRef]
28. Peng, J.; Zhang, D.; Liu, Y. An improved snake model for building detection from urban aerial images. *Pattern Recognit. Lett.* **2005**, *26*, 5875–5895. [CrossRef]
29. Ahmadi, S.; Zoej, M.J.V.; Ebadi, H.; Moghaddam, H.A.; Mohammadzadeh, A. Automatic urban building boundary extraction from high resolution aerial images using an innovative model of active contours. *Int. J. Appl. Earth Obs. Geoinf.* **2010**, *12*, 1501–1557. [CrossRef]
30. Garcin, L.; Descombes, X.; Men, H.L.; Zerubia, J. Building detection by markov object processes. In Proceedings of the International Conference on Image Processing, Thessaloniki, Greece, 7–10 October 2001; Volume 562, pp. 565–568.
31. Kass, A. Snake: Active contour models. *Int. J. Comput. Vis.* **1988**, *1*, 321–331. [CrossRef]
32. Zhou, J.Q. Spatial relation-aided method for object-oriented extraction of buildings from high resolution image. *J. Appl. Sci.* **2012**, *30*, 511–516.

33. Tan, Q. Urban building extraction from vhr multi-spectral images using object-based classification. *Acta Geod. Cartogr. Sin.* **2010**, *39*, 618–623.
34. Wu, H.; Cheng, Z.; Shi, W.; Miao, Z.; Xu, C. An object-based image analysis for building seismic vulnerability assessment using high-resolution remote sensing imagery. *Nat. Hazards* **2014**, *71*, 151–174. [CrossRef]
35. Benarchid, O.; Raissouni, N.; Adib, S.E.; Abbous, A.; Azyat, A.; Achhab, N.B.; Lahraoua, M.; Chahboun, A. Building extraction using object-based classification and shadow information in very high resolution multispectral images, a case study: Tetuan, Morocco. *Can. J. Image Process. Comput. Vis.* **2013**, *4*, 1–8.
36. Mariana, B.; Lucian, D. Comparing supervised and unsupervised multiresolution segmentation approaches for extracting buildings from very high resolution imagery. *ISPRS J. Photogramm. Remote Sens.* **2014**, *96*, 67–75.
37. Tao, C.; Tan, Y.; Cai, H.; Du, B.; Tian, J. Object-oriented method of hierarchical urban building extraction from high-resolution remote-sensing imagery. *Acta Geod. Cartogr. Sin.* **2010**, *39*, 394–395.
38. Guo, Z.; Du, S. Mining parameter information for building extraction and change detection with very high-resolution imagery and gis data. *Mapp. Sci. Remote Sens.* **2017**, *54*, 38–63. [CrossRef]
39. Liu, Z.J.; Wang, J.; Liu, W.P. Building extraction from high resolution imagery based on multi-scale object oriented classification and probabilistic hough transform. In Proceedings of the 2005 IEEE International Geoscience and Remote Sensing Symposium (IGARSS'05), Seoul, Korea, 25–29 July 2005; pp. 250–253.
40. Krizhevsky, A.; Sutskever, I.; Hinton, G.E. Imagenet classification with deep convolutional neural networks. In Proceedings of the International Conference on Neural Information Processing Systems, Lake Tahoe, NV, USA, 3–6 December 2012; pp. 1097–1105.
41. Badrinarayanan, V.; Kendall, A.; Cipolla, R. Segnet: A deep convolutional encoder-decoder architecture for scene segmentation. *IEEE Trans. Pattern Anal. Mach. Intell.* **2015**, *39*, 2481–2495. [CrossRef] [PubMed]
42. Huang, Z.; Cheng, G.; Wang, H.; Li, H.; Shi, L.; Pan, C. Building extraction from multi-source remote sensing images via deep deconvolution neural networks. In Proceedings of the Geoscience and Remote Sensing Symposium, Beijing, China, 10–15 July 2016; pp. 1835–1838.
43. Saito, S.; Yamashita, T.; Aoki, Y. Multiple object extraction from aerial imagery with convolutional neural networks. *Electron. Imaging* **2016**, *2016*, 1–9.
44. Zhong, Z.; Li, J.; Cui, W.; Jiang, H. Fully convolutional networks for building and road extraction: Preliminary results. In Proceedings of the Geoscience and Remote Sensing Symposium, Beijing, China, 10–15 July 2016; pp. 1591–1594.
45. Xu, Y.; Wu, L.; Xie, Z.; Chen, Z. Building extraction in very high resolution remote sensing imagery using deep learning and guided filters. *Remote Sens.* **2018**, *10*, 144. [CrossRef]
46. Cao, J.; Chen, Z.; Wang, B. Deep convolutional networks with superpixel segmentation for hyperspectral image classification. In Proceedings of the Geoscience and Remote Sensing Symposium, Beijing, China, 10–15 July 2016; pp. 3310–3313.
47. Zhao, W.; Jiao, L.; Ma, W.; Zhao, J.; Zhao, J.; Liu, H.; Cao, X.; Yang, S. Superpixel-based multiple local cnn for panchromatic and multispectral image classification. *IEEE Trans. Geosci. Remote Sens.* **2017**, *55*, 4141–4156. [CrossRef]
48. Liu, Y.; Cao, G.; Sun, Q.; Siegel, M. Hyperspectral classification via deep networks and superpixel segmentation. *Int. J. Remote Sens.* **2015**, *36*, 3459–3482. [CrossRef]
49. Gao, J.; Wang, Q.; Yuan, Y. Embedding structured contour and location prior in siamesed fully convolutional networks for road detection. In Proceedings of the IEEE International Conference on Robotics and Automation, Singapore, 29 May–3 June 2017; pp. 219–224.
50. Zhou, J.T.; Zhao, H.; Peng, X.; Fang, M.; Qin, Z.; Goh, R.S.M. Transfer hashing: From shallow to deep. *IEEE Trans. Neural Netw. Learn. Syst.* **2018**, *PP*, 1–11. [CrossRef] [PubMed]
51. Peng, X.; Feng, J.; Xiao, S.; Yau, W.Y.; Zhou, J.T.; Yang, S. Structured autoencoders for subspace clustering. *IEEE Trans. Image Process.* **2018**, *27*, 5076–5086. [CrossRef] [PubMed]
52. Huang, X.; Zhang, L. Morphological building/shadow index for building extraction from high-resolution imagery over urban areas. *IEEE J. Sel. Top. Appl. Earth Obs. Remote Sens.* **2012**, *5*, 1611–1672. [CrossRef]
53. Rongming, H.U.; Huang, X.; Huang, Y. An enhanced morphological building index for building extraction from high-resolution images. *Acta Geod. Cartogr. Sin.* **2014**, *43*, 514–520.

54. Huang, X.; Yuan, W.; Li, J.; Zhang, L. A new building extraction postprocessing framework for high-spatial-resolution remote-sensing imagery. *IEEE J. Sel. Top. Appl. Earth Obs. Remote Sens.* **2017**, *10*, 654–668. [CrossRef]
55. Lin, X.; Zhang, J. Object-based morphological building index for building extraction from high resolution remote sensing imagery. *Acta Geod. Cartogr. Sin.* **2017**, *46*, 724–733.
56. Jiménez, L.I.; Plaza, J.; Plaza, A. Efficient implementation of morphological index for building/shadow extraction from remotely sensed images. *J. Supercomput.* **2017**, *73*, 482–489. [CrossRef]
57. Ghandour, A.; Jezzini, A. Autonomous building detection using edge properties and image color invariants. *Buildings* **2018**, *8*, 65. [CrossRef]
58. Cardona, E.U.; Mering, C. Extraction of buildings in very high spatial resolution's geoeye images, an approach through the mathematical morphology. In Proceedings of the Information Systems and Technologies, Nashville, TN, USA, 12–13 November 2016; pp. 1–6.
59. Liow, Y.T.; Pavilidis, T. Use of shadows for extracting buildings in aerial images. *Comput. Vis. Graph. Image Process.* **1989**, *49*, 242–277. [CrossRef]
60. Shi, W.Z.; Mao, Z.Y. Building extraction from high resolution remotely sensed imagery based on shadows and graph-cut segmentation. *Acta Electron. Sin.* **2016**, *69*, 11–13.
61. Wang, L. Development of a multi-scale object-based shadow detection method for high spatial resolution image. *Remote Sens. Lett.* **2015**, *6*, 596–598.
62. Raju, P.L.N.; Chaudhary, H.; Jha, A.K. Shadow analysis technique for extraction of building height using high resolution satellite single image and accuracy assessment. *ISPRS Int. Arch. Photogramm. Remote Sens. Spat. Inf. Sci.* **2014**, *XL-8*, 1185–1192. [CrossRef]
63. Pesaresi, M.; Gerhardinger, A.; Kayitakire, F. A robust built-up area presence index by anisotropic rotation-invariant textural measure. *IEEE J. Sel. Top. Appl. Earth Obs. Remote Sens.* **2009**, *1*, 180–192. [CrossRef]
64. Pesaresi, M.; Gerhardinger, A. Improved textural built-up presence index for automatic recognition of human settlements in arid regions with scattered vegetation. *IEEE J. Sel. Top. Appl. Earth Obs. Remote Sens.* **2011**, *4*, 162–166. [CrossRef]
65. Hu, L.; Zheng, J.; Gao, F. A building extraction method using shadow in high resolution multispectral images. In Proceedings of the Geoscience and Remote Sensing Symposium, Vancouver, BC, Canada, 24–29 July 2011; pp. 1862–1865.
66. Fraser, C. *3D Building Reconstruction from High-Resolution Ikonos Stereo-Imagery*; Automatic Extraction Of Man-Made Objects From Aerial And Space Images (iii); Balkema: London, UK, 2001.
67. Gilani, S.; Awrangjeb, M.; Lu, G. An automatic building extraction and regularisation technique using lidar point cloud data and orthoimage. *Remote Sens.* **2016**, *8*, 27. [CrossRef]
68. Uzar, M.; Yastikli, N. Automatic building extraction using lidar and aerial photographs. *Boletim De Ciências Geodésicas* **2013**, *19*, 153–171. [CrossRef]
69. Awrangjeb, M.; Fraser, C. Automatic segmentation of raw lidar data for extraction of building roofs. *Remote Sens.* **2014**, *6*, 3716–3751. [CrossRef]
70. Shaker, I.F.; Abdelrahman, A.; Abdelgawad, A.K.; Sherief, M.A. Building extraction from high resolution space images in high density residential areas in the great cairo region. *Remote Sens.* **2011**, *3*, 781–791. [CrossRef]
71. Sportouche, H.; Tupin, F.; Denise, L. Extraction and three-dimensional reconstruction of isolated buildings in urban scenes from high-resolution optical and sar spaceborne images. *IEEE Trans. Geosci. Remote Sens.* **2011**, *49*, 3932–3946. [CrossRef]
72. Grigillo, D.; Fras, M.K.; Petrovič, D. *Automated Building Extraction from Ikonos Images in Suburban Areas*; Taylor & Francis, Inc.: London, UK, 2012; pp. 5149–5170.
73. Hu, X.; Ye, L.; Pang, S.; Shan, J. Semi-global filtering of airborne lidar data for fast extraction of digital terrain models. *Remote Sens.* **2015**, *7*, 10996–11015. [CrossRef]
74. Pang, S.; Hu, X.; Wang, Z.; Lu, Y. Object-based analysis of airborne lidar data for building change detection. *Remote Sens.* **2014**, *6*, 10733–10749. [CrossRef]
75. Siddiqui, F.U.; Awrangjeb, M. A novel building change detection method using 3d building models. In Proceedings of the International Conference on Digital Image Computing: Techniques and Applications, Sydney, Australia, 29 November–1 December 2017; pp. 1–8.

76. Yang, B.; Huang, R.; Li, J.; Tian, M.; Dai, W.; Zhong, R. Automated reconstruction of building lods from airborne lidar point clouds using an improved morphological scale space. *Remote Sens.* **2016**, *9*, 14. [CrossRef]
77. Tian, J.; Cui, S.; Reinartz, P. Building change detection based on satellite stereo imagery and digital surface models. *IEEE Trans. Geosc. Remote Sens.* **2013**, *52*, 406–417. [CrossRef]
78. Siddiqui, F.U.; Awrangjeb, M.; Teng, S.W.; Lu, G. A new building mask using the gradient of heights for automatic building extraction. In Proceedings of the International Conference on Digital Image Computing: Techniques and Applications, Gold Coast, Australia, 30 November–2 December 2016; pp. 1–7.
79. Mnih, V. *Machine Learning for Aerial Image Labeling*; University of Toronto: Toronto, ON, Canada, 2013.
80. Simonyan, K.; Zisserman, A. Very deep convolutional networks for large-scale image recognition. *arXiv* **2014**, arXiv:1409.1556.
81. Jia, Y.; Shelhamer, E.; Donahue, J.; Karayev, S.; Long, J.; Girshick, R.; Guadarrama, S.; Darrell, T. Caffe: Convolutional architecture for fast feature embedding. In Proceedings of the 22nd ACM international conference on Multimedia, Orlando, FL, USA, 3–7 November 2014; pp. 675–678.
82. Hermosilla, T.; Ruiz, L.A.; Recio, J.A.; Estornell, J. Evaluation of automatic building detection approaches combining high resolution images and lidar data. *Remote Sens.* **2011**, *3*, 1188–1210. [CrossRef]
83. Zhang, Z.; Zhang, X.; Xin, Q.; Yang, X. Combining the pixel-based and object-based methods for building change detection using high-resolution remote sensing images. *Acta Geod. Cartogr. Sin.* **2018**, *47*, 102–112.
84. Lin, X.; Zhang, J. Extraction of human settlements from high resolution remote sensing imagery by fusing features of right angle corners and right angel sides. *Acta Geod. Cartogr. Sin.* **2017**, *46*, 838–839.
85. Buckland, M.; Gey, F. The relationship between recall and precision. *J. Am. Soc. Inf. Sci.* **1994**, *45*, 12–19. [CrossRef]
86. Zhou, Z. *Machine Learning*; Tsinghua University Press: Beijing, China, 2016.
87. Envi Feature Extraction Module User's Guide. Available online: http://www.harrisgeospatial.com/portals/0/pdfs/envi/Feature_Extracyion_Module.pdf (accessed on 1 December 2008).
88. Deng, S.B.; Chen, Q.J.; Du, H.J. *Envi Remote Sensing Image Processing Method*; Higher Education Press: Beijing, China, 2014.
89. Canny, J. A computational approach to edge detection. In *Readings in Computer Vision*; Elsevier: New York, NY, USA, 1987; pp. 184–203.
90. Zhang, Z.; Schwing, A.G.; Fidler, S.; Urtasun, R. Monocular object instance segmentation and depth ordering with cnns. In Proceedings of the The IEEE International Conference on Computer Vision (ICCV), Santiago, Chile, 13–16 December 2015; pp. 2614–2622.

© 2018 by the authors. Licensee MDPI, Basel, Switzerland. This article is an open access article distributed under the terms and conditions of the Creative Commons Attribution (CC BY) license (http://creativecommons.org/licenses/by/4.0/).

Article

Extracting Building Boundaries from High Resolution Optical Images and LiDAR Data by Integrating the Convolutional Neural Network and the Active Contour Model

Ying Sun [1,2], Xinchang Zhang [3,*], Xiaoyang Zhao [4] and Qinchuan Xin [1,2,*]

1. Department of Geography and Planning, Sun Yat-Sen University, Guangzhou 510275, China; sunying23@mail.sysu.edu.cn
2. Guangdong Key Laboratory for Urbanization and Geo-simulation, Guangzhou 510275, China
3. School of Geographical Sciences, Guangzhou University, Guangzhou 510006, China
4. Guangzhou Urban Planning and Design Survey Research Institute, Guangzhou 510060, China; zhaoxiaoyang@gzpi.com.cn
* Correspondence: eeszxc@mail.sysu.edu.cn (X.Z.); xinqinchuan@gmail.com (Q.X.); Tel.: +86-20-8411-5103 (X.Z.)

Received: 18 July 2018; Accepted: 11 September 2018; Published: 12 September 2018

Abstract: Identifying and extracting building boundaries from remote sensing data has been one of the hot topics in photogrammetry for decades. The active contour model (ACM) is a robust segmentation method that has been widely used in building boundary extraction, but which often results in biased building boundary extraction due to tree and background mixtures. Although the classification methods can improve this efficiently by separating buildings from other objects, there are often ineluctable salt and pepper artifacts. In this paper, we combine the robust classification convolutional neural networks (CNN) and ACM to overcome the current limitations in algorithms for building boundary extraction. We conduct two types of experiments: the first integrates ACM into the CNN construction progress, whereas the second starts building footprint detection with a CNN and then uses ACM for post processing. Three level assessments conducted demonstrate that the proposed methods could efficiently extract building boundaries in five test scenes from two datasets. The achieved mean accuracies in terms of the $F1$ score for the first type (and the second type) of the experiment are 96.43 ± 3.34% (95.68 ± 3.22%), 88.60 ± 3.99% (89.06 ± 3.96%), and 91.62 ±1.61% (91.47 ± 2.58%) at the scene, object, and pixel levels, respectively. The combined CNN and ACM solutions were shown to be effective at extracting building boundaries from high-resolution optical images and LiDAR data.

Keywords: building boundary extraction; convolutional neural network; active contour model; high resolution optical images; LiDAR

1. Introduction

Information regarding the spatiotemporal variation of buildings is important for various applications, such as geodatabase updating, environment management, and urban planning and development. Accompanying the revolutionary development of aerial and space remote sensing technology, identifying and extracting building boundaries from remote sensing data, such as high resolution optical images and recently airborne light detection and ranging (LiDAR) data, is a research frontier in the field of photogrammetry and remote sensing [1–4].

Among the tremendous efforts that have been made to extract building boundaries from remote sensing data [5], the active contour model (ACM) is a widely used method [6,7]. ACM, also referred to as the snake model, is a closed curve extracting method based on the idea of minimizing energy guided by external constraint forces such as lines or edges. ACM could generate smooth and closed object contours with various shapes [8]. Most existing ACMs could be categorized into edge-based and region-based ACMs. In the edge-based models, the contour is guided by the edge information [6]. The edge-based models are sensitive to the initial contour, as they focus on the image pixels, and the ACM contour often docks at the pseudo edges generated by textures [9]. Kabolizade, Ebadi and Ahmadi [10] used an improved snake model for building extraction. Compared with traditional ones, the snake model in their work performed efficiently, as they added a new height similarity energy and regional similarity energy, as well as gradient vector flow. However, their work depends on the initial contour selected. To solve this, Liasis and Stavrou [11] used Hue, Saturation and Value color space as well as the Red, Green, and Blue representation to extract the building boundaries from satellite images by using an ACM. A new energy term is encoded in this work for curve initialization, which leads to higher extraction accuracy. Another solution for curve initialization is to use region-based models which attract the contour by a region descriptor from the global or region context. Chan and Vese [12] presented a region-based active contour model that used a piecewise smooth function. The region-based models are not sensitive to the initial contour, although they are inefficient for the images in which the objects have inhomogeneity textures (i.e., intensity inhomogeneity). Li et al. [13] developed robust a region-scalable fitting (RSF) model that is capable of dealing with intensity inhomogeneity. However, one major limitation of the above-mentioned ACM methods is that confusion caused by trees and ground surfaces could result in errors on identified buildings. To avoid the influence of irrelevant confusing objects, Yan et al. [14] introduced a building model construction framework based on the snake model. They first derived non-terrain objects from LiDAR data and separated buildings from trees, and then extracted and refined the buildings by the snake model. In their work, they made use of a novel graph reduction method to extend the dynamic programming to 2-D planar topology snake model. Bypina and Rajan [15] used the object-based method to extract buildings from very high resolution satellite images, where scene objects are segmented by the Chan-Vese model, and tree objects are removed based on normalized difference vegetation index (NDVI). In practice, separating the buildings from other ground objects such as trees is often difficult by using only a vegetation index.

An effective building footprints detection method could provide helpful information to avoid the effects of other terrain objects, and improve the extraction of building boundaries accordingly. Methods such as the classic hierarchical stripping classification and machine-learning-based classification have been developed to detect building footprints [16–18]. In the classic hierarchical stripping approach, building footprints are separated from vegetation footprints, other off-terrain footprints, and terrain footprints progressively [19]. Awrangjeb and Fraser [20] proposed a method for automatic segmentation of LiDAR data. The ground and the non-ground footprints are separated based on a "building mask". The building roof footprints are then segmented from the non-ground cluster of points and refined by rules. In the method of Wang et al. [21], the building boundaries are detected by a four-step method. The thresholding method is applied to separate footprints with high heights from others. Oriented boundaries are detected by an edge-detection algorithm. Building and non-building objects are classified by two shape measures finally. When extracting building footprints, the hierarchical stripping classification is operationally complicated due to multiple-step operation and manual interaction.

In the past few decades, researches have used the machine learning approaches, such as Artificial Neural Networks (ANN) [22,23], Support Vector Machine (SVM) [24,25], AdaBoost [26] and Random Forests (RF) [27], to extract building footprints. The machine learning approaches could establish a model that detects building footprints by learning the classification rules automatically using training data [28]. Lodha et al. [29,30] employed SVM and AdaBoost classifiers for LiDAR data classification. Du et al. [31] presents a semantic building classification method by using RF classifier from a large number of imbalanced samples. The RF classifiers are improved in two aspects: one is the voting distribution ranked rule for imbalanced samples, and the other is the feature importance measurement. Structured prediction methods, such as Conditional Random Field (CRF), are also used. Niemeyer et al. [32] integrated a RF classifier into a CRF framework, in which the CRF probabilities for the classes are computed using a unary potential and a pairwise potential. The RF approach is more reliable when compared to the linear models for the CRF computation. Overall, the performances of the traditional methods are often dependent on the derived handcraft features. Recently, deep learning has shown a great ability in high level feature extraction or object detection. Vakalopoulou et al. [33] proposed a convolutional neural network (CNN) for deep feature learning. The deep features and additional spectral information were then fed to a SVM classifier for automated building detection, and the result was refined by Markov Random Field. However, they only used CNN for deep features extraction; accordingly, the procedure of feature extraction cannot optimize the classification adaptively. Erhan et al. [34] developed a saliency-inspired neural network for object detection. The network contains several convolutional layers, pooling layers, and full connected layers. Although the abstract features derived from the convolutional layers are helpful to classify the categories of objects in an image, the pooling layers in the architecture reduces the image resolution. Accordingly, the details of the object are lost, and the specific outline of the object cannot be detected well. In essence, classic CNN is more suitable for patch-based image category classification rather than pixel-wise classification. Fully convolutional networks (FCNs) add upsampling layers and convert the full connected layer into the convolutional layers, which could up-sample the feature maps to the original size. Li et al. [35] compared the performance between the fully convolutional network [36] model and shallow models in building detection. A qualitative and quantitative analysis showed that FCN gives better results than shallow models. Although FCN improves the pixel-wise classification, the results are not sensitive enough to the details, and the shapes of the building boundaries are still blurred. Compared with FCN, the symmetrical encoder-decoder network SegNet [37] improves the boundary delineation, and is easy to incorporate into any end-to-end architecture, such as FCN. Although CNN shows robust ability in object classification, it suffers from the "salt and pepper" artifacts inevitably, which in turn affects the detected object boundary.

Recent work has also explored CNN for contour extraction. Maninis et al. [38] proposed an architecture called *convolutional oriented boundaries* for multiscale oriented contours producing. However, the model is designed for natural images. Remote sensing images are often complex scenes, which are not guaranteed to work. Rupprecht et al. [39] developed a deep active contour model. In their work, they predicted the vector point of the contour by a CNN. Nevertheless, they also need an initial curve for image patch deriving, which is costly and time-consuming.

To reduce the influence of other ground objects and "salt and pepper" artifacts, we developed an automatic building boundary extraction method from high-resolution optical images and LiDAR data by integrating CNN and ACM together. We conducted two types of experiments: the first was to extract the building boundaries directly by integrating ACM into CNN construction progress; the second was to use CNN for initial building footprint detection, and apply ACM for the post process.

2. Materials and Methods

2.1. Study Materials

Two different datasets are used in our experiment. The first (hereinafter referred to as the Potsdam dataset) is the ISPRS benchmark data of Potsdam that covers a historical city with large buildings. The dataset contains 38 patches, and each provides high-resolution orthorectified aerial photograph and digital surface models (DSM) with pixel size 6000 × 6000 at the spatial resolution of 5 cm. The aerial photograph has 4 channels: red, green, blue, and near-infrared bands. NDSM is derived based on automatic filtering. The dataset was classified into six land cover classes, of which five classes were merged into non-buildings. Among the 38 patches, 24 patches were labeled by the benchmark test organizers and were used for the training of the CNN, whereas 3 patches (Potsdam 2_13, Potsdam

6_15 and Potsdam 7_13) were used for validation (Figure 1). The ground truths of the three patches are obtained by manual labelling.

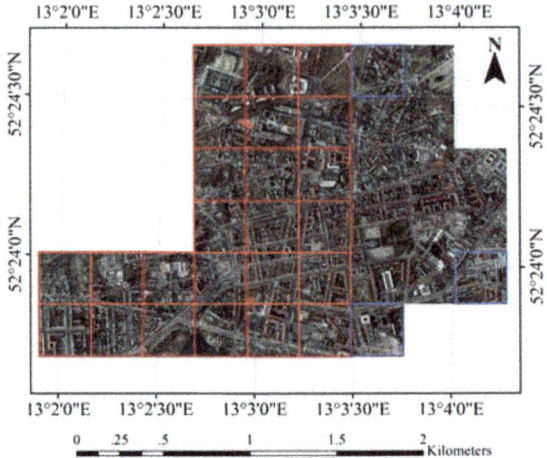

Figure 1. The true color composite image is shown for the Potsdam dataset, where the scenes marked in red are used for the training of the convolutional neural network, and the ones marked in blue are used for validation.

The second dataset (hereinafter referred to as the Marion dataset) that covered Marion in Indiana, USA was downloaded from the Indiana Spatial Data Portal (ISDP). The dataset (Figure 2) includes orthophotography (RGBI) and LiDAR/elevation data. The ground sampling distance of the optical image is about 0.15 m, and the LiDAR data is about 1 point/m^2. We choose seven blocks for CNN training from the Marion County with the size of 10,000 × 10,000 each. We label the images as buildings and non-buildings using the vector data of Open Street Map, as well as by manual labeling. NDSM is derived from the original LiDAR data. The CNN networks are trained by the composite images of RGB+IR+NDSM. The validation data in the Potsdam and Marion datasets have a window size of 2000 × 2000 pixels and 1200 × 1800 pixels, respectively.

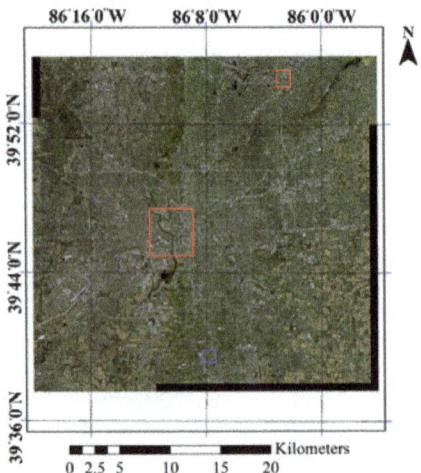

Figure 2. The same as Figure 1 but showing the training and validation data for the Marion dataset.

2.2. Preliminaries

CNN: the encoder-decoder architecture, such as SegNet that is capable of performing semantic pixel labeling of an image, is employed for building footprints detection. For the task of building footprint detection, we can predict the probability that each pixel belongs to a building or non-building in the image by using SegNet. SegNet is a supervised approach with a convolutional-deconvolutional structure. It has a set of convolutional stages, and typically includes fine layers, including the convolutional layer, the activation function layer, the pooling layer, the batch normalization layer, and the up-sample layer. The convolutional layer is the core component in the convolutional stage, and applies a series of filters for feature extraction. The batch normalization layer aims to avoid the vanishing gradients or the explosive gradients. The activation function layer controls the activation level of a neuron for the forward signal transform. A rectified linear unit (ReLU) is often used for non-linear mapping of the input features. The pooling layer generalizes the input features by applying a non-overlapping window to achieve the down-sampled feature maps. The up-sample layer is to resample the feature maps which were down-sampled by the pooling layers to original image sizes. The feature maps are fed into the softmax for pixel-wise classification. A detailed description on SegNet may be found in [37]. The final classification map for a given image can be obtained by calculating the category corresponding to the maximum probability of each pixel.

Active contour model: the ACM method that accounts for both edge and region [40] is employed for the building boundary refinement. Given an image $I(x,y) : \Omega \to R$, $\Omega \to R^n$ is the image domain. Suppose a closed contour $C \to \Omega$, which separates the image into two regions Ω_1 and Ω_2, where Ω_1 and Ω_2 denote the exterior and interior of C, respectively. For a given pixel $x \in \Omega$, the energy function of the ACM is defined as follows:

$$E(C, \overline{f_1}, \overline{f_2}) = \mu \int_C g(|\nabla I[C(s)]|) ds + \sum_{i=1}^{2} \lambda_i \iint_{\Omega_i} K_\sigma(x-y)|I(y) - f_i(x)|^2 dy dx \qquad (1)$$

where, the first term is the edge energy. $g(x) = \frac{1}{1+(x+K)^2}$ is the edge function, and K is the contrast coefficient of the edge function g which is greater than 0. The second term is the RSF energy. The positive parameters μ and λ_i are the weights of the two terms, respectively. $f_i(x)$ is the approximate image intensity inside or outside the contour C. $I(y)$ is the intensity of a local region centered at pixel x, and σ is the size of the region. The bigger that σ is, the higher the calculation complexity of the model.

We employ the variational level set method for the above model solution. The closed contour $C \to \Omega$ is presented by the level set function $\phi \in \Omega$. An arbitrary rectangle is chosen for the initialization of contour C, and the value of level set function ϕ is as follows:

$$\begin{cases} \phi(x,y) > 0 \text{ outside the contour } C \\ \phi(x,y) = 0 \text{ on the contour } C \\ \phi(x,y) < 0 \text{ inside the contour } C \end{cases} \qquad (2)$$

Moreover, we introduced the regularization Heaviside function $H(\phi)$, as well as its derivative $\delta(\phi)$, and added the level set regularization term to Equation (1).

2.3. Building Boundary Extraction Based on CNN and ACM

We developed two strategies for CNN and ACM combination in this study. For the first (CNN_ACM_1), we integrated ACM into CNN construction progress, while the second solution (CNN_ACM_2) starts with CNN for building footprints detection, and then uses ACM for post processing. Figure 3 shows the frame work of the first solution. The optical images and NDSM are fed into the encoder-decoder architecture for deep feature learning. Meanwhile, ACM is used to extract the boundaries features to improve the boundaries perception. The ACM hand-crafted features and CNN deep features are concatenated before the softmax classifier for the final classification.

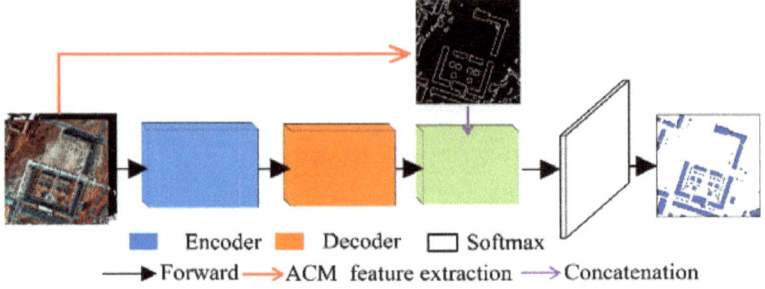

Figure 3. The architecture of the CNN_ACM_1 building boundary extraction method.

Figure 4 illustrates the framework of the second solution. CNN is first applied to detect the candidate building footprints, which are then clustered into subsets for individual building patch generation. Each building boundary is refined by ACM and mosaicked into a whole scene. Details on these processes as follows.

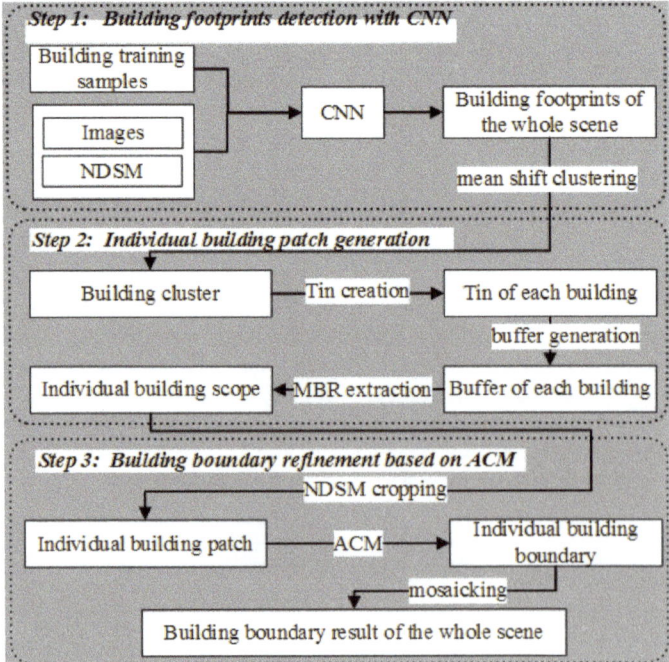

Figure 4. The flowchart of the CNN_ACM_2 building boundary extraction method.

CNN could misclassify pixels, resulting in apparent salt and pepper artifacts; as such, ACM is used to refine the extracted building footprints. To reduce the dimensionality of the ACM searching space, we generate individual building patches from the CNN classification results for feeding into the ACM model. Figure 5 illustrates the detailed procedures to generate individual building patches. Given the remote sensing data, building footprints are first identified based on the mean shift clustering method (Figure 5b). The triangulated irregular network is then established for each individual building footprint using Delaunay triangulation, and the areas of the triangulated irregular network are delineated (Figure 5c). A buffered area (the buffer distance varies from 5–10 m depending on the building sizes in the scene) of the triangulated irregular network (marked with the black curve in Figure 5d) is built as some of the buildings that are not completely detected in CNN, and small footprints less than a priori minimum building area are then deleted. The minimum bounding rectangle (MBR) of the triangulated irregular network area is finally generated for the building patch cropping (Figure 5e, the red rectangle). In the ACM boundary extraction, the edges in the high resolution optical images are often located at the texture changes; however, they appear at the places where the elevation changes in NDSM. Comparatively, the contrast between building objects and ground surfaces is stronger in NDSM than in high resolution optical images, and thus, we employed NDSM for further ACM refinement (Figure 5f). After the boundary extraction, all building patches are mosaicked based on the cropping position to the original scenes.

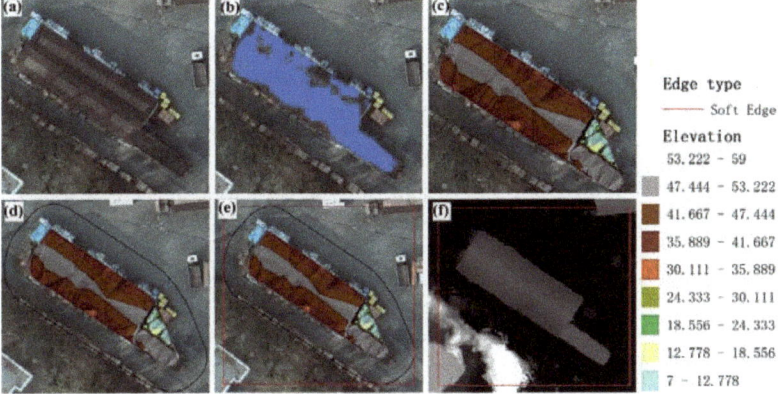

Figure 5. Individual building patch generation. (**a**) The high resolution optical images, (**b**) building footprints detected by CNN and clustered together for an individual building, (**c**) Tin generated based on the individual building footprints, (**d**) the buffer area of the Tin domain (marked with black curve), (**e**) MBR of the buffer (the red rectangle), and (**f**) individual NDSM building patch cropped by the MBR.

2.4. Experiment Setup

Our CNN architecture is running on NVIDIA TITAN X based on Caffe, and the ACM algorithm and the RF classification algorithm are implemented by Matlab R2014a. The remote sensing images in this study are processed by ArcGIS 10.4.1 and ENVI 5.3. The building samples from the ISPRS benchmark dataset and Open street map (OSM) were used for training. High resolution optical images and NDSM are cropped into small patches of 300 × 300 pixels. For the Potsdam and Marion datasets, 8400 and 8092 patches are used for CNN model training, respectively. The trained CNN are then used for mapping building footprints.

To understand the algorithm robustness, the proposed methods are compared with the methods that use CNN [37] or ACM [40], as well as the state of the art classification method, RF [27]. The training and inference manners of RF and CNN are quite different. The stratified random sampling strategy is used for RF method, and the samples are only from the test images. For the ACM method, the entire scene was fed into the ACM model for building boundary extraction. The detected building footprints in the raster format were converted to the vector format. Small objects, i.e., less than the minimum building area, e.g. often cars, small trees, or the salt and pepper noise caused by classification, are removed. All the building boundary results are post-processed using the DP algorithm [41].

2.5. Assessment

Method assessments were conducted at the scene, object, and pixel levels. Detected buildings are split or merged based on the topological relations, as identified by the topological clarification method proposed by Rutzinger et al. [42]. The metrics of *Completeness* (*Comp*), *Correctness* (*Corr*), and *F1*-score (*F1*) were derived as follows:

$$Comp = TP/(TP+FN)$$
$$Corr = TP/(TP+FP)$$
$$F1 = 2 \times \frac{Comp \times Corr}{Comp+Corr} \quad (3)$$

where, *TP*, *FP* and *FN* have different definitions in the three levels, and they are described in more detail below.

At the scene level, we establish correspondences between buildings in the detected results and ground reference by their overlapping rate (Equations (4)). The overlapping rate is derived as follows:

$$R_{\text{overlap}} = A_{\text{overlap}}/A_{\text{ref}} \quad (4)$$

where, A_{overlap} is the overlapping area of the detected building and the corresponding building in the ground reference and A_{ref} is the area of the building in ground reference.

At the scene level, the detected results are categorized based on five different critical thresholds for the overlapping rates (i.e., T_{overlap} = 10%, 30%, 50%, 70%, and 90%). The detected buildings with the overlapping rates larger than the critical threshold are labeled as *TP*, the reference buildings with the overlapping rates lower than the critical threshold are considered as *FN*, and the detected buildings with the overlapping rates lower than the critical threshold are considered as *FP*.

At the object level, we only evaluated each detected building which has an overlap with ground reference data set (i.e., the *TPs* in scene level). The object level metrics give estimates of a single building. Object level *TP* denotes the overlapping area between the detected building and the reference building, *FN* denotes the undetected area of the reference building, and *FP* denotes the falsely detected area of the detected building. With the defined *TP*, *FN*, and *FP*, the metrics of *Comp*, *Corr*, and *F1*-score are first derived for each individual building, and then averaged for all the objects across the scene.

To perform assessments at the pixel level, both the detected results and the reference data are converted to the raster formats and then compared with each other. At the pixel level, the pixel correctly detected as building is referred as *TP*. *FN* denotes the building pixel that is not detected, and *FP* denotes the pixel that is not a building in the reference data, but which was misclassified as building.

The three-level assessment shows the performance of our method in different ways. The scene-based assessment is based on the overlapping area, indicating the accuracy of the whole scene. The object-based metrics can evaluate how a building object can be extracted. Pixel-based metrics are easily done by comparing the detect images and ground truth. However, pixel-based assessment may be distorted owing to the problems of building boundaries [42]. The different metrics are indicative to the algorithm accuracies from different aspects, but should not be compared across different levels.

3. Results

3.1. Building Boundary Extraction Results

Figure 6 shows visual comparisons among methods. ACM often misclassifies tall trees as buildings, and fails to extract buildings of low height due to background confusion. RF can better extract building footprints than ACM, but it frequently generates classification results with apparent "salt and pepper" artifacts. CNN outperforms both ACM and RF in distinguishing trees from buildings, whereas CNN could misclassify the buildings with heterogeneous textures. The methods of both CNN_ACM_1 and CNN_ACM_2 obtain reasonable results, as compared using the algorithms above.

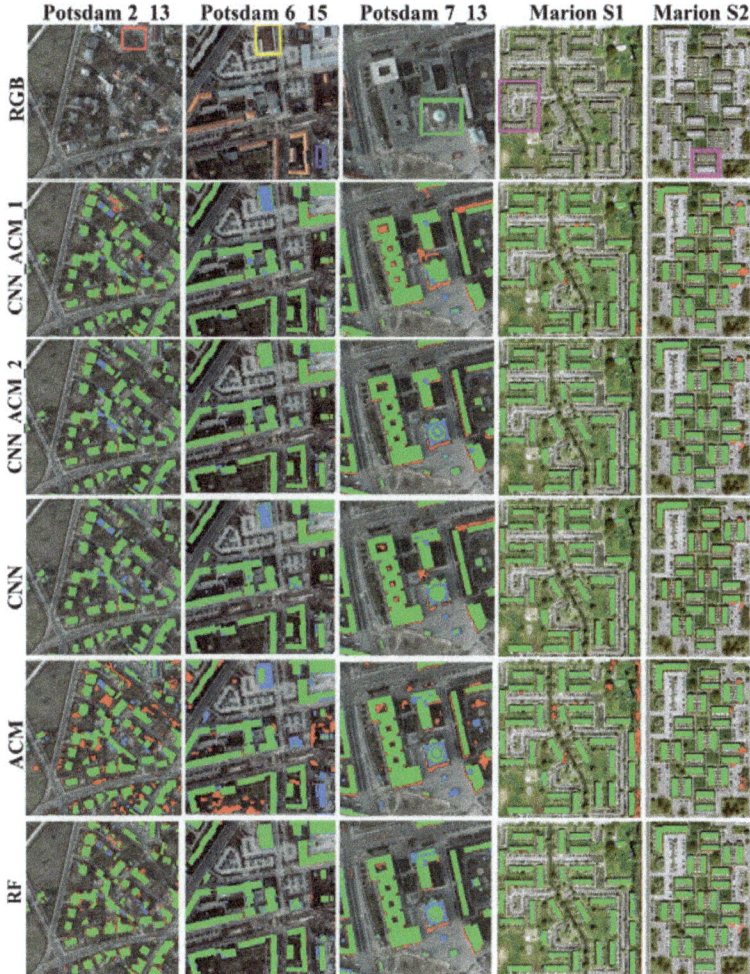

Figure 6. The detected buildings in five test scenes with five different methods. Areas in the green color denote *TP*, areas in the blue color denote *FN*, and areas in the red color denote *FP* at the object level.

As marked with red rectangles in Figure 6, buildings with inconsistent roof texture are rarely extracted correctly in CNN, whereas the use of ACM clearly refines the building boundaries. Figure 7 shows the details of the marked building in Potsdam 2_13. CNN_ACM_2 tracks the boundary fairly well, whereas CNN_ACM_1 can detect the building, but the detected boundary is not accurate enough. ACM underestimates the building and some building footprints are not detected. Both CNN and RF have the salt-and-pepper artifacts. For buildings with vegetation on top of the roof (marked with yellow rectangles in Figure 6), CNN_ACM_2 could provide good results, while CNN_ACM_1 and CNN failed to extract the roof areas covered by vegetation (see details in Figure 7, Potsdam 6_15). Results detected by RF still have the salt and pepper artifacts. The building missed by the other methods as marked by blue rectangles in Figure 6 could be detected well using CNN_ACM_1. For the tower with complex structure in Potsdam 7_13 (marked with green rectangle in Figure 6), CNN_ACM_1 yields a more complete result than other methods. The buildings in Marion dataset have simple structures and similar spectrum. All methods except ACM and RF successfully extracted the building boundaries.

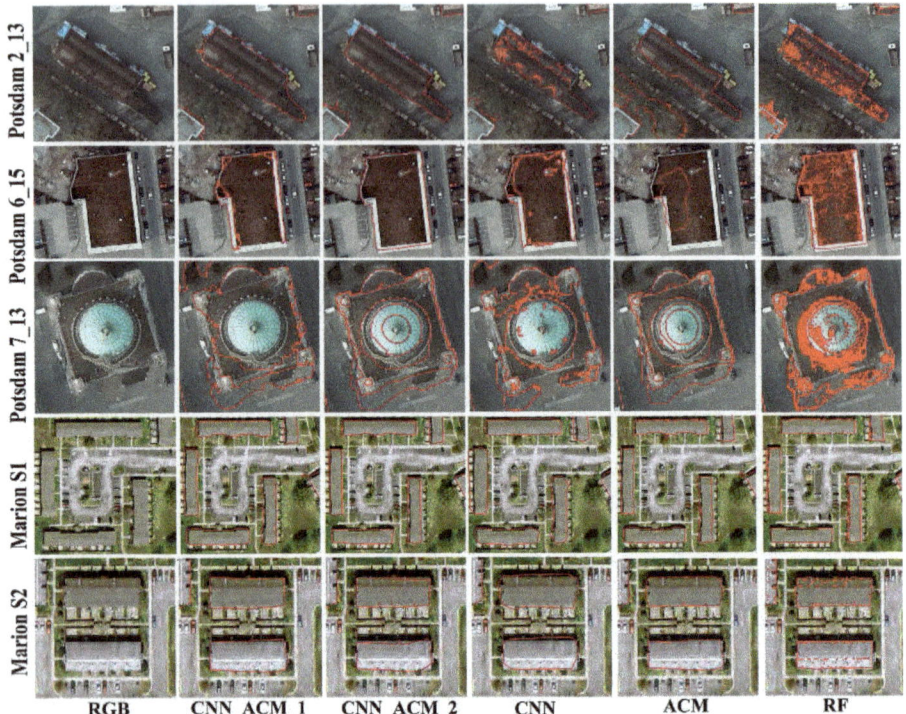

Figure 7. The zoom-ups of the marked buildings in Figure 6 with five different methods.

3.2. Performance Assessment

Figure 8 presents the assessment results of the proposed building boundary extraction methods for five test scenes at the scene level (see the details in Tables A1 and A2). The overlapping thresholds are used to determine whether the detected building is a *TP* at the scene level. This means that if the overlapping rate $R_{overlap}$ of a building is lower than the threshold, it will be considered as an undetected one. Obviously, the methods could detect more *TPs* and achieve higher accuracies using low overlapping thresholds than high overlapping thresholds. For the Potsdam dataset, CNN_ACM_1 achieves the accuracies higher than 90.41% when the overlapping threshold is less than or equal to 30%. For 50–70%, the scene level accuracies are almost all above 82.05%, except Potsdam 6_15 at $T_{overlap}$ = 70%. While for the highest threshold (90%), the average accuracy of the three scenes is 73.22%. When using CNN_ACM_2, similar accuracies were obtained, except for a slight drop in Potsdam 6_15. In the Marion dataset, the accuracies are higher than those of Potsdam, as few buildings are missed in both scenes. CNN_ACM_1 obtains the accuracies of above 98.00% for the overlapping threshold less than or equal to 70%. The accuracies are above 95.00% when assessed by the threshold of $T_{overlap}$ = 90%. CNN_ACM_2 obtains higher accuracies in Marion S1 than CNN_ACM_1, and slightly lower accuracies in Marion S2.

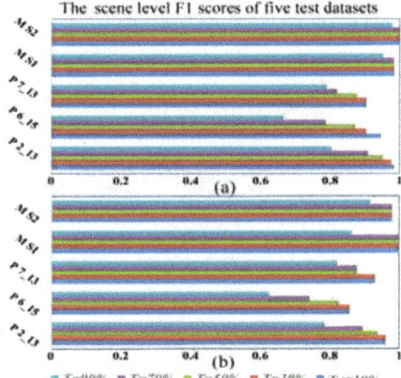

Figure 8. The scene level $F1$ scores of the five test images. (**a**) The accuracies of the method CNN_ACM_1, (**b**) the accuracies of the method CNN_ACM_2. The abbreviation of P denotes Potsdam, the abbreviation of M for Marion, T for the overlapping threshold.

Figure 9 shows the assessment of the extracted building boundaries at the pixel and object levels (see the details in Tables A3 and A4). At the object level, the mean values of *Comp*, *Corr*, and *F1* for all the detected buildings overlapped with ground truth are derived and shown in Figure 9a,b. *Comp* represents the similarity between overlapping area $A_{overlap}$ and ground truth, while *Corr* represents the similarity between overlapping area $A_{overlap}$ and detect results. *F1* can be regarded as a weighted average of *Comp* and *Corr*. For the method of CNN_ACM_1, we can see that the detected buildings have good area similarity compared with ground reference objects: the mean *F1* scores are above 82.98% for all the five test scenes, among which Marion S1 achieves 94.35%. For the methods of CNN_ACM_2, the mean *F1* scores of all the assessed buildings are above 84.15%, and the highest accuracy (93.96%) is also obtained for Marion S1. The accuracies at the pixel level (Figure 9c,d) can be perceived as a kind of average of scene and object level assessment. The average *F1* score of the five test scenes at the pixel level is 91.62% for CNN_ACM_1, and 91.72% for CNN_ACM_2.

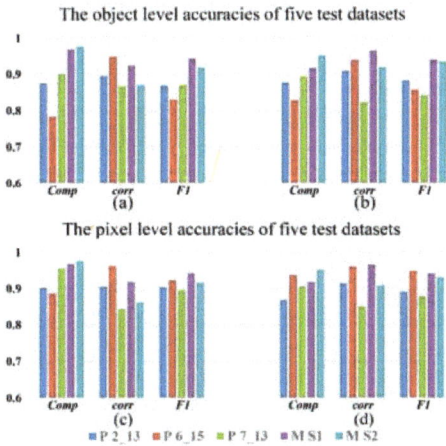

Figure 9. The three metrics of the five test scenes at the object level and the pixel level. (**a**) The object level accuracies of the method CNN_ACM_1, (**b**) the object level accuracies of the method CNN_ACM_2, (**c**) The pixel level accuracies of the method CNN_ACM_1, and (**d**) the pixel level accuracies of the method CNN_ACM_2. The abbreviations of P, M and T are the same as Figure 8.

3.3. Comparative Analysis

Figure 10 compares the assessment results of different building boundary extraction methods across two datasets. The horizontal axis denotes the assessment level, namely, the object level, the pixel level, and the scene level with five different overlapping thresholds. The vertical axis denotes the accuracies of *F*1 scores.

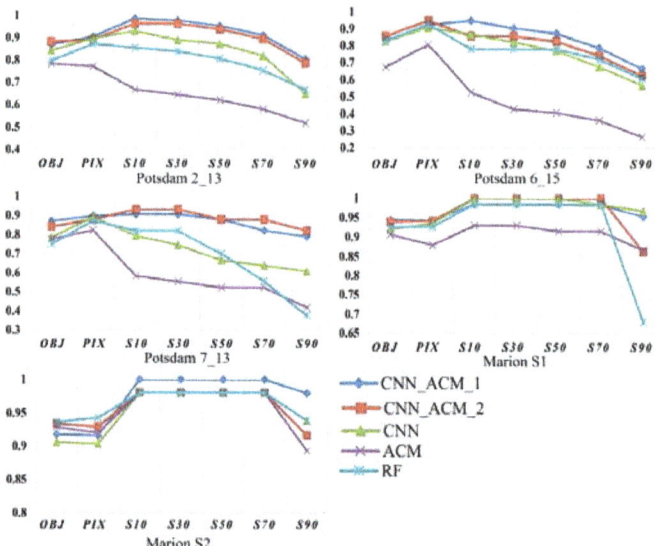

Figure 10. Assessments using the two datasets are compared for the building boundary extraction methods, including the proposed methods, CNN, RF, and ACM. The abbreviation of OBJ denotes results at the object level, the abbreviation of PIX for pixel-based assessment, S10 for scene-based assessment with the overlapping threshold of 10%, and so on.

For the scene of Potsdam 2_13, CNN_ACM_1 performs the best in the five scene level assessments, and the method of CNN_ACM_2 comes second. This means that CNN_ACM_1 can detect more buildings which overlap with ground truths than other methods. At the object level, CNN_ACM_1 also works the best. Higher object-level accuracy implies that the detected buildings have better area similarity with ground truth. The other methods, CNN, ACM and RF, all work worse than our proposed method on all the three levels. In Potsdam 6_15, CNN_ACM_1 performs the best in all the five scene level assessments. This is because CNN_ACM_1 detects several small buildings which other methods do not extract. However, the detected building boundaries are poorer than CNN_ACM_2, as shown in Figure 6. At the object level and pixel level, CNN_ACM_2 undoubtedly achieves the best results. The accuracy of CNN_ACM_1 is slightly higher than that of CNN and RF, and ACM is the worst. In Potsdam 7_13, the opposite result is obtained. CNN_ACM_2 detects more buildings, but the building shapes are worse than with CNN_ACM_1. In Marion S1, CNN_ACM_2 and CNN performs best in the scene level assessments. CNN_ACM_1 and RF miss a small building, and their accuracies are a bit worse. ACM also obtains the worst accuracy. For Marion S2, the accuracy of RF is as good as CNN_ACM_1, except $T_{overlap} = 90\%$. The other three methods show the same ability in the scene level. CNN_ACM_1 achieves the best object level accuracy, and RF obtains the highest pixel level accuracy, respectively. Overall, our proposed methods are effective for buildings under various scenes. CNN_ACM_1 obtains the best results at the scene level, and CNN_ACM_2 is good at the object level. CNN and RF only attain satisfactory results in simple building types.

4. Discussion

In practice, most building footprints can be detected by CNN, which shows a powerful ability in distinguishing buildings and vegetation. However, salt-and-pepper artifacts remain inside a building or on the building boundaries in the classification results. Accordingly, the completeness of a building needs to be improved to some extent. As reported in Section 3, the introduction of ACM improves the accuracies obviously when the footprints of a building are partly missed in CNN classification. On the whole, the integrated solution of CNN_ACM_1 works the best, except in the case of buildings with vegetation on the roof, as it can detect more building areas than other methods. CNN_ACM_2 also performs well on the building boundary refinement, which benefits from the excellent edge extraction capability of ACM, as the contour of ACM can stop at the relative reliable building edges. Moreover, the individual building patch generation process reduces the calculation range of ACM. The method of RF can obtain good results in simple scenarios. However, it has a more severe salt and pepper effect than CNN. The method of ACM is often influenced by other ground objects such as trees. In terms of the performance of the proposed methods in the two datasets, the results for the Marion dataset are better than Potsdam in almost all the three assessment levels. Buildings with diverse shapes and different spectral in Potsdam make it harder for accurate extraction, while the simple structures and spectral characteristics of buildings in Marion resulted in high accuracy.

Although the proposed models perform well, further improvements are needed. First, the generalization ability of the network should be improved. CNN_ACM_1 shows poor handling capacity in case of buildings with vegetation on the roof. This is mainly due to the different data distribution of the training data and the test scene, although they have the same data sources. The reason that RF can detect this kind of building is attributed to the sampling strategies: it selects samples from the very classification images. Second, a softer and more effective building boundary regularization method is required. The DP regularization algorithm reduces the building extraction results to some extent.

5. Conclusions

We developed a method for building boundary extraction using CNN and ACM. Two kinds of strategies are designed. The first employs ACM for boundary feature extraction, which is then fed to the CNN architecture. The second starts building footprints detection with CNN classification, and then clusters the footprints to obtain subsets of candidate buildings, from which the buffer of every building is constructed and the MBR is derived. Next, the NDSM of the scene are cropped by the MBRs. Finally, the cropped NDSMs are fed to the ACM for building boundary refinement, and mosaicked into a whole scene based on their original positions. The benefits of our method are as follows: (1) the proposed solution can reduce the influence of vegetation and salt and pepper artifacts. (2) It can extract buildings which are similar to the ground surfaces, which are missed in the other methods. When testing two datasets with various building shapes, we obtained better results than other three methods in the five test scenarios. In the future, we hope to extend our method to other complex building types, such as the archaeological buildings.

Author Contributions: Y.S. was responsible for the conceptualization and the methodology, and wrote the original draft; X.Z. (Xinchang Zhang) acquired the funding and supervised the study; X.Z. (Xiaoyang Zhao) contributed materials, performed the experiments of Section 3.1 and contributed to the figures; Q.X. reviewed and edited the draft, and contributed to the article's organization.

Funding: This research was funded by the National Natural Science Foundation of China (grant Nos. 41801351 41431178, and 41875122), the Natural Science Foundation of Guangdong Province, China (grant No. 2016A030311016), the National Administration of Surveying, Mapping and Geoinformation of China (grant No. GZIT2016-A5-147), and the Key Projects for Young Teachers at Sun Yat-sen University (grant No. 17lgzd02).

Acknowledgments: The authors wish to thank the study material providers. The Potsdam data were produced by International Society for Photogrammetry and Remote Sensing: http://www2.isprs.org/commissions/comm3/wg4/2d-sem-label-potsdam.html. The Marion data were obtained from the Indiana Spatial Data Portal (ISDP): http://gis.iu.edu/datasetInfo/index.php. We also would like to thank the anonymous reviewers for their constructive comments.

Conflicts of Interest: The authors declare no conflict of interest.

Appendix A

Table A1. Accuracies of CNN_ACM_1 at the scene level.

Scenes	Metrics	Overlapping Threshold				
		10%	30%	50%	70%	90%
Potsdam 2_13	Comp	0.9701	0.9552	0.9104	0.8358	0.6716
	Corr	1.0000	1.0000	1.0000	1.0000	1.0000
	F1 score	0.9848	0.9771	0.9531	0.9106	0.8036
Potsdam 6_15	Comp	0.9730	0.8919	0.8378	0.7027	0.5405
	Corr	0.9231	0.9167	0.9118	0.8966	0.8696
	F1 score	0.9474	0.9041	0.8732	0.7879	0.6667
Potsdam 7_13	Comp	0.9048	0.9048	0.8571	0.7619	0.7143
	Corr	0.9048	0.9048	0.9000	0.8889	0.8824
	F1 score	0.9048	0.9048	0.8780	0.8205	0.7895
Marion S1	Comp	0.9697	0.9697	0.9697	0.9697	0.9091
	Corr	1.0000	1.0000	1.0000	1.0000	1.0000
	F1 score	0.9846	0.9846	0.9846	0.9846	0.9524
Marion S2	Comp	1.0000	1.0000	1.0000	1.0000	0.9600
	Corr	1.0000	1.0000	1.0000	1.0000	1.0000
	F1 score	1.0000	1.0000	1.0000	1.0000	0.9796

Table A2. Accuracies of CNN_ACM_2 at the scene level.

Scenes	Metrics	Overlapping Threshold				
		10%	30%	50%	70%	90%
Potsdam 2_13	Comp	0.9403	0.9403	0.8955	0.8209	0.6567
	Corr	0.9844	0.9844	0.9836	0.9821	0.9778
	F1 score	0.9618	0.9618	0.9375	0.8943	0.7857
Potsdam 6_15	Comp	0.8919	0.8919	0.8378	0.7027	0.5405
	Corr	0.8250	0.8250	0.8158	0.7879	0.7407
	F1 score	0.8571	0.8571	0.8267	0.7429	0.6250
Potsdam 7_13	Comp	0.9524	0.9524	0.8571	0.8571	0.7619
	Corr	0.9091	0.9091	0.9000	0.9000	0.8889
	F1 score	0.9302	0.9302	0.8780	0.8780	0.8205
Marion S1	Comp	1.0000	1.0000	1.0000	1.0000	0.7576
	Corr	1.0000	1.0000	1.0000	1.0000	1.0000
	F1 score	1.0000	1.0000	1.0000	1.0000	0.8621
Marion S2	Comp	1.0000	1.0000	1.0000	1.0000	0.8800
	Corr	0.9615	0.9615	0.9615	0.9615	0.9565
	F1 score	0.9804	0.9804	0.9804	0.9804	0.9167

Table A3. Accuracies of the proposed method at the object level.

Scenes	CNN_ACM_1			CNN_ACM_2		
	Mean_Comp	Mean_Corr	Mean_F1	Mean_Comp	Mean_Corr	Mean_F1
Potsdam 2_13	0.8752	0.8949	0.8693	0.8769	0.9086	0.8822
Potsdam 6_15	0.7827	0.9481	0.8298	0.8278	0.9386	0.8567
Potsdam 7_13	0.9009	0.8669	0.8701	0.8948	0.8226	0.8415
Marion S1	0.9681	0.9235	0.9435	0.9170	0.9646	0.9396
Marion S2	0.9756	0.8704	0.9173	0.9514	0.9181	0.9333

Table A4. Accuracies of the proposed method at the pixel level.

Scenes	CNN_ACM_1			CNN_ACM_2		
	Comp	*Corr*	*F1*	*Comp*	*Corr*	*F1*
Potsdam 2_13	0.9021	0.9054	0.9038	0.8678	0.9140	0.8903
Potsdam 6_15	0.8866	0.9626	0.9230	0.9369	0.9601	0.9483
Potsdam 7_13	0.9555	0.8438	0.8962	0.9058	0.8509	0.8775
Marion S1	0.9679	0.9187	0.9427	0.9184	0.9654	0.9413
Marion S2	0.9755	0.8621	0.9153	0.9511	0.9078	0.9290

References

1. Awrangjeb, M. Using point cloud data to identify, trace, and regularize the outlines of buildings. *Int. J. Remote Sens.* **2016**, *37*, 551–579. [CrossRef]
2. Laefer, D.F.; Hinks, T.; Carr, H.; Truong-Hong, L. New advances in automated urban modelling from airborne laser scanning data. *Recent Pat. Eng.* **2011**, *5*, 196–208. [CrossRef]
3. Awrangjeb, M.; Lu, G.; Fraser, C. Automatic building extraction from LiDAR data covering complex urban scenes. *Int. Arch. Photogramm. Remote Sens. Spat. Inf. Sci.* **2014**, *40*, 25. [CrossRef]
4. Von Schwerin, J.; Richards-Rissetto, H.; Remondino, F.; Spera, M.G.; Auer, M.; Billen, N.; Loos, L.; Stelson, L.; Reindel, M. Airborne LiDAR acquisition, post-processing and accuracy-checking for a 3D WebGIS of Copan, Honduras. *J. Archaeol. Sci. Rep.* **2016**, *5*, 85–104. [CrossRef]
5. Tomljenovic, I.; Höfle, B.; Tiede, D.; Blaschke, T. Building extraction from airborne laser scanning data: An analysis of the state of the art. *Remote Sens.* **2015**, *7*, 3826–3862. [CrossRef]
6. Kass, M.; Witkin, A.; Terzopoulos, D. Snakes: Active contour models. *Int. J. Comput. Vis.* **1988**, *1*, 321–331. [CrossRef]
7. Ahmadi, S.; Zoej, M.V.; Ebadi, H.; Moghaddam, H.A.; Mohammadzadeh, A. Automatic urban building boundary extraction from high resolution aerial images using an innovative model of active contours. *Int. J. Appl. Earth Obs. Geoinf.* **2010**, *12*, 150–157. [CrossRef]
8. Chan, T.F.; Vese, L.A. *Image Segmentation Using Level Sets and the Piecewise-Constant Mumford-Shah Model*; UCLA CAM Report 00-14; Kluwer Academic Publishers: Alphen aan den Rijn, The Netherlands, 2000.
9. He, L.; Peng, Z.; Everding, B.; Wang, X.; Han, C.Y.; Weiss, K.L.; Wee, W.G. A comparative study of deformable contour methods on medical image segmentation. *Image Vis. Comput.* **2008**, *26*, 141–163. [CrossRef]
10. Kabolizade, M.; Ebadi, H.; Ahmadi, S. An improved snake model for automatic extraction of buildings from urban aerial images and LiDAR data. *Comput. Environ. Urban Syst.* **2010**, *34*, 435–441. [CrossRef]
11. Liasis, G.; Stavrou, S. Building extraction in satellite images using active contours and colour features. *Int. J. Remote Sens.* **2016**, *37*, 1127–1153. [CrossRef]
12. Chan, T.F.; Vese, L.A. Active contours without edges. *IEEE Trans. Image Process.* **2001**, *10*, 266–277. [CrossRef] [PubMed]
13. Li, C.; Kao, C.-Y.; Gore, J.C.; Ding, Z. Minimization of region-scalable fitting energy for image segmentation. *IEEE Trans. Image Process.* **2008**, *17*, 1940–1949. [CrossRef] [PubMed]
14. Yan, J.; Zhang, K.; Zhang, C.; Chen, S.-C.; Narasimhan, G. Automatic construction of 3-D building model from airborne LiDAR data through 2-D snake algorithm. *IEEE Trans. Geosci. Remote Sens.* **2015**, *53*, 3–14. [CrossRef]
15. Bypina, S.K.; Rajan, K. Semi-automatic extraction of large and moderate buildings from very high-resolution satellite imagery using active contour model. In Proceedings of the IEEE International Geoscience and Remote Sensing Symposium, Milan, Italy, 26–31 July 2015; pp. 1885–1888. [CrossRef]
16. Dai, Y.; Gong, J.; Li, Y.; Feng, Q. Building segmentation and outline extraction from UAV image-derived point clouds by a line growing algorithm. *Int. J. Digit. Earth* **2017**, *10*, 1077–1097. [CrossRef]
17. Rottensteiner, F.; Sohn, G.; Gerke, M.; Wegner, J.D.; Breitkopf, U.; Jung, J. Results of the ISPRS benchmark on urban object detection and 3D building reconstruction. *ISPRS J. Photogramm. Remote Sens.* **2014**, *93*, 256–271. [CrossRef]

18. Mongus, D.; Lukač, N.; Žalik, B. Ground and building extraction from LiDAR data based on differential morphological profiles and locally fitted surfaces. *ISPRS J. Photogramm. Remote Sens.* **2014**, *93*, 145–156. [CrossRef]
19. Shan, J.; Sampath, A. Building extraction from LiDAR point clouds based on clustering techniques. In *Topographic Laser Ranging and Scanning: Principles and Processing*; Toth, C.K., Shan, J., Eds.; CRC Press: Boca Raton, FL, USA, 2008; pp. 423–446.
20. Awrangjeb, M.; Fraser, C.S. Automatic segmentation of raw LiDAR data for extraction of building roofs. *Remote Sens.* **2014**, *6*, 3716–3751. [CrossRef]
21. Wang, R.; Hu, Y.; Wu, H.; Wang, J. Automatic extraction of building boundaries using aerial LiDAR data. *J. Appl. Remote Sens.* **2016**, *10*, 016022. [CrossRef]
22. Fukushima, K.; Miyake, S.; Ito, T. Neocognitron: A neural network model for a mechanism of visual pattern recognition. *IEEE Trans. Syst. Man Cybern.* **1983**, *SMC-13*, 826–834. [CrossRef]
23. Lari, Z.; Ebadi, H. Automatic extraction of building features from high resolution satellite images using artificial neural networks. In Proceedings of the ISPRS Conference on Information Extraction from SAR and Optical Data, with Emphasis on Developing Countries, Istanbul, Turkey, 16–18 May 2007.
24. Vapnik, V.N. An overview of statistical learning theory. *IEEE Trans. Neural Netw.* **1999**, *10*, 988–999. [CrossRef] [PubMed]
25. Turker, M.; Koc-San, D. Building extraction from high-resolution optical spaceborne images using the integration of support vector machine (SVM) classification, Hough transformation and perceptual grouping. *Int. J. Appl. Earth Obs. Geoinf.* **2015**, *34*, 58–69. [CrossRef]
26. Freund, Y.; Schapire, R.E. A decision-theoretic generalization of on-line learning and an application to boosting. *J. Comput. Syst. Sci.* **1997**, *55*, 119–139. [CrossRef]
27. Ho, T.K. The random subspace method for constructing decision forests. *IEEE Trans Pattern Anal. Mach. Intell.* **1998**, *20*, 832–844. [CrossRef]
28. Guo, B.; Huang, X.; Zhang, F.; Sohn, G. Classification of airborne laser scanning data using JointBoost. *ISPRS J. Photogramm. Remote Sens.* **2015**, *100*, 71–83. [CrossRef]
29. Lodha, S.K.; Kreps, E.J.; Helmbold, D.P.; Fitzpatrick, D.N. Aerial LiDAR Data Classification Using Support Vector Machines (SVM). In Proceedings of the 3rd International Symposium on 3D Data Processing, Visualization and Transmission (3DPVT 2006), Chapel Hill, NC, USA, 14–16 June 2006; pp. 567–574. [CrossRef]
30. Lodha, S.K.; Fitzpatrick, D.M.; Helmbold, D.P. Aerial lidar data classification using AdaBoost. In Proceedings of the Sixth International Conference on 3-D Digital Imaging and Modeling, Montreal, QC, Canada, 21–23 August 2007; pp. 435–442. [CrossRef]
31. Du, S.; Zhang, F.; Zhang, X. Semantic classification of urban buildings combining VHR image and GIS data: An improved random forest approach. *ISPRS J. Photogramm. Remote Sens.* **2015**, *105*, 107–119. [CrossRef]
32. Niemeyer, J.; Rottensteiner, F.; Soergel, U. Contextual classification of lidar data and building object detection in urban areas. *ISPRS J. Photogramm. Remote Sens.* **2014**, *87*, 152–165. [CrossRef]
33. Vakalopoulou, M.; Karantzalos, K.; Komodakis, N.; Paragios, N. Building detection in very high resolution multispectral data with deep learning features. In Proceedings of the IEEE International Geoscience and Remote Sensing Symposium, Milan, Italy, 26–31 July 2015; pp. 1873–1876. [CrossRef]
34. Erhan, D.; Szegedy, C.; Toshev, A.; Anguelov, D. Scalable object detection using deep neural networks. *arXiv* **2014**, arXiv:1312.2249.
35. Li, Y.; He, B.; Long, T.; Bai, X. Evaluation the performance of fully convolutional networks for building extraction compared with shallow models. In Proceedings of the IEEE International Geoscience and Remote Sensing Symposium, Fort Worth, TX, USA, 23–28 July 2017; pp. 850–853. [CrossRef]
36. Long, J.; Shelhamer, E.; Darrell, T. Fully convolutional networks for semantic segmentation. In Proceedings of the IEEE Conference on Computer Vision and Pattern Recognition, Boston, MA, USA, 7–12 June 2015; pp. 3431–3440.
37. Badrinarayanan, V.; Kendall, A.; Cipolla, R. Segnet: A deep convolutional encoder-decoder architecture for image segmentation. *IEEE Trans. Pattern Anal. Mach. Intell.* **2017**, *39*, 2481–2495. [CrossRef] [PubMed]
38. Maninis, K.-K.; Pont-Tuset, J.; Arbeláez, P.; Van Gool, L. Convolutional oriented boundaries: From image segmentation to high-level tasks. *IEEE Trans. Pattern Anal. Mach. Intell.* **2018**, *40*, 819–833. [CrossRef] [PubMed]

39. Rupprecht, C.; Huaroc, E.; Baust, M.; Navab, N. Deep active contours. *arXiv* **2016**, arXiv:1607.05074.
40. Jing, Y.; An, J.; Liu, Z. A novel edge detection algorithm based on global minimization active contour model for oil slick infrared aerial image. *IEEE Trans. Geosci. Remote Sens.* **2011**, *49*, 2005–2013. [CrossRef]
41. Douglas, D.H.; Peucker, T.K. Algorithms for the reduction of the number of points required to represent a digitized line or its caricature. *Cartogr. Int. J. Geogr. Inf. Geovisual.* **1973**, *10*, 112–122. [CrossRef]
42. Rutzinger, M.; Rottensteiner, F.; Pfeifer, N. A comparison of evaluation techniques for building extraction from airborne laser scanning. *IEEE J. Sel. Top. Appl. Earth Obs. Remote Sens.* **2009**, *2*, 11–20. [CrossRef]

© 2018 by the authors. Licensee MDPI, Basel, Switzerland. This article is an open access article distributed under the terms and conditions of the Creative Commons Attribution (CC BY) license (http://creativecommons.org/licenses/by/4.0/).

Article

A Boundary Regulated Network for Accurate Roof Segmentation and Outline Extraction

Guangming Wu [1], Zhiling Guo [1], Xiaodan Shi [1], Qi Chen [1,2], Yongwei Xu [1], Ryosuke Shibasaki [1] and Xiaowei Shao [1,*]

[1] Center for Spatial Information Science, University of Tokyo, Kashiwa 277-8568, Japan; huster-wgm@csis.u-tokyo.ac.jp (G.W.); guozhilingcc@csis.u-tokyo.ac.jp (Z.G.); shixiaodan@csis.u-tokyo.ac.jp (X.S.); qichen@csis.u-tokyo.ac.jp (Q.C.); xyw@csis.u-tokyo.ac.jp (Y.X.); shiba@csis.u-tokyo.ac.jp (R.S.)
[2] Faculty of Information Engineering, China University of Geosciences (Wuhan), Wuhan 430074, China
* Correspondence: shaoxw@csis.u-tokyo.ac.jp; Tel.: +81-04-7136-4390

Received: 16 June 2018; Accepted: 26 July 2018; Published: 30 July 2018

Abstract: The automatic extraction of building outlines from aerial imagery for the purposes of navigation and urban planning is a long-standing problem in the field of remote sensing. Currently, most methods utilize variants of fully convolutional networks (FCNs), which have significantly improved model performance for this task. However, pursuing more accurate segmentation results is still critical for additional applications, such as automatic mapping and building change detection. In this study, we propose a boundary regulated network called BR-Net, which utilizes both local and global information, to perform roof segmentation and outline extraction. The BR-Net method consists of a shared backend utilizing a modified U-Net and a multitask framework to generate predictions for segmentation maps and building outlines based on a consistent feature representation from the shared backend. Because of the restriction and regulation of additional boundary information, the proposed model can achieve superior performance compared to existing methods. Experiments on an aerial image dataset covering 32 km^2 and containing more than 58,000 buildings indicate that our method performs well at both roof segmentation and outline extraction. The proposed BR-Net method significantly outperforms the classic FCN8s model. Compared to the state-of-the-art U-Net model, our BR-Net achieves 6.2% (0.869 vs. 0.818), 10.6% (0.772 vs. 0.698), and 8.7% (0.840 vs. 0.773) improvements in F1 score, Jaccard index, and kappa coefficient, respectively.

Keywords: roof segmentation; outline extraction; convolutional neural network; boundary regulated network; very high resolution imagery

1. Introduction

In the field of remote sensing, for applications such as urban planning, land use analysis, and automatic updating or generation of maps, automatic extraction of building outlines is a long-standing problem. Recent years, based on the rapid development of imaging sensors and operating platforms, a dramatic increase in the availability and accessibility of very high resolution (VHR) remote sensing imagery has made this problem increasingly urgent [1]. Extracting building outlines directly from images containing various backgrounds is very challenging because of the complexity of color, luminance, and texture conditions. A two-step approach that first segments building roofs and then generates outlines according to the segmentation results is more appropriate for this problem.

Based on the scale, resolution, and precision level of extracted data, various methods and algorithms have been proposed for segmenting VHR images [2]. These methods have achieved acceptable precision levels that solve the aforementioned problem to some extent.

However, for additional applications, such as building change detection and automatic mapping, more accurate and robust methods are required.

According to the sources of the data, existing methods can be categorized as three groups: (1) image only [3]; (2) Light Detection and Ranging (LiDAR) point cloud only [4]; and (3) combination of both image and point cloud [5,6]. Based on the algorithms for segmentation, these methods can also be divided into two groups: (1) non-classification-based methods; and (2) classification-based methods. For non-classification-based methods, segmentation is performed by: (a) analyzing pixels values or histograms to determine a threshold [7]; (b) detecting edges utilizing edge detectors [8]; or (c) utilizing region information [9,10]. Classification-based methods produce segmentations of an image by classifying every pixel. Classification-based methods will first learn a pattern according to ground truth data and then apply it to new images. Because these patterns can be adjusted based on the ground truth data, learning-based methods have achieved superior performance in terms of generalization and precision [11–13].

Prior to the introduction of convolutional neural networks (CNNs), classification-based methods extract features from image by utilizing hand-crafted descriptors [14–17] and produce classification result by utilizing various classifiers [18–20]. Because the type and parameters of a descriptor are manually selected and optimized, an optimal solution typically requires significant trial-and-error testing, which is labor intensive and lacks generalization ability. Rather than utilizing hand-crafted descriptors, CNN methods automatically extract features and perform classification by utilizing convolutional, subsampling, and fully-connected layers [21]. Because the feature extraction patterns are learned directly from the data, CNNs have superior generalization capability and precision [22].

Since AlexNet overwhelmingly won the Large Scale Visual Recognition Challenge 2010 (LSVRC-2010) and 2012 [23], and based on the availability of open-source large-scale annotated datasets [24–26], CNN-based algorithms have become the gold standard in many computer vision tasks, such as image classification, object detection, and image segmentation. Initially, researchers mainly applied patch-based CNN methods to detecting or segmenting buildings in aerial or satellite images [27] and significantly improved classification performance. However, owing to extreme memory costs and low computational efficiency, fully convolutional networks (FCNs) [28] have recently attracted more attention in this area. Instead of utilizing small patches and fully-connected layers to predict the class of a pixel, FCN methods utilize sequential convolutional, subsampling, and upsampling operations to generate pixel-to-pixel translations between input and output images. Because no patches or fully-connected layers are required, FCN methods greatly reduce memory costs and the number of parameters, which significantly improves processing efficiency [29]. The classical FCN simply performs single (FCN32s) or multiple (FCN16s and FCN8s) instances of upsampling of subsampled layers to generate predictions for input images of consistent height and width. Because of the information loss caused by the subsampling and upsampling operations, the prediction results of FCN models often have blurred edges and low precision.

To overcome the limitations of the basic FCN model, some novel FCN-based methods have been introduced to improve model performance. In place of the traditional upsampling operations, the SegNet [30] adopts an unsampling operation that records pooling indices during the pooling stage and then applies them during upsampling. The DeconvNet [31] method introduces a novel deconvolution layer that can produce upsampled results utilizing convolution transpose operations. Both unsampling and deconvolution partially solve the information loss caused by upsampling operations, which leads to superior performance. Other methods, such as U-Net [32] and FPN [33], adopt skip connections that utilize both the lower and upper layers to generate a final output, resulting in superior performance. The MC-FCN [34] method utilizes multi-constraints to prevent bias and improve precision.

These methods have improved the traditional FCN model through various innovative techniques and achieved state-of-the art performance. However, these techniques either focus on replacing bilinear upsampling with more information-preserving methods (SegNet and DeconvNet) or adding skip-connections/constraints (U-Net and MC-FCN) to achieve better utilization of the feature representation capability of hidden layers. Another critical issue in FCN-based still exists. Regardless of how these models generate predictions, for each pixel, its value is solely dependent on the features of the upper layer within its localized receptive field (e.g., a 5 × 5 kernel), meaning the global shape information (e.g., linear relationships between points and right-angle relationships between lines) of building polygons are ignored. Additionally, when capturing aerial images, it is inevitable to include noisy data, such as portions of buildings that are shadowed by surrounding trees. In such cases, the more accurately a model can recognize boundary pixels, the greater the distance between predictions and the ground truth will be.

In light of this issue, we propose a novel deep CNN architecture called the boundary regulated network (BR-Net) to utilize both local and global information for better roof segmentation and more accurate outline extraction. The BR-Net model adopts a modified U-Net structure as a shared backend and simultaneously produces predictions for both segmentation and outlines. In the proposed BR-Net, the optimizer has two main tasks. It must ensure that both the segmentation and outlines of the prediction results are as close as possible to those of the ground truth. In this manner, in every iteration, parameters are updated by considering both segmentation and outlines, which prevents parameters from focusing on surrounding pixels and utilizes a wider range of global information. Experiments on a VHR imagery dataset (see details in Section 2.1) demonstrate the effectiveness of the proposed BR-Net model. In comparative experiments, the values of precision, recall, overall accuracy, F1 score, Jaccard index [35] and kappa coefficient [36] achieved by the proposed method are 0.857, 0.885, 0.952, 0.869, 0.772, and 0.840, respectively. For all evaluation metrics other than recall, the proposed BR-Net outperforms U-Net and significantly outperforms classic FCN8s. Furthermore, sensitivity analysis indicates that other techniques, such as batch normalization (BN) [37] and leaky rectified linear units (LeakyReLUs) [38], can be easily integrated into our BR-Net model to enhance model performance for segmentation and outline extraction. The main contribution of this paper is that we propose a novel boundary regulated network that improves the performance of the state-of-the-art method (e.g., U-Net) for performing segmentation and outline extraction on VHR aerial imagery. The introduction of boundary regulation provides new insight for improving model performance.

The materials and methods are presented in Section 2, where the configuration of the network models are also described. In Section 3, the results of comparisons between four methods and sensitivity analysis of BR-Net are introduced. Discussion and conclusions regarding our study are presented in Sections 4 and 5, respectively.

2. Materials and Methods

2.1. Data

To evaluate the performance of different methods, a study area that covers 32 km^2 in Christchurch, New Zealand is chosen for this study. The aerial image dataset and corresponding building outlines (polygons in .shp format) are downloaded from Land Information of New Zealand (https://data.linz.govt.nz/layer/53413-nz-building-outlines-pilot/). The spatial resolution of the aerial images is 0.075 m. The original images are captured during the flying seasons of 2015 and 2016. Later, they are converted into orthophotos and divided into tiles by the provider. The size of each tile is 3200 × 4800 pixels (240 × 360 m^2). Prior to conducting our experiments, we merge the 370 tiles within the study area into a single mosaic. Additionally, for the purpose of accurate roof segmentation, we manually adjust vectorized building outlines to ensure that all building polygons are strictly aligned with their corresponding roofs.

As shown in Figure 1, the study area is largely covered by residential or manufacturing buildings with sparsely distributed patches of grassland. Prior to conducting our experiments, the study area is evenly divided into two areas for training (Figure 1, left) and testing (Figure 1, right). The training and testing areas contain 28,786 and 26,747 building objects, respectively.

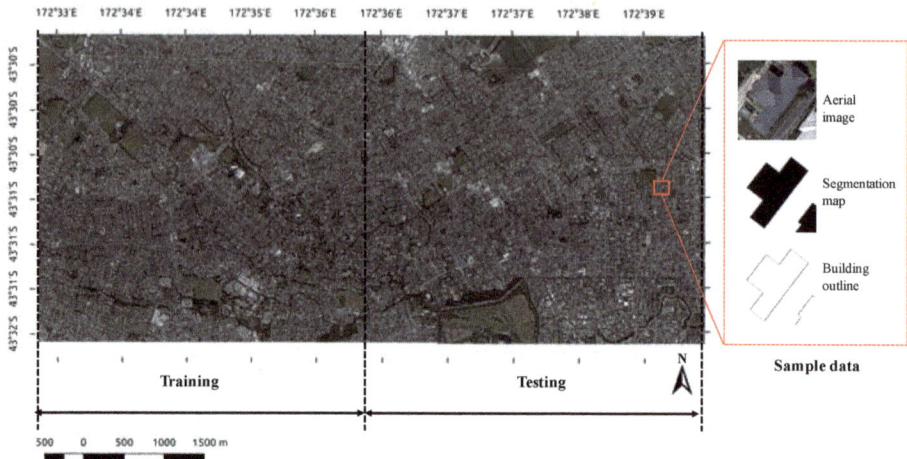

Figure 1. Aerial imagery of the study area ranging from 172°33′ E to 172°40′ E and 43°30′ S to 43°32′ S.

2.2. Methodology

Figure 2 presents the workflow for our study. The aerial imagery from the study area is processed by utilizing a data preprocessing framework to extract proper training and testing data (see details in Section 2.2.1). Then, the training data are further divided into two portions: 70% of the data are utilized for direct model training and the remaining 30% are utilized for cross validation. Through training and cross validation, hyper-parameters, such as number of iterations (or epochs) and value of learning rate, are optimized and determined. Then, the model trained by optimized hyper-parameters is utilized for generating predictions from the testing data. The performance of the model is evaluated based on commonly used evaluation metrics. For evaluating segmentation performance in this study, we chose precision, recall, overall accuracy, Jaccard index, and kappa coefficient. To compare the raw performance of different methods, all evaluation metrics are computed without any post-processing operations, such as conditional random fields [39] or morphological operations [40]. The final outlines of the buildings are extracted from the segmentation maps by utilizing the Canny operator [41].

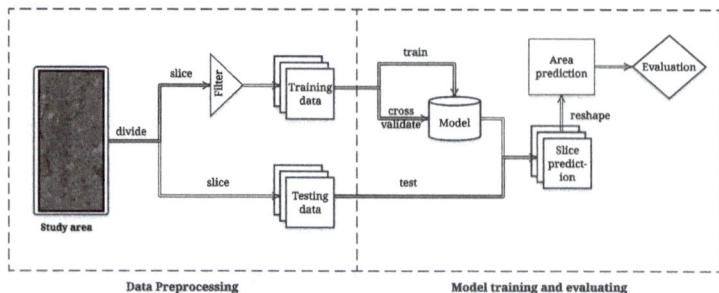

Figure 2. Workflow for our study. The proposed BR-Net method is trained and cross validated utilizing the training data. Later, evaluation of model performance is conducted by utilizing the testing data.

2.2.1. Data Preprocessing

The aerial imagery from the study area is divided into training and testing regions. Later, the aerial imagery from both regions is processed by a sliding window of 224 × 224 pixels (with stride of 224 pixels) to generate image slices. In deep learning, particularly for classification tasks, biased data typically leads to overfitting and poor generalization [42]. To avoid this issue, thresholding is applied to the slices generated from the training region to filter out image slices with low building coverage rates (e.g., building coverage rate ≤ 15%). After data preprocessing, the number of samples in training, validation and testing data are 27,912, 1952 and 71,688, respectively.

2.2.2. Boundary Regulated Network

The classic FCN model, which utilizes fully convolutional layers to perform pixel-to-pixel translations from inputs to outputs, is first proposed by Long et al. in 2015. By removing fully-connected layers, the FCN model greatly reduces the total number of parameters and significantly improves model performance. Advanced FCN-based models improve model performance by utilizing novel techniques, such as unsampling (SegNet), deconvolution (DenconvNet), skip connections (U-Net), and multi-constraints (MC-FCN). Although these FCN-based models are already very powerful, they still have some limitations:

- For these models, the prediction value of each pixel is solely based on the features within a localized receptive field (e.g., a 3 × 3 kernel). Therefore, global information (e.g., linear relationships between points and right angle relationships between lines) of building polygons cannot be utilized by these models.
- When capturing aerial imagery, it is inevitable to obtain noisy data, such as portions of buildings that are shadowed by surrounding trees. If the models are successfully trained to strictly segment the image solely by surrounding pixels, the hidden part of building polygon will be ignored.

To overcome these limitations, the proposed BR-Net model adopts multitask learning for segmentation and outline extraction to utilize both local and global information of images. During the training phase, the optimizer has two main tasks. It must ensure that both the segmentation and outline extraction prediction results are as consistent as possible with the corresponding ground truth. In this manner, during every iteration, the boundary information can restrict and regulate the parameter updating. It will prevent mapping pattern of model from biasing toward segmentation map of surrounding pixels.

Figure 3 presents the network architecture of the proposed BR-Net model. This model is composed of two parts: (1) an optimized U-Net-style FCN as a shared backend; and (2) a dual prediction framework for generating segmentation and outline extraction results. In the shared backend, there are several convolution, nonlinear activation, subsampling, and skip-connection operations.

The convolution operation is an element-wise multiplication performed via kernels. The size of the kernel determines the range of receptive field. In contrast to a rectified linear unit (ReLU) [43], which sets all values less than zero to zero, the output will be handled by a LeakyReLU with an alpha value of 0.1. To accelerate deep network training, avoid bias and prevent gradient vanishing, BN layers are heavily applied following convolutional layers. In this study, max-pooling [44] is chosen for subsampling the height and width of intermediate features. To achieve a consistent size between inputs and outputs, sequential bilinear upsampling [45] and skip-connection operations are implemented. A skip-connection is a concatenating operation across a single axis.

For multitask prediction, both segmentation and outline predictions are generated from the same output from the shared backend. For each prediction, a single kernel convolution operation followed by a sigmoid operation is required.

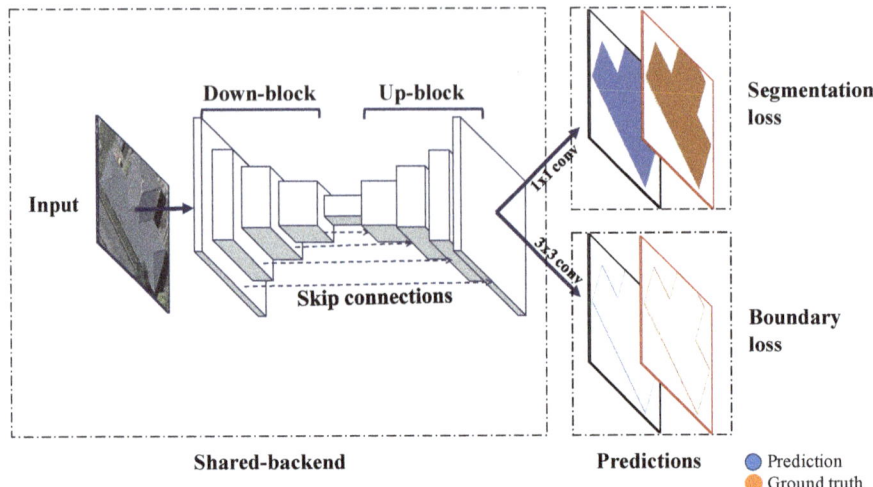

Figure 3. The network architecture of the proposed BR-Net model. The BR-Net model adopts a modified U-Net structure as a shared backend and performs multitask predictions for roof segmentation and outline extraction.

The binary cross entropy [46] between a prediction and the corresponding ground truth is utilized to compute the losses for segmentation ($Loss_{seg}$) and outline ($Loss_{bou}$). Each loss can be calculated as

$$Loss = -\frac{1}{h \times w} \sum_{i=1, j=1}^{h,w} g_{i,j} \times \log(y_{i,j}) + (1 - g_{i,j}) \times \log(1 - y_{i,j}) \quad (1)$$

where h and w represent the height and width of the prediction (y) and corresponding ground truth (g). The value of $y_{i,j}$ is the predicted probability of the pixel category.

Therefore, the total loss of the BR-Net can be formulated as

$$Loss_{final} = (1 - \alpha) \times Loss_{seg} + \alpha \times Loss_{bou} \quad (2)$$

where α is the weight of the boundary loss ($Loss_{bou}$). In this study, the value of α is set to 0.5.

With final loss being minimized by an Adam optimizer [47] in every iteration, the BR-Net model learns a mapping pattern that can produce predictions for both segmentation and outlines utilizing a single input.

2.3. Experimental Setup

2.3.1. Architecture of the BR-Net

The architecture of the BR-Net consists of a shared backend and multitask prediction model. The shared backend consists of four sequential down-blocks, one central conv-block, and four sequential up-blocks. The central conv-block is a 3 × 3 convolutional layer with 384 kernels followed by a LeakyReLU activation function and BN layer. Four skip connections are placed between the 2nd BN layer among the down-blocks and corresponding upsampling layer among the up-blocks. The initial input of the model is an RGB image slice of 224 × 224 pixels. The output of each block serves as the input for the next block.

Figure 4a presents the structure of a down-block. The h, w, and d represent the height, width, and depth of an input, respectively. k represents the number of kernels that are utilized for convolution operations. Each down-block has two convolutional layers followed by two LeakyReLU activation functions, two BN layers, and a max-pooling layer. For each input, a down-block generates an output with half the width and height. The numbers of kernels in the four down-blocks are [24, 48, 96, 192].

Figure 4b presents the structure of an up-block. The h, w, and d represent the height, width and depth of an input, respectively. k and k' represent the dimension of the corresponding BN layer among the down-blocks and the number of kernels utilized for convolution operations, respectively. In an up-block, there is a single bilinear upsampling layer, a skip connection layer, and three convolutional layers followed by LeakyReLU activation functions and BN layers. An up-block doubles the width and height of its input. The numbers of kernels in the four up-blocks are [192, 96, 48, 24].

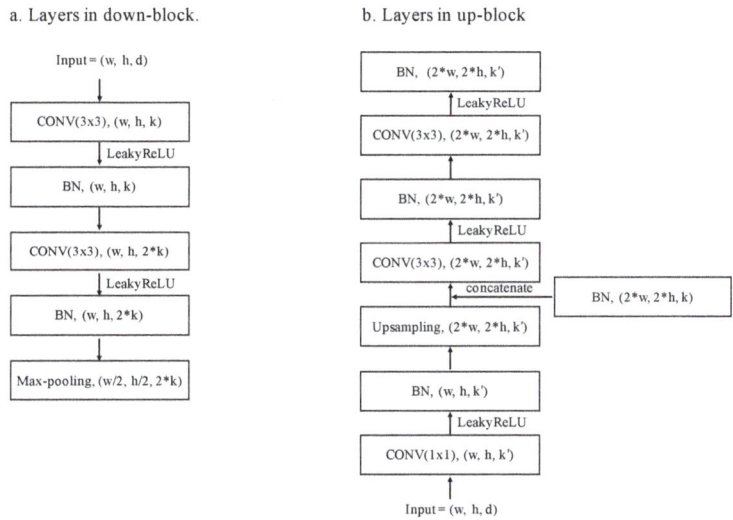

Figure 4. Layers in down-blocks and up-blocks of the shared backend.

The output of the shared backend is a 3D matrix with consistent width and height of the input image. A single 1×1 convolutional kernel followed by a sigmoid activation function is applied to the output to generate predictions for segmentation maps. Similarly, single 3×3 convolutional kernel with sigmoid activation function is used for generating outlines. The losses of different tasks are then calculated by computing the binary cross entropy between the predictions and ground truth.

2.3.2. Integration of Different Components

To further analyze the importance and significance of different components, including BN, LeakyReLU, and the proposed multitask training loss function, various combinations of the three components are tested in a comparison experiment. As shown in Table 1, BR-Net models with different combinations of components (with and without BN after each convolution operation, and with and without nonlinear activation of ReLU/LeakyReLU functions (see details in Figure 4)) are trained and validated utilizing the same training and testing data.

Table 1. Component combinations of BR-Net models.

Combinations	BN	ReLU	LeakyReLU
− BN / ReLU		*	
+ BN / ReLU	*	*	
− BN / LeakyReLU			*
+ BN / LeakyReLU	*		*

3. Results

The best FCN variant (FCN8s) and classic U-Net model are adopted as baseline models in our comparisons. These models, as well as the proposed BR-Net model, are trained and evaluated utilizing the same dataset and processing platform.

3.1. Hyper-Parameter Optimization

Figure 5 shows the trends of model performances under training rates of 5×10^{-3}, 1×10^{-3}, 2×10^{-4}, 4×10^{-5} and 8×10^{-6}. In general, too large ($>1 \times 10^{-3}$) or too small ($<4 \times 10^{-5}$) learning rate leads to poor performance. Three different methods (FCN8s, U-Net and BR-Net) show similar trends over various learning rates:

- As shown in Figure 5a, FCN8s model achieves the best performance with the learning rate of 2×10^{-4}. For major metrics, FCN8s model shows similar values using learning rate between 4×10^{-5} and 2×10^{-4}.
- As shown in Figure 5b, U-Net model shows the highest values of major metrics with the learning rate of 2×10^{-4}. Under learning rates from 2×10^{-4} to 1×10^{-3}, the performances of U-Net model are almost identical.
- As shown in Figure 5c, similar to FCN8s and U-Net methods, the BR-Net model reaches its best performance with the learning rate of 2×10^{-4}.

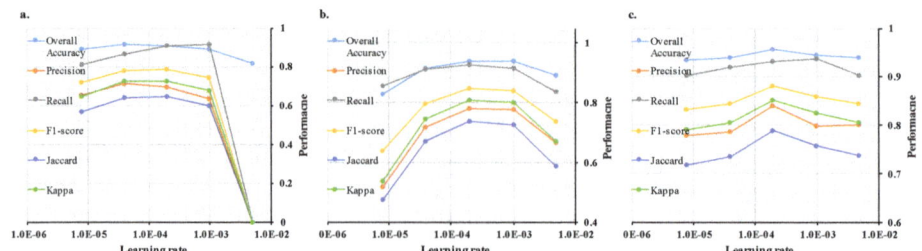

Figure 5. Model performances using learning rates of 5×10^{-3}, 1×10^{-3}, 2×10^{-4}, 4×10^{-5} and 8×10^{-6}: (**a**) performances of FCN8s under various learning rates; (**b**) performances of U-Net under various learning rates; and (**c**) performances of BR-Net under various learning rates.

3.2. Qualitative Result Comparisons

3.2.1. Result Comparisons at Region Level

Figure 6 reveals that the BR-Net method is superior to U-Net and significantly outperformed the FCN8s method in the region-level comparison. In residential regions, such as the top-left and bottom-right regions, all three methods are capable of building recognition and segmentation. The FCN8s model presents significantly more false positives than the other methods. The U-Net model presents fewer false positives than FCN8s, but still failed to discriminate roads when compared to the BR-Net model. In non-residential regions, such as the top-right, central, and bottom-left regions, the U-Net and BR-Net models present a significantly smaller number of false positives than FCN8s.

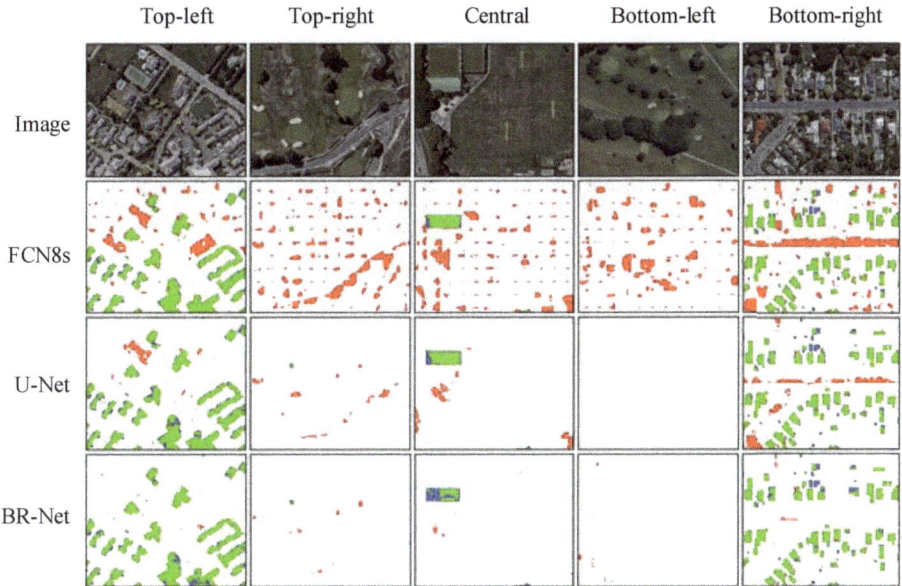

Figure 6. Results of roof segmentation of regions by FCN8s, U-Net, and the proposed BR-Net. The five regions are located in the top-left, top-right, central, bottom-left, and bottom-right portions of the testing area. Each region contains 2240 × 2240 pixels. The green, red, blue, and white channels in the results represent true positive, false positive, false negative, and true negative predictions, respectively.

Figure 7 presents the outline extraction results of the FCN8s, U-Net, and BR-Net methods. In residential regions (e.g., top-left and bottom-right regions), the majority of building outlines are extracted by all three models. However, the results from the FCN8s model contain more false positive polygons and lines compared to the other two methods. Compared to U-Net, BR-Net presents fewer false positives in adjacent areas between buildings and roads. Similar to the residential regions, in the non-residential regions in the top-right, central, and bottom-left portions of the test area, the FCN8s method generates a relatively large number of false positives.

3.2.2. Result Comparisons at Single-House Level

To further explore the improvements in our method compared to other methods, several representative samples are selected for additional comparison.

Figure 8 presents eight representative groups of segmentation results generated by FCN8s, U-Net, and BR-Net. In general, U-Net and BR-Net perform better than FCN8s with slightly fewer false negatives (d and c) and significantly fewer false positives (a, b, e, f, and h), respectively. Compared to the U-Net model, BR-Net model generates fewer false negatives within buildings (a, d, f, and g) and fewer false positives around building edges (b, c, and e).

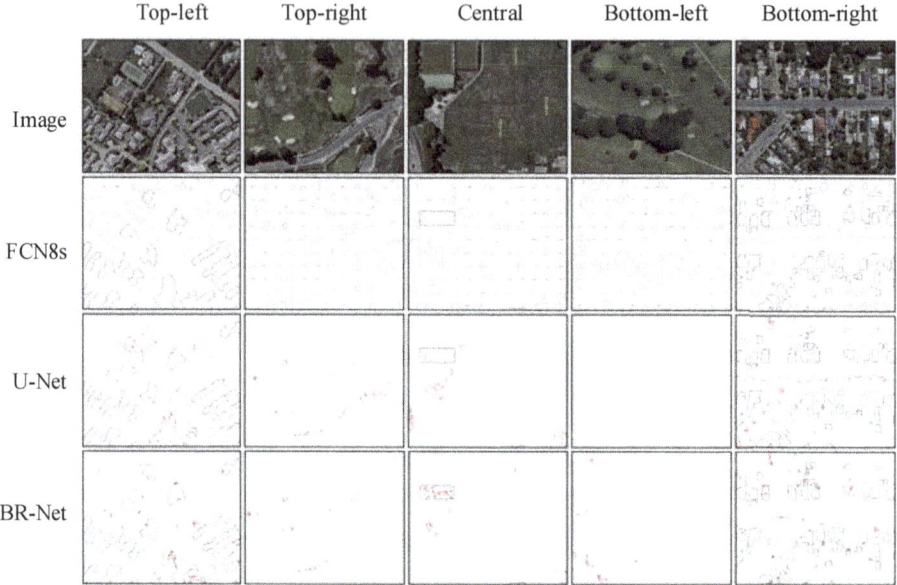

Figure 7. Results of outline extraction from different regions by FCN8s, U-Net, and the proposed BR-Net. The five regions are located in the top-left, top-right, central, bottom-left, and bottom-right portions of the testing area. Each region contains 2240 × 2240 pixels. The green, red, blue, and white channels in the results represent true positive, false positive, false negative, and true negative predictions, respectively.

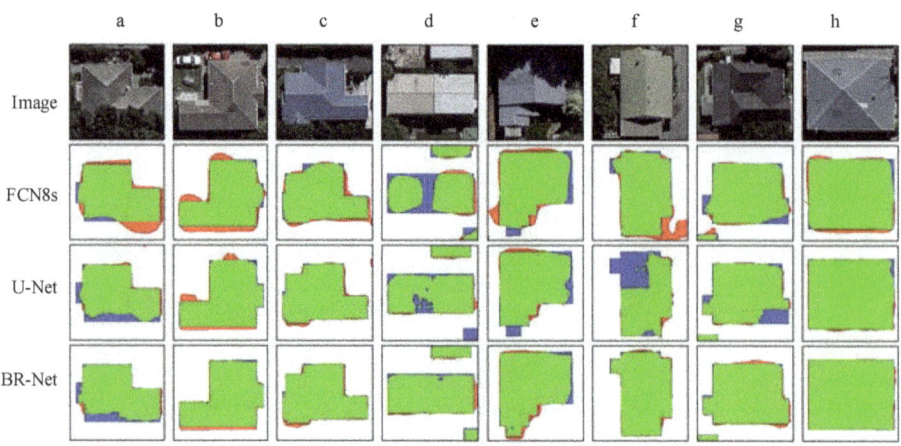

Figure 8. Representative results of single-building-level segmentation by FCN8s, U-Net, and BR-Net. The green, red, blue, and white channels in the results represent true positive, false positive, false negative, and true negative predictions, respectively.

Figure 9 presents eight representative groups of outline extraction results from FCN8s, U-Net, and BR-Net. In general, all three methods can extract the major parts of buildings. For aerial images captured in good imaging conditions, both BR-Net and U-Net can generate near-perfectly aligned building outlines, whereas the polygon shapes in the FCN8s results are slightly twisted (c and h). For aerial images captured in shadowy condition, the BR-Net model produces results that are close to the actual shapes of buildings, instead of only the unobstructed parts of building (a, e, and g). It should be noted that, when both FCN8s and U-Net produce broken polygons, the proposed BR-Net model can still generate acceptable outlines (d and f).

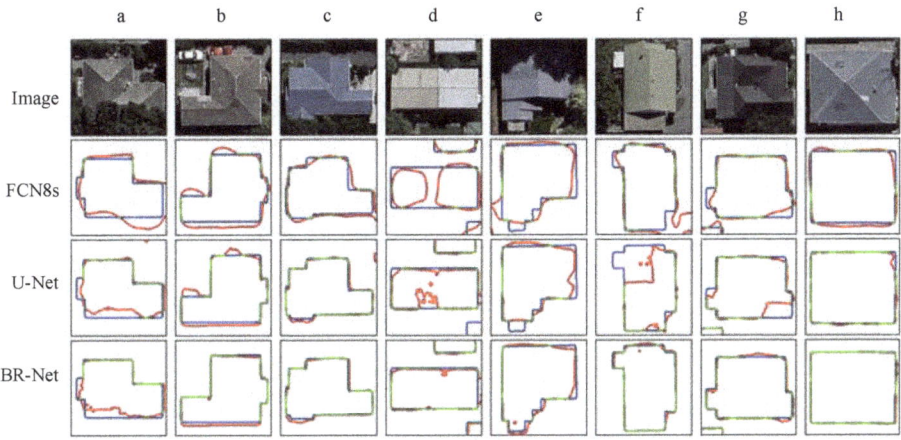

Figure 9. Representative results of single-building-level outline extraction by FCN8s, U-Net and, BR-Net. The green, red, blue, and white channels in the results represent true positive, false positive, false negative, and true negative predictions, respectively.

3.3. Quantitative Result Comparisons

In this study, two imbalanced metrics of precision and recall, and four general metrics of overall accuracy, F1 score, Jaccard index, and kappa coefficient are utilized for quantitative evaluations of roof segmentation results. Figure 10 presents comparative results between FCN8s, U-Net, BR-Net for the testing area.

For the imbalanced metrics of precision and recall, the BR-Net method achieves significantly higher values of precision (0.857 vs. 0.742 for U-Net and 0.620 for FCN8s), which indicates that our method performs well in terms of suppressing false positives. This result is consistent with the observations in Figure 6. However, compared to the recall value of 0.922 for FCN8s and U-Net, BR-Net achieves a slightly lower value of 0.885. Compared to the U-Net method, the BR-Net method shows 15.5% (0.857 vs. 0.742) improvement of precision and 4.0% (0.885 vs. 0.922) decline of recall. The improvement in precision (15.5%) significantly outweighs the decline in recall (4.0%).

For the four general metrics, the BR-Net model achieves the highest values for overall accuracy, F1 score, Jaccard index, and kappa coefficient. For overall accuracy, BR-Net achieves improvements of approximately 2.8% (0.952 vs. 0.926) over U-Net and 8.1% (0.952 vs. 0.881) over FCN8s. For F1 score, BR-Net achieves improvements of approximately 6.2% (0.869 vs. 0.818) over U-Net and 17.9% (0.869 vs. 0.737) over FCN8s. Compared to the FCN8s method, the BR-Net method achieves improvements of 30.1% (0.772 vs. 0.589) and 26.3% (0.840 vs. 0.665) for Jaccard index and kappa coefficient, respectively. Compared to the U-Net method, the BR-Net method achieves improvements of 10.6% (0.772 vs. 0.698) and 8.7% (0.840 vs. 0.773) for Jaccard index and kappa coefficient, respectively.

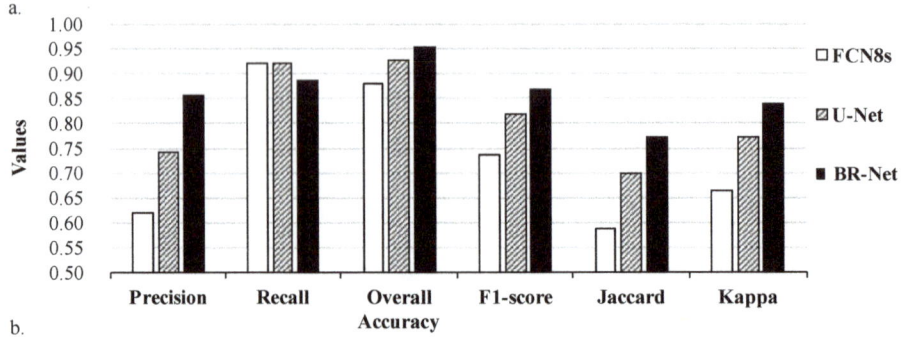

Methods	Precision	Recall	Overall Accuracy	F1-score	Jaccard	Kappa
FCN8s	0.620	**0.922**	0.881	0.737	0.589	0.665
U-Net	0.742	**0.922**	0.926	0.818	0.698	0.773
BR-Net	**0.857**	0.885	**0.952**	**0.869**	**0.772**	**0.840**

Figure 10. Comparison of segmentation performances of FCN8s, U-Net, and BR-Net across the entire testing area. (**a**) Bar chart for performance comparison. The x- and y-axis represent the evaluation metrics and corresponding values, respectively. (**b**) Table of performance comparisons of methods. For each evaluation metric, the highest values are highlighted in **bold**.

3.4. Sensitivity Analysis of Components

The sensitivity of the components for BN and nonlinear activation of ReLU/LeakyReLU functions is analyzed in this section.

Figure 11 presents representative roof segmentation results from BR-Net with different combinations of components. Compared to the basic BR-Net model (−BN/ReLU), adding BN (+BN/ReLU) or replacing the ReLU activation function with a LeakyReLU activation function (−BN/LeakyReLU), or combining both batch normalization and LeakyReLU (+BN/LeakyReLU) slightly reduces the number false positives (e and h) and false negatives (a, b, d, and g), which leads to better overall performance for roof segmentation. The performance improvements resulting from adding BN and replacing the activation function are quite similar.

Figure 12 presents representative results of single-house-level outline extraction from BR-Net with different combinations of components. Similar to the roof segmentation results, the BR-Net model with the addition of BN (+BN/ReLU) or replacement of the ReLU activation function with a LeakyReLU activation function (−BN/LeakyReLU), or combining both BN and LeakyReLU (+BN/LeakyReLU), produces better building contours for both shadowed (a, c, d, and g) and non-shadowed (b, e, f, and h) images. However, the differences between the BR-Net models of +BN/ReLU, −BN/LeakyReLU, and +BN/LeakyReLU are not significant.

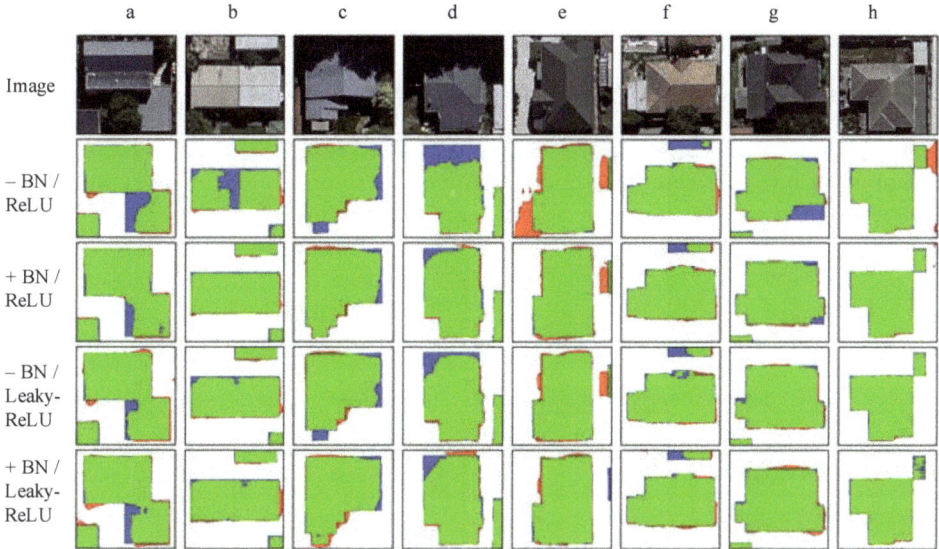

Figure 11. Representative results of single-building-level roof segmentation from BR-Net with various combinations of components. The green, red, blue, and white channels in the results represent true positive, false positive, false negative, and true negative predictions, respectively.

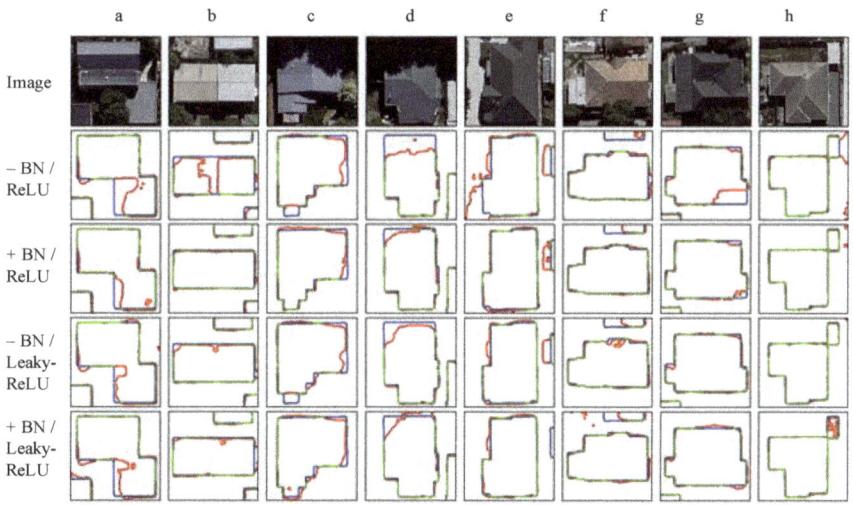

Figure 12. Representative results of single-building-level outline extraction from BR-Net with various combinations of components. The green, red, blue, and white channels in the results represent true positive, false positive, false negative, and true negative predictions, respectively.

The evaluation results of BR-Net with various combinations of components are presented in Figure 13.

In Figure 13a, for all evaluation metrics other than recall, the BR-Net model with the addition of BN (+BN/ReLU) or replacement of ReLU with LeakyReLU (−BN/LeakyReLU), or combining BN and

LeakyReLU (+BN/LeakyReLU), produces slightly higher values than the basic model (−BN/ReLU). Compared to the basic model, the model utilizing LeakyReLU (−BN/LeakyReLU) produces a higher value of recall.

In Figure 13b, the BR-Net model with BN and LeakyReLU (+BN/LeakyReLU) produces the highest values for five out of six evaluation metrics, namely precision, overall accuracy, F1 score, Jaccard index, and kappa coefficient. Compared to the basic model, the increases in these metrics are 4.3% (0.857 vs. 0.822), 0.5% (0.952 vs. 0.947), 1.2% (0.869 vs. 0.859), 2.1% (0.772 vs. 0.756), and 1.6% (0.840 vs. 0.827), respectively. However, the model with BN and LeakyReLU results in the lowest value of recall with a decrease of approximately 2.0% (0.885 vs. 0.903) compared to the base model.

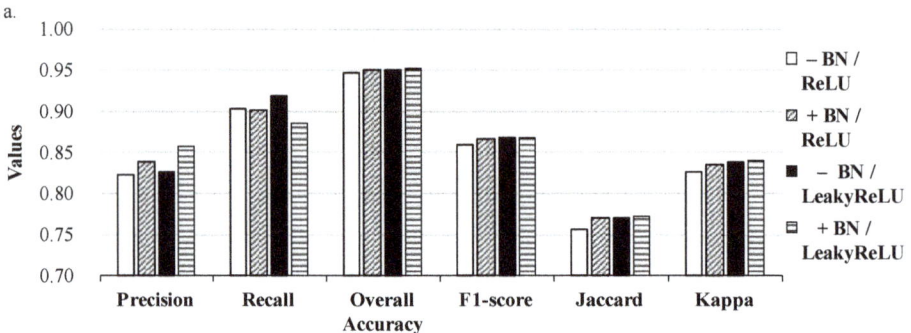

Figure 13. Comparison of segmentation performances of BR-Net models with various combinations of components. (a) Bar chart for performance comparison. The x- and y-axis represent the evaluation metrics and corresponding values, respectively. (b) Table of performance comparisons of methods. For each evaluation metric, the highest values are highlighted in **bold**.

3.5. Computational Efficiency

The FCN8s, U-Net, and BR-Net models were implemented in PyTorch (https://pytorch.org/) and tested on a 64-bit Ubuntu system equipped with an NVIDIA GeForce GTX 1070 GPU (https://www.nvidia.com/en-us/geforce/products/10series/geforce-gtx-1070-ti/) and 8 GB of memory. During training, the Adam stochastic optimizer [47] with a learning rate of 2×10^{-4} and betas of (0.9, 0.999) was utilized. To conduct fair comparisons between the different methods, the batch size and iteration number for training were fixed as 24 and 10,000, respectively.

The computational efficiencies of the different methods during different stages are listed in Table 2. During the training stage, the FCN8s model processes approximately 29.3 frames per second (FPS), while the fastest model (U-Net) reached 91.7 FPS. For the BR-Net models, adding BN or replacing ReLU with LeakyReLU will decrease training speed. During the testing stage, as there is no need for gradient calculation or parameter updating, all models are 3–4 times faster. Similar to the training stage, the U-Net model is faster than all BR-Net models. However, the differences in their computational efficiencies become smaller. Compared to the BR-Net model with the best performance

(+BN/LeakyReLU), the U-Net model achieves 16.2% (91.7 vs. 80.2) and 12.3% (280.6 vs. 249.9) higher FPS during the training and testing stages, respectively.

Table 2. Comparison of computational efficiency of FCN8s, U-Net, and BR-Net with various combinations of components.

Stage	FCN8s	U-Net	BR-Net (−BN/ReLU)	BR-Net (+BN/ReLU)	BR-Net (−BN/LeakyReLU)	BR-Net (+BN/LeakyReLU)
Training (FPS)	29.3	91.7	88.1	80.2	86.6	78.9
Testing (FPS)	130.2	280.6	276.5	252.5	274.1	249.9

4. Discussion

4.1. Regarding the Proposed BR-Net Model

In the field of remote sensing, deep CNN models are first applied to detecting buildings in rural area [48] or informal settlements [49]. Because of limitations in terms of heavy memory costs and low computational efficiency, these patch-based CNN models are not capable of performing roof segmentation over large areas. In 2016, Maggiori et al. first adopted an FCN for segmenting large-scale aerial images [50,51]. With the development of new computer vision algorithms, more advanced FCN-based models, such as SegNet, U-Net, and MC-FCN, have been introduced and optimized for roof segmentation tasks.

In this paper, we propose a novel boundary regulated network termed BR-Net to improve capability of roof segmentation and outline extraction through combination of both local and global information of images. Existing advanced FCN-based models enhance the performance of the classic FCN model by either focusing on replacing the simple bilinear upsampling operation with more information-preserving methods (e.g., unsampling in SegNet and deconvolution in DeconvNet) or making better usage of the feature representation capability of hidden layers (e.g., skip-connections in U-Net and multi-constraints in MC-FCN). In contrast to other advanced FCN-based models, the proposed BR-Net model adopts a shared backend utilizing a modified U-Net and a dual prediction framework for the generation of segmentation and outline extraction results. Because of the multitask learning, BR-Net can utilize both local information from surrounding pixels to segment buildings and global information from polygons to generate outline. Comparative results from the testing area demonstrated that the proposed BR-Net model further improves the capability of FCN-based methods (FCN8s and U-Net) and achieves state-of-the-art performance on this task. Additionally, other techniques, such as BN and LeakyReLU activation, can be easily integrated into BR-Net to achieve superior performance.

4.2. Accuracies, Uncertainties, and Limitations

Compared to classic FCNs (FCN8s) and the state-of-the-art fully convolutional model (U-Net), BR-Net achieved the highest values for five out of six evaluation metrics (precision, overall accuracy, F1 score, Jaccard index, and kappa coefficient). The BR-Net model achieves a value of 0.857 for the precision, whereas U-Net and FCN8s only achieve values of 0.742 and 0.620, respectively. However, BR-Net shows slightly lower recall than FCN8s and U-Net (0.885 of BR-Net vs. 0.922 of FCN8s and U-Net). The increment of the precision as well as the decline of recall from BR-Net might due to the regulation of boundary information that avoid making prediction solely by surrounding pixels. Since the improvement in precision significantly outweighs the decline in recall, the proposed BR-Net model is superior to FCN8s and U-Net at roof segmentation and outline extraction tasks.

From the sensitivity analysis of different components, adding BN after each convolutional operation or replacing the traditional ReLU activation function with a LeakyReLU or combining both BN and LeakyReLU is able to improve the performance of the basic BR-Net model (see details in Figure 13).

As shown in Table 3, compared to U-Net, even the basic BR-Net model (−BN/LeakyReLU) achieves higher values for all evaluation metrics other than recall. Adding boundary loss to U-Net leads to better performance (basic BR-Net vs. U-Net). In comparison to optimized BR-Net, negative BR-Net shows smaller values of major metrics including precision, overall accuracy, f1-score, Jaccard index and kappa (see Rows 4 and 5 of Table 3). Removing boundary loss from optimized BR-Net leads to weaker performance (negative BR-Net vs. optimized BR-Net). These results demonstrate that our proposed boundary loss is a critical factor for improving model performance.

Table 3. Comparison of segmentation performances of U-Net, basic BR-Net, negative BR-Net and optimized BR-Net. The highest values for different metrics are highlighted in **bold**.

Methods	Precision	Recall	Overall Accuracy	F1-score	Jaccard	Kappa
U-Net	0.742	0.922	0.926	0.818	0.698	0.773
basic BR-Net [1]	0.822	0.903	0.947	0.859	0.756	0.827
negative BR-Net [2]	0.768	**0.951**	0.936	0.845	0.739	0.806
optimized BR-Net [3]	**0.857**	0.885	**0.952**	**0.869**	**0.772**	**0.840**

[1] BR-Net (−BN/ReLU); [2] BR-Net (+BN/LeakyReLU), without boundary loss; [3] BR-Net (+BN/LeakyReLU).

During our computational efficiency analysis, we observed a significant increasing in computational cost when utilizing the multitask framework, BN, or LeakyReLU in the training stage. The differences in processing speed became much smaller in testing stage. This decrease in computational efficiency may become a problem when applying our method to very large datasets, such as automatic mappings of provinces or entire countries. Additionally, compared to the performances of FCN8s and U-Net, the performance of BR-Net is lower by approximately 4.0% (0.885 vs. 0.922) in terms of recall. The balance between precision and recall must be studied further. Additionally, even for the optimized BR-Net model, there is still a certain amount of false positives in its prediction results (see top-right and bottom-left regions in Figure 6), which prevents its further application for more precise outline extraction and vectorization.

5. Conclusions

In this paper, we propose a novel boundary regulated network for accurate roof segmentation and outline extraction from VHR aerial images. The proposed BR-Net model has the ability to perform automatic segmentation and outline extraction from RGB images. Its performance is verified through several experiments on a VHR dataset covering approximately 32 km^2. With its unique design of boundary restriction and regulation, the proposed method achieved significantly better performance than FCN8s and U-Net. In comparison to U-Net, BR-Net achieved gains of 6.2% (0.869 vs. 0.818), 10.6% (0.772 vs. 0.698), and 8.7% (0.840 vs. 0.773) in F1 score, Jaccard index, and kappa coefficient, respectively. Sensitivity analysis demonstrated that adding BN or utilizing LeakyReLU, or combining BN and LeakyReLU, can further improve model performance. In future studies, we will further optimize our network architecture to achieve better performance with less computational cost.

Author Contributions: G.W., X.S. (Xiaowei Shao), and R.S. conceived and designed the experiments. G.W. performed the experiments. G.W., Z.G., and X.S. (Xiaowei Shao) analyzed the data. X.S. (Xiaodan Shi), Q.C., and Y.X. contributed reagents/materials/analysis/tools. G.W. wrote the paper. All authors read and approved the submitted manuscript.

Funding: This work was partially supported by the Japan Society for the Promotion of Science (JSPS) Grant (No. 16K18162); National Natural Science Foundation of China, Project Number 41601506; and China Postdoctoral Science Foundation, Project Number 2016M590730.

Acknowledgments: We would like to thank the National Topographic Office of New Zealand for kindly sharing their data.

Conflicts of Interest: The authors declare no conflict of interest.

Abbreviations

The following abbreviations are used in this manuscript:

CNN   Convolutional Neural Network
BN   Batch Normalization
ReLU   Rectified Linear Unit
FCN   Fully Convolutional Networks
FPS   Frames Per Second
BR-Net Boundary Regulated Network

References

1. Ma, L.; Li, M.; Ma, X.; Cheng, L.; Du, P.; Liu, Y. A review of supervised object-based land-cover image classification. *ISPRS J. Photogramm. Remote Sens.* **2017**, *130*, 277–293. [CrossRef]
2. Li, M.; Zang, S.; Zhang, B.; Li, S.; Wu, C. A review of remote sensing image classification techniques: The role of spatio-contextual information. *Eur. J. Remote Sens.* **2014**, *47*, 389–411. [CrossRef]
3. Chen, R.; Li, X.; Li, J. Object-based features for house detection from rgb high-resolution images. *Remote Sens.* **2018**, *10*, 451. [CrossRef]
4. Xu, B.; Jiang, W.; Shan, J.; Zhang, J.; Li, L. Investigation on the weighted ransac approaches for building roof plane segmentation from lidar point clouds. *Remote Sens.* **2016**, *8*, 5. [CrossRef]
5. Huang, Y.; Zhuo, L.; Tao, H.; Shi, Q.; Liu, K. A novel building type classification scheme based on integrated LiDAR and high-resolution images. *Remote Sens.* **2017**, *9*, 679. [CrossRef]
6. Gilani, S.A.N.; Awrangjeb, M.; Lu, G. An automatic building extraction and regularisation technique using lidar point cloud data and orthoimage. *Remote Sens.* **2016**, *8*, 258. [CrossRef]
7. Sahoo, P.K.; Soltani, S.; Wong, A.K. A survey of thresholding techniques. *Comput. Vis. Graph. Image Process.* **1988**, *41*, 233–260. [CrossRef]
8. Kanopoulos, N.; Vasanthavada, N.; Baker, R.L. Design of an image edge detection filter using the Sobel operator. *IEEE J. Solid-State Circuits* **1988**, *23*, 358–367. [CrossRef]
9. Wu, Z.; Leahy, R. An optimal graph theoretic approach to data clustering: Theory and its application to image segmentation. *IEEE Trans. Pattern Anal. Mach. Intell.* **1993**, *15*, 1101–1113. [CrossRef]
10. Tremeau, A.; Borel, N. A region growing and merging algorithm to color segmentation. *Pattern Recognit.* **1997**, *30*, 1191–1203. [CrossRef]
11. Gómez-Moreno, H.; Maldonado-Bascón, S.; López-Ferreras, F. Edge detection in noisy images using the support vector machines. In *International Work-Conference on Artificial Neural Networks*; Springer: Berlin/Heidelberg, Germany, 2001; pp. 685–692.
12. Zhou, J.; Chan, K.; Chong, V.; Krishnan, S.M. Extraction of Brain Tumor from MR Images Using One-Class Support Vector Machine. In Proceedings of the 2005 IEEE 7th Annual International Conference of the Engineering in Medicine and Biology Society (EMBS 2005), Shanghai, China, 17–18 January 2006; pp. 6411–6414.
13. Xie, S.; Tu, Z. Holistically-Nested Edge Detection. In Proceedings of the IEEE International Conference on Computer Vision, Santiago, Chile, 13–16 December 2015; pp. 1395–1403.
14. Viola, P.; Jones, M. Rapid Object Detection Using a Boosted Cascade of Simple Features. In Proceedings of the 2001 IEEE Computer Society Conference on Computer Vision and Pattern Recognition (CVPR 2001), Kauai, HI, USA, 8–14 December 2001; Volume 1, p. I.
15. Lowe, D.G. Object Recognition from Local Scale-Invariant Features. In Proceedings of the Seventh IEEE International Conference on Computer Vision, Kerkyra, Greece, 20–27 September 1999; Volume 2, pp. 1150–1157.
16. Ojala, T.; Pietikainen, M.; Maenpaa, T. Multiresolution gray-scale and rotation invariant texture classification with local binary patterns. *IEEE Trans. Pattern Anal. Mach. Intell.* **2002**, *24*, 971–987. [CrossRef]

17. Dalal, N.; Triggs, B. Histograms of Oriented Gradients for Human Detection. In Proceedings of the 2005 IEEE Computer Society Conference on Computer Vision and Pattern Recognition (CVPR 2005), San Diego, CA, USA, 20–25 June 2005; Volume 1, pp. 886–893.
18. Inglada, J. Automatic recognition of man-made objects in high resolution optical remote sensing images by SVM classification of geometric image features. *ISPRS J. Photogramm. Remote Sens.* **2007**, *62*, 236–248. [CrossRef]
19. Aytekin, Ö.; Zöngür, U.; Halici, U. Texture-based airport runway detection. *IEEE Geosci. Remote Sens. Lett.* **2013**, *10*, 471–475. [CrossRef]
20. Dong, Y.; Du, B.; Zhang, L. Target detection based on random forest metric learning. *IEEE J. Sel. Top. Appl. Earth Obs. Remote Sens.* **2015**, *8*, 1830–1838. [CrossRef]
21. LeCun, Y.; Bengio, Y. Convolutional networks for images, speech, and time series. *Handb. Brain Theory Neural Netw.* **1995**, *3361*, 1995.
22. Ciresan, D.; Giusti, A.; Gambardella, L.M.; Schmidhuber, J. Deep neural networks segment neuronal membranes in electron microscopy images. In Proceedings of the Advances in Neural Information Processing Systems, Lake Tahoe, NV, USA, 3–6 December 2012; pp. 2843–2851.
23. Krizhevsky, A.; Sutskever, I.; Hinton, G.E. Imagenet classification with deep convolutional neural networks. In Proceedings of the Advances in Neural Information Processing Systems, Lake Tahoe, NV, USA, 3–6 December 2012; pp. 1097–1105.
24. Everingham, M.; Van Gool, L.; Williams, C.K.; Winn, J.; Zisserman, A. The pascal visual object classes (voc) challenge. *Int. J. Comput. Vis.* **2010**, *88*, 303–338. [CrossRef]
25. Lin, T.Y.; Maire, M.; Belongie, S.; Hays, J.; Perona, P.; Ramanan, D.; Dollár, P.; Zitnick, C.L. Microsoft coco: Common objects in context. In *European Conference on Computer Vision*; Springer: Cham, Switzerland, 2014; pp. 740–755.
26. Deng, J.; Dong, W.; Socher, R.; Li, L.J.; Li, K.; Li, F.-F. Imagenet: A Large-Scale Hierarchical Image Database. In Proceedings of the 2009 IEEE Conference on Computer Vision and Pattern Recognition (CVPR 2009), Miami, FL, USA, 20–25 June 2009; pp. 248–255.
27. Guo, Z.; Shao, X.; Xu, Y.; Miyazaki, H.; Ohira, W.; Shibasaki, R. Identification of village building via Google Earth images and supervised machine learning methods. *Remote Sens.* **2016**, *8*, 271. [CrossRef]
28. Long, J.; Shelhamer, E.; Darrell, T. Fully Convolutional Networks for Semantic Segmentation. In Proceedings of the IEEE Conference on Computer Vision and Pattern Recognition, Boston, MA, USA, 7–12 June 2015; pp. 3431–3440.
29. Kampffmeyer, M.; Salberg, A.B.; Jenssen, R. Semantic Segmentation of Small Objects and Modeling of Uncertainty in Urban Remote Sensing Images Using Deep Convolutional Neural Networks. In Proceedings of the IEEE Conference on Computer Vision and Pattern Recognition Workshops, Las Vegas, NV, USA, 26 June–1 July 2016; pp. 1–9.
30. Badrinarayanan, V.; Kendall, A.; Cipolla, R. Segnet: A deep convolutional encoder-decoder architecture for image segmentation. *IEEE Trans. Pattern Anal. Mach. Intell.* **2017**, *39*, 2481–2495. [CrossRef] [PubMed]
31. Noh, H.; Hong, S.; Han, B. Learning Deconvolution Network for Semantic Segmentation. In Proceedings of the IEEE International Conference on Computer Vision, Santiago, Chile, 13–16 December 2015; pp. 1520–1528.
32. Ronneberger, O.; Fischer, P.; Brox, T. U-Net: Convolutional networks for biomedical image segmentation. In *International Conference on Medical Image Computing and Computer-Assisted Intervention*; Springer: Cham, Switherland, 2015; pp. 234–241.
33. Lin, T.Y.; Dollár, P.; Girshick, R.; He, K.; Hariharan, B.; Belongie, S. Feature pyramid networks for object detection. *CVPR* **2017**, *1*, 4.
34. Wu, G.; Shao, X.; Guo, Z.; Chen, Q.; Yuan, W.; Shi, X.; Xu, Y.; Shibasaki, R. Automatic Building Segmentation of Aerial Imagery Using Multi-Constraint Fully Convolutional Networks. *Remote Sens.* **2018**, *10*, 407. [CrossRef]
35. Polak, M.; Zhang, H.; Pi, M. An evaluation metric for image segmentation of multiple objects. *Image Vis. Comput.* **2009**, *27*, 1223–1227. [CrossRef]
36. Carletta, J. Assessing agreement on classification tasks: The kappa statistic. *Comput. Linguist.* **1996**, *22*, 249–254.

37. Ioffe, S.; Szegedy, C. Batch Normalization: Accelerating Deep Network Training by Reducing Internal Covariate Shift. In Proceedings of the International Conference on Machine Learning, Lille, France, 6–11 July 2015; pp. 448–456.
38. Maas, A.L.; Hannun, A.Y.; Ng, A.Y. Rectifier nonlinearities improve neural network acoustic models. In Proceedings of the 30th International Conference on Machine Learning, Atlanta, GA, USA, 16–21 June 2013; Volume 30, p. 3.
39. Li, E.; Femiani, J.; Xu, S.; Zhang, X.; Wonka, P. Robust rooftop extraction from visible band images using higher order CRF. *IEEE Trans. Geosci. Remote Sens.* **2015**, *53*, 4483–4495. [CrossRef]
40. Plaza, A.; Martínez, P.; Pérez, R.; Plaza, J. Spatial/spectral endmember extraction by multidimensional morphological operations. *IEEE Trans. Geosci. Remote Sens.* **2002**, *40*, 2025–2041. [CrossRef]
41. Canny, J. A computational approach to edge detection. In *Readings in Computer Vision*; Elsevier: New York, NY, USA, 1987; pp. 184–203.
42. Goodfellow, I.; Bengio, Y.; Courville, A.; Bengio, Y. *Deep Learning*; MIT Press: Cambridge, MA, USA, 2016; Volume 1.
43. Nair, V.; Hinton, G.E. Rectified Linear Units Improve Restricted Boltzmann Machines. In Proceedings of the 27th International Conference on Machine Learning (ICML-10), Haifa, Israel, 21–24 June 2010; pp. 807–814.
44. Nagi, J.; Ducatelle, F.; Di Caro, G.A.; Cireşan, D.; Meier, U.; Giusti, A.; Nagi, F.; Schmidhuber, J.; Gambardella, L.M. Max-Pooling Convolutional Neural Networks for Vision-Based Hand Gesture Recognition. In Proceedings of the 2011 IEEE International Conference on Signal and Image Processing Applications (ICSIPA), Kuala Lumpur, Malaysia, 16–18 November 2011; pp. 342–347.
45. Novak, K. Rectification of digital imagery. *Photogramm. Eng. Remote Sens.* **1992**, *58*, 339–344.
46. Shore, J.; Johnson, R. Properties of cross-entropy minimization. *IEEE Trans. Inf. Theory* **1981**, *27*, 472–482. [CrossRef]
47. Kingma, D.P.; Ba, J. Adam: A method for stochastic optimization. *arXiv* **2014**, arXiv:1412.6980.
48. Guo, Z.; Chen, Q.; Wu, G.; Xu, Y.; Shibasaki, R.; Shao, X. Village Building Identification Based on Ensemble Convolutional Neural Networks. *Sensors* **2017**, *17*, 2487. [CrossRef] [PubMed]
49. Mboga, N.; Persello, C.; Bergado, J.R.; Stein, A. Detection of Informal Settlements from VHR Images Using Convolutional Neural Networks. *Remote Sens.* **2017**, *9*, 1106. [CrossRef]
50. Maggiori, E.; Tarabalka, Y.; Charpiat, G.; Alliez, P. Convolutional neural networks for large-scale remote-sensing image classification. *IEEE Trans. Geosci. Remote Sens.* **2017**, *55*, 645–657. [CrossRef]
51. Maggiori, E.; Tarabalka, Y.; Charpiat, G.; Alliez, P. Fully Convolutional Networkss for Remote Sensing Image Classification. In Proceedings of the 2016 IEEE International on Geoscience and Remote Sensing Symposium (IGARSS), Beijing, China, 10–15 July 2016; pp. 5071–5074.

© 2018 by the authors. Licensee MDPI, Basel, Switzerland. This article is an open access article distributed under the terms and conditions of the Creative Commons Attribution (CC BY) license (http://creativecommons.org/licenses/by/4.0/).

MDPI
St. Alban-Anlage 66
4052 Basel
Switzerland
Tel. +41 61 683 77 34
Fax +41 61 302 89 18
www.mdpi.com

Remote Sensing Editorial Office
E-mail: remotesensing@mdpi.com
www.mdpi.com/journal/remotesensing